D0202511

Insect Ultrastructure

Volume 1

Insect Ultrastructure

Volume 1

Edited by

Robert C. King
Northwestern University
Evanston, Illinois, USA

and

Hiromu Akai
Sericultural Experiment Station
Yatabe, Ibaraki, Japan

Plenum Press • New York and London

Library of Congress Cataloging in Publication Data

Main entry under title:

Insect ultrastructure.

Includes bibliographical references and index.
1. Insects—Anatomy. 2. Insects—Cytology. 3. Ultrastructure (Biology). I. King, Robert C. II. Akai, Hiromu, 1930- .
QL494.I49 595.7′04 82-5268
ISBN 0-306-40923-2 (v. 1) AACR2

A Division of Plenum Publishing Corporation
233 Spring Street, New York, N.Y. 10013

Printed in the United States of America

Contributors

DOREEN E. ASHURST — Department of Anatomy
St. George's Hospital Medical School
Cranmer Terrace
Tooting, London SW17 ORE, England

ROBERT BERDAN — Department of Zoology
University of Western Ontario
London, Ontario, Canada N6A 5B7

JOSEPH D. CASSIDY, O.P. — Department of Biological Sciences
University of Illinois at Chicago Circle
Chicago, Illinois 60680, USA

M. DONALD CAVE — Department of Anatomy
University of Arkansas for Medical Sciences
Little Rock, Arkansas 72205, USA

STANLEY CAVENEY — Department of Zoology
University of Western Ontario
London, Ontario, Canada N6A 5B7

BERTIL DANEHOLT — Department of Medical Cell Genetics
Medical Nobel Institute
Karolinska Institutet
S-10401 Stockholm, Sweden

BARRY K. FILSHIE — Commonwealth Scientific and Industrial Research Organization
Division of Entomology
Canberra, Australia

VICTORIA FOE — Department of Biochemistry and Biophysics
University of California Medical School
San Francisco, California 94143, USA

HUGH FORREST
Department of Zoology
University of Texas
Austin, Texas 78712, USA

DIANE K. FRISTROM
Department of Genetics
University of California
Berkeley, California 94720, USA

PREBEN BACH HOLM
Department of Physiology
Carlesberg Laboratory
Copenhagen, Valby, Denmark

ERWIN HUEBNER
Department of Zoology
University of Manitoba
Winnipeg, Manitoba R3T 2N2, Canada

FOTIS C. KAFATOS
Department of Cellular and Developmental Biology
The Biological Laboratories
Harvard University
16 Divinity Avenue
Cambridge, Massachusetts 02138, USA

ROBERT C. KING
Department of Ecology and Environmental Biology
Northwestern University
Evanston, Illinois 60201, USA

CHARLES LAIRD
Department of Zoology
University of Washington
Seattle, Washington 98195, USA

NANCY J. LANE
Agricultural Research Council
Unit of Invertebrate Chemistry and Physiology
Department of Zoology
University of Cambridge
Downing Street
Cambridge CB2 3EJ, England

GRACE DANE MAZUR
Department of Cellular and Developmental Biology
The Biological Laboratories
Harvard University
16 Divinity Avenue
Cambridge, Massachusetts 02138, USA

CHARLES NOIROT
Laboratory of Zoology
University of Dijon
6, Boulevard Gabriel
21100 Dijon, France

Cécile Noirot-Timothée — Laboratory of Zoology
University of Dijon
6, Boulevard Gabriel
21100 Dijon, France

Søren Wilken Rasmussen — Department of Physiology
Carlesberg Laboratory
Copenhagen, Valby, Denmark

Jerome C. Regier — Department of Biochemistry and Molecular and Cellular Biology
Northwestern University
Evanston, Illinois 60201, USA

Wayne L. Rickoll — Department of Genetics
University of California
Berkeley, California 94720, USA

André Rousset — Laboratoire de Biologie des Insectes
Université Paul Sabatier
Toulouse, 31062, France

Klaus Sander — Institute für Biologie I (Zoologie)
Albert-Ludwigs-Universität
Albertstrasse 21a
D 7800 Freiburg i.Br., Federal Republic of Germany

D. Spencer Smith — Department of Zoology and the Hope Entomological Collection
Oxford University
Oxford, England

Annette Szöllösi — Laboratoire d'Histophysiologie Fondamentale et Appliquée
Université Pierre et Marie Curie
12 rue Cuvier
75005 Paris, France

William H. Telfer — Department of Biology
University of Pennsylvania
Philadelphia, Pennsylvania 19104, USA

Linda Wilkinson — Department of Zoology
University of Washington
Seattle, Washington 98195, USA

Dieter Zissler — Institut für Biologie I (Zoologie)
Albert-Ludwigs-Universität
Albertstrasse 21a
D 7800 Freiburg i.Br., Federal Republic of Germany

Preface

Fourteen years have passed since the publication of David Spencer Smith's *Insect Cells: Their Structure and Function*. Here the results of a decade of electronmicroscopic studies on insect cells were summarized in an organized and integrated fashion for the first time, and the ultrastructural characteristics of different specialized cells and tissues were abundantly illustrated in the 117 plates this monograph contained.

In the intervening period great progress has been made in the field of Insect Ultrastructure. Organelles not even mentioned in Smith's book, such as synaptonemal complexes, clathrin baskets, fusomes, and retinular junctions, have been identified and functions proposed for them. There have also been many technical advances that have profoundly influenced the direction of subsequent research. A spectacular example would be the development by Miller and Beatty of the chromosomal spreading technique which allowed for the first time ultrastructural studies on segments of chromosomes containing genes in various stages of replication and transcription. Then there is the freeze-fracture procedure first described by Moor and his colleagues. This technique permitted an analysis of intercellular junctions that was impossible with the conventional sectioning methods. The results greatly clarified our understanding of the channels for ion movement and the permeability barriers between cells and also the membrane changes that occur during the embryonic differentiation and metamorphosis of various types of insect cells.

Along with such scientific milestones involving technical methods there have also been instances where hypotheses have been advanced which explained for the first time findings of unclear significance. An example would be the explanation by Yves Bouligand that a helicoidal packing of microfibrils was responsible for the arced patterns seen in sectioned insect cuticles. Finally, advances made in the genetics, developmental biology, biochemistry, and physiology of a limited number of insect species have been integrated with the results of ultrastructural studies to supply functional interpretations for the intricate organization of various subcellular organelles.

The purpose of the volume that follows is to provide the interested reader with a series of up-to-date, well illustrated, reviews of selected aspects of

Insect Ultrastructure by authorities in the field. The topics fall into three groups, *i.e.*, those dealing with gametes, with developing cells, and with the differentiation and functioning of specialized tissues and organs.

The 16th International Congress of Entomology held in Kyoto, Japan in August of 1980 brought together an extremely diverse group of biologists with very different trainings, research interests and philosophies. In particular it provided a golden opportunity for experts from many countries to meet and discuss various aspects of Insect Ultrastructure. Many of the chapters that follow arose as a result of discussions held during the Kyoto meetings. We deeply appreciate the efforts of the many people who contributed to the success of the Kyoto Congress, and we wish to especially thank Dr. Shoziro Ishii, President of the Congress, Drs. Masatoshi Kobayashi and Yoshio Waku, Members of the Organizing Committee, Dr. Kazuo Hazama, Director of the Sericultural Experiment Station, Dr. Seijiro Morohoshi, President of the Tokyo University of Agriculture and Technology, and Mr. Genkichi Hara, Managing Director of the Kajima Foundation.

Work is underway upon a second volume of similar size that will cover a wide variety of topics. Under the general heading of the "Ultrastructure of Developing Cells" there will be chapters on oogenesis in telotrophic ovaries, on silkworm embryogenesis, and on the high resolution mapping of the giant polytene chromosomes of the larval salivary gland cells of *Drosophila*. Under the heading of the "Ultrastructure of the Development, Differentiation, and Functioning of Specialized Tissues and Organs" there will be chapters on muscle, fat body, digestive and excretory organs, endocrine glands and their target cells, wax glands, silk glands, photoreceptors, and glial cells. Finally there will be chapters dealing with the pathological changes occurring in cells infected by parasites, with tumor-forming cells and with the cell-mediated defense systems of insects. This volume should appear about one year after the first volume is published.

R. C. King and H. Akai
April 1982

References

Bouligand, Y., 1972, Twisted fibrous arrangements in biological materials and cholesteric mesophases, *Tissue Cell* **4**:189–217.

Miller, O. L., Jr., and Beatty, B. R., 1969, Visualization of nucleolar genes, *Science* **164**:955–957.

Moor, H., Mühlethaler, K., Waldner, H., and Frey-Wyssling, A., 1961, A new freezing ultramicrotome, *J. Biophys. Biochem. Cytol.* **10**:1–13.

Smith, D. S., 1968, *Insect Cells: Their Structure and Function*, Oliver and Boyd, Edinburgh.

Contents

I. THE ULTRASTRUCTURE OF GAMETES

III. THE ULTRASTRUCTURE OF THE DEVELOPMENT, DIFFERENTIATION, AND FUNCTIONING OF SPECIALIZED TISSUES AND ORGANS

Chapter 10

Fine Structure of the Cuticle of Insects and Other Arthropods

Barry K. Filshie

Chapter 11

The Structure and Development of Insect Connective Tissues

Doreen E. Ashhurst

Chapter 12

The Structure and Development of the Tracheal System

Charles Noirot and Cécile Noirot-Timothée

Chapter 13

Structural and Functional Analysis of Balbiani Ring Genes in the Salivary Glands of Chironomus tentans

Bertil Daneholt

Chapter 14

Insect Intercellular Junctions: Their Structure and Development

Nancy J. Lane

Contents of Volume 2

I

The Ultrastructure of Gametes

1

The Formation of Clones of Interconnected Cells during Gametogenesis in Insects

ROBERT C. KING, JOSEPH D. CASSIDY, O.P., AND ANDRÉ ROUSSET

1. Introduction

Higher insects have evolved a highly effective mechanism for loading their unfertilized eggs with long-lived messenger RNAs, ribosomes, transfer RNAs, enzymes, and less complex organic molecules. These are utilized later for synthesizing the proteins needed by the young embryo. The mechanism generates endopolyploid nurse cells that produce the required compounds and a system of canals through which these products are exported to the oocyte.

Oocyte–nurse cell syncytia are formed in all insects characterized by meroistic polytrophic ovarioles (Bonhag, 1958; Engelmann, 1970; King, 1970; Mahowald, 1972; Telfer, 1975). A vast number of species fall into this category, including most Lepidoptera, Hymenoptera and Diptera, and in these insects the ovarioles contain a moniliform collection of egg chambers in which each oocyte is one member of a clone of interconnected cells. The other cells of the cluster, the nurse cells, grow and simultaneously transfer cytoplasmic macromolecules to the oocyte (see section 15 for further discussion).

In *Drosophila melanogaster* hundreds of tandem copies of the genes coding for ribosomal RNAs are localized at specific chromosomal sites (reviewed by

ROBERT C. KING • Department of Ecology and Environmental Biology, Northwestern University, Evanston, Illinois 60201, USA.
JOSEPH D. CASSIDY, O.P. • Department of Biological Sciences, University of Illinois at Chicago Circle, Chicago, Illinois 60680, USA.
ANDRÉ ROUSSET • Laboratoire de Biologie des Insectes. Université Paul Sabatier, Toulouse 31062, France.

Chooi, 1976). The nurse cell chromosomes undergo endomitosis and are replicated thousands of times. Since there are fifteen nurse cells, the pool of rRNA genes at the disposal of the oocyte undergoes a 10^6-fold amplification during oogenesis (Dapples and King, 1970). The nurse cells synthesize about 20 billion ribosomes and transfer them through the canal system to the oocyte (Klug *et al.*, 1970).

2. *The Origin of an Oocyte–Nurse Cell Syncytium*

Brown and King (1964) analyzed the pattern of interconnections in *Drosophila* egg chambers after making serial reconstructions of sections viewed with the light microscope. They found that the oocyte and one nurse cell were always connected to each other and to three other cells as well. Two nurse cells were each connected to three other cells, four were connected to two cells, and eight were each connected to only one other cell. They suggested that the lineage of the cells could be determined by the pattern of their interconnections. Since the number of "ring canals" was equal to the number of spindles formed during the four consecutive divisions, they proposed that the interconnected cells arose as the result of incomplete cytokinesis.

Koch and King (1966) made an electron microscopic study of serially sectioned *Drosophila* germaria to identify various developmental stages in the production of egg chambers. They found an apical, mitotically active area in the germarium that contained in its most anterior region a few single cells followed posteriorly by clusters of two, four, eight, and sixteen interconnected cells. The volumes of the cystocytes in this developmental sequence varied in a characteristic fashion. Single cells were the largest, and the cystocytes of sixteen-cell clusters were the smallest in the series. To explain these findings, Koch and King suggested that each germarium contains a few stemline oogonia. Each divides, forming two daughter cells, and these separate. The apical one continues to behave as a stem cell; whereas the other functions as a "cystoblast." The cystoblast and the cells derived from it do not double their birth sizes before entering the next division cycle, as does the stem cell, and the sister "cystocytes" do not separate, but form a system of permanent, interconnecting ring canals. Therefore, as the divisions continue, these cystocytes become smaller and smaller, and the ring canal system becomes more complex (Figure 1A, B).

It follows from the above arguments that the apical regions of germaria in adult ovaries contain two types of single cells that are derived from the germ line. These stem cells and cystoblasts differ in at least four ways. (1) Stem cells double their birth volumes before dividing; whereas the cystocytes derived from cystoblasts divide before reaching double their birth size and, therefore, become smaller with each division. (2) Stem cells proliferate indefinitely; whereas the progeny cells of cystoblasts cease dividing after a specific number of mitoses. (3) Stem cells undergo complete cleavage; whereas cytokinesis of a cystoblast and of each of its progeny cells is incomplete, resulting in an interconnected chain of cystocytes. (4) Only cystoblasts and their progeny move posteriorly through the germarium.

3. *Incomplete Cytokinesis during Spermatogenesis*

In *Drosophila melanogaster* the early stages of both oogenesis and spermatogenesis show striking similarities, since the gonial cells of both sexes give rise to cystoblasts, and these undergo four consecutive divisions to generate a cluster of sixteen interconnected cells. The clones of sixteen spermatocytes then proceed through the meiotic divisions to produce clusters of sixty-four spermatids (reviewed by Lindsley and Tokuyasu, 1980). Thus, each cystoblast generates one oocyte in the female, but sixty-four sperm in the male. According to Hihara (1976), the pattern of interconnections shown by primary spermatocytes in *D. melanogaster, D. tumiditarsus,* and *D. sordidula* is the same as that described by Brown and King (1964) in *D. melanogaster* oocytes.

Thus, in both males and females gametocytes undergo, just before meiosis, a unique series of mitoses that are followed by incomplete cytokinesis. In *D. melanogaster* the number of mitoses is the same in both sexes. However, this equality between the sexes is not universal. For example, in *Bombyx mori* there are three cystocyte divisions in the female to yield an oocyte and seven nurse cells, and six in the male to produce sixty-four primary spermatocytes (King and Akai, 1971b).

4. *Mitotic Synchrony and the 2^n Rule*

Studies have been made of the distribution of dividing cells in the germaria of wild type *D. melanogaster* (Bucher, 1957; Johnson and King, 1972; Grell, 1973). All of the cystocyte divisions were found to take place in the anterior third of each germarium. Here a few isolated metaphases were seen, representing stem line oogonia and cystoblasts. The rest of the division figures were found in groups of two, four, and eight and presumably represented dividing cystocytes. Thus, interconnected sister cystocytes divide in synchrony, and their number (N) after their final division is given by $N = 2^n$, where n equals the number of consecutive mitoses preceding meiosis. N is species specific. For example, since the final number of cystocytes per cluster during oogenesis is eight, sixteen, and thirty-two, respectively, in *Bombyx mori, Drosophila melanogaster,* and *Habrobracon juglandis, n* is three, four, and five in these species.

Certain female lacewings* do not obey the 2^n rule. For example, in *Chrysopa perla* (Rousset, 1978a, b) egg chambers contain 12–14 cystocytes (see section 13). Over 60% of the chambers contain an oocyte and only eleven nurse cells. The first three mitoses proceed as in *Drosophila* to generate eight third generation cystocytes (Figure 1A). First and second generation cystocytes divide in synchrony. At M_4 cells 3, 4, 7, and 8 divide; the rest do not. The result is a central group made up of four third generation cystocytes and two lateral

*The Neuroptera (lacewings), Megaloptera (alderflies), and Raphidioptera (snakeflies) have been grouped into a superorder, the Neuropterida (Kristensen, 1981). However, megalopterans and raphidiopterans have telotrophic meroistic ovarioles (Matsuzaki and Ando, 1977; Buning, 1979; 1980).

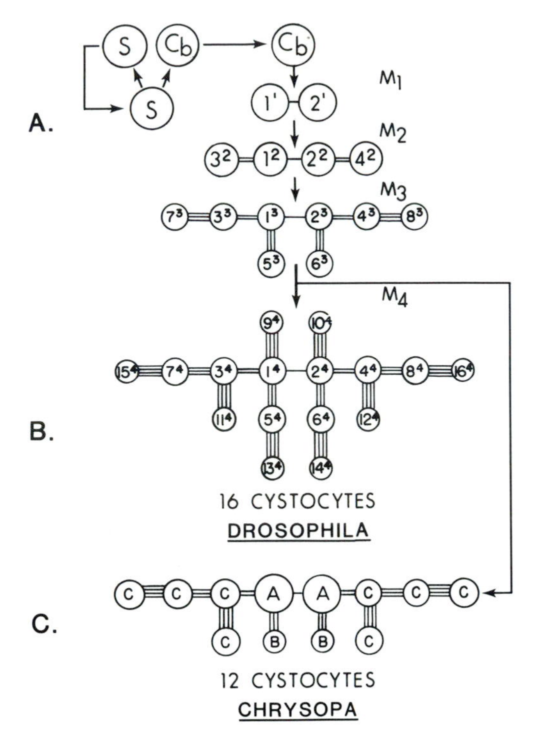

Figure 1. In this diagram, cells (represented by open circles) are given superscript designations (1,2,3,4) depending on whether they are first, second, third, or fourth, generation cystocytes. The connecting canals are labeled according to the division at which they were produced. In the twelve-cell ovarian cyst of *Chrysopa perla*, the A cells correspond to cells 1^3 and 2^3 in the sixteen-cell cyst of *Drosophila melanogaster*. B cells correspond to cells 5^3 and 6^3, and C cells correspond to cells 3^4, 4^4, 7^4, 8^4, 11^4, 12^4, 15^4, and 16^4. Cb, cystoblast; M, mitosis; S, stemline oogonium. See text for further discussion.

groups, each made up of four fourth generation cystocytes (Figure 1C). In *Chrysopa*, as in *Drosophila*, the oocyte differentiates from one of the two central cells (labeled A in Figure 1C). *Chrysopa* males do obey the 2^n rule, since cysts containing sixteen spermatocytes have been observed in *C. ventralis* and *C. prasina* (Suomalainen, 1952).

5. Arrested Cleavage and the Formation of Ring Canals

The mechanism that arrests cleavage remains to be worked out in detail. Normally, cleavage is initiated by a furrow that forms because of microfilaments that slide past one another during the closure of the contractile ring (Schroeder, 1975; Arnold, 1976; Fujiwara *et al.*, 1978). This is a transitory organelle that assumes the form of a continuous annulus beneath the plasma membrane of the cleavage furrow during late anaphase and telophase. The furrow advances centripetally until it is obstructed by the interzonal microtubules of the mitotic spindle. These are compressed until they form a tight bundle. Cytokinesis is temporarily arrested and the daughter cells remain connected by the resulting cytoplasmic bridge. The bridge contains not only the microtubules, but a dense, disc-shaped midbody (Figure 2). The bridge elongates and eventually breaks to one side or the other of the midbody (Byers and Abramson, 1968). During a study of spermatogenesis in the commercial silkmoth, *Bombyx mori*, King and Akai (1971a) proposed that the cleavage furrows are arrested in cystocytes because material from the midbodies some-

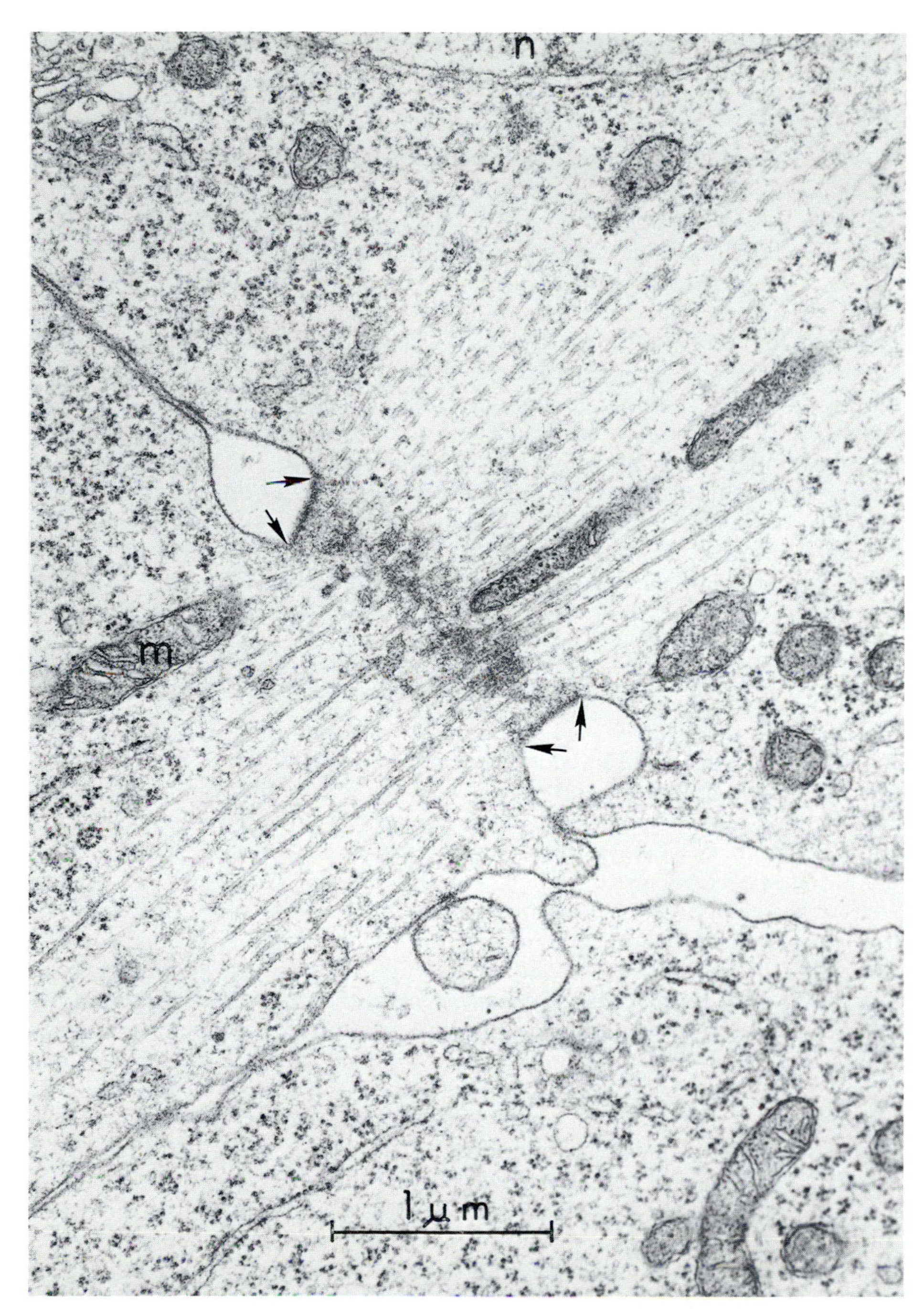

Figure 2. An electron micrograph of a sectioned testis from a fourth instar *Bombyx mori* larva. A cytoplasmic bridge joins two spermatocytes undergoing incomplete cleavage. The bridge contains a disc of dense, fibrous material, the midbody. Arrows show the edges of the contractile ring that occupies the neck of the cleavage furrow. Many of the interzonal microtubules of the spindle have been longitudinally sectioned. Ribosomes are excluded from the area rich in microtubules. m, mitochondrion; n, nucleus.

how crosslinks the component microfilaments of the contractile ring, and therefore they are prevented from constricting the orifice further.

6. Stability of the Plasmalemma Near Canal Rims

Koch and King (1969) utilized electron micrographs of selected serial sections through a normal *Drosophila* germarium to make three-dimensional models of the oldest and youngest sixteen-cell clusters it contained. The oocyte in the oldest cluster had a surface area more than double that of the largest pro-oocyte in the youngest sixteen-cell cyst. Yet the four canals were grouped together in areas of similar size in the two cells. Since the portion of the plasma membrane containing the canal rims seems stable while more distant areas expand, it appears that the oocyte increases its surface areas by adding new membrane to areas remote from the canal zone.

In the model of the young cluster it was found that the volumes of cells 1–8 were larger than those of cells 9–16. This observation shows that the division of third generation cystocytes is unequal and that the cell which gets all the previously formed canals is invariably larger than its sister.

7. The Differentiation of Canal Rims

Koch and King (1969) paid particular attention to the morphology of the ring canals in cells 1 and 2 and noted that as time passed the canal rims increased in diameter and thickness, and deposits developed along their inner circumferences, with different cytochemical properties from the rims themselves.

Developmental changes of the ultrastructure of canal rims are even more striking in germarial cystocytes of *Apis mellifera*. Cystocyte divisions take place in the region near the terminal filament, and here one finds clones of first, second, third, or fourth generation cystocytes. A newly formed canal is surrounded by a dense ring about 0.5 μm high, 0.05 μm thick, and 2.5–3 μm in diameter. The ring represents a thickening of the plasmalemma containing a dense collection of circumferentially arranged microtubules and microfilaments. Ribosomes and mitochondria are excluded from the polyfusome (see section 9) which extends between the canals of the adjoining cells. Each fusome contains microtubules, which may be single or grouped in bundles, and they form a randomly oriented, overlapping collection of tubes embedded in a densely packed mass of pale, shorter fibrils.

In young clusters of *Apis* cystocytes, pro-oocytes can be distinguished from nurse cells by the synaptonemal complexes that form in their nuclei. The polyfusome dissolves, cytoplasmic organelles begin to move through the canals, and an inner coating is deposited on each canal rim. Eventually this coating forms an inner ring about 1.5 μm high that projects above and below the outer ring (Figure 3A). As the inner ring grows, its top and bottom extend inward, so that a vertical section through the ring resembles a vertical section through an automobile tire lying on its side. The inner ring is composed of circumferentially arranged, parallel rows of microtubules and filaments (Figure 3B). The outer ring must grow in circumference, since rings of 5 μm diameter are commonly observed in young oocytes.

8. The Origin of the Branched System of Canals

During *Drosophila* oogenesis the cystocytes form a branching chain of cells (see Figure 1). To generate a branched chain of sister cells, the division planes must be oriented in such a way that during a given division cycle one sister cell always retains all previously formed canals, the other none. Since the positions of the division planes are determined in turn by the positions of the centriole pairs (Rappaport, 1971), there must be a mechanism that controls the orientations of centriole pairs during the consecutive cystocyte divisions. A possible mechanism was suggested by Koch *et al.* (1967). If during the interval between mitoses, a mother and a newly formed daughter centriole always moved in opposite directions roughly one quarter the circumference of the cell, and if the directions of the movements were always perpendicular to the long axes of the previous spindles, the cleavage furrows would always develop at right angles to the furrows of the previous generation. In a given division cycle this behavior would result in one sister cell retaining all the previously formed canals, the other none. The main problem with this hypothesis was that it could not specify the precise mechanism used for centriole navigation.

9. Fusomes and Polyfusomes

Telfer (1975) called attention to the fact that during the cystocyte divisions the cytoplasmic residues of the spindles appear to join together to form a tube-shaped gel that extended through the canal system. Giardina (1901), who first described this structure in the ovary of the beetle *Dytiscus*, called it the *residue fusoriale*. Hirschler (1945) invented the less cumbersome term *fusome* to refer to similar material, but he also applied it to a hypothetical organelle that he believed governed the pattern of divisions that a given clone of animal cells could undergo. Telfer wisely adopted fusome as a term to refer only to the tubular gel extending through adjacent ring canals that contained the residues of the spindles and persisted during the cystocyte divisions.

We have found it more convenient to use the term fusome to refer only to the gel derived from a single spindle and polyfusome to designate the continuous structure representing the fusomes formed at two or more cystocyte divisions that have fused secondarily. A typical polyfusome is shown in Figure 4. This is from a clone of *Chrysopa* cystocytes such as diagrammed in Figure 1C. This polyfusome resulted from the fusion of eleven fusomes.

Polyfusomes also form during the production of spermatocyte clones. The situations in the drone honeybee and the Cecropia moth have been described by MacKinnon and Basrur (1970) and Mandelbaum (1980), respectively.

Maziarski published a cytological study of the ovary of the wasp, *Vespa*

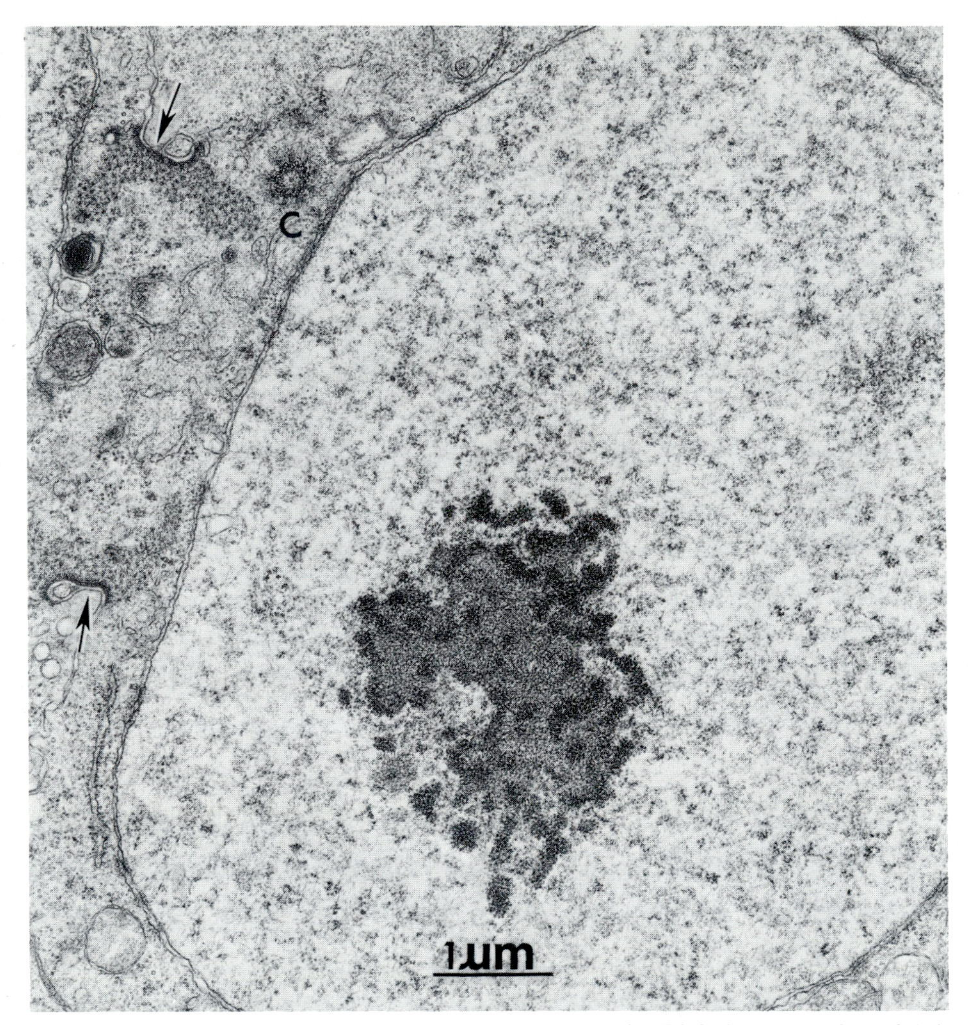

Figure 3A. A nurse cell from a cystocyte cluster from the posterior third of the germarium of *Apis mellifera*. Two ring canals are included in this section. The left one has been cut vertically, and the filaments and tubules making up the inner ring have been cross sectioned. About seventy-five microtubules with associated microfilaments can be counted in each rim. The dense outer ring (arrows) is attached to the plasma membrane. The upper canal has been cut tangentially through the rim, and the inner ring can be seen to contain fiber bundles stacked in parallel rows. c, centriole.

(*Vespula*) *vulgaris*, in 1913. Telfer (1975) called attention to an illustration of a section through dividing cells in the germarium. Each of the five spindles included in the section had one of its poles embedded in a fibrous strand that connected the cells. Telfer assumed that this strand was a polyfusome, and he pointed out that such an arrangement would insure that the sister cell closest to the fusome would receive all previously formed canals, while the more distant cystocyte would receive none.

Let us diagrammatically reconstruct the generation of a polyfusome during the cystocyte divisions in *Drosophila*. In Figure 5A a cystoblast in

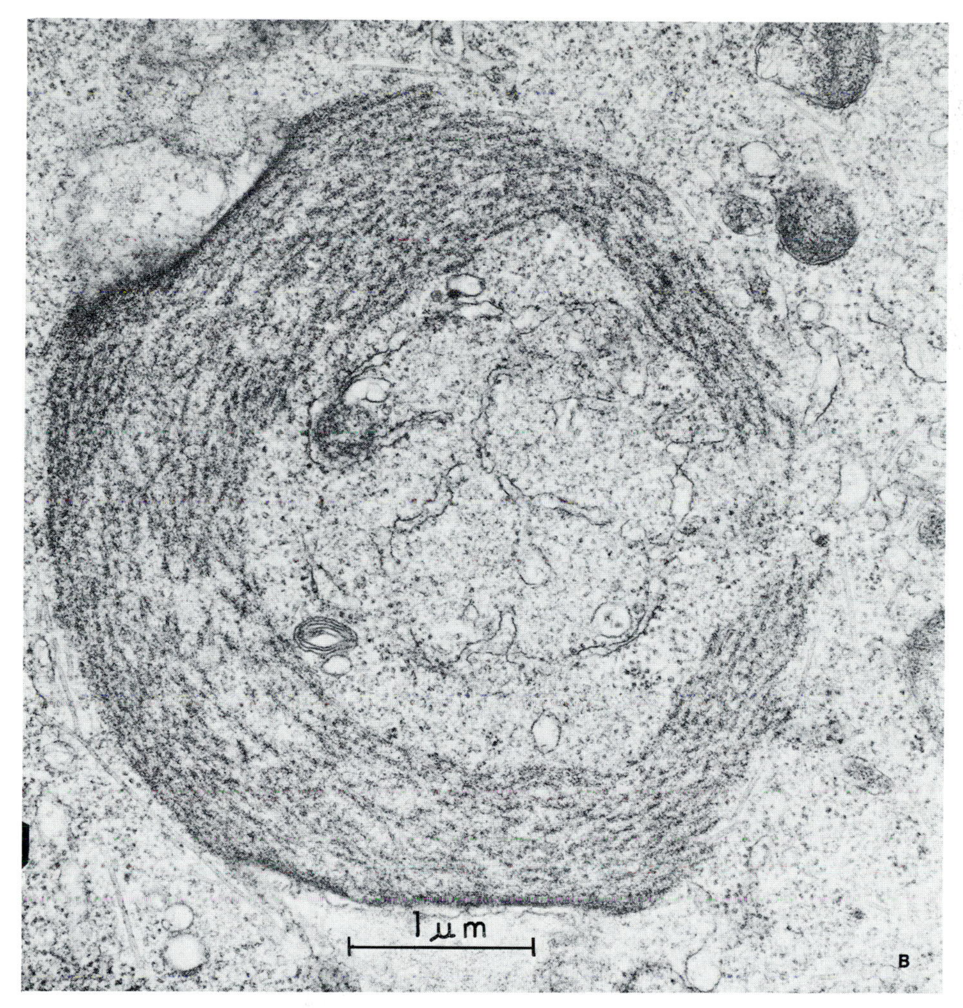

Figure 3B. A horizontally sectioned, *Apis* ring canal showing the circumferential arrangement of loosely packed tubules. Their maximum diameters are similar to the outer diameters of the microtubules in the neighboring cytoplasm.

mitosis is drawn. Centrioles 1 and 2 are at the spindle poles. At the next stage (Figure 5B) two first generation cystocytes have formed. They are joined by a ring canal containing fusome 1. In Figure 5C the fusome has elongated, trapping centrioles 1 and 2. Centrioles 1 and 2 generate centrioles 3 and 4, and these migrate to opposite sides of each nucleus. The nuclei break down and spindles form. As a result, each spindle has one pole attached to the fusome. Four second generation cystocytes are shown in Figure 5D. Two new fusomes extend through the two new canals. In Figure 5E these have elongated and fused with the original fusome. Now centrioles 3 and 4 are also trapped in the polyfusome. Centrioles 1, 2, 3, and 4 generate centrioles 5, 6, 7, and 8, and these move to the opposite side of each nucleus. In Figure 5F the nuclei have broken down,

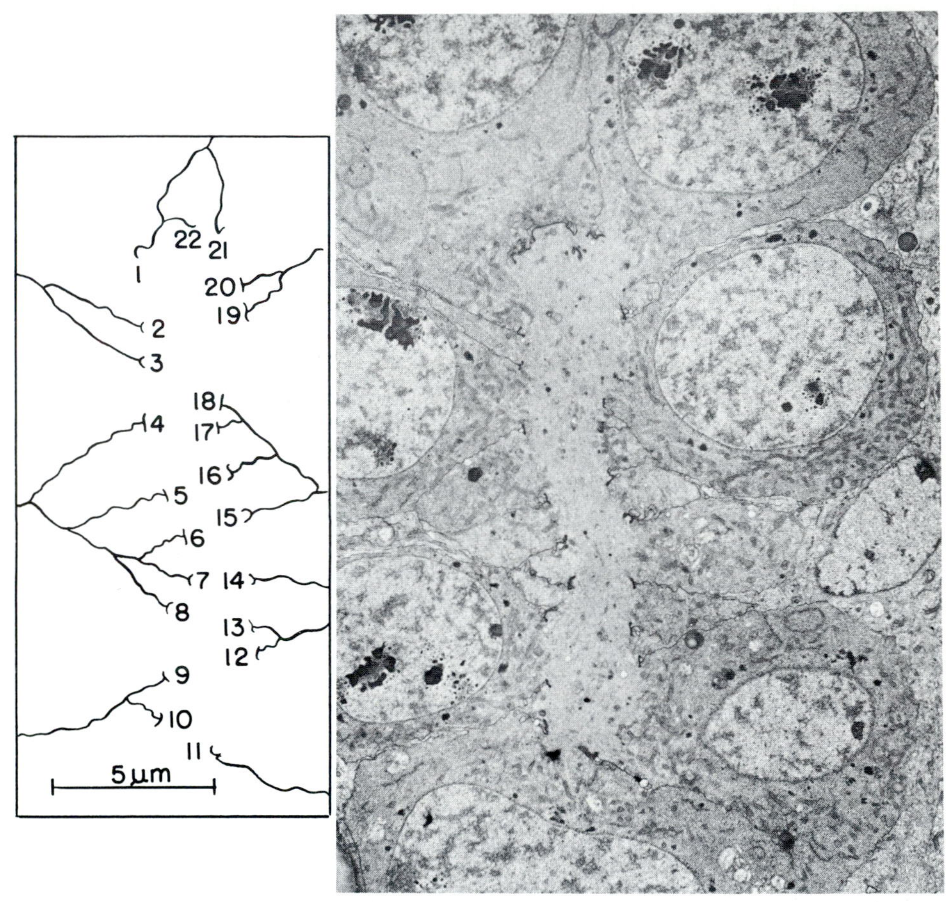

Figure 4. A longitudinal section through a clone of twelve cystocytes from *Chrysopa perla.* The cells are clustered about a central polyfusome. Only eight cystocyte nuclei have been included in the section, but all canals are present, since twenty-two sectioned rims are visible (note left drawing). The polyfusome appears paler than the adjacent cytoplasm because ribosomes are excluded from it. Reproduced from Rousset (1978a).

and the chromosomes are arranged on spindles, which again have one pole embedded in the polyfusome. Eight third generation cystocytes now form, and four new fusomes extend through the four new canals (Figure 5G). The new fusomes fuse to the polyfusome, trapping centrioles 5–8 (Figure 5H). Centrioles 1–8 generate centrioles 9–16. These migrate to the opposite sides of the nuclei and form spindles during mitosis (Figure 5I). In Figure 5J, the sixteen interconnected fourth generation cystocytes are shown. Eight new fusomes extend through the eight new canals. The cells grow, and their nuclei are pushed aside as these fusomes elongate and fuse with the polyfusome (Figure 5K). Note that throughout this proposed scheme the mother centriole always is embedded in the fusome, while the daughter centriole is at the opposite spindle pole. This implies that only the daughter centriole can move out of the fusome.

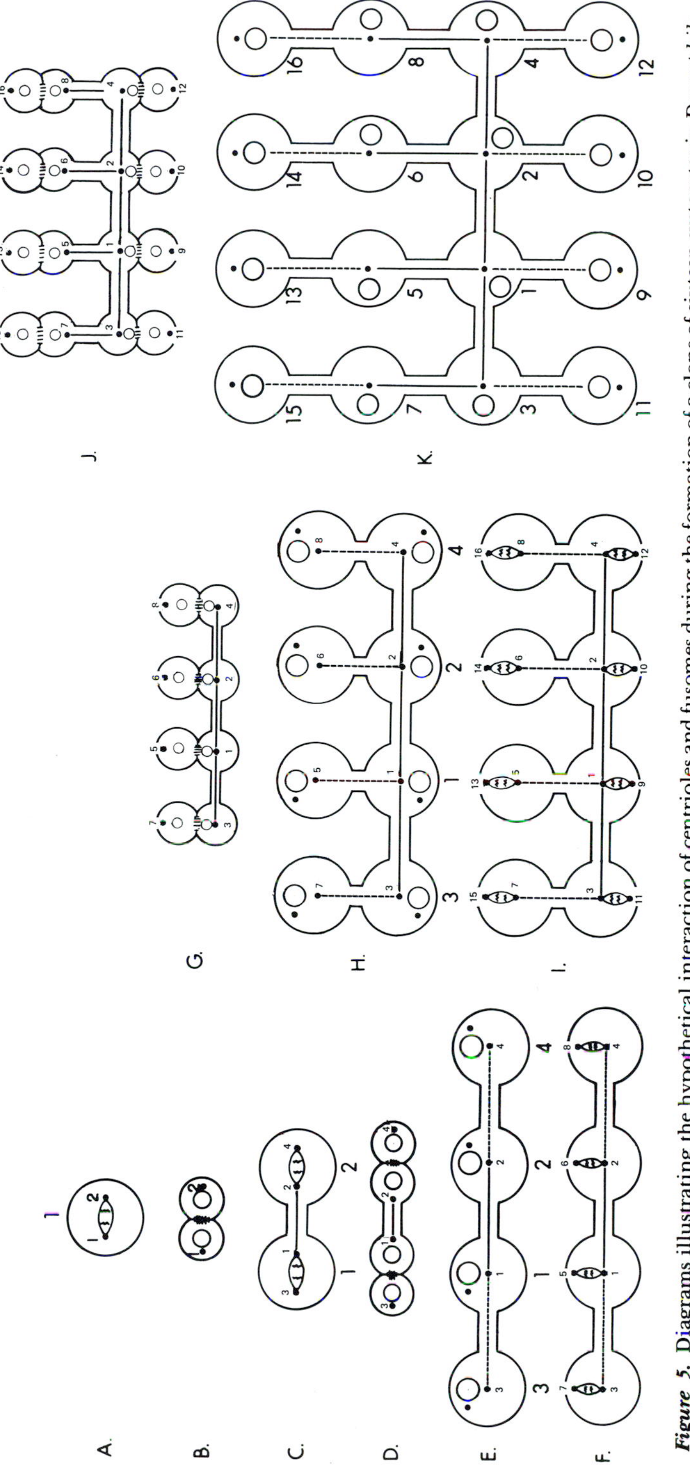

Figure 5. Diagrams illustrating the hypothetical interaction of centrioles and fusomes during the formation of a clone of sixteen cystocytes in *Drosophila melanogaster*. See text for further discussion.

The two-dimensional diagrams give a false impression, in that they show the spindles all oriented in space in parallel and in the same plane (Figure 5C, F, I). The information we have from one serially sectioned cluster shows that this is not the case. Figure 6A illustrates the actual orientation of the cystocytes, and Figure 6B tabulates the relative positions in space of the spindles as estimated from the positions of the fusomes joining the sister cells. Note that the spindle orientations of cells occupying identical positions in the lineage may be the same or different. For example, during the second division, the two spindles were actually at right angles to each other, while during the third division the first and third spindles were parallel, and the second and fourth were at right angles to these, but parallel to each other. During the fourth division the spindles marking the future positions of the 2–10 and the 8–16 canals were parallel to each other but at right angles to the other six spindles, which were parallel to each other. We concluded from these observations that sister spindles are positioned in space independently of each other.

According to the sequence of events outlined in Figure 5, a chain of cystocytes will start to form branches once internal cells are formed. The earliest internal cells have two canals, and a polyfusome extends through them. The cell nucleus can reside anywhere in the cell not occupied by the fusome. For example, if the polyfusome crosses the cell from left to right, the nucleus can be in front of, behind, above, or below it. If one centriole is always adjacent to the nucleus, but anchored to the fusome, and the other moves to the opposite side of the nucleus, then when the nucleus breaks down and the spindle forms, it can project in front of, behind, above, or below the polyfusome, but never can be parallel to it. Thus spindles formed simultaneously need not be oriented symmetrically in space, and Figure 6B illustrates that spindle asymmetries are observed.

According to the proposal of Koch *et al.* (1967) and illustrated in their Figure 7, the orientations of the spindles during one series of cystocyte mitoses

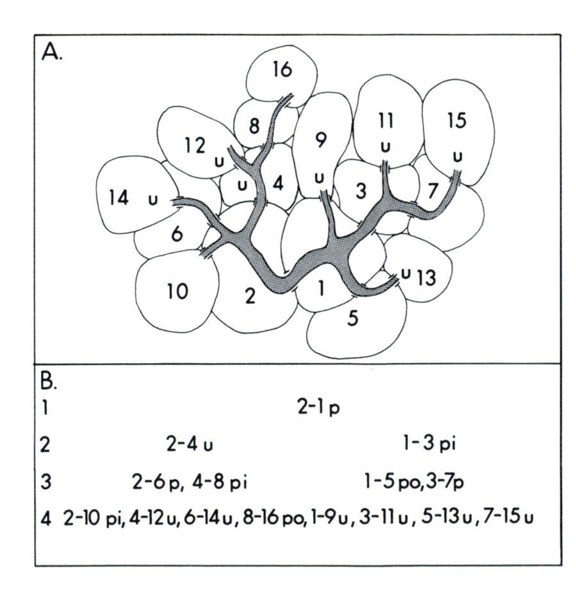

Figure 6. (A) The orientation in space of interconnected sister cells in a newly formed clone in a *Drosophila* germarium. Note, when attempting to visualize the sixteen cells in three dimensions, that those cells labeled with u's extend above the plane of the page. (B) The orientation of the fifteen spindles formed during the four consecutive divisions. p, planar (left–right); pi, planar inward; po, planar outward; u, upward; d, downward. The paired numbers refer to the canals marking the position of the previous spindles. For example, 2-4u refers to the upward-pointing canal joining cells 2 and 4.

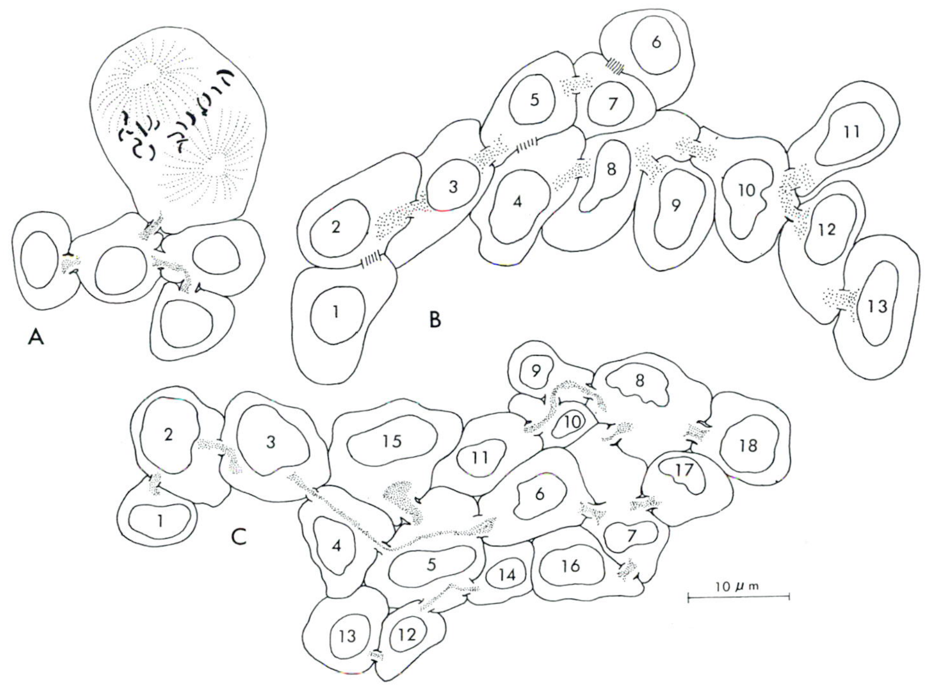

Figure 7. (A) A schematic drawing of five interconnected cells in a tumor from an *fs231 Drosophila*. The upper cell is polyploid and is in metaphase. (B) A clone of thirteen cells. Note that three pairs of cells (1–2, 4–5, 6–7) are joined by canals containing midbodies. (C) A clone of eighteen interconnected cells. It contains four-canal cells (No. 7) and a five-canal cell (No. 5). Reproduced from King (1979).

would always be at right angles to the orientations of the spindles of the previous division. The present proposal is much less rigid and specifies that a spindle is oriented at right angles to the plane of the previous spindle only in two-canal cells. Thus, in one-canal cells the spindle can be parallel to the plane of the previous spindle, and this has been observed. For example, Rousset (1978a) has reported that during the second cystocyte division in *Chrysopa perla* the two spindles are symmetrically oriented at angles of about 150° relative to the axis of the first spindle. It is at the next division, when branches in the chain first occur, that spindles oriented 90° to the previous spindles are observed.

The diagram shown in Figure 5K predicts that in a newly formed sixteen-cell cluster, all one-canal cells should have a centriole distal to the canal, while each of the other cells should have its centriole located in the polyfusome. In Figure 3A the centriole lies between a canal rim and the nucleus, as one would expect for a cystocyte containing at least two canals.

10. Drosophila Mutations That Influence the Cystocyte Divisions

Insights into the genetic control of oogenesis have been gained through studies of pathological changes occurring in the ovaries of *Drosophila* females

Table 1. *Drosophila* Mutants Belonging to the Ovarian Tumor Class

Mutation	Discovered by	Year	Genetic locus	Cytological locus	Mutation included in deficiency	Alleles available for study
fs1621	M. Gans[a]	1975	1-11.7[b]	between 4F1 & 5A1[b]	Df(1)C159[b]	1
fs231	M. Gans[a]	1973	1-22.7[c]	within 7F[d]	Df(1)RA2 & Df(1)KA14[d]	3[e,f]
fu	C. B. Bridges[g]	1912	1-59.5	between 17B6 and 17F1[h]	Df(1)N19[i]	16[j]
fes	C. B. Bridges[g]	1929	2-5±	between 22F and 24D[g]	none	1
nw	C. B. Bridges[g]	1916	2-83	—	—	mutant lost

[a] Gans *et al.*, 1975.
[b] Gollin and King, 1981.
[c] King *et al.*, 1978.
[d] Gans, personal communication.
[e] King and Buckles, 1980.
[f] King *et al.*, 1981.
[g] Lindsley and Grell, 1968.
[h] King, 1970.
[i] Cramer and Roy, 1980.
[j] Wurst and Hanratty, 1979.

homozygous for certain female sterile genes (see reviews by King, 1970; King and Mohler, 1975; and Mahowald and Kambysellis, 1980). Five recessive mutations are known that result in the production of ovarian tumors (see Table 1). The *fu* and *fs1621* mutations can be grouped together because the ovaries of newly eclosed homozygotes are generally normal, and tumors develop as the females age (Smith and King, 1966; Gollin and King, 1981). Ovarian tumors appear earlier if the environmental temperature is raised. In the case of *fes, nw,* and *fs231,* ovarian tumors first form during the pupal period, and the ovaries are completely tumorous when the homozygous females eclose (Koch and King, 1964; King, 1970; King *et al.*, 1978). The number of tumors per ovariole and the number of cells per tumor increase as the mutant females age.

For each of the X-linked mutations, deficiencies are available that lack the wild type alleles (Table 1, column 6), and therefore it is possible to study the ovarian pathology of fs/fs^- females. Since fs/fs^- ovaries are more abnormal than *fs/fs* ovaries, *fs1621, fs231,* and *fu* are hypomorphic mutants. In the case of *fu, fes,* and *nw,* the ovaries behave autonomously in transplantation experiments (King and Bodenstein, 1965). The five mutations listed in Table 1 do not sterilize males.

10.1. The Abnormal Cytokinesis of fes Germarial Cystocytes

The tumors in *fes* ovaries contain hundreds to thousands of cells that resemble cystocytes in that they are similar in size, are mitotically active, and are sometimes interconnected. The mitotic figures in hundreds of *fes* germaria were observed in Feulgen-stained ovarian whole mounts (Johnson and King, 1972). Metaphases were found throughout the *fes* germarium (not just in the anterior third as is normally the case), and the number of isolated metaphase figures was fifteen to twenty times higher than in wild type. Clusters of two metaphases were twice as abundant in *fes* as wild type, and clusters of four were equally abundant. Clusters of eight were seen about six times more often in wild type than in *fes*, but clusters of three, five, six, seven, nine, ten, and eleven metaphases (which are never observed in wild type germaria) occurred in *fes*. Clustered cells in mitotic synchrony were assumed to be connected. More than twice as many dividing cells were found in the average *fes* germarium than in wild type, showing that the average *fes* cystocyte undergoes at least one additional division before leaving the germarium. Since *fes* germaria are made up of large numbers of single cystocytes and clusters containing two, three, or four cystocytes, Johnson and King concluded that a sizable fraction of mutant cystocytes undergoes complete cleavage and that cystocytes that separate from interconnected sister cells do not receive the cue to stop dividing. The tumors are the result of the continuing division of these cells.

10.2. The Behavior of Cystocytes in fs231 Tumors

In an *fs231* female the average number of ovarioles is twenty, and each ovariole contains an average of four tumors. Since these tumors together contain about 4000 cystocytes and each female has two ovaries, there are at least 160,000 cystocytes per fly. A normal female lays about 1000 eggs in its lifetime, and since each egg represents sixteen cystocytes, the total production of cystocytes would be 16,000. It follows that the ovaries of *fs231* females generate about ten times as many cystocytes as those of wild type females. If one assumes that the number of stem line oogonia is the same in *fs231* and wild type females, then the average mutant cystocyte must undergo three or four supernumerary divisions and a total of at least seven consecutive mitoses.

King (1979) analyzed reconstructions made from electron micrographs of serially sectioned regions of three adjacent tumors in an *fs231* ovary. A total of 559 cells was analyzed, and 1.6% were in mitotic stages. About 34% of the cells were not connected to any other cells, while the remainder were present in clusters made up of between two and eighteen cells. The majority of interconnected cells were in clusters of two, three, or four. The tissue segment sampled contained 265 canals, and the ratio between the number of ring canals observed and the number of cells sampled was 0.47. This ratio gives an accurate, direct estimate of the average frequency of incomplete cytokinesis. Thus, in *fs231* tumors, cystocytes undergo complete cytokinesis 53% of the time, and this constitutes one of the primary defects of the mutant.

The distributions of interconnected tumor cells among clusters ranging in size between two and eighteen cells are illustrated in Table 2. The arrangements of the interconnected cells were abnormal, since the chains of cystocytes showed relatively few branches. For example, there was a seven-cell and an eight-cell clone with no side branches, two seven-cell clusters and one nine-cell cluster each with only one side branch, a thirteen-cell clone with only three branches, and an eighteen-cell cluster with only three cells that had more than two canals each.

The five-cell clone shown in Figure 7A demonstrates that interconnected cells are sometimes out of mitotic synchrony. The thirteen-cell clone (Figure

Table 2. The Three-Dimensional Interrelationships of Clones of Interconnected Cells[a]

total		cells/cluster	total		cells/cluster
49		2	1		7
17		3	1		8
15	14 1	4	1		8
1		4	1		8
3		5	1		9
3	2 1	5	2		9
1		6	1		9
1		6	1		13
1		7	1		18
1		7			

[a]Cells with (x) canals: (1) 233; (2) 111; (3) 18; (4) 4; (5) 1; total cells 367.

The normal patterns of connections found in clusters of 8 and 16 ovarian cystocytes are shown below the table. m = mitotic cells. Cells containing 4 or more canals are shown by solid circles. Reproduced from King, 1979.

7B) again suggests that interconnected cells are out of synchrony. In this cluster, midbodies mark the positions of the most recent spindles. Obviously, three cells have divided more recently than the other seven. Note that cells 5 and 7 are connected and have both just divided, since midbodies connect them to cells 4 and 6, respectively. However, cell 3, which is also connected to cell 5, has not yet divided. The position of the midbody connecting the 4–5 pair implies that a spindle had formed with its long axis parallel to that of the major chain, an orientation that normally does not occur. As a result, cell 5 received two and cell 4 received one of the three old canals.

As cells of a normal sixteen-cystocyte clone grow, the canal rims increase in diameter and develop a protein coating along their inner circumference. In cells containing three or four canals (cells 1–4, Figure 5K), the canal rims differ in morphology because they are of different ages. In the *fs231* sample (Table 2) there were eighteen, three-canal cells, four four-canal cells, and one five-canal cell. In these, all rims looked the same as those seen in one- and two-canal cells. Therefore, it is clear that the canal rims of cystocytes in the *fs231* tumors remain immature.

As the distribution of single cells and groups of interconnected cells was worked out for *fs231* tumors, particular attention was paid to the morphology of the fusomes that extended through the canals of interconnected cells, especially those cells with multiple canals. Data are available for a total of 103 cells with two or more canals. In only nineteen of these cases was the cell traversed by a polyfusome. The longest polyfusomes we observed extended through a chain of only four cells. In the five cells with more than three canals, a multipolar fusome was not observed. It follows that in *fs231* cystocytes, fusomes are abnormal in that they seldom fuse with neighboring fusomes.

10.3. The Differentiation of Pseudonurse Cells in Tumorous Ovaries

In a normal sixteen-cell cluster, the nurse cell nuclei undergo many rounds of DNA replication and increase greatly in volume. At first, the homologous chromosomes in each nurse cell are conjoined, and all chromosomes adhere at their centromeric regions. However, as endomitosis proceeds, the chromosomes uncoil, lengthen, and fall apart, so that by the time the maximum DNA content is reached, all fifteen nuclei in each chamber contain jumbled masses of Feulgen-positive threads (King, 1970, Figure VI-1).

Ovarian tumors sometimes contain cells, which, because of their large nuclei, resemble nurse cells (Figure 8A). They are called pseudonurse cells (pnc) because (1) they are usually not attached to an oocyte and therefore are not serving as "nurses" and (2) their nuclei are morphologically different from wild type nurse cells. This is because the replicating chromatids in pnc nuclei remain in register to form giant polytene chromosomes. Pseudonurse cells were first described in the ovaries of *fes* females by King *et al.* (1957). Temperature shift experiments showed that *fes* flies kept for a number of days after eclosion at 18°C have a larger number of pnc in their ovaries than do female flies maintained at temperatures between 21° and 25°C (King *et al.*, 1961; King, 1969a, Figure 2). Lowering the environmental temperature also stimulates pnc to differentiate within the tumorous ovaries of females homozygous for *fs231* (Dabbs and King, 1980, Figure 4) or *fs1621* (Gollin and King, 1981).

Within the ovarioles of *fs231* females reared at 23°C there occur chambers containing (1) only tumor cells, (2) only pnc, and (3) both tumor cells and pnc (King *et al.*, 1978). In chambers containing only pnc, clusters of pnc are fifteen times more abundant than single pnc, suggesting that the probability of pnc differentiation increases once a cell becomes a member of a cluster (Dabbs and King, 1980). However, chambers containing a single nurse cell have been observed about one hundred times, so it is clear that cystocytes can differentiate into pnc in the absence of ring canals.

Oocytes were never observed in *fs231* females reared at 23°C. However, if females were kept at 18°C and then shifted to 23°C, chambers containing an oocyte and large numbers of pnc were observed. However, in those experiments in which oocytes were observed, chambers containing only pnc were one hundred times more frequent than chambers containing pnc and an oocyte. We interpret these results to show that in *Drosophila*, ovarian cystocytes are programmed to enter the nurse cell developmental pathway once they stop dividing and that additional cues must be supplied to switch a cystocyte to the oocyte developmental pathway.

When pseudonurse cells are found in the ovaries of *fs231* females they are

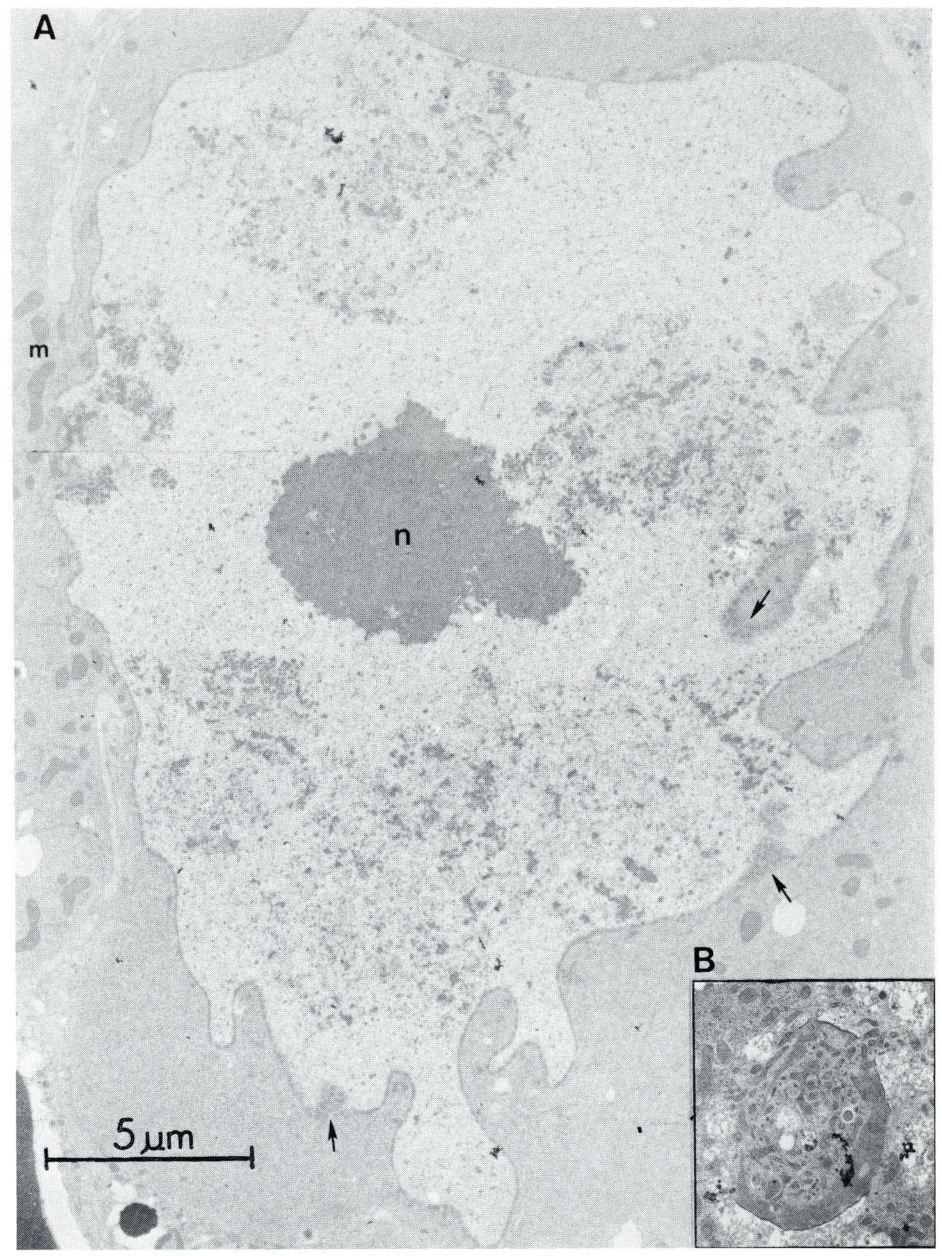

Figure 8. (A) The nucleus of a pseudonurse cell from an *fs231* ovary. Arrows point to areas where the nuclear envelope has been cut tangentially revealing nuclear poles. m, mitochondrion; n, nucleolus. (B) The ring canal connecting this cell to a sister pseudonurse cell.

generally present in chambers containing no tumor cells. Clusters containing two, four, eight, and sixteen pseudonurse cells are the most common (Dabbs and King, 1980, Figure 6). Canals connecting pnc have diameters three to four times larger than those between tumor cells. Rims show the coating of fibrous material characteristic of normal canals (Figure 8B). The nuclei in a given cluster are similar in volume and morphology, and we assume that such cells are members of an interconnected clone of cells. Among the 1400 pnc clusters observed by Dabbs and King, about one fourth contained "forbidden numbers" of cells (numbers other than one, two, four, eight, sixteen, or thirty-two). Clusters containing forbidden numbers of pnc arise presumably because one cell in a clone undergoes complete cytokinesis or because not all cells of the clone divide simultaneously (as is the case for tumor cells, Figure 7A).

Dabbs and King paid particular attention to the pattern of pseudonurse cell differentiation in *fs231* females that had been shifted to a lower temperature (23° to 18°C). In instances where tumors (t) and pseudonurse cell chambers (pcc) were present in the same ovariole, pnc were located near the germarium (g). Typical sequences were g-pcc-t-t-t or g-pcc-pcc-t-t. Rare chambers containing both tumor cells and pnc were always located more posteriorly. From this they concluded that most pnc in the ovaries of down-shifted, mutant females were the progeny of germarial cystoblasts activated to produce pnc by cold treatment and that tumor cells in the vitellarium were rarely activated by the same treatment.

In pnc clusters, eight- and sixteen-cell clones outnumber two-, four- and thirty-two-cell clones. So the cue to stop dividing and to differentiate as a pnc generally comes after the third or fourth division. Perhaps a compound necessary for pnc differentiation is synthesized slowly by the clone, and a time interval equivalent to two or three division cycles is required for a critical concentration to be reached within the clone. This suggestion also implies that the cold-shocked cystoblast and its progeny have started to differentiate by synthesizing a nurse cell factor before they have finished their consecutive series of mitoses.

11. Evidence for Differentiation Prior to Mitotic Arrest in Chrysopa

The idea that cystocytes can enter a specific developmental pathway before they have stopped dividing is supported by studies on *Chrysopa perla* (Rousset, 1978a,b). Here, differentiation of primary nurse cells occurs at the four-cell stage. The cleavages producing these cells are unequal, with the result that cells 1^2 and 2^2 are larger than cells 3^2 and 4^2. Cells 1^2 and 2^2 resemble their parents, but the nucleoli of cells 3^2 and 4^2 are more compact. Cells 1^2 and 2^2 cleave unequally again to produce two large (1^3, 2^3) and two small (5^3, 6^3) cells. In contrast, cells 3^2 and 4^2 divide twice, to produce eight cells of about the same size. The result is a twelve-cell clone containing two large A cells, two small B cells, and eight C cells of intermediate size (Figure 1C). C cells differentiate as nurse cells, one A cell becomes the oocyte, and the remaining cells eventually enter the nurse cell development pathway.

12. Synaptonemal Complex Formation by Cystocytes

In a clone containing sixteen genetically identical cells, why should one differentiate into an oocyte and the rest into nurse cells? Early speculations produced a hypothetical, nonreplicating particle, the "oocyte determinant," that was transmitted at each division to but one daughter cell. After the final division of the series, only the cell possessing the particle would enter the oocyte-specific developmental pathway. In *Drosophila melanogaster* such a determinant, whether located in the nucleus or cytoplasm of a cystoblast, would be randomly distributed during the four ensuing divisions, and therefore one would expect that any one of the sixteen fourth-generation cystocytes could become an oocyte. However, Brown and King (1964) showed that the oocyte always contained four ring canals, and so an oocyte determinant had to be ruled out, unless it had properties leading to its being distributed nonrandomly.

When reared at low temperature, *fes* females produce chambers that can contain anywhere from one to forty cystocytes; the modal number is ten. P.A. Smith and P. Murphy (cited in King, 1969b) worked out the distribution of ring canals among the *fes* cystocytes found in thirteen ovarian chambers possessing seven to fifteen cells. Only three chambers contained oocytes, and these possessed four canals each. All cells with one, two, or three canals showed the nuclear characteristics of pseudonurse cells. One oocyte was a member of an eight-cell cluster. These data demonstrated that an oocyte could differentiate in a chamber containing half the normal number of cells, so long as one had four canals, and that no oocyte would develop in a chamber regardless of the number of cystocytes it contained, if none had four ring canals.

However, in a normal chamber where there are two four-canal cells, why don't both enter the oocyte developmental pathway? Actually, both do. Koch *et al.* (1967) showed that immediately after a sixteen-cell cluster was formed *both* four-canal cells synthesized synaptonemal complexes in their nuclei. Since synaptonemal complexes form during zygonema, are completed during pachynema, and are responsible for the synapsis of homologous chromosomes during meiotic prophase (reviewed in King, 1970, Chapter V; Rasmussen and Holm, 1980), both cells 1^4 and 2^4 start meiosis. Since cells 5^4–16^4 never form synaptonemal complexes, it is clear that these pro-nurse cells fail to receive the cue to enter meiotic prophase. Sometimes cells 3^4 and 4^4 also form synaptonemal complexes (Rasmussen, 1974; Carpenter, 1975), so it appears that specific cells connected directly with 1^4 and 2^4 can also receive the stimulus to enter meiosis.

Thus, in *Drosophila melanogaster* the first visible step in the differentiation of a clone of cystocytes is the formation of synaptonemal complexes by a small group of cells. These pro-oocytes are directly connected to each other, and each possesses three or four canals. Cells that contain one or two canals are pro-nurse cells, and these never begin meiosis. Obviously, both pro-oocytes with three canals belatedly switch to the nurse cell developmental pathway, and so does one four-canal cell. On the other hand, in the Lepidoptera it appears that all cystocytes form synaptonemal complexes. This is true for the eight-cell cysts in the ovaries of cecropia moth larvae (Mandelbaum, 1980). In *Bombyx mori* it is not until the end of pachynema that the oocyte can be distinguished from the seven nurse cells (Rasmussen, 1976). In the nurse cells, the synaptonemal complexes subsequently detach from the chromosomes and aggregate to form polycomplexes. In the oocyte, the synaptonemal complexes persist after the polycomplexes in the nurse cells have disappeared.

However, in the Neuroptera and Hymenoptera some, but not all, cystocytes form synaptonemal complexes, as in *Drosophila*. In *Chrysopa perla*, both type A and type B cells (figure 1C) form synaptonemal complexes, but C cells never do (Rousset, 1977). While the majority of the cystocytes in a germarial clone never form synaptonemal complexes, they are seen in the nuclei of certain adjacent, interconnected cells in *Habrobracon juglandis* (Cassidy and King, 1972) and in *Apis mellifera* (unpublished observations).

In *Drosophila melanogaster* egg chambers, the oocyte invariably contains four ring canals, but do cystocytes with four canals invariably differentiate into oocytes? The answer is no, at least in the case of *fs231* homozygotes. King (1979) found four four-canal cells and one five-canal cell among a population of 559 tumor cells analyzed from a *fs231* female. Synaptonemal complexes were not present in the nuclei of any of these five cells, and calculations indicated that cells with more than three canals occur in the ovaries of *fs231* homozygotes thousands of times more often than oocytes. Therefore, while normal oocytes always have four canals, in clones of tumor cells there are cells with four or more canals that fail to differentiate into oocytes. However, in the case of *fs231*, the four- and five-canal cells were abnormal in that their multiple fusomes had not fused.

13. The Developmental Significance of Fusomes

The sixteen interconnected cystocytes in a *Drosophila* egg chamber are the mitotic products of a single cell. Mitosis should insure the nuclear equivalence of all cells, and the particulate cytoplasmic components should be distributed initially in a random fashion between all sister cells. However, there is a cytoplasmic organelle that is distributed nonrandomly during the series of divisions. After the fourth division, *the polyfusome* extends into each cystocyte, but the relative amount of fusomal material per cell is proportional to the number of canals a cystocyte possesses. The polyfusomes meet the requirements placed upon oocyte determinants, since pro-oocytes would have received the highest concentrations of fusomal materials.

Much of what we know concerning early normal oogenesis in *Drosophila melanogaster* and the cytological abnormalities brought about by genes like *fs231* and *fes* can be placed in a logical framework by assuming that the wild type alleles of these genes are active in germarial cystoblasts and their progeny. The products of these genes could facilitate the formation of fusomes, and these in turn could cause cystocyte cleavages to be arrested. The mutant alleles could produce defective fusomes that often allow complete cleavage. Normal fusomes fuse, centrioles interact with fusomes (according to the scheme outlined in Figure 5), and certain orientations of spindles are prevented. Mutant fusomes often fail to fuse with their neighbors, spindles then can assume any orientation, and linear chains of cystocytes result. Furthermore, the normal polyfusome can be viewed as an intercellular organelle through which chemical or physical stimuli are transmitted. We propose that in the two-, four-, and eight-cell stage, mitoses are synchronized by such signals, and at the sixteen-cell stage the signals initiate the synthesis of synaptonemal complexes in the central cells, and mitosis is then shut off throughout the clone. In the tumor mutants, the defective polyfusomes allow asynchronous mitoses, and cells continually detach from clones. Single cells and cells in clones that contain no tetrapolar fusomes continue dividing, and tumors result. Oocytes rarely differentiate because the multiple fusomes in four-canal cells break down prematurely without fusing.

Normal polyfusomes extend into all the cystocytes of a clone, and therefore it is reasonable to suggest that they can function to specify the maximum number of divisions allowed within a given clone. Where four is the maximum, the cue could be the production of a tetrapolar fusome (or that amount of fusomal material) in any cell of the clone. In clones of spermatocytes this may be the only function of the fusome.

In *Chrysopa perla,* 61% of the egg chambers contain twelve cells, 31% contain thirteen, and 8% contain fourteen cystocytes. There are always eight C cells, but there may be an additional A cell or an additional B cell or both (Rousset, 1978a). From these data one must conclude that mitotic synchrony is not controlled as rigidly at M_4 in *Chrysopa* as in *Drosophila*. Perhaps C cells respond to a weaker signal than A cells or B cells.

In *Chrysopa perla,* as in *Drosophila melanogaster,* the most extensive synaptonemal complexes are formed in the two central cystocytes. In both species, less extensive synaptonemal complexes also form in two other cystocytes, each joined to a central cell. In *D. melanogaster* the lateral cells containing synaptonemal complexes are 3^4 and 4^4 (Figure 1), cells with three canals each and hence richer in fusomal material than any of the twelve remaining cystocytes. However, in *C. perla* the B cells, which also form synaptonemal complexes, correspond to 5^3 and 6^3, and cystocytes 3^4 and 4^4 are primary nurse cells. If one wishes to assign a role to the polyfusome during *Chrysopa* oogenesis, one can suggest that it keeps cystocytes 1^2 and 2^2 from entering the nurse cell developmental pathway. Cystocytes 3^2 and 4^2 become C cells and transmit "C-ness" to their offspring. Once cystocytes have entered the nurse cell developmental pathway, they cannot switch to the oocyte route. However, cystocytes (like B cells) that have begun to differentiate as oocytes can later switch to the nurse cell pathway of differentiation.

14. Cystocyte Clones in the Collembola

The Collembola have meroistic polytrophic ovarioles in which the cystocyte clusters form nonbranching chains (Palévody 1971; 1973; 1976). In *Isoto-*

murus palustris the number of cystocytes per chain is usually sixteen. The chain in *Folsomia candida* usually contains eight cystocytes, but nine have also been observed. All eight cells form synaptonemal complexes. Eventually, when the oocyte becomes morphologically distinct from the seven nurse cells, it is always the second or third cystocyte, starting from the end. Thus, cystocytes with two canals, one formed at the second and one formed at the third mitosis, are potential oocytes. In the winged, holometabolous insects with meroistic polytrophic ovarioles the general rule is to have branching chains of cystocytes, if there are eight, sixteen, or thirty-two, and the oocyte is always a cystocyte containing a canal formed at the first division. *Folsomia* cystocytes have typical ring canals, but no polyfusomes have been observed in the eight-cell chains. So the living representatives of this ancient order* of hexapods form cystocyte clones in a very different way from advanced insects. Polyfusomes are not present, and as one would expect, cystocyte chains do not branch and oocytes do not differentiate from cells placed centrally in the chains.

15. The Determination of the Oocyte and Its Interaction with Nurse Cells

Koch and King (1966, Figure 1) observed that the *Drosophila* germarium could be subdivided into three morphologically distinct regions. The mitosis of germ cells took place in the most anterior region. It contained stem cells, cystoblasts and first, second, third, and fourth generation cystocytes. Region 2 was the area of mesodermal cell invasions. These cells detached from the tunica propria and moved centripetally between the sixteen-cell clusters as they entered region 2. In this way a follicular envelope was produced about each cluster. The posterior sixteen-cell cluster resided in region 3. The follicle cells covering its posterior hemisphere had formed a cuboidal epithelium, and eventually the more anterior follicle cells also became cuboidal. It was at the boundary between regions 2 and 3 that an interleafing of follicle cells resulted in the production of an interfollicular stalk (as shown in Brown and King, 1964, Figure 13). Under optimal conditions, a cluster of sixteen cystocytes enters the vitellarium about once every 12 hours (King, 1970, p. 50), and it carries a layer of about eighty-five follicle cells with it (King and Vanoucek, 1960; Rizzo and King, 1977). These are replaced by the division of profollicle cells in region 2 (Johnson and King, 1972).

In chambers within the vitellarium the oocyte is always the most posteriorly located of the sixteen cystocytes. However, Koch and King (1966) found that in germarial region 2, the two four-canal cells had no preferred orientation. Therefore they concluded that the cell clusters were simply pushed through region 2 by the cells that had proliferated in region 1, and that the posterior migration of potential oocytes occurred in region 3. Compound chambers that

*In earlier phylogenies the Collembola were placed among the apterygote insects. However, taxonomists now separate the Collembola from the true insects. For example, in Kristensen (1981) the class Hexapoda is subdivided into two subclasses, the more primitive Entognatha (containing the Collembola, Protura, and Diplura) and the Insecta. Proturans and diplurans have panoistic ovarioles.

contain two sixteen-cell clusters occur rarely in normal ovaries, but their frequency is greatly increased in the ovaries of certain female sterile mutants (*fused* is an example). In compound chambers the oocytes generally reside at opposite poles (as shown in King, 1970, Figure III-6). These bipolar chambers presumably arise when two clusters are pinched off the germarium simultaneously. The observation that the anterior oocyte comes to lie against the stalk connecting the bipolar chamber to the germarium, while the posterior oocyte lies against the stalk attaching the bipolar chamber to chamber 2, supports the hypothesis that oocytes are attracted by and move toward neighboring stalk cells.

Autoradiographic studies with ^{3}H-thymidine demonstrate that all sixteen cells in a clone of newly formed fourth generation cystocytes replicate their DNA before entering germarial region 2 (Chandley, 1966). Synaptonemal complexes form in cells 1^4 and 2^4 (Figure 1B) during the time they pass through germarial region 2. Eventually a synaptonemal complex extends the entire length of each bivalent, and therefore the pro-oocytes have completed leptonema and zygonema and have entered the pachytene stage of meiosis (Carpenter, 1975). Cytoplasmic organelles start to appear in the ring canals, indicating that the polyfusome is dispersing. The fusome connecting the two pro-oocytes is the last to dissolve. No DNA synthesis occurs in the sixteen-cell clusters during their passage through region 2, but uptake of ^{3}H-thymidine does occur in region 3 (Calvez, 1979). Oocyte nuclei maintain their DNA level at 4C value throughout oogenesis (Mulligan and Rasch, 1980). However, each nurse cell in the first chamber in the vitellarium contains the 8C amount (Mulligan *et al.*, 1982). Therefore, all fifteen nurse cells begin their cycle of endomitotic DNA replications in germarial region 3.

In the posterior portion of region 2, one of the two four-canal cells loses its synaptonemal complexes (Rasmussen, 1974; Carpenter, 1975) and enters the cycle of endomitoses characteristic of nurse cells. The other cell continues to develop as an oocyte and retains its synaptonemal complexes during the previtellogenic stages of oogenesis in the vitellarium (Koch *et al.*, 1967). The oocytes in each posterior chamber in germaria and in anterior chambers in vitellaria are characterized by multiple centrioles (Koch and King, 1969). Mahowald and Strassheim (1970) concluded from an analysis of the location of centrioles in sixteen-cell clusters at different positions in germaria that most of the centrioles in the oocyte were derived from the nurse cells. The movement of cytoplasmic organelles into the oocyte begins at about the same time the other pro-oocyte switches to the nurse cell developmental pathway, and therefore the switchover coincides with the breakdown of the fusomes. The fact that both four-canal cells start to develop as oocytes, but only one finishes suggests that only this cell receives a decisive futher stimulus. The divergence of the two pro-oocytes takes place in the posterior region of germarial region 2. Since this is where the follicle cells first surround the sixteen-cystocyte cluster, it may be that the first four-canal cell to come in contact with a follicle cell is the one chosen to remain an oocyte.

Woodruff and Telfer (1973; 1974) have shown that the polarized movement of macromolecules that takes place within vitellogenic egg chambers of the Cecropia moth is caused by an electric current generated by potential differences between the oocyte and the nurse cells. The equilibrium potential of the nurse cells is several millivolts more negative than that of the oocyte, and therefore negatively charged molecules manufactured by the nurse cells can be carried through the ring canals and into the oocyte by electrophoresis. Earlier studies (for example: Bier, 1963; Pollack and Telfer, 1969; Hughes and Berry, 1970; Paglia *et al.*, 1976) had shown that molecules of nurse cell RNA were transported to the oocyte, and evidently electrophoresis is the mechanism employed. The electrophoretic movement of negatively charged molecules would be expected to continue from nurse cells to oocyte irrespective of the orientation of the oocyte relative to sister nurse cells in the ovariole, and hence in *Drosophila* streams of negatively charged molecules must move simultaneously in opposite directions in bipolar compound chambers. In *Drosophila*, the earliest transfer of cytoplasmic organelles such as centrioles or mitochondria (see King *et al.*, 1968, Figure 17) from nurse cell to oocyte may signal the turning-on of an ion pump similar to that functioning in *Hyalophora cecropia*. In this moth the ion pump is thought to reside in that region of the nurse cell plasmalemma which borders the oolemma (Jaffe and Woodruff, 1979).

In their most recent paper, Woodruff and Telfer (1980) demonstrate that positively charged proteins injected into the Cecropia oocyte are transferred through ring canals to adjacent nurse cells. So there may be a two-way traffic of charged macromolecules, and therefore the oocyte may be able to regulate the activity of the nurse cells by synthesizing appropriate postively charged, effector molecules. Such oocyte regulation could explain why during the development of the egg chamber in the vitellarium the nurse cells closest to the oocyte show the greatest nuclear growth (Jacob and Sirlin, 1959; Brown and King, 1964; Dapples and King, 1970; Mulligan *et al.*, 1982).

16. Summary

In the gonads of a multitude of insect species a series of mitoses occur, each followed by incomplete cytokinesis. These mitoses immediately precede meiosis and generate clones of interconnected germ cells. In the testis, each cystoblast begets equivalent cells, and each clone of spermatocytes proceeds through the meiotic divisions to produce a cluster of spermatids. However, a divergence becomes visible in the destiny of ovarian sibling cells, since one member of the clone completes meiosis (the oocyte) and the others (the nurse cells) do not. The cell that eventually becomes the oocyte always occupies a central position in the clone, and cells directly connected to it sometimes also develop as oocytes, before switching to the nurse cell development pathway. Nurse cells undergo many endomitotic cycles of DNA replication, whereas oocytes stop at 4C.

Dividing cystocytes are joined by a system of interconnected fusomes that extend through the ring canals. We suggest that fusomes arrest the cleavage furrows, and that polyfusomes synchronize the mitoses of sibling cells, influence the positioning of mitotic spindles in space, and restrict the number of consecutive mitotic divisions the clone undergoes. In ovarian cystocytes, the polyfusome may also function as an oocyte determinant. The polyfusome eventually dissolves, and then a polarized flow of negatively charged molecules from the nurse cells to the oocyte takes place.

Drosophila females homozygous for certain recessive mutations produce ovarian tumors that result from the uncontrolled division of cystocytes that undergo complete cytokinesis. In mutant ovaries, oocytes are rarely seen, but large numbers of pseudonurse cells can differentiate. We suggest that the underlying cause of aberrant oogenesis in these mutants is the production of defective fusomes.

ACKNOWLEDGMENTS

The authors are grateful to Drs. Susanne M. Gollin, Pamela K. Mulligan, and Annette Szöllösi for their critical comments on early drafts of this chapter. The paper contains previously unpublished results of research at Northwestern University supported by a grant by the National Science Foundation (PCM7907597).

References

Arnold, J. M., 1976, Cytokinesis in animal cells: New answers to old questions. In *The Surface in Animal Embryogenesis and Development*, edited by G. Poste and G. L. Nicholson, pp. 55–80, Elsevier, Amsterdam.

Bier, K., 1963, Synthese, interzellulärer Transport, und Abbau von Ribonukleinsäure im Ovar der Stubenfliege *Musca domestica*, *J. Cell Biol.* **16**:436–440.

Bonhag, P. F., 1958, Ovarian structure and vitellogenesis in insects, *Annu. Rev. Entomol.* **3**: 137–160.

Brown, E. H., and King, R. C., 1964, Studies on the events resulting in the formation of an egg chamber in *Drosophila melanogaster*, *Growth* **28**:41–81.

Bucher, N., 1957, Experimentelle Untersuchungen über die Beziehungen zwischen Keimzellen und somatischen Zellen im Ovar von *Drosophila melanogaster*, *Rev. Suisse Zool.* **64**:91–188.

Büning, J., 1979, The telotrophic-meroistic ovary of Megaloptera. I. The ontogenetic development, *J. Morphol.* **162**:37–66.

Büning, J., 1980, The ovary of *Raphidia flavipes* is telotrophic and of the *Sialis* type (Insecta, Raphidioptera), *Zoomorphologie* **97**:127–131.

Byers, B., and Abramson, D. H., 1968, Cytokinesis in HeLa: Post-telophase delay and microtubule associated motility, *Protoplasma* **66**:413–435.

Calvez, C., 1979, Duration of egg-chamber growth in young imagines of *Drosophila melanogaster* Meig., *Develop. Growth Differ.* **21**:383–390.

Carpenter, A. T. C., 1975, Electron microscopy of meiosis in *Drosophila melanogaster* females. I. Structure, arrangement, and temporal change of the synaptonemal complex in wild-type, *Chromosoma* **51**:157–182.

Cassidy, J. D., and King, R. C., 1972, Ovarian development in *Habrobracon juglandis* (Ashmead) (Hymenoptera: Braconidae). I. The origin and differentiation of the oocyte–nurse cell complex, *Biol. Bull.* **143**:483–505.

Chandley, A. C., 1966, Studies on oogenesis in *Drosophila melanogaster* with ^{3}H-thymidine label, *Exp. Cell Res.* **44**:201–215.

Chooi, W. Y., 1976, RNA transcription and ribosomal protein assembly in *Drosophila melanogaster*. In *Handbook of Genetics*, vol. 5, *Molecular Genetics*, edited by R. C. King, pp. 219–265, Plenum, New York.

Cramer, L., and Roy, E., 1980, New duplications and deficiencies, *Drosophila Inform. Serv.* **55:** 200–203.

Dabbs, C. K., and King, R. C., 1980, The differentiation of pseudonurse cells in the ovaries of *fs231* females of *Drosophila melanogaster* Meigen (Diptera: Drosophilidae), *Int. J. Insect Morphol. Embryol.* **9:**215–229.

Dapples, C. C., and King, R. C., 1970, The development of the nucleolus of the ovarian nurse cell of *Drosophila melanogaster, Z. Zellforsch.* **103:**34–47.

Engelmann, F., 1970, *The Physiology of Insect Reproduction,* Pergamon Press, Oxford.

Fujiwara, K., Porter, M. E., and Pollard, T. D., 1978, Alpha-actinin localization in the cleavage furrow during cytokinesis, *J. Cell Biol.* **79:**268–275.

Gans, M., Audit, C., and Masson, M., 1975, Isolation and characterization of sex-linked female sterile mutants in *Drosophila melanogaster, Genetics* **81:**683–704.

Giardina, A., 1901, Origine dell' oöcite e delle cellule nutrici nei *Dytiscus, Int. Mschr. Anat. Physiol.* **18:**417–484.

Gollin, S. M., and King, R. C., 1981, Studies of *(1)1621,* a mutation producing ovarian tumors in *Drosophila melanogaster, Develop. Genet.* **2:**203–218.

Grell, R., 1973, Recombination and DNA replication in the *Drosophila melanogaster* oocyte, *Genetics* **73:**87–108.

Hihara, F., 1976, A morphological study of spermatogenesis in *Drosophila,* with special reference to encysted germ cells, *Annot. Zool. Jpn.* **49:**48–54.

Hirschler, J., 1945, Gesetzmässigkeiten in den Ei-Nährzellen-verbanden, *Zool. Jb. Abt. allg. Zool. Physiol.* **61:**141–236.

Hughes, M., and Berry, S. J., 1970, The synthesis and secretion of ribosomes by nurse cells of *Antheraea polyphemus, Develop. Biol.* **23:**651–664.

Jacob, J., and Sirlin, J. L., 1959, Cell function in the ovary of *Drosophila.* I. DNA, *Chromosoma* **10:**210–228.

Jaffe, L. F., and Woodruff, R. I., 1979, Large electrical currents traverse developing Cecropia follicles, *Proc. Nat. Acad. Sci. USA* **76:**1328–1332.

Johnson, J. H., and King, R. C., 1972, Studies on *fes,* a mutation affecting cystocyte cytokinesis, in *Drosophila melanogaster, Biol. Bull.* **143:**525–547.

King, R. C., 1969a, The hereditary ovarian tumors of *Drosophila melanogaster, Nat. Cancer Inst. Monograph* No. 31, pp. 323–345.

King, R. C., 1969b, Control of oocyte formation by *female sterile (fes) Drosophila melanogaster, Nat. Cancer Inst. Monograph* No. 31, pp. 347–349.

King, R. C., 1970, *Ovarian Development in Drosophila melanogaster,* Academic Press, New York.

King, R. C., 1979, Aberrant fusomes in the ovarian cystocytes of the *fs(1)231* mutant of *Drosophila melanogaster* Meigen (Diptera: Drosophilidae), *Int. J. Insect Morphol. Embryol.* **8:**297–309.

King, R. C., and Akai, H., 1971a, Spermatogenesis in *Bombyx mori.* I. The canal system joining sister spermatocytes. *J. Morphol.* **134:**47–55.

King, R. C., and Akai, H., 1971b, Spermatogenesis in *Bombyx mori.* II. The ultrastructure of synapsed bivalents, *J. Morphol.* **134:**181–194.

King, R. C., and Bodenstein, D., 1965, The transplantation of ovaries between genetically sterile and wild type *Drosophila melanogaster, Z. Naturforsch.* **20b:**292–297.

King, R. C., and Buckles, B. D., Jr., 1980, Three mutations blocking early steps in *Drosophila* oogenesis: *fs(4)34, fs(2)A16,* and *fs(1)231M, Drosophila Inform. Serv.* **55:**74.

King, R. C., and Mohler, J. D., 1975, The genetic analysis of oogenesis in *Drosophila melanogaster,* in *Handbook of Genetics,* vol. 3, *Invertebrates of Genetic Interest,* edited by R. C. King, pp. 757–791, Plenum, New York.

King, R. C., and Vanoucek, E. G., 1960, Oogenesis in adult *Drosophila melanogaster.* X. Studies on the behavior of the follicle cells, *Growth* **24:**333–338.

King, R. C., Burnett, R. G., and Staley, N. A., 1957, Oogenesis in adult *Drosophila melanogaster.* IV. Hereditary ovarian tumors, *Growth* **21:**239–261.

King, R. C., Koch, E. A., and Cassens, G. A., 1961, The effect of temperature upon the hereditary ovarian tumors of the *fes* mutant of *Drosophila melanogaster, Growth* **25:**45–65.

King, R. C., Aggarwal, S. K., and Aggarwal, U., 1968, The development of the female *Drosophila* reproductive system, *J. Morphol.* **124:**143–166.

King, R. C., Bahns, M., Horowitz, R., and Larramendi, P., 1978, A mutation that affects female and male germ cells differentially in *Drosophila melanogaster* Meigen (Diptera: Drosophilidae), *Int. J. Insect Morphol. Embryol.* **7**:359–375.

King, R. C., Riley, S. F., Cassidy, J. D., White, P. E., and Paik, Y. K., 1981, Giant polytene chromosomes from ovaries of a *Drosophila* mutant, *Science* **212**:441–443.

Klug, W. S., King, R. C., and Wattiaux, J. M., 1970, Oogenesis in the *suppressor*[2] *of Hairy-wing* mutant of *Drosophila melanogaster.* II. Nucleolar morphology and *in vitro* studies of RNA and protein synthesis, *J. Exp. Zool.* **174**:125–140.

Koch, E. A., and King, R. C., 1964, Studies on the *fes* mutant of *Drosophila melanogaster, Growth* **28**:325–369.

Koch, E. A., and King, R. C., 1966, The origin and early differentiation of the egg chamber of *Drosophila melanogaster, J. Morphol.* **119**:283–304.

Koch, E. A., and King, R. C., 1969, Further studies on the ring canal system of the ovarian cystocytes of *Drosophila melanogaster, Z. Zellforsch.* **102**:129–152.

Koch, E. A., Smith, P. A., and King, R. C., 1967, The division and differentiation of *Drosophila* cystocytes, *J. Morphol.* **121**:55–70.

Kristensen, N. P., 1981, Phylogeny of insect orders, *Annu. Rev. Entomol.* **26**:135–157.

Lindsley, D. L., and Grell, E. H., 1968, *Genetic Variations of Drosophila melanogaster,* Carnegie Inst. of Washington, Publ. No. 627, Carnegie Institution, Washington, D.C.

Lindsley, D. L., and Tokuyasu, K. T., 1980, Spermatogenesis. In *The Genetics and Biology of Drosophila,* vol. 2d, edited by M. Ashburner and T. R. F. Wright, pp. 225–294, Academic Press, London.

MacKinnon, E. A., and Basrur, P. K., 1970, Cytokinesis in the gonocysts of the drone honey bee (*Apis mellifera* L.), *Can. J. Zool.* **48**:1163–1166.

Mahowald, A. P., 1972, Oogenesis. In *Developmental Systems: Insects,* vol. 1, edited by; S. J. Counce and C. H. Waddington, pp. 1–47, Academic Press, New York.

Mahowald, A. P., and Kambysellis, M. P., 1980, Oogenesis. In *The Genetics and Biology of Drosophila,* vol. 2d, edited by M. Ashburner and T. R. F. Wright, pp. 141–224, Academic Press, London.

Mahowald, A. P., and Strassheim, J. M., 1970, Intercellular migration of centrioles in the germarium of *Drosophila melanogaster, J. Cell Biol.* **45**:306–320.

Mandelbaum, I., 1980, Intercellular bridges and the fusome in the germ cells of the cecropia moth, *J. Morphol.* **166**:37–50.

Matsuzaki, M., and Ando, H., 1977, Ovarian structures of the adult alderfly, *Sialis mitsuhashii* Okamoto (Megaloptera: Sialidae), *Int. J. Insect Morphol. Embryol.* **6**:17–29.

Maziarski, S., 1913, Sur la persistance des résidus fusoriaux pendant les nombreuses générations cellulaires au cours de l'ovogénèse de *Vespa vulgaris* L., *Arch. Zellforsch.* (Leipzig) **10**:507–532.

Mulligan, P. K., and Rasch, E. M., 1980, The determination of genome size in male and female germ cells of *Drosophila melanogaster* by DNA-Feulgen cytophotometry, *Histochemistry* **66**:11–18.

Mulligan, P. K., Rasch, E. M., and Rasch, R. W., 1982, Cytophotometric determination of DNA content in nurse and follicle cells from ovaries of *Drosophila melanogaster, Histochemistry* (in press).

Paglia, L. M., Berry, S. J., and Kastern, W. H., 1976, Messenger RNA synthesis, transport, and storage in silkmoth ovarian follicles, *Develop. Biol.* **51**:173–181.

Palevody. C., 1971, L'ovogenèse chez les Collemboles: Structure et évolution de l'ovaire, *C. R. Acad. Sc. Paris, D,* **272**:3165–3168.

Palévody, C., 1973, Differenciation du noyau de l'ovocyte au cours de la prophase méiotique chez les Collemboles (Insectes Aptérygotes). Étude ultrastructurale, *C. R. Acad. Sci. Paris, D,* **277**:2201–2204.

Palévody, C., 1976, L'ovogenèse chez les Collemboles Isotomides. Cytologie et approche physiologique. These Doct. Sci., Univ. Paul Sabatier de Toulouse, Lab. Entomol.

Pollack, S. B., and Telfer, W. H., 1969, RNA in cecropia moth ovaries: Sites of synthesis, transport and storage, *J. Exp. Zool.* **170**:1–24.

Rappaport, R., 1971, Cytokinesis in animal cells, *Int. Rev. Cytol.* **31**:169–215.

Rasmussen, S. W., 1974, Studies on the development and ultrastructure of the synaptinemal complex in *Drosophila melanogaster, C. R. Trav. Lab. Carlsberg* **39**:443–468.

Rasmussen, S. W., 1976, The meiotic prophase in *Bombyx mori* females analyzed by three-dimensional reconstruction of synaptonemal complexes, *Chromosoma* **54**:245–293.

Rasmussen, S. W., and Holm, P. B., 1980, Mechanics of meiosis, *Hereditas* **93**:187–216.

Rizzo, W. B., and King, R. C., 1977, Oogenesis in the *female sterile (1)42* mutant of *Drosophila melanogaster*, *J. Morphol.* **152**:329–340.

Rousset, A., 1977, Formation et développement d'hétérochromatine extrachromosomienne dans le noyau des cellules germinales femelles de *Chrysopa perla* (L.) (Neuroptera), *C. R. Acad. Sci. Paris, D.*, **285**:65–68.

Rousset, A., 1978a, La formation des cystes dans l'ovaire de *Chrysopa perla* (L.) (Insecta, Neuroptera). Étude ultrastructurale, *Int. J. Insect Morphol. Embryol.* **7**:45–57.

Rousset, A., 1978b, La différenciation des cellules germinales dans l'ovariole de *Chrysopa perla* (L.) (Insecta, Neuroptera), *Int. J. Insect Morphol. Embryol.* **7**:59–71.

Schroeder, T. E., 1975, Dynamics of the contractile ring. In *Molecules and Cell Movement*, edited by S. Inoué and R. E. Stephens, pp. 305–334, Raven Press, New York.

Smith, P. A., and King, R. C., 1966, Studies on *fused*, a mutant gene producing ovarian tumors in *Drosophila melanogaster*, *J. Nat. Cancer Inst.* **36**:455–463.

Suomalainen, H. O. T., 1952, Localization of chiasmata in the light of observations on the spermatocytes of certain Neuroptera, *Ann. Zool. Fenn.* **15**(3):1–105.

Telfer, W. H., 1975, Development and physiology of the oocyte-nurse cell syncytium, *Adv. Insect Physiol.* **11**:223–319.

Woodruff, R. I., and Telfer, W. H., 1973, Polarized intercellular bridges in ovarian cells of the Cecropia moth, *J. Cell Biol.* **58**:172–188.

Woodruff, R. I., and Telfer, W. H., 1974, Electrical properties of ovarian cells linked by intercellular bridges, *Ann. N.Y. Acad. Sci.* **238**:408–419.

Woodruff, R. I., and Telfer, W. H., 1980, Electrophoresis of proteins in intercellular bridges, *Nature* **286**:84–86.

Wurst, G. G., and Hanratty, W. P., 1979, Studies on the developmental characteristics of *fused* mutants of *Drosophila melanogaster*, *Can. J. Genet. Cytol.* **21**:335–346.

2

Relationships between Germ and Somatic Cells in the Testes of Locusts and Moths

ANNETTE SZÖLLÖSI

1. Introduction

The development of the insect male gamete has often been considered heretofore from a rather restrictive point of view: i.e., from descriptive accounts of the complex morphological changes occuring in the germ cell during its transformation into the mature sperm. Spermatogenesis is indeed a fascinating case of cell specialization. Besides, it is unique in that it can take place in an already fully differentiated organism and seems to proceed following a very autonomous pattern. Hence, a study of the whole process is readily accessible to the cytologist with only a superficial knowledge of the physiology of the "host-organism," and the anatomy of the sperm has actually been thoroughly investigated in almost all orders of insects. In contrast, the somatic cells of the gonad are rarely mentioned, and have received little attention from the electron microscopists. However, tissues and cells with which the developing gametes are associated or *via* which they are related to the rest of the body, obviously are important for the survival of the germ line. Because this fact was neglected until recently, our knowledge of these somatic cells is still fragmentary, and the approaches to this subject are essentially descriptive. All considerations on the possible functions of the various gonadal somatic cells are mainly speculative, and they will be presented briefly as (sometimes fragile) working hypotheses designed to give direction to future explorations.

ANNETTE SZÖLLÖSI • Laboratoire d'Histophysiologie Fondamentale et Appliquée, Université Pierre et Marie Curie, 12 rue Cuvier, 75005 Paris, France.

In this chapter, I shall refer primarily to the male gonad of locusts (*Locusta migratoria* and other acridids) and of moths (*Anagasta kuehniella* and *Platysamia cynthia*). Three kinds of germ–somatic cell relationships will be considered. (1) The first kind of relationship concerns the close topographical association that occurs regularly and obligatorily between a somatic cell and a germ cell at specific stages of its development. To this type belongs the association between primary gonia and the apical cell, and that between spermatids and cyst cells. Also, under this same heading, a quite peculiar case of phagocytosis of ripe sperm in the genital duct will be reviewed. (2) The second relationship concerns another role played by the somatic cells. By surrounding the developing gemetes, they isolate them from the rest of the body. This isolation of the male germline is probably of prime importance for its survival, and it depends upon the existence of a "blood–testis" barrier located in the enveloping somatic layers of the gonad. (3) The third relationship, up to now demonstrated only in Lepidoptera, consists in the establishment of potential cell-to-cell communications between the germ cell and the somatic cell by way of gap junctions.

2. *General Anatomy of the Insect Testis*

2.1. The "Locust Type" and the "Lepidopteran Type"

The gross anatomy of the testis differs considerably from one insect to another, sometimes even within a given order, owing to the unequal development of the different somatic tissues or internal fluids (reviewed by Matsuda, 1976). From this point of view, the differences between locusts and Lepidoptera are extreme (Figure 1a,b). In locusts, the gonad consists of a series of 150 to 200 follicles that open individually into the sperm duct through an efferent duct. The two gonads are fused in a single ovoid organ, and all follicles are held together by fat body and by a network of tracheae that penetrate between the follicles but never enter them. However, in other insects, the follicles in turn are enclosed within a sheath. In Lepidoptera, the sheath, penetrated by tracheae, is very thick and is fused with the somatic layer that delimits the four follicles of the testis. These follicles open at the same site into the sperm duct. The two testes are separated in some species, such as Saturnids, while in many others they fuse and twist one around the other during the pupal stage.

In spite of these anatomical differences, the basic organization of the male germ line within the follicle is similar in all insects. The site of the follicle, distal to the efferent duct, is defined as the apex. There, primary gonia surround an apical structure represented in locusts and Lepidoptera by a single cell. More basally, the germ cells are found enclosed within cysts. The formation of cysts is realized during the third larval instar in *Locusta migratoria* and in *Anagasta kuehniella*. In the cysts, the germ cells remain interconnected by cytoplasmic bridges, and they develop synchronously as a clone until late spermiogenesis. In moths as well as in locusts, a germ cell undergoes six mitotic divisions producing sixty-four spermatocytes, which enter meiosis,

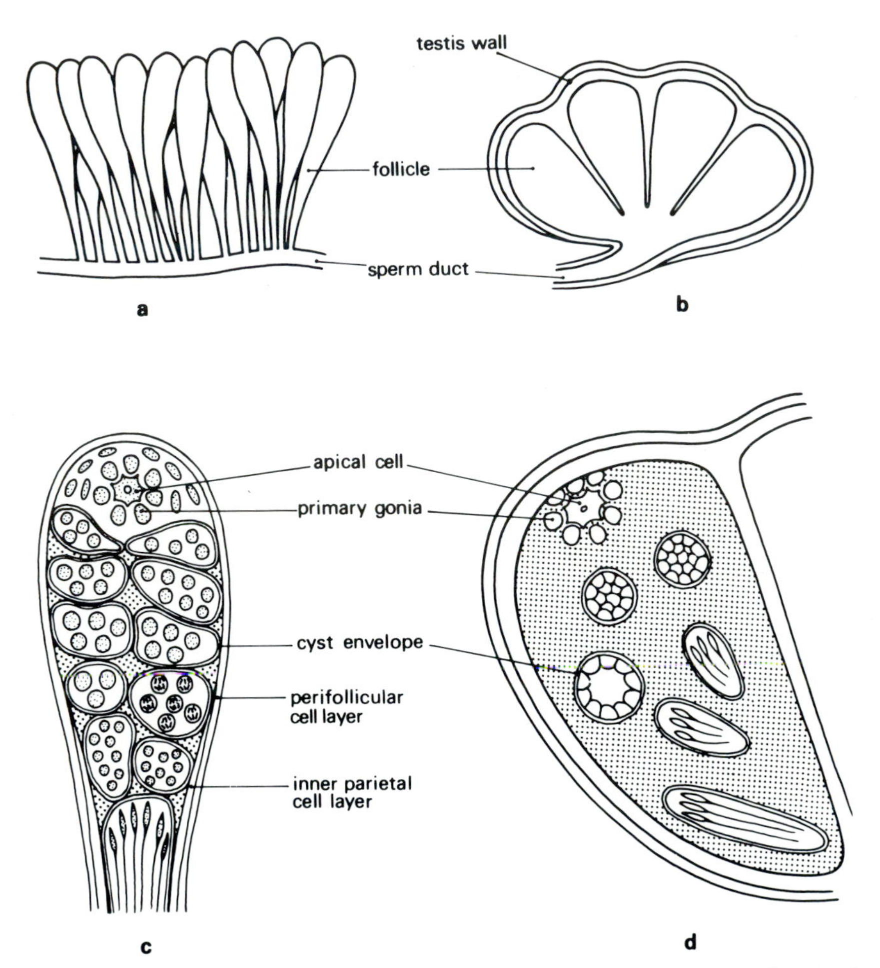

Figure 1. Diagrams presenting the general anatomy of the testis in locusts (a) and moths (b) and the arrangement of the germ cells within the follicles of locusts (c) and moths (d).

giving rise to 256 spermatids. There is almost no space between the germinal cysts in the locust follicle (Figure 1c), and gap junctions are occasionaly found between two adjacent cysts. In contrast, the lepidopteran cysts are isolated from one another, and they are bathed in the abundant fluid that fills the four follicles (Figure 1d). The oldest cysts are located in the basal region of the follicle.

2.2. *Somatic Cells and Envelopes in the Testis*

Except for the cyst cells, the various somatic envelopes of the gonad have received different names from one insect to an other. Tissues "homologous" in terms of their location relative to the germ cells can differ greatly in their structure and possibly in their function. They will therefore be considered separately.

2.2.1. The Follicular Wall of the Locust Testis

The organization of the follicular wall is identical in all locusts. This thin envelope is composed of two different tissues (Szöllösi and Marcaillou, 1977): (1) The perifollicular layer (PFL), which consists of flat cells enclosed between two lamellae, each composed of a carbohydrate-rich amorphous material and containing collagen fibrils (Figure 2). The fibrils are oriented circularly so that their periodicity can be observed only in cross sections of the follicle. A distinctive feature of the perifollicular cells is their high content of microtubules circularly oriented in relation to the major axis of the follicle. Microfilaments showing the same orientation are interspersed among the microtubules (Figure 3). The spaces between adjacent cells are of a constant width of about 20nm.

(2) The inner parietal cell layer (IPL). The thickness of the PFL decreases from the apex to the base, where, under the internal cellular lamella, another tissue appears, made of cells displaying complex overlapping processes and thus exhibiting a tortuous system of intercellular spaces. The IPL becomes conspicuous at the level of the round spermatids and is visible even with the light microscope from this region to the efferent duct. The IPL nuclei were studied in thirty species of grasshoppers (Kiknadzé and Istomina, 1980). They are polyploid (4n to 16n), and their chromosomes demonstrate a highly condensed core surrounded by lampbrushlike peripheral loops. According to the authors, all chromosomes transcribe RNA, and the ^{3}H-uridine label is found mainly over the peripheral fibrils and loops. In the follicle of adult locusts, a space separates the IPL from the subjacent cyst envelopes, providing the route for the sperm bundles in their descent towards the sperm ducts. In contrast to the perifollicular cells, the IPL cells are associated by extensive septate junctions (Figures 2, 4, and 5), and most of the septa are in a plane oriented parallel to the long axis of the follicle.

2.2.2. The Testis Wall in Lepidoptera

The overall organization of the testis wall varies little between different species of Lepidoptera. In *Anagasta kuehniella,* the wall consists of an external layer that surrounds the gonad and of an internal layer that extends inward, separating the four chambers one from the other (Szöllösi *et al.*, 1980). A basement lamella lies over the outer surface of the testis, while another one marks the inner limit of the wall. In the external layer, most cells show a few ER cisterns, large lipid droplets, and scattered glycogen particles. Cells of the inner layer have a denser cytoplasm showing large glycogen-rich areas, lipid droplets, and pigment granules (Figure 6). A quite peculiar network of channels gives a "spongy" aspect to the innermost region of this layer (Figure 7). These dilated intercellular spaces contain a flocculent material. Large tracheae

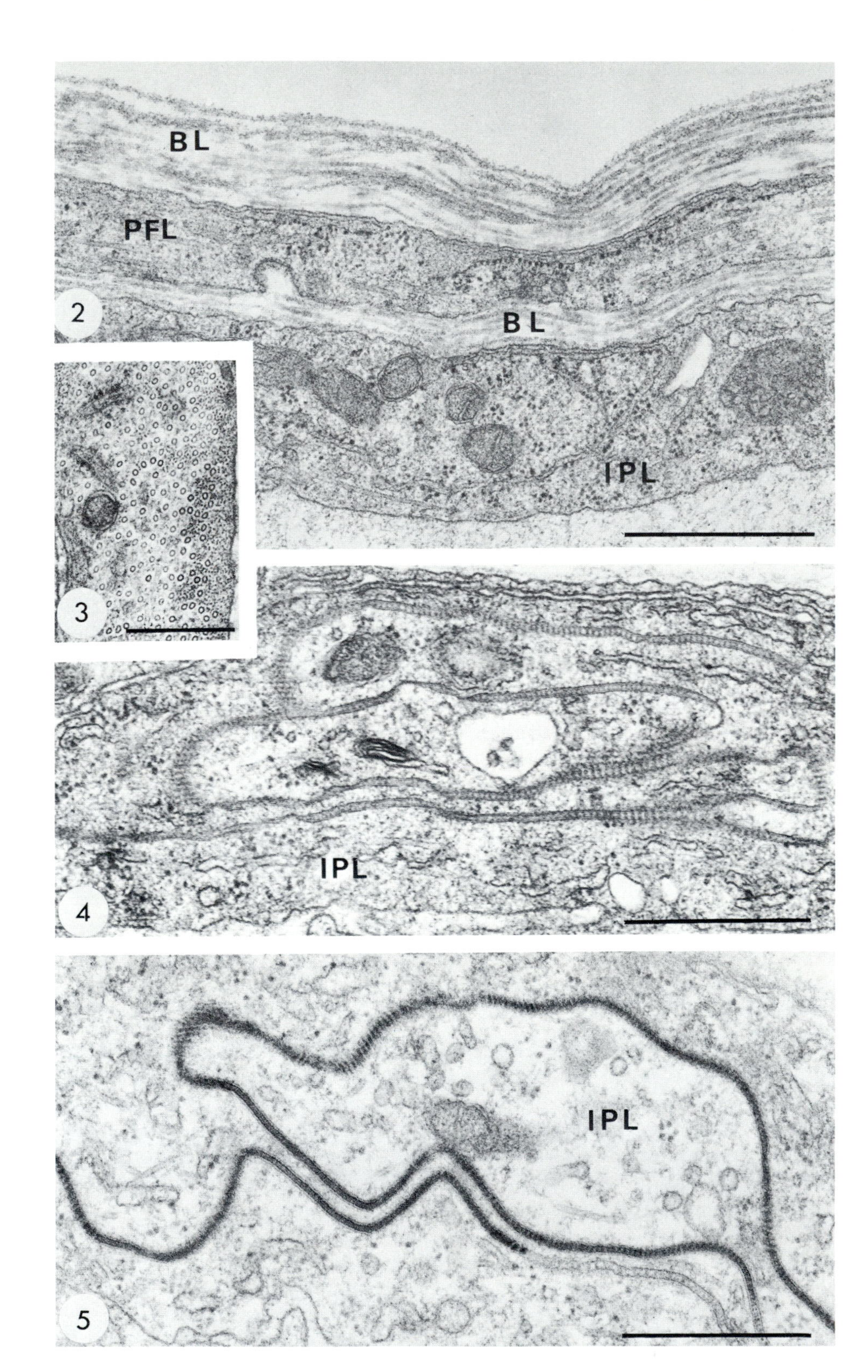
BL
PFL
2
BL
IPL
3
IPL
4
IPL
5

are found in the external layer of the wall, while tracheolae penetrate its inner layer and reach the cavity of the follicle.

2.3. The Cyst Envelope

The cyst envelopes are composed of a few cells. In *Locusta*, there are probably two cells per cyst, but this number has been determined only at the light microscopial level and cannot be considered as accurate. In *Anagasta kuehniella* about twenty cells per cyst of primary spermatocytes were observed. In the cyst cell, the usual organelles are located around an ovoid nucleus, and long slender cytoplasmic processes extend from the perinuclear region to surround the germ cells. Although the cyst envelope is extremely thin, a sinuous system of intercellular spaces is created between adjacent cells by their overlapping processes. Gap junctions, septate junctions, and desmosomes are observed between the cyst cells. The septate junctions are well developed in the lepidopteran cyst envelope. In contrast, they are poorly represented and loosely organized in those of locusts (Figure 8), (Figure 7 in Lane and Skaer, 1980). By freeze fracture, an additional type of membrane differentiation has been found in the cyst wall that might represent junctional structures and will be described later (section 4.6).

3. Specific Associations between Germ and Somatic Cells

3.1. Primary Gonia and the Apical Cell

Since the first description of a "Keimstelle" by Spichardt in 1886, apical cells or groups of cells have been observed in representatives of almost all orders of insects, and numerous interpretations of their role have been proposed (reviewed by Roosen-Runge, 1977). Some new information has been gained recently on their fine structure and life cycle in Lepidoptera (Leclerck-Smekens, 1978) and locusts (Szöllösi and Marcaillou, 1979). The apical cell, already differentiated at the time of hatching, persists throughout the reproductive life of the insect, that is, throughout the period during which spermatogonia enter into mitotic divisions. The apical cell degenerates in lepidopterans, whose imaginal life is short, but not in locusts.

In locusts, the apical cell is star shaped; the spherical, centrally located nucleus is surrounded by a cytoplasm rich in organelles and by radiating

←

Figures 2–5. The testis wall of *Locusta*.
Figure 2. The perifollicular cell layer (PFL) is encompassed by two basal lamellae (BL) containing collagen fibrils. In the basal region of the follicle a second somatic layer, the inner partial cell layer (IPL), lies under the PFL (Scale bar = 1 μm.)
Figure 3. Tubules and filaments of the PFL, circularly oriented, are cross sectioned in a longitudinal section of the follicle. (Scale bar = 0.5 μm.)
Figure 4. The septate junction system in the IPL shows a tortuous pattern. (Scale bar = 0.5 μm.)
Figure 5. Septate junctions of the IPL in negative contrast after Lanthanum impregnation. (Scale bar = 0.5 μm.) Figures 2, 3, and 4 are reproduced from Szöllösi and Marcaillou (1977).

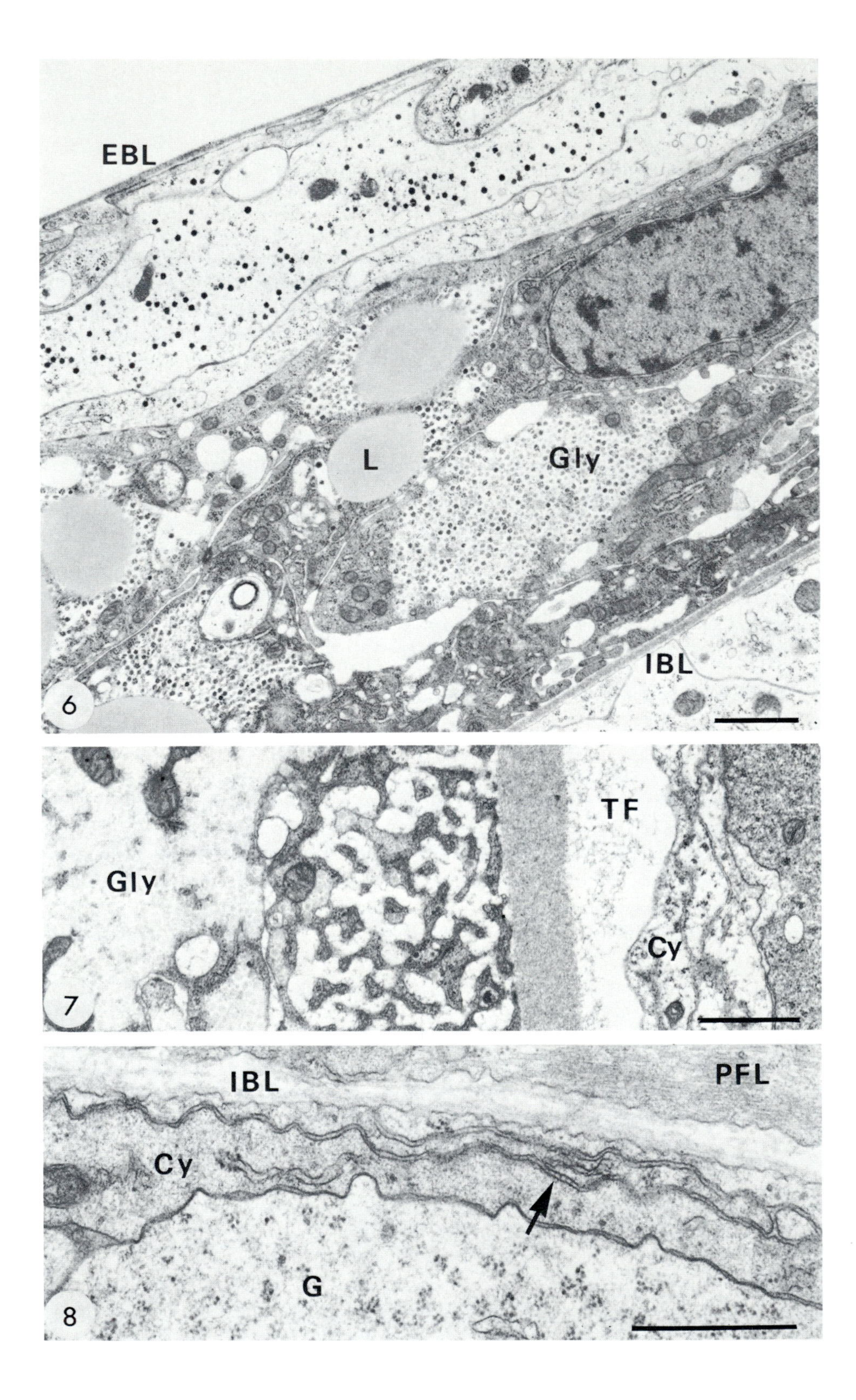
EBL
L
Gly
IBL
6
Gly
TF
Cy
7
IBL
PFL
Cy
G
8

cytoplasmic processes that penetrate between the germ cells and partially surround them (Figure 9). Within the nucleus, the chromatin is highly dispersed, which is consistent with an intense synthetic activity, and a high rate of nuclear ^{3}H-uridine incorporation was reported in the apical cell of the locust *Melanoplus* (Muckenthaller, 1964). A clear nucleus is also described in the apical cell of the moth *Euproctis chrysorrhea* (Leclerck-Smekens, 1978). In moths and locusts, the cytoplasm of the apical cell contains numerous mitochondria, lysosomes, and residual bodies originating from autophagic processes and phagocytosis of exogenous material (whole spermatogonia or parts thereof). In both types of insects, the endocytotic activity of the apical cell has been experimentally demonstrated by the use of an electron-dense tracer (Szöllösi and Marcaillou, 1979; Szöllösi *et al.*, 1980). The most intriguing feature discovered by electron microscopy in the locust apcial cell is the impressive amount of smooth endoplasmic reticulum (SER) this cell contains (Figure 10). The cytomembranes are organized in whorls and stacks of parallel cisterns or in a network of tubular channels occupying the entire periphery of the cell. SER is also reported to occur in the apical cell of the moth *Euproctis* (Leclerck-Smekens, 1978), but is poorly developed and is not organized in whorls.

Although the phagocytotic activity of the apical cell is now clearly demonstrated in locusts and Lepidoptera, the significance of this activity remains enigmatic. Germ cell loss occurs at all stages of development (Roosen-Runge, 1973), and surely some defective gonia could be removed by this means from the apical region before entering the cycle of mitotic divisions. However, we have no way to distinguish any "defect" occurring before an engulfed spermatogonium truly begins to degenerate. The high metabolic activity of the apical cell is suggested by several morphological features (particularly in locusts), but we can only speculate as to its function. Is the cell a secretory one? Does it produce some specific factor(s) acting on the adjacent germ cells? Does it exert a regulatory role on the spermatogenetic process? In the testis of *Drosophila*, a regulatory function has been attributed to a "hub" of apical cells (Hardy *et al.*, 1979; Lindsley and Tokuyasu, 1980). But these cells do not exhibit a secretory activity and are thought to support mechanically the future germ cells and progenitor cyst cells in such a way that orderly generations of spermatogenetic cysts can be formed. The shape of the single apical cell in locusts and moths is compatible with a similar organizing role. The radiating cytoplasmic processes of the locust apical cell are in contact with peripheral somatic cells intercalated between the germ cells (Figure 11), but

Figures 6–8. Testis wall and cyst envelope.

Figure 6. The testis wall of the moth *Anagasta* consists of external clear cells and of internal dark ones. (Scale bar = 1 μm.) Reproduced from Szöllösi *et al.* (1980).

Figure 7. Internal region of the testis wall in the moth *Platysamia*. A network of intercellular spaces gives this region a spongy aspect. (Scale bar = 1 μm.)

Figure 8. A cyst envelope in the testis of *Locusta*. The cells of the cyst envelope (Cy) are associated by a few septate junctions (arrow). (Scale bar = 1 μm.) EBL, external basal lamella; G, spermatogonium; Gly, glycogen; IBL, internal basal lamella; L, lipid; PFL, perifollicular cell layer; TF, testicular fluid.

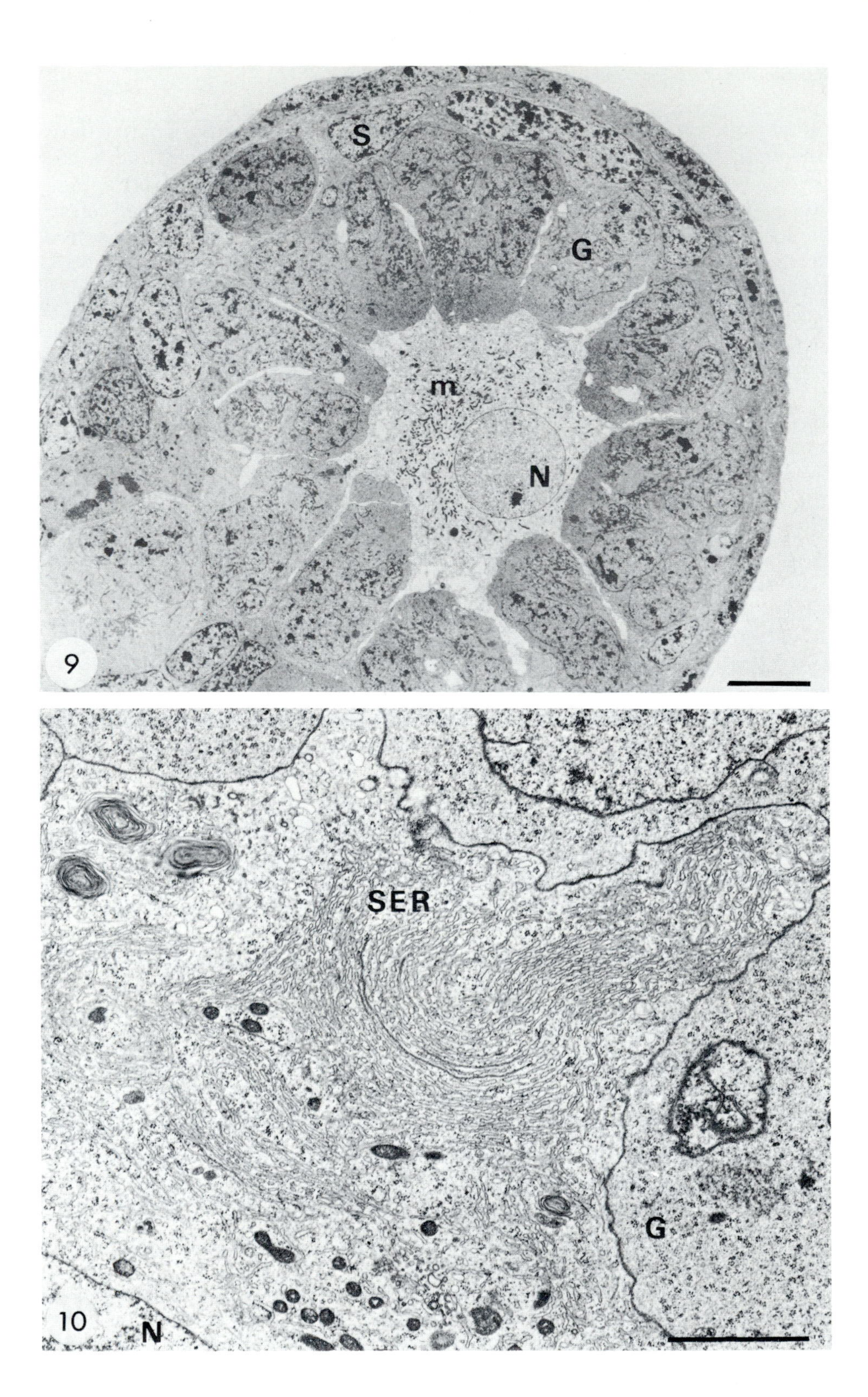

S
G
m
N
9
SER
G
10
N

in locusts, the fate of these cells has not been traced, and it is not certain that they represent "progenitor" cyst cells.

The germ cells surrounding the apical one(s) have received various designations: primary gonia, prospermatogonia, or stem cells. They show a large surface of contact with the apical cell but are not associated with it *via* any junctional differentiations. There are no cytoplasmic bridges between apical and germ cells, and all "membrane interruptions" reported in the literature must certainly be attributed to artifactual ruptures of the membranes. When a primary gonium (or stem cell) divides in the adult testis, one daughter cell remains in contact with the apical cell and continues to function as a stem line spermatogonium, while the other becomes surrounded by cyst cells and undergoes a specified number of mitoses each followed by incomplete cytokinesis (King and Akai, 1971). In locusts, during the larval life, a primary spermatogonium can also divide into two daughter cells, both of which retain their contact with the apical cell. Cytokinesis is delayed for a while, and the spermatogonia remain interconnected by a cytoplasmic bridge. (Figure 12). These canals are similar to those connecting the germ cells within a cyst.

Whatever the immediate fate of the daughter cells may be, it is clear that the apical region constitutes a proliferative center on which the entire process of spermatogenesis depends. It was therefore tempting to try and destroy the apical cell. This was attempted on Lepidoptera (Leclerck-Smekens, 1978), but in my view, a selective destruction has not been realized so far and needs to await further technical improvement. Perhaps, a laser microcautery of the locust apical cell could be performed with minimal damage to the surrounding germ cells, since the apical cell nucleus is clearly visible at the apical blind end of the living follicle under the dissecting microscope.

3.2. Formation of the Sperm Bundle

In all insects, the spermatids are gathered in bundles while they elongate, and their heads are oriented toward one of the cyst cells (Phillips, 1970). At present nothing is known about this step of spermiogenesis and, in particular, about the orientation of the spermatids toward one of the cyst cells. Is this cell "chosen" at random or does it differ in some manner from the other(s)? By the end of sperm differentiation the bundles may or may not disintegrate; in Lepidoptera, the cyst envelopes are destroyed when the mature sperm leave the testis. From this moment on, apyrene (sterile) sperm become dispersed and are stored individually in the seminal vesicles, while eupyrene (fertile) sperm remain loosely associated and are secondarily embedded within a secretory material of the sperm duct (Thibout, 1980). In locusts, it is in the cyst

Figures 9 and 10. The apical cell of *Locusta*.
Figure 9. The star shaped apical cell is surrounded by primary spermatogonia (G) and by somatic cell(s) (S). (Scale bar = 10 μm.)
Figure 10. The smooth endoplasmic reticulum (SER) of the apical cell is abundant and organized in stacks of cisternae. (Scale bar = 1 μm.) m, mitochondria; N, nucleus of the apical cell. Reproduced from Szöllösi and Marcaillou (1979).

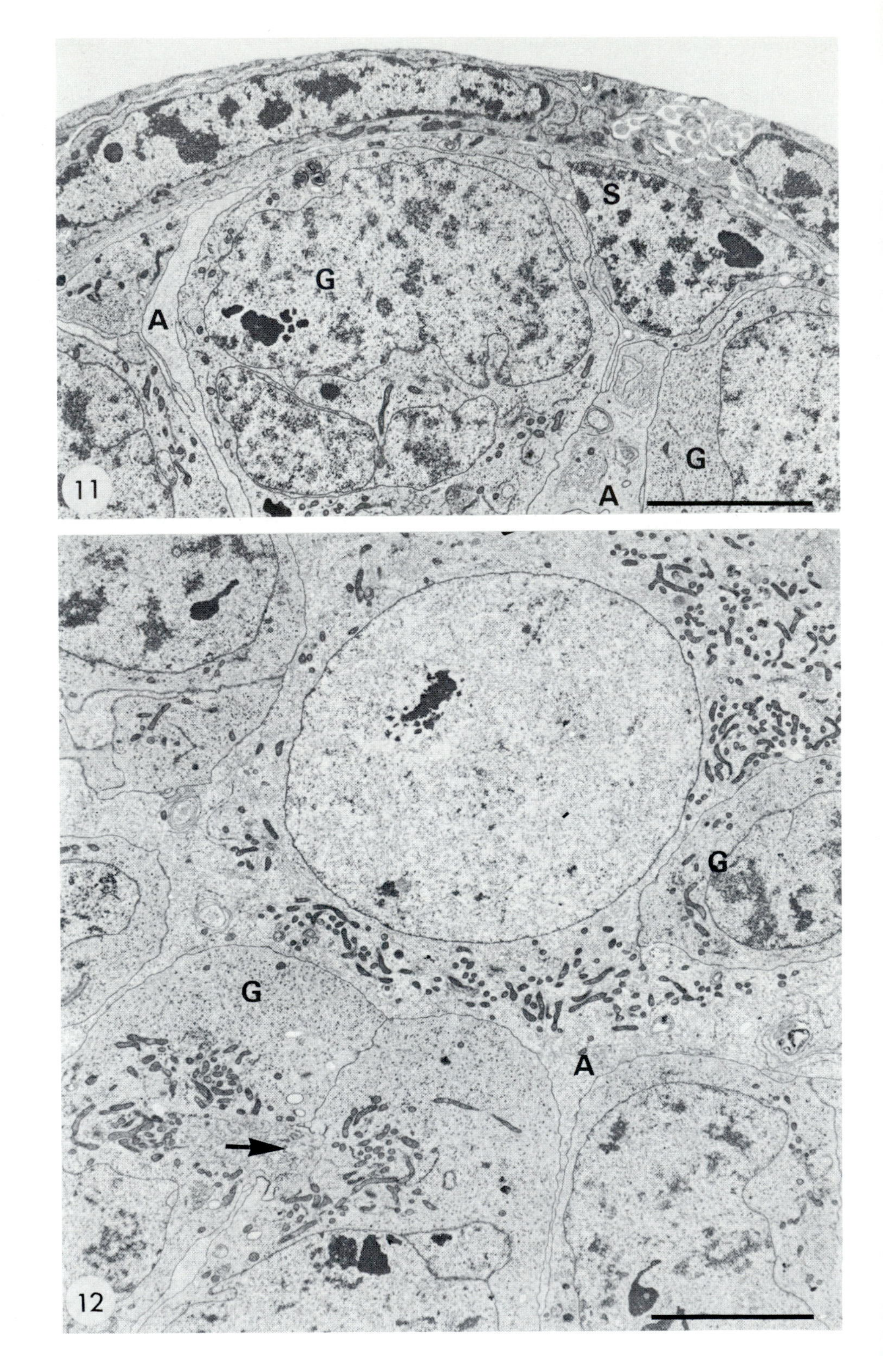

S
G
A
G
A
11
G
G
A
12

and during spermiogenesis that the acrosomes of the spermatids become tightly linked by a cap of glycoprotein (Figure 13) (Szöllösi, 1974; 1975). In this case, the sperm of the original bundle remain firmly associated and are transmitted to the female in the form of "spermatodesms" (Cholodkovsky, 1913). Although the formation of the cap takes place in the immediate vicinity of one of the cyst cells, nothing is seen in this cell that might shed light on its role in the process. So, whether the material of the cap is produced or transformed by the cyst cell remains an open question. Under certain experimental conditions, this particular step of spermiogenesis can be disturbed and the formation of the cap prevented, but in these cases we are ignorant as to whether or not such alterations are mediated by the cyst cell (Cantacuzène *et al.*, 1972; Szöllösi, 1976a,b).

3.3. Mature Sperm and Spermiophagic Cells in the Sperm Duct

In locusts, phagocytosis of mature sperm takes place in the anterior part of the sperm ducts (Cantacuzène, 1971). In the larva, the future sperm duct is represented by a cord of undifferentiated cells showing no lumen, and it is only at the time of the imaginal molt that it truly becomes a duct. At that time, the peripheral cells organize into a simple epithelium, microvilli differentiate at their apex, and they develop an extensive secretory apparatus. Not all cells are integrated into the epithelium lining the duct, and a number of them remain in the center of the lumen, forming a plug of cells along which the sperm bundles have to pass during their descent towards the seminal vesicles (Figure 14). These internal cells retain some larval characteristics such as large nuclei and clear cytoplasms, and they become able to phagocytize mature sperm. A few days after the imaginal molt, when the spermatodesms are liberated from the testes and pass through the sperm duct, some individual sperm come in contact with the plasma membrane of the spermiophagic cells and are engulfed within their cytoplasm. Sperm or sperm fragments are internalized, nuclei and tails are lysed, and, in some areas, only the accessory fibers of the axoneme are observed, either in vacuoles or directly in the cytoplasm of the spermiophagic cells (Figure 15). A similarly located spermiophagic tissue has been found in all studied species of Acridids, but does not exist in other Orthoptera. Perhaps this tissue is present only in those species that possess spermatodesms. Counts performed in transverse sections of the bundles stored in the seminal vesicle reveal that the number of sperm per bundle is always slightly less than the theoretical number of 256. The seminal vesicle never contains isolated sperm. Although these observations might be explained by the death of cells at any stage of the spermatogenetic process,

←

Figures 11 and 12. The apical region of the locust follicle.

Figure 11. The radiating processes of the apical cell (A) are seen in contact not only with primary spermatogonia (G), but also with interspersed somatic cells (S), which might represent future cyst cells. (Scale bar = 4 μm.)

Figure 12. After an incomplete cleavage, two primary spermatocytes (G) (bottom left) remain interconnected by a cytoplasmic bridge (arrow). (Scale bar = 4 μm.) Figure 12 is reproduced from Szöllösi and Marcaillou (1979).

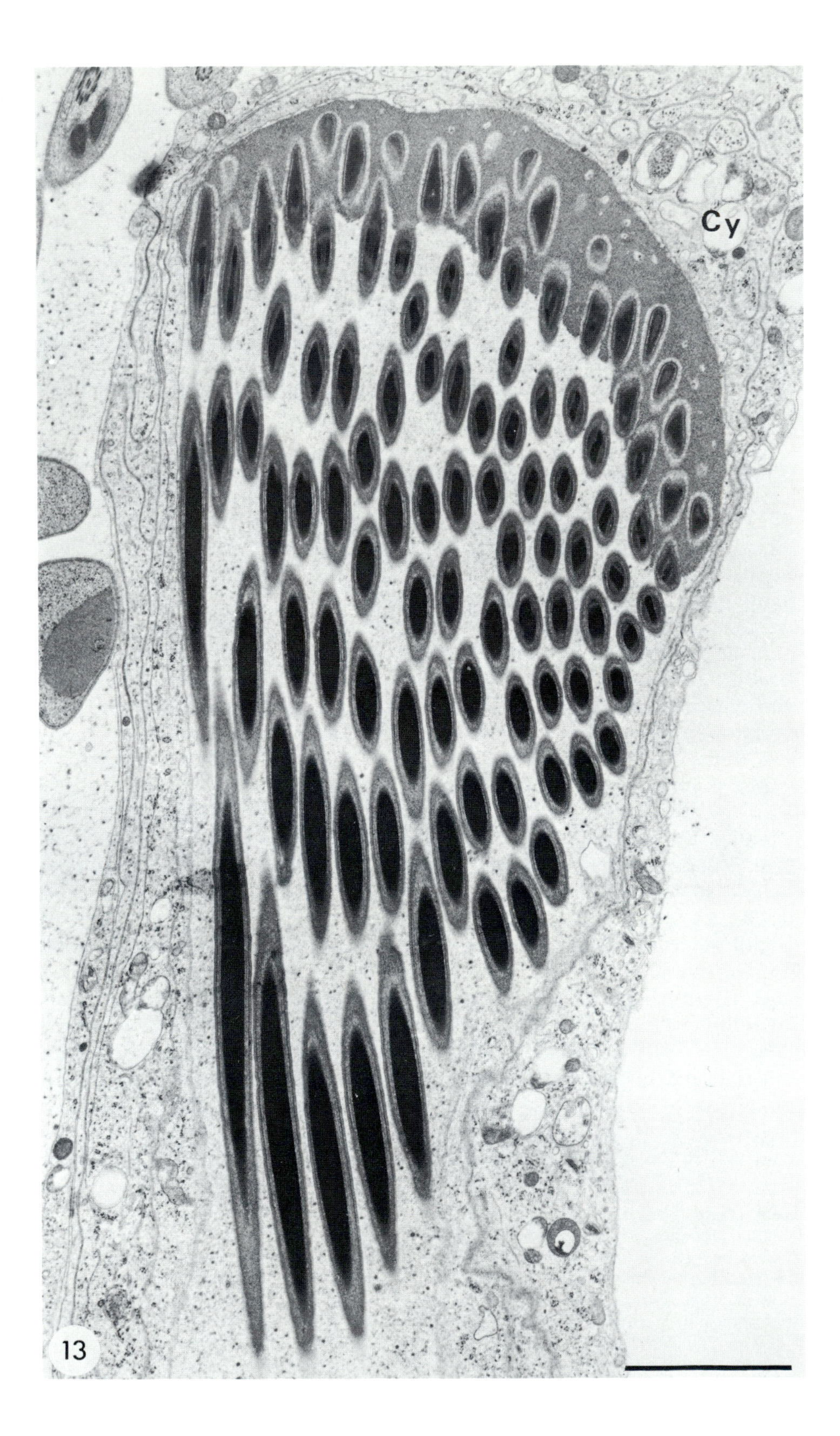

Cy
13

it is clear that the spermiophagic cells eliminate only mature sperm that have been detached (probably accidentally) from the spermatodesms. The existence of a well-defined phagocytic tissue is probably a peculiar feature of the locust gonad, but some phagocytotic phenomena were also observed in the reproductive tract of other insects (Riemann and Thorson, 1976). They may be more widespread, but in a less visible form. It is tempting to consider the elimination of isolated sperm as a regulatory process, but this is highly speculative since the functional significance of persisting sperm associations is not known. Whether they are free or associated in spermatodesms, the sperm of all insects are transmitted to the female enclosed within the spermatophore, a structure that is elaborated by the various secretions of the male accessory glands.

4. The Blood–Testis Barrier

4.1. Historical Perspective

The concept of a "blood-testis barrier" (BTB) in mammals arose at the beginning of the century from the observation that dyes injected into the blood stained many tissues but did not enter the seminiferous tubules. Interest in this area was revived by Kormano (1967), who tested the distribution in the body of a variety of substances injected into the blood stream, and by the physiologists, who found that there were striking differences in the composition of blood plasma and testicular lymph when compared with the fluid of the seminiferous tubule (reviewed by Setchell and Waites, 1975; Howards *et al.*, 1976; Setchell, 1980).

The physiological data suggested that the BTB was located in the seminiferous tubule itself, and this was further demonstrated by examination of testes of animals injected with electron-opaque tracers of various sizes (Dym and Fawcett, 1970; Fawcett *et al.*, 1970). Great importance was attributed to the existence of this barrier, especially when it became clear that autoimmune accidents leading to azoospermia were accompanied by its breakdown. It was also presumed that the BTB was responsible for the high concentrations in the seminiferous tubule of substances needed for spermatogenesis, such as inhibin or androgen-binding protein (Setchell, 1980). The existence of a BTB was demonstrated for the first time in an insect by Marcaillou and Szöllösi (1975) and then studied at the ultrastructural level in locusts (Szöllösi and Marcaillou, 1977; Jones, 1978) and Lepidoptera (Szöllösi *et al.*, 1980).

The term blood-testis barrier has been and still is so widely used in all studies of the mammalian seminiferous tubule that we decided to retain this term for insects as well. It must be realized, though, that in every case studied the barrier is not between the blood and the testis, but inside the testis itself,

---

Figure 13. Formation of a spermatodesm in the locust testis. A glycoprotein substance is deposited around the acrosomes of the maturing sperm. This process takes place in the vicinity of a cyst cell (Cy). It results in the formation of a cap which transforms the bundle into a spermatodesm. (Scale bar = 2 μm.) Reproduced from Szöllösi (1974).

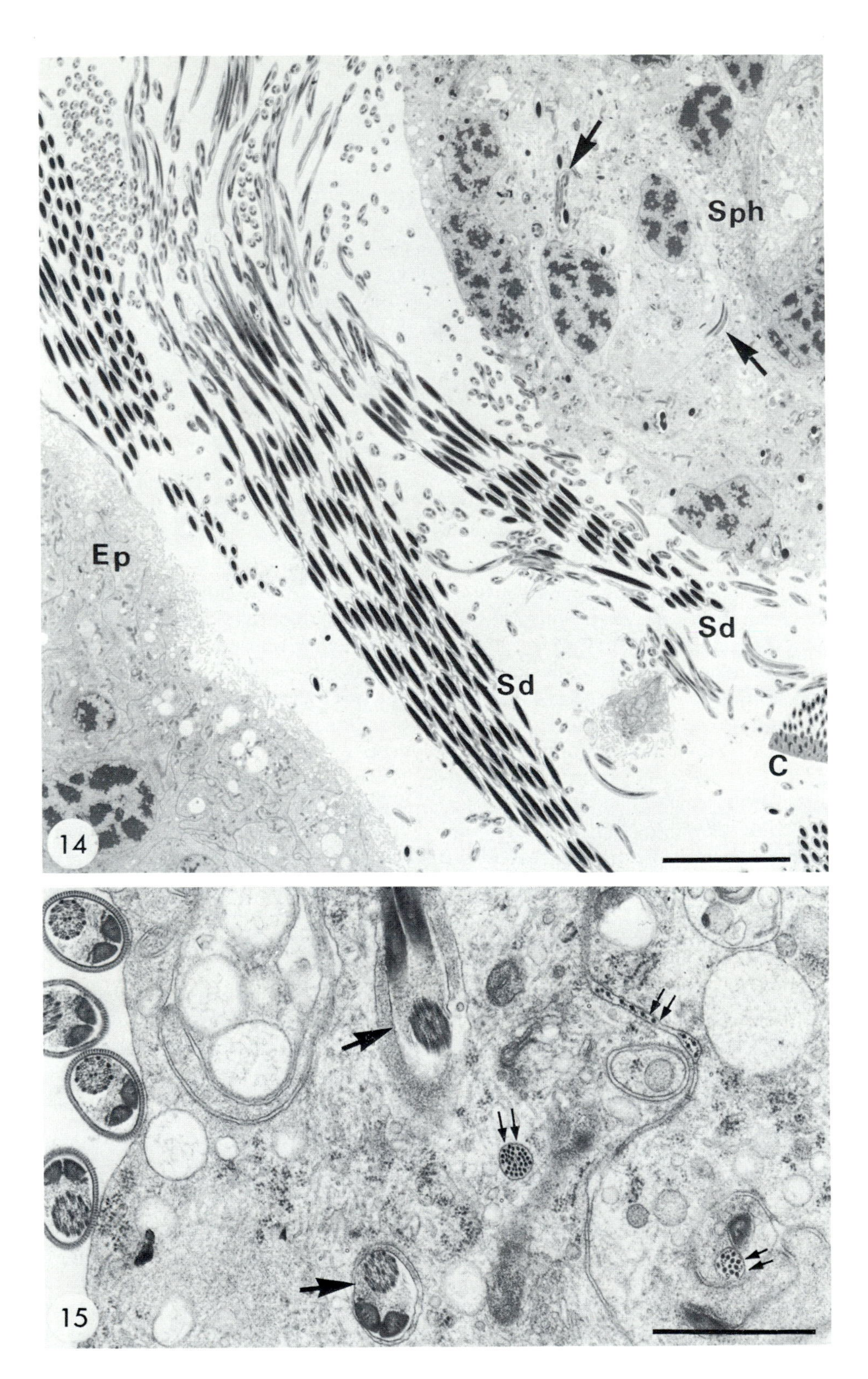
Sph
Ep
Sd
Sd
C
14
15

protecting only the germ cells. Indeed, the proper term would be "blood-germ cell barrier," as used by Jones (1978).

4.2. Cytological Techniques

The barrier mentioned in the following description is a barrier to macromolecules demonstrated mainly by the use of horseradish peroxidase (HRP), a globular protein of known molecular weight (40,000 d) and diameter (60 Å). The penetration of such a tracer into an organ does not tell us what really happens *in vivo* to blood macromolecules; it only indicates that there is a *possibility* for entry of large molecules into certain regions of the organ, and a *restriction* or *barrier* to that entry in other parts. In insects, we have no data on the composition of the cyst fluid in which the sperm differentiate and we do not know if, as in mammals, there are differences in the concentration of ions and small molecules between the hemolymph and the cyst fluid. Our data refer exclusively to macromolecules.

4.3. "Open" and "Closed" Compartments in Testes of Locusts and Moths

By the use of horseradish peroxidase added to the physiological medium, it can be demonstrated that the locust follicle consists of an "apical" compartment that contains the youngest germ cells and is readily penetrated by the tracer and of a "basal" one where the spermatids differentiate and that is tighly closed (Marcaillou and Szöllösi, 1975; Szöllösi and Marcaillou, 1977; Jones, 1978) (Figure 16). All along the follicle, the intercellular spaces of the perifollicular layer provide open channels through which the tracer reaches the internal lamella. In the apical compartment, the tracer infiltrates the spaces separating adjacent cysts and penetrates them, reaching the young germ cells (Figure 19). In the basal compartment, HRP has free access to the inner parietal layer (IPL) where it penetrates for some distance before being stopped. It is found in places within the intercellular spaces and even in several interseptae of the long septate junctions present in this layer. However, it never crosses the IPL, and consequently it never reaches the germ cells. The border between the open and closed compartments is located in the region where the germ cells have entered meiotic prophase.

In the lepidopteran testis, there are two structures that potentially can play the role of barrier to exogenous macromolecules: the thick wall of the organ and the cyst envelopes. In *Anagasta kuehniella*, a species whose sper-

---

Figures 14 and 15. The locust spermiophagic cells.

Figure 14. A plug of these cells (Sph) is located in the lumen of the spermiduct. Isolated sperm are engulfed by the cells (arrows). (Scale bar = 10 μm.) C, cap of a spermatodesm; Ep, epithelium of the spermiduct; Sd, spermatodesm.

Figure 15. A higher magnification of the spermiophagic cells shows engulfed, degenerating sperm that have lost their characteristic membrane ornamentation (arrow). Isolated accessory fibers of the axonemes (double arrow) are found in vacuoles or in the intercellular spaces of the tissue. (Scale bar = 1 μm.)

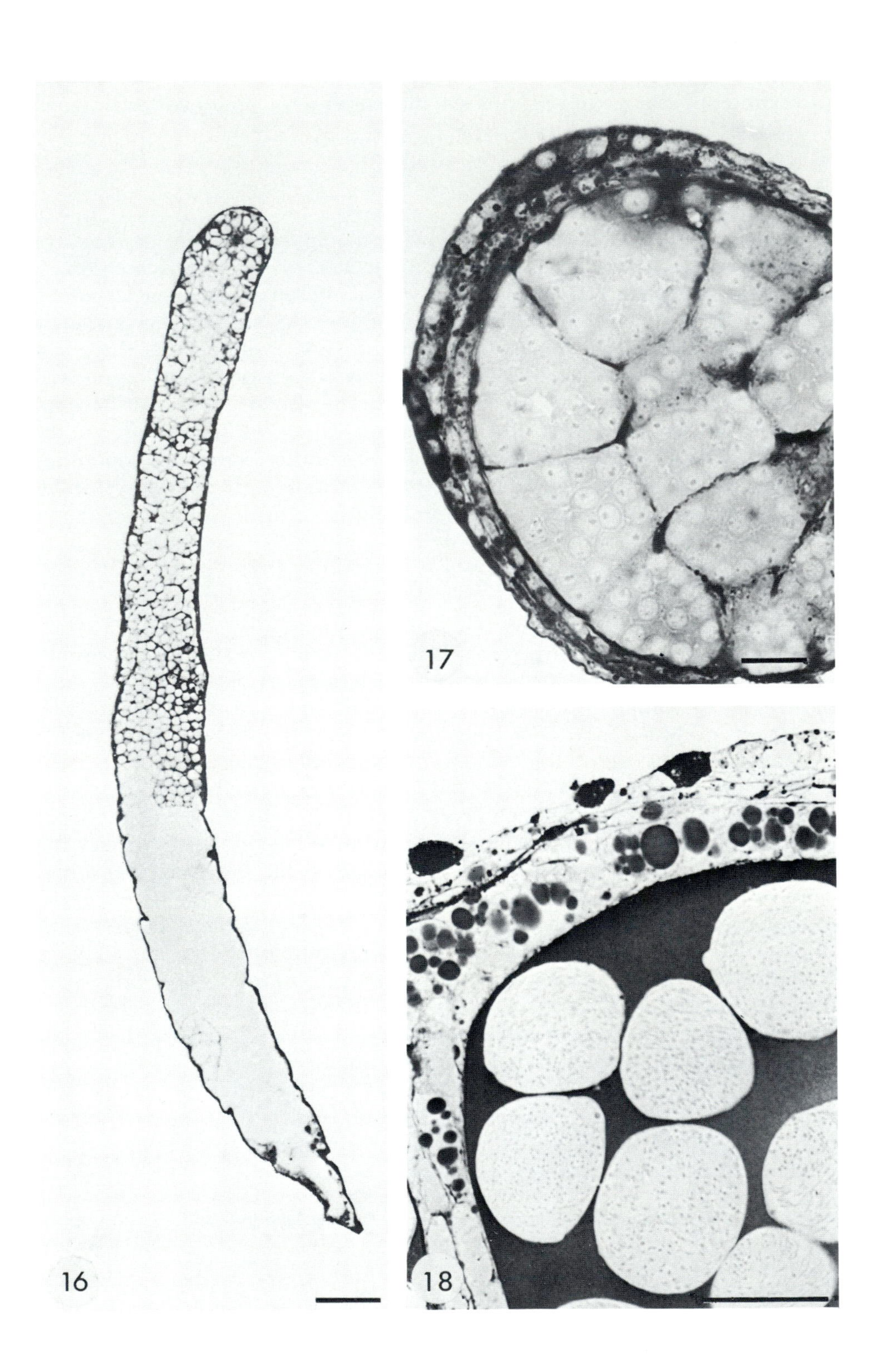
16
17
18

matogenesis is continuous, HRP readily penetrates across the wall of the testis. Larval, pupal, and adult testes are equally permeable to the tracer. HRP is found in all the intercellular spaces and, particularly, in the system of anastomosed intercellular channels that characterizes the inner layer of the wall. It is also captured by pinocytotic vesicles of the wall cells, but nevertheless reaches the inner basal lamina and the testicular fluid, whether this fluid is sparse as in young larvae or more abundant as in older stages (Figures 17 and 18). The primary gonia that surround the apical cell and are bathed directly by the testicular fluid are therefore readily reached by the tracer. The apical cell takes up HRP, as already mentioned in section 3.1, while the primary spermatogonia and the cells of the cyst envelopes demonstrate only a small number of micropinocytotic vesicles containing the exogenous protein. The tracer is never observed within the cysts, among the germ cells (Figures 18, 20) (Szöllösi *et al.*, 1980).

Regarding the distribution of "open" and "closed" compartments, the situation in lepidoptera appears therefore quite different from that found in locusts. The "open" compartment is very restricted and the cysts appear "closed" as soon as they form. Thus, HRP does not reach the dividing spermatogonia and the young spermatocytes as it does in locusts (and in mammals). On the other hand, while tracer experiments showed that HRP is efficiently blocked within the IPL in the basal region of the locust follicle, the much thicker wall of the lepidopteran testis does not play this role. This does not mean that the composition of the testicular fluid is identical to that of hemolymph, but it implies that the only effective barrier between the germ cells and their environment lies in the cyst envelope. This point will be discussed in more detail in sections 4.6 and 5.2.

4.4. Establishment of the Blood–Testis Barrier in Locusts

Observations performed on locust follicles in the last larval instar showed that in *Schistocerca* (Jones, 1978), as well as in *Locusta* (Marcaillou and Szöllösi, 1975; Szöllösi and Marcaillou, 1977), the tracers are found in the spermatogonial cysts but are excluded in those containing cells in the diplotene stage. It was also demonstrated that in the lifetime of the insect the onset of meiosis coincides with the appearance of the blood–testis barrier. The barrier is established in the third larval instar in *Schistocerca* (Jones, 1978), while it appears only during the fourth instar in *Locusta*. In both cases how-

←

Figures 16–18. Penetration of horseradish peroxidase into the testis as seen by light microscopy. ***Figure 16.*** In the *Locusta* follicle, the apical compartment is permeable, while the basal one is not. (Scale bar = 100 μm.) Reproduced from Szöllösi and Marcaillou (1977).
Figure 17. In the young larva of the moth *Anagasta* the cysts are tightly packed, and there is little testicular fluid between them. This fluid is impregnated by the tracer. (Scale bar = 50 μm.)
Figure 18. In the last larval instar of the moth, the cysts are dispersed in an abundant testicular fluid. At all stages, the tracer penetrates through the testis wall but does not enter into the cysts. (Scale bar = 50 μm.) Figures 17 and 18 are reproduced from Szöllösi *et al.* (1980).

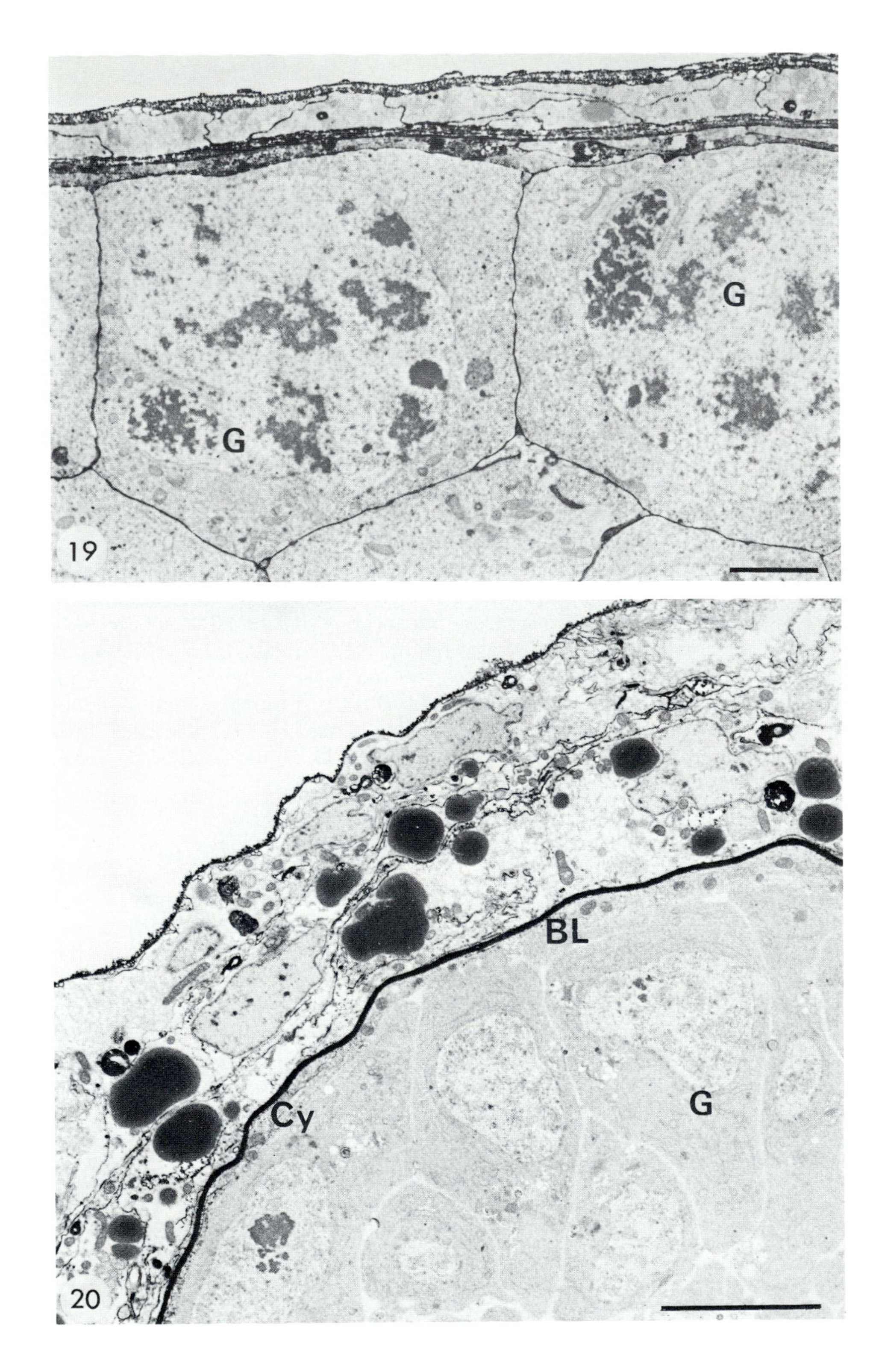
G
G
19
BL
Cy
G
20

ever, this is the time when the first wave of germ cells enters meiotic prophase (zygotene or pachytene stages). The sequence of events appears then to be: "molt/barrier formation/meiosis," and Jones (1978) postulated that the molting hormone, as well as other substances such as juvenile hormone and testicular factor, might be involved in barrier formation. *In vitro* experiments performed on follicles of young larvae of *Schistocerca* led Jones to conclude that ecdysterone is able to induce a premature formation of the barrier in the follicles of second instar grasshoppers and that this effect is antagonized by juvenile hormone. Testis extracts could also induce premature barrier formation, and therefore, the testis itself was supposed to store some stimulating factor.

These interesting experiments are difficult to interpret until a more precise knowledge of the structures involved in the blood–testis barrier is gained. It must be recalled that under normal conditions, the onset of meiosis coincides not only with the establishment of the barrier but also with the appearance of a conspicuous differentiation of the follicular wall: a system of septate junctions in the inner parietal cell layer. In Jones's experiments the meiotic divisions are not stimulated by the hormonal treatment, and it is the spermatogonial region that becomes closed. Could ecdysterone have stimulated the development of IPL in a more apical region? The short duration of the experiments (6 hr) makes this highly improbable. It is more likely that the hormone is acting on the cyst envelopes or on the germ cells themselves. The young cysts might have developed some structures rendering them prematurely impermeable. On the other hand, interference of the hormone with the germ cells and their ability to actively capture an exogenous protein should not be excluded (see section 4.5). Whatever the exact situation, it is almost certain that the closure produced experimentally differs from that observed *in situ* in the basal region of the locust follicle. Consequently, further investigations are needed before we can conclude about an hormonal control on the formation of the blood–testis barrier.

4.5. Permeability of the Apical Compartment in the Locust Testis

In locusts, while meiotic cells and maturing sperm are sequestered in a closed compartment, the spermatogonia, located in the apical compartment are *almost* always in direct contact with the macromolecules used in the experiments. However, HRP penetration into the apical compartment varies during the fifth larval instar (Marcaillou and Szöllösi, 1975; Marcaillou *et al.*, 1978). During the first four to five days of the fifth instar, the tracer reaches

Figures 19 and 20. Penetration of horseradish peroxidase in the testis as seen by electron microscopy.

Figure 19. In the apical compartment of the locust follicle, the tracer reaches the spermatogonia (G). (Scale bar = 2 μm.) Reproduced from Marcaillou *et al.* (1978).

Figure 20. In the testis of *Anagasta*, the tracer crosses the wall and impregnates the internal basal lamella (BL) but does not enter the young cyst. G, spermatogonia; Cy, cyst envelope. (Scale bar = 5 μm.) Reproduced from Szöllösi *et al.* (1980).

all of the germ cells, including those located at the center of the follicle. One day later, penetration is restricted to the peripheral cysts, and then it ceases for a period of about 24 hr. After this, the follicle becomes again slightly permeable at its periphery, and by the last (ninth) day of the instar, all germ cells are reached by the tracer even though the follicle has enlarged considerably.

The temporary closure of the apical compartment does not depend on transitory specialized structures such as newly formed junctions. Perhaps the success or failure for an exogenous protein to penetrate could be an expression of different metabolic states of the testicular tissues. Although the pinocytotic vesicles observed in the cyst cells and germ cells never give rise to large inclusions similar to those found in a vitellogenic oocyte, it is possible that blood proteins entering into the testis could rapidly be degraded and utilized by the dividing gonia and by spermatocytes in their growth phase. This active capture would occur most of the time in the apical compartment of the follicle but would be inhibited on the sixth or seventh day of the last instar, a time when high concentrations of circulating ecdysteroids are present in the animal (Marcaillou *et al.*, 1978).

Thus, the closure of the spermatogonial compartment observed after incubating young follicles of *Schistocerca* with ecdysterone (Jones, 1978) could correspond to this inhibition rather than to a premature establishment of the blood–testis barrier. Clearly, further studies are needed to determine what the influence of ecdysterone is on the testis permeability. Intriguing questions remain: why is the spermatogonial compartment rendered impenetrable to HRP for a few hours during the fifth instar, and how is this achieved?

4.6. Junctions and Membrane Differentiations of the Somatic Envelopes of the Testis

Testicular regions that are permanently closed demonstrate morphological structures that certainly contribute to the isolation of the germ cells. This situation is quite different from that described in the preceeding section (4.5).

In the locust follicle, negative staining and freeze-fracture studies confirm the existence of extensive septate junctions in the inner parietal cell layer (Figures 5, 21). In *Locusta*, no additional junctions of the "tight" type are found in this layer. Therefore, the IPL septate junctions represent the only system capable of restricting the passage of the exogenous tracer across this

Figures 21–23. Freeze-fracture preparations of the somatic envelopes of the testis.
Figure 21. In the inner parietal cell layer of the follicle of *Locusta*, the membranes show parallel rows of protoplasmic-face (PF) particles typical of septate junctions. Several planes can be seen in this micrograph, which correspond to the complex interdigitations between adjacent cells. (Scale bar = 0.5 μm.)
Figure 22. In the cyst envelope of *Locusta*, the membranes demonstrate rows of fused particles forming short PF ridges. Two or three individual ridges sometimes run parallel and close to one another, but more generally are isolated. (Scale bar = 0.25 μm.)
Figure 23. In the cyst envelope of *Anagasta*, the membranes show a network of numerous, branching PF ridges, EF, exoplasmic face of the membrane. (Scale bar = 0.25 μm.)

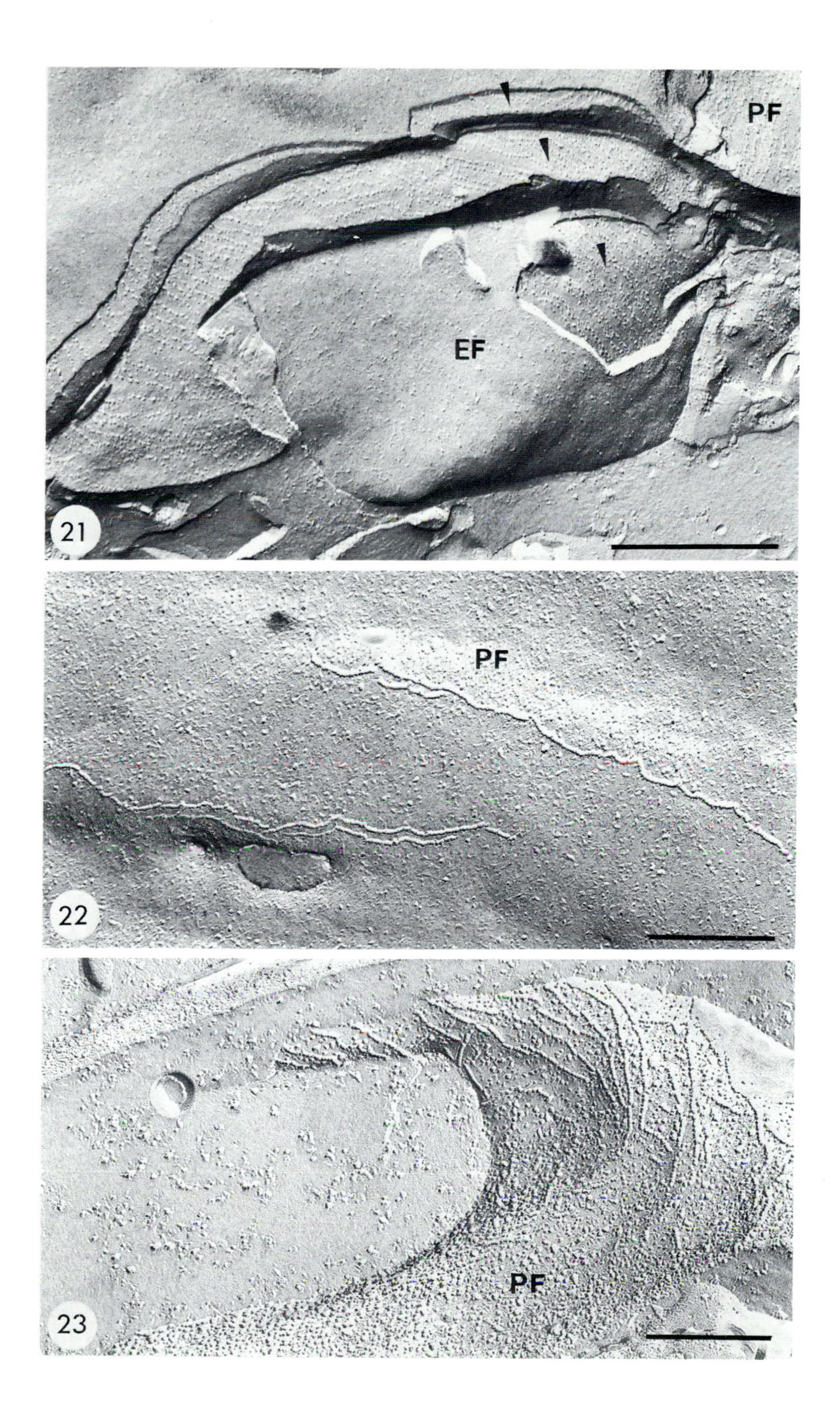
PF
EF
21
22
PF
23
PF

layer. Many authors have suggested that invertebrate septate junctions act as a transepithelial permeability barrier [reviewed by Lane and Skaer (1980) and by Noirot-Timothée and Noirot (1980)], and this role has been clearly demonstrated in *Hydra* (Wood and Kuda, 1980). This view has been questioned on the basis of experiments performed with ionic lanthanum that showed that this tracer usually penetrates the septate junctions (Lane and Skaer, 1980). In the locust follicle, ionic lanthanum itself penetrates very slowly into the IPL, and it does not traverse the whole septate junctions. This can probably be explained by the fact that interdigitations between IPL cells greatly increase the length of the transepithelial pathway. The entry of HRP appears still more drastically restricted than that of ionic lanthanum. This could be due to the larger size of the molecule, even though the molecular size is not the only parameter limiting the passage of a substance across septate junctions. The presence of acidic polysaccharides in the interseptal matrix has been demonstrated. These substances, by their anionic groups, can bind cations whose presence in the intercellular space in turn can restrict the passage of charged molecules of any size (Noirot-Timothée and Noirot, 1980; Wood and Kuda, 1980). In the cyst envelopes, the cells are associated by rare and loosely organized septate junctions. In this tissue, freeze fracture demonstrates another kind of membrane differentiation, consisting of short rows of fused particles forming ridges on the protoplasmic face (PF) of the membranes (Figure 22) and corresponding furrows of the exoplasmic face (EF). Similar structures have been observed in a variety of tissues where they have been considered as nonjunctional particle arrays (Lane, 1979). The appearance of these short ridges reminds one of focal tight junctions, and, in the testis of *Schistocerca*, they actually have been considered as belonging to the tight junctional system (Lane and Skaer, 1980). According to the authors, these ridges and furrows constitute the morphological basis of the barrier in the testis. However, in this organ, we have no proof that there is a constriction of the intercellular space at the site of these membrane differentiations. Consequently, it is not clearly established that they represent junctional areas, and they could just as well be nonjunctional structures, as was suggested in the case of other tissues. On the other hand, because the ridges are very short and randomly dispersed on the membrane, it is doubtful that they could be very effective anyway in restricting the passage of molecules between the cells.

In the cyst envelopes of Lepidoptera, septate junctions are well developed, particularly in the older cysts. Besides, a system of parallel and branching PF ridges and EF furrows was found in *Bombyx mori* by Toshimori *et al.* (1979), who considered it as a true tight junctional system insuring the isolation of the maturing germ cells. In *Anagasta kuehniella*, anastomosed ridges also are found in the cyst envelopes. These ridges, like those of the locust cyst envelopes, are made of fused particles about 10 nm in diameter, but they are much more numerous than in the locust. In places, they form a true network (Figure 23). Ridges coexist with parallel rows of aligned particles characteristic of septate junctions, and the two systems can be observed side by side in some areas of the membrane.

In summary, two different situations are found regarding the blood–testis barrier in the two types of insects so far studied. In locusts, the impermeability of the basal compartment is due mainly to the septate junctions present in the follicle wall. In Lepidoptera, in contrast, the wall is permeable, and the only site for the barrier lies in the cyst envelopes. In this tissue, septate and tight junctions coexist, and for the time being, it is not possible to decide whether only one or both of these structures act in isolating the germ line.

5. Communication between Germ Cells and Somatic Cells

5.1. Historical Perspective

Adjacent animal cells can "communicate" by means of gap junctions, through which ions and small molecules can rapidly move from one cell to another. These junctions are found mostly between cells of the same type (homocellular junctions). However, they can also occur between cells of different origin, an example being the heterocellular gap junctions recently discovered between germ cells and somatic cells.

In the ovary, the presence of gap junctions and the parallel existence of electrotonic and metabolic coupling between the oocyte and the surrounding follicular cells have recently been demonstrated in mammals (Gilula *et al.*, 1978; Moor *et al.*, 1980) and in insects (Wollberg *et al.*, 1976; Woodruff, 1979; Huebner, 1981). In the testis, minute gap junctions have also been detected between the germ cells and the Sertoli cells in the seminiferous tubule of the rat (McGinley *et al.*, 1979; Russell, 1980). *In vitro* experiments indicate that a coupling also exists between these two cell types (Palombi *et al.*, 1980). In insects, gap junctions between male germ cells and somatic cells have been detected in *Anagasta kuehniella* (Szöllösi and Marcaillou, 1980), and recently in two other moths, *Platysamia cynthia* and *Hyalophora cecropia*.

5.2. Differentiation of Gap Junctions during the Development of Moth Germ Cells

Numerous gap junctions appear between the germ cells and those of the cyst envelopes as soon as the cysts are formed, and they persist until the spermatids begin to elongate. Between secondary gonia and cyst cells the junctional areas appear as rigid and flat regions of contact between the two cells (Figures 24, 25). The length of such junctional profiles varies widely and can reach up to 2 μm. At the surface of the spermatocytes in early meiotic prophase, the junctions become hemispherical. They are located in regions where the germ cell bulges into the adjacent cyst cell, indenting it to various degrees. Thin filaments associated with the junction become clearly visible at this stage. In the cytoplasm of the germ cell, the filaments radiate toward the junctional area, and they seem to attach to an electron-dense material lining the membrane (Figure 26). These filaments are very similar to those reported in gap junctions of granulosa cells (Zamboni, 1974). In cysts of late spermatocytes, gap junctions are found either as buttonlike structures or as caps at the extrem-

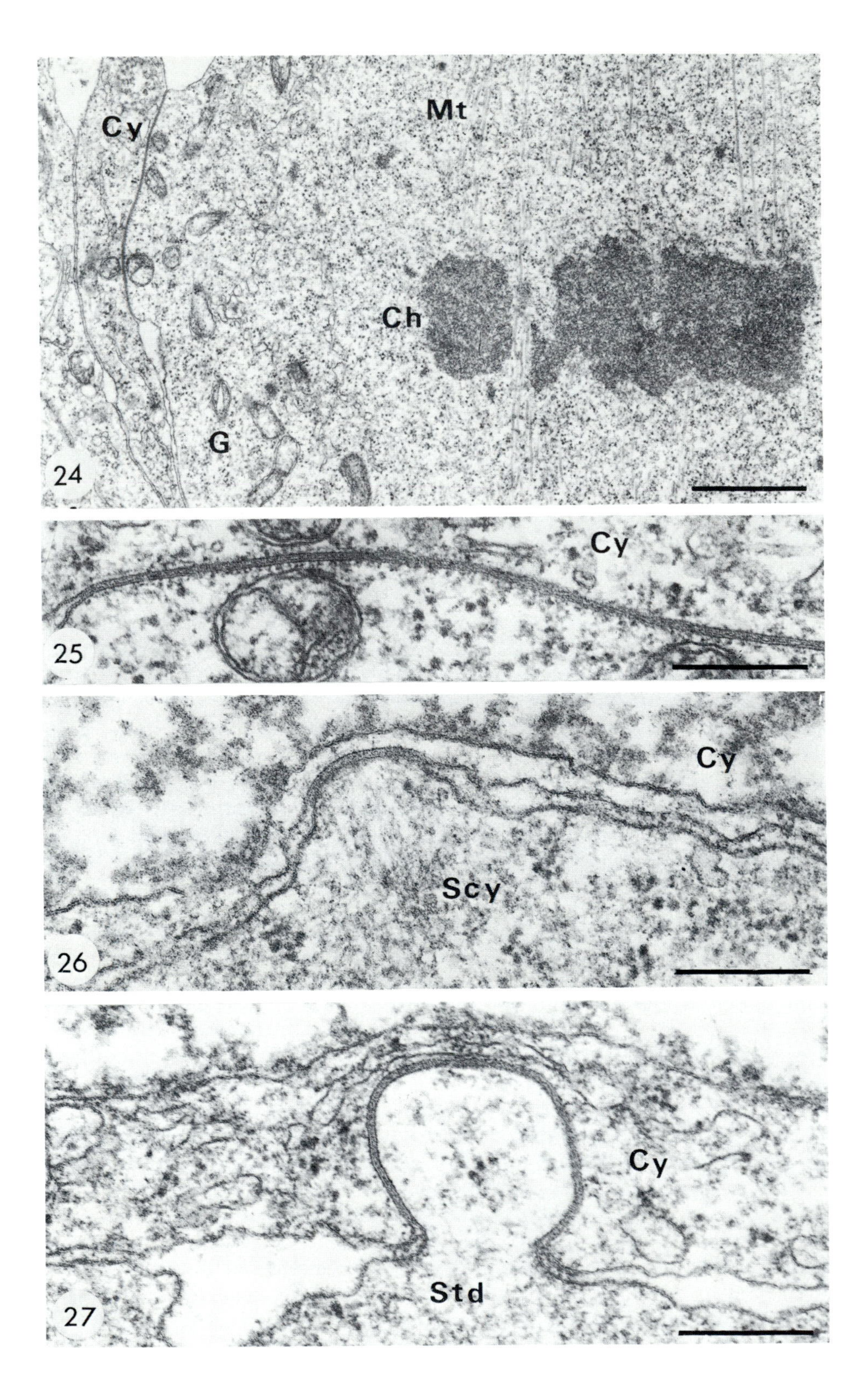
Cy
Mt
Ch
G
24
25
Cy
26
Cy
Scy
27
Cy
Std

ity of cytoplasmic protrusions of the germ cell. At the surface of young spermatids, the junctional areas can be observed to be pinched off from the germ cell surface (Figure 27), and "gap vesicles" are found in the cytoplasm of the cyst cell. These vesicles could correspond to the definitive elimination of the germ–somatic cell gap junctions at this stage of the spermatid differentiation. Thus, from the moment a germ cell is enclosed within a cyst to that when it becomes an elongated spermatid, it always appears to be linked to an adjacent somatic cell by means of gap junctions.

Several remarks should be made before discussing the possible role of these junctions: (1) The total number of junctions per cyst must be quite large, since five to twenty junctions are observed in a single thin section of a cyst containing spermatocytes. (2) The junctions are located exclusively at the surface of the peripheral germ cells. (3) Due to incomplete cytokinesis, all germ cells remain interconnected by large cytoplasmic bridges until they reach the stage of elongating spermatids. (4) The presence of a gap junction between two cells does not in itself indicate that these cells are permanently coupled, but their presence probably means that such a coupling does occur at least at certain stages of development. This assumption needs to be confirmed by further investigations.

5.3. Functions of the Gap Junctions between Germ Cells and Somatic Cells

The precocious closure of the lepidopteran cysts observed in experiments using HRP is quite puzzling. This situation is in contrast with that found in locusts, where the same tracer has free access to all secondary gonia, at least during the larval life. The impermeability of the young cysts raises distinct questions about the role in spermatogenesis of a macromolecular factor (MF) found in the blood of diapausing saturnids. Kambysellis and Williams (1971 a, b) postulated that the postdiapause reinitiation of meiosis is under the double control of a hormone (ecdysterone) and of MF. The hormone would act on the testicular wall, making it permeable to the MF, which would in turn reinitiate the meiotic process. Since the young cysts of saturnids are impermeable to peroxidase, MF also may be unable to reach the germ cells. The first site of action of the MF would then be the cyst cell, and meiosis would be induced through a stimulus received by cells of the somatic envelope.

Figures 24–27. Gap junctions between germ and somatic cells in the testis of *Anagasta*.
Figure 24. Between the dividing spermatogonium (G) and the cell of the cyst envelope (Cy), the junctional area is large and flat. Ch, chromosome; Mt, microtubule. (Scale bar = 1 μm.)
Figure 25. Detail: Periodic densities are observed in the intercellular space of the junction, corresponding to the connecting gap particles. (Scale bar = 0.25 μm.)
Figure 26. Between a spermatocyte (Scy) and a cyst cell (Cy) the gap junction bulges into the cyst cell. In the germ cell, microfilaments radiate towards the junction. (Scale bar = 0.25 μm.)
Figure 27. The junctional area appears as pinched off from the surface of the spermatid (Std). Since gap junctions are no longer found after this stage, this constriction probably results in the elimination of the junction. (Scale bar = 0.25 μm.)

The discovery of gap junction between the germ cells and the enveloping cyst cells supports this hypothesis. It has been recently demonstrated that a hormonal stimulation can be transmitted between two cells belonging to different cell types by means of cAMP delivered from one cell to another through gap junctions (Lawrence *et al.*, 1978). A mechanism of this kind may be involved in the development of the lepidopteran cyst. Stimulated by some macromolecule (MF), the cyst cell could produce a small molecule acting as a secondary messenger (e.g., cAMP), which could in turn be rapidly transfered to the germ cell through gap junctions.

Further investigations along this line should include electrophysiological studies which may shed light on the role of the somatic envelopes in regulating spermatocyte meiosis.

References

Cantacuzène, A. M., 1971, Origine et caractères ultrastructuraux des cellules spermiophages du criquet migrateur *Locusta migratoria* (R. et F.) Orthoptère Caelifère, *J. Microsc.* (Paris) **10**:179–190.

Cantacuzène, A. M., Lauverjat, S., and Papillon, M., 1972, Influence de la température d'élevage sur les caracteres histologiques de l'appareil genital de *Schistocerca gregaria, J. Insect Physiol.* **18**:2077–2093.

Cholodkovsky, N. A., 1913, Uber die Spermatodosen der Locustiden, *Zool. Anz.* **41**:615–619.

Dym, M., and Fawcett, D. W., 1970, The blood-testis barrier in the rat and the physiological compartmentation of the seminiferous epithelium, *Biol. Reprod.* **3**:308–326.

Fawcett, D. W., Leak, L. V., and Heidger, P. M., 1970, Electron microscopic observations on the structural components of the blood-testis barrier, *J. Reprod. Fertil. Suppl.* **10**:105–122.

Gilula, N. B., Epstein, M. L., and Beers, W. H., 1978, Cell-to-cell communication and ovulation. A study of the cumulus-oocyte complex, *J. Cell Biol.* **78**:58–75.

Hardy, R. W., Tokuyasu, K. T., Lindsley, D. L., and Garavito, M., 1979, The germinal proliferation center in the testis of *Drosophila melanogaster, J. Ultrastruct. Res.* **69**:180–190.

Howards, S. S., Jessee, S. J., and Johnson, A. L., 1976, Micropuncture studies of the blood-seminiferous tubule barrier, *Biol. Reprod.* **14**:264–269.

Huebner, E., 1981, Oocyte-follicle cell interaction during normal oogenesis and atresia in an insect, *J. Ultrastruct. Res.* **74**:95–104.

Jones, R. T., 1978, The blood/germ cell barrier in male *Schistocerca gregaria*: The time of its establishment and factors affecting its formation, *J. Cell. Sci.* **31**:145–164.

Kambysellis, M. P., and Williams, C. M., 1971a, *In vitro* development of insect tissues. I. A macromolecular factor prerequisite for silkworm spermatogenesis, *Biol. Bull.* **141**:527–540.

Kambysellis, M. P., and Williams, C. M., 1971b, *In vitro* development of insect tissues. II. The role of ecdysone in the spermatogenesis of silkworms. *Biol. Bull.* **141**:541–552.

Kiknadzé, I. L., and Istomina, A. G., 1980, Endomitosis in grasshoppers. I. Nuclear morphology and synthesis of DNA and RNA in the endopolyploid cells of the inner parietal layer of the testicular follicle, *Eur. J. Cell Biol.* **21**:122–123.

King, R. C., and Akai, H., 1971, Spermatogenesis in *Bombyx mori*. I. The canal system joining sister spermatocytes, *J. Morphol.* **134**:47–55.

Kormano, M., 1967, Dye permeability and alkaline phosphatase activity of testicular capillaries in the postnatal rat, *Histochemie* **9**:327–338.

Lane, N. J., 1979, Intramembranous particles in the form of ridges, bracelets or assemblies in arthropod tissues, *Tissue Cell* **11**:1–18.

Lane, N. J., and Skaer, H. B., 1980, Intercellular junctions in insect tissues, *Adv. Insect Physiol.* **15**:35–213.

Lawrence, T. S., Beers, W. H., and Gilula, N. B., 1978, Transmission of hormonal stimulation by cell to cell communication, *Nature* (London) **272**:501–505.

Leclerck-Smekens, M., 1978, Reproduction d'*Euproctis chrysorrhea* L. (Lépidoptère Lymantridae). II. Aspects cytologiques et rôle de la cellule apicale (cellule de Verson) des testicules, *Acad. R. Belg. Cl. Sci.* **64**:318–322.

Lindsley, D. L., and Tokuyasu, K. T., 1980, Spermatogenesis. In *The Genetics and Biology of Drosophila,* vol. 2d, edited by M. Ashburner and T. R. F. Wright, pp. 225–249, Academic Press, London.

McGinley, D. M., Posalaky, Z., Porvaznik, M., and Russell, L., 1979, Gap junctions between sertoli and germ cells of rat seminiferous tubules, *Tissue Cell* **11**:741–754.

Marcaillou, C., and Szöllösi, A., 1975, Variations de perméabilité du follicule testiculaire chez le criquet *Locusta migratoria migratoriodes* (Orthoptère) au cours du dernier stade larvaire, *C. R. Acad. Sci. Paris* **281**:2001–2004.

Marcaillou, C., Szöllösi, A., Porcheron, P., and Dray, F., 1978, Uptake of Horseradish peroxidase by the testis of *Locusta migratoria* during the last larval instar; relation with variations of ecdysteroid levels in haemolymph, *Cell Tissue Res.* **188**:63–74.

Matsuda, R., 1976, *Morphology and Evolution of the Insect Abdomen,* Pergamon Press, Oxford.

Moor, R. M., Smith M. W., and Dawson, R. M. C., 1980, Measurement of intercellular coupling between oocytes and cumulus cells using intracellular markers, *Exp. Cell Res.* **126**:15–29.

Muckenthaler, F. A., 1964, Autoradiographic study of nucleic acid synthesis during spermatogenesis in the grasshopper *Melanoplus differentialis, Exp. Cell Res.* **35**:531–597.

Noirot-Timothée, C., and Noirot, C., 1980, Septate and scalariform junctions in arthropods, *Int. Rev. Cytol.* **63**:97–140.

Palombi, F., Ziparo, E., Galdieri, M., Russo, M. A., and Stefanini, M., 1980, Ultrastructural evidence for germ cells-Sertoli cells interaction in cultures of rat seminiferous epithelium, *Biol. Cellulaire* **39**:249–252.

Phillips, D. M., 1970, Insect sperm: Their structure and morphogenesis, *J. Cell Biol.* **44**:243–277.

Riemann, J. G., and Thorson, B. J., 1976, Ultrastructure of the *vasa deferentia* of the mediterranean flour moth, *J. Morphol.* **149**:483–506.

Roosen-Runge, E. C., 1973, Germinal cell loss in normal metazoan spermatogenesis, *J. Reprod. Fertil.* **35**:339–348.

Roosen-Runge, E. C., 1977, *The Process of Spermatogenesis in Animals,* Cambridge University Press, Cambridge.

Russell, L. D., 1980, Sertoli germ cell interrelations: A review, *Gamete Res.* **3**:179–202.

Setchell, B. P., 1980, The functional significance of the blood-testis barrier, *J. Androl.* **1**:3–10.

Setchell, B. P., and Waites, G. M. H., 1975, The blood-testis barrier. In *Handbook of Physiology,* section 7, *Endocrinology,* vol. V, *Male Reproductive System,* edited by D. W. Hamilton and R. O. Greep, pp. 143–172, American Physiological Society, Washington.

Spichardt, C., 1886, Beitrag zur Entwicklung der männlichen Genitalien und ihrer Ausführungsgänge bei Lepidopteren. *Vehr. Naturlist. Verein Rheinland,* Bonn. **43**:1–43.

Szöllösi, A., 1974, Ultrastructural study of the spermatodesm of *Locusta migratoria migratorioides* (R. et F.): Acrosome and cap formation, *Acrida* **3**:175–192.

Szöllösi, A., 1975, Electron microscope study of spermiogenesis in *Locusta migratoria* (Insect Orthoptera), *J. Ultrastruct. Res.* **50**:322–346.

Szöllösi, A., 1976a, Influence of infra-optimal breeding temperature on spermiogenesis of the locust *Locusta migratoria.* I. Abnormalities in differentiation of the cytoplasmic organelles, *J. Ultrastruct. Res.* **54**:202–214.

Szöllösi, A., 1976b, Influence of infra-optimal breeding temperature on spermiogenesis of the locust *Locusta migratoria.* II. Abnormalities in differentiation of the nucleus, *J. Ultrastruct. Res.* **54**:215–223.

Szöllösi, A., and Marcaillou, C., 1977, Electron-microscope study of the blood-testis barrier in an insect: *Locusta migratoria, J. Ultrastruct Res.* **59**:158–172.

Szöllösi, A., and Marcaillou, C., 1979, The apical cell of the locust testis: An ultrastructural study, *J. Ultrastruct. Res.* **69**:331–342.

Szöllösi, A., and Marcaillou, C., 1980, Presence of gap junctions between germ and somatic cells in the testis of an insect, the moth *Anagasta kuehniella, Cell Tissue Res.* **213**:137–147.

Szöllösi, A., Riemann, J., and Marcaillou, C., 1980, Localization of the blood-testis barrier in the testis of the moth *Anagasta kuehniella*, *J. Ultrastruct. Res.* **72:**189–199.

Thibout, E., 1980, Evolution and role of apyrene sperm cells of lepidopterans: Their activation and denaturation in the leek moth *Acrolepiopsis assectella*. In *Advances in Invertebrate Reproduction*, edited by W. H. Clark and T. S. Adams, pp. 231–242, Elsevier, Amsterdam.

Toshimori, D., Iwashita, T., and Oura, C., 1979, Cell junctions in the cyst envelope in the silkworm testis, *Bombyx mori*, *Cell Tissue Res.* **202:**63–74.

Wollberg, Z., Cohen, E., and Kalina, M., 1976, Electrical properties of developing oocytes of the migratory locust, *Locusta migratoria*, *J. Cell Physiol.* **88:**145–158.

Wood, R. L., and Kuda, A. M., 1980, Formation of junctions in regenerating hydra: Septate junctions, *J. Ultrastruct. Res.* **70:**104–117.

Woodruff, R. I., 1979, Electrotonic junctions in *Cecropia* moth ovaries, *Develop. Biol.* **69:**281–295.

Zamboni, L., 1974, Fine morphology of the follicle wall and follicle cell oocyte association, *Biol. Reprod.* **10:**125–149.

3

The Meiotic Prophase in *Bombyx mori*

SØREN WILKEN RASMUSSEN AND
PREBEN BACH HOLM

1. Introduction

The first major step toward an understanding and a characterization of the ultrastructure of meiosis was performed about twenty-five years ago. In 1956, Moses reported in a combined light and electron microscopic study of crayfish spermatocytes, that homologous chromosomes at pachynema were held in register by a bipartite structure. The same year, Fawcett independently described similar structures in spermatocytes of pigeons, cats, and humans. The term synaptinemal complex [later changed to synaptonemal complex (King, 1970)] was coined for this structure.

Since then, numerous studies have revealed that synaptonemal complexes are present during the prophase of meiosis in all taxonomic groups of eukaryotes investigated so far (Moses, 1968; Sotelo, 1969; Westergaard and von Wettstein, 1972; Gillies, 1975). The synaptonemal complex (SC) consists of two banded or amorphous-filamentous lateral components (LC) about 30–40 nm in diameter, held together by a central region 100–120 nm in diameter. The central region has a medially located central component 20–30 nm wide, with an amorphous or scalariform substructure.

Basically, formation of the SC involves the following steps (Rasmussen and Holm, 1980): At leptonema, each chromosome organizes an LC, the telomeres become attached to the nuclear envelope, and after the congression of the telomeres, whereby a chromosome bouquet is formed, chromosome pairing and SC formation start. In some organisms, SC formation is initiated

---

SØREN WILKEN RASMUSSEN AND PREBEN BACH HOLM • Department of Physiology Carlesberg Laboratory, Copenhagen, Valby, Denmark.

preferentially from the telomeres, while in others complex formation may also occur interstitially. Complete pairing is attained at pachynema, where the two homologues of each bivalent are held in register by a continuous SC. This condition is maintained during the pachytene stage. At early diplonema, the bulk of the SCs are eliminated from the bivalents, and only a few SC fragments remain associated with the homologues. It has been proposed that such retained fragments of the SC participate in the organization of chiasmata (Westergaard and von Wettstein, 1968; 1970; Solari, 1970).

Recombination-deficient organisms such as *Drosophila* males, and females carrying the $c(3)G^{17}$ mutation (Gowen, 1933), do not form SCs (Meyer, 1964; Smith and King, 1968; Rasmussen, 1973; 1975b), which demonstrates that homologous pairing with SC formation is a necessary prerequisite for crossing over.

However, the presence of an SC is not in itself sufficient for crossing over to occur. *Bombyx* females do not exhibit crossing over (Sturtevant, 1915) but apparently have normal SC formation (Miya *et al.*, 1970; Rasmussen, 1976), as is the case for meiotic mutants in *Drosophila* partially deficient in recombination (Carpenter, 1979b). Crossing over and chiasma formation are also absent in haploid organisms despite extensive nonhomologous SC formation (Gillies, 1974). Hence, the primary function of the SC in crossing over is presumably to provide a structural framework, which ensures that homologous DNA regions are brought into proximity and which maintains this condition for a certain period of time.

On the basis of electron microscopic analyses of *Drosophila* oocytes, Carpenter (1975; 1979a) proposed that in addition to homologous pairing with the SC, small electron-dense nodes—recombination nodules (RN)—attached to the central region of the SC were necessary prerequisites for crossing over. This proposal has since then received strong support from a number of investigations (reviewed by Carpenter, 1979b).

Ultrastructural investigations of meiosis in insects are numerous and have to a large extent contributed to the understanding of the basic scheme of meiosis. Furthermore, a number of the most intriguing modifications of the meiotic processes have been reported from this taxonomic group. It is, however, not the aim of this paper to summarize the available information on the ultrastructure of meiosis in insects, mainly because this has already been done quite extensively in previously published reviews on the SC (Moses, 1968; Sotelo, 1969; Westergaard and von Wettstein, 1972; Gillies, 1975; Moens, 1978). Instead, we have tried to extract the common features of these processes exemplified by meiosis in the domesticated silkworm, *Bombyx mori*.

This organism constitutes a favorable medium for ultrastructural analyses of meiosis for several reasons: Crossing over and chiasma formation in *Bombyx* are limited to the male sex (Sturtevant, 1915; Maeda, 1939). In addition, triploid and tetraploid males as well as females are viable and easily available (Astaurov, 1967), a unique situation in animals. Finally, the silkworm has been intensively studied both at the genetic and at the developmental level (Tanaka, 1953; Tazima, 1964; 1978; Tazima *et al.*, 1975).

The work described below comprises the first example of an ultrastructural investigation of meiosis in which the entire meiotic prophase has been studied by serial sectioning and three-dimensional reconstruction in diploid females (Rasmussen, 1976; 1977a) as well as males (Holm and Rasmussen, 1980). These investigations are supplemented by analyses of chromosome pairing in triploid (Rasmussen, 1977b) and tetraploid *Bombyx* females (Rasmussen and Holm, 1979).

2. *The Chromosome Complement of Bombyx mori*

Three-dimensional reconstructions of LCs and SCs at pachynema allow a precise determination of the chromosome number of individual nuclei. In all the reconstructed pachytene nuclei of diploid males and females, the expected fifty-six chromosomes were found paired into twenty-eight bivalents, each with a continuous SC from telomere to telomere. In contrast, aneuploidy was frequent in triploid and tetraploid females; in triploids in the form of supernumerary chromosome fragments, and in tetraploids as incomplete chromosome complements (Rasmussen, 1977b; Rasmussen and Holm, 1979).

The data given in Table 1 show that the total LC length is very similar in diploid oocytes and spermatocytes, particularly at early pachynema. The LCs increase in length during pachynema, with tetraploid females as a possible exception. The data in Table 1 also show that an increase in ploidy level unexpectedly is accompanied by a decrease of the LC length of the genome.

Identification and classification of the twenty-eight *Bombyx* bivalents are met with difficulties, mainly because centromeric heterochromatin is absent and other ultrastructural markers are scarce, but also owing to the relatively small differences in length between the individual bivalents. Only the two longest bivalents can be unequivocally identified, bivalent 1 by a knob and bivalent 2 by the nucleolus organizer region. For the remaining fifty-two

Table 1. Total Lateral Component Length in μm of One Genome in Diploid, Triploid, and Tetraploid Oocytes and Diploid Spermatocytes of *Bombyx mori*

Organism	Late zygonema	Early pachynema	Mid pachynema	Late pachynema	Pachynema–diplonema
2n oocytes	—	196 (4)[a]	212 (6)		239 (1)
3n oocytes	—	137 (4)	171 (7)		—
4n oocytes	—	150 (7)	133 (11)		—
2n spermatocytes	202 (8)	198 (16)	258 (8)	256 (6)	347 (3)

[a]The numbers in parentheses are the number of nuclei reconstructed.

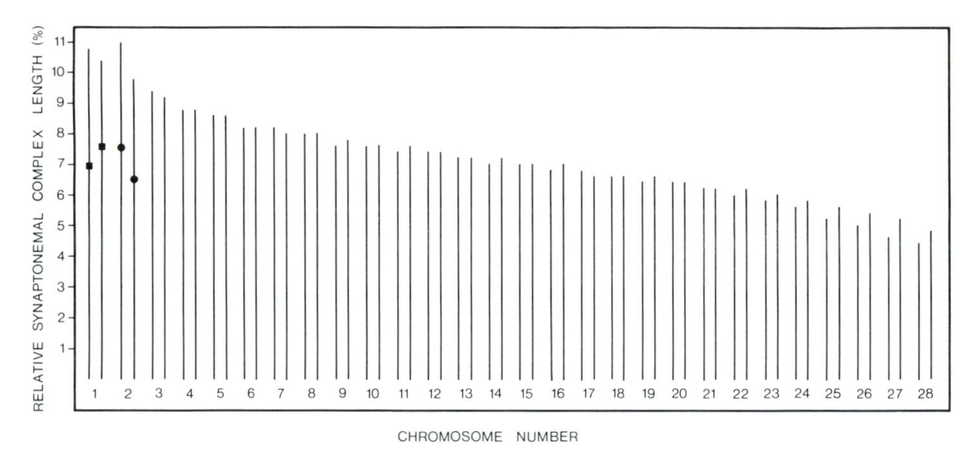

Figure 1. An idiogram showing the mean relative length of synaptonemal complexes in the twenty-eight bivalents at pachynema in diploid *Bombyx* spermatocytes (leftmost lines) and oocytes (rightmost lines). The position of the knob on bivalent 1 is denoted by a solid square and the position of the nucleolus organizer region on bivalent 2 by a solid circle.

chromosomes, classification is restricted to a simple ranking according to LC length. Obviously, this restrains the reliability of the classification, and it is to be expected that the number assigned to a certain bivalent will be different in different nuclei.

A precise comparison of the male and female pachytene karyotypes is thus not feasible. It is apparent, however, that at the diploid level the similarity in total LC length among the two sexes is matched by a corresponding similarity in the length of the individual bivalents (Figure 1). Minor differences exist for bivalents 1 and 2, the latter being the longer in spermatocytes in contrast to the situation in the oocyte. The position of the knob and of the nucleolus organizer region varies slightly. In spermatocytes, the knob divides bivalent 1 into two arms with relative lengths of 1 and 1.8, compared to a ratio of 1 to 2 in the oocyte. For bivalent 2, the ratio between the two arms is 1 to 2.1 in the male, and 1 to 2.4 in the female.

In spermatocytes, as well as in oocytes, bivalents other than numbers 1 and 2 occasionally carry morphological markers in the form of various chromatin condensations. These are especially prominent at mid and late pachynema, but although of potential use, their presence is too erratic to be of any value in the classification of *Bombyx* chromosomes at the electron microscopic level. In this context, it is of interest that the Z and W chromosomes in the heterogametic female sex, form a normal bivalent with a continuous SC that neither in pairing behavior nor in ultrastructure differs from the autosomes, and an unequivocal identification of the ZW bivalent has so far not been possible in the electron microscope. According to Murakami *et al.*, (1978), the Z chromosome is one of the longest chromosomes, being statistically the third longest.

3. Meiosis in Diploid *Bombyx mori*

3.1. Chromosome Pairing and Synaptonemal Complex Formation

3.1.1. The Normal Course of Synapsis

In *Bombyx* males and females, homologous chromosomes are fortuitously distributed within the nucleus prior to entering the meiotic prophase, and how the homologous chromosomes are brought together at the beginning of zygonema is not yet understood. Two processes initiated at leptonema may, however, facilitate the initial association of the homologues, namely, the attachment of the telomeres to the nuclear envelope, whereby the freedom of telomere movement is reduced to a two-dimensional space, and the congression of the attachment sites, which brings into proximity all telomeres, thus also homologous telomeres.

In *Bombyx* oocytes and spermatocytes, initiation of pairing and SC formation preceeds the completion of continuous LCs and are with few exceptions confined to the telomere regions. Following this initial matching of homologous telomeres, SC formation proceeds from telomere to telomere in a zipper-like fashion if not impeded by interlockings (see section 3.1.2).

It is a general observation in normal diploid organisms that chromosome pairing and SC formation during zygonema is highly specific. This is also the case in diploid *Bombyx* oocytes and spermatocytes (Figure 2), eventually resulting in complete bivalent formation at pachynema (Rasmussen and Holm, 1980).

3.1.2. Interlocking of Chromosomes

Interlocking of chromosomes and bivalents during SC formation at zygonema was first discovered in *Bombyx* oocytes, where 75% of the analyzed late zygotene nuclei contained one or more interlockings (Rasmussen, 1976). Since then, interlockings have been found in human spermatocytes (Rasmussen and Holm, 1978) and *Bombyx* spermatocytes (Holm and Rasmussen, 1980). In addition, interlockings are present in the basidiomycete, *Coprinus cinereus* (Holm and Rasmussen, unpublished observations) and in *Zea maize* (Gillies, 1981).

The frequency of interlockings has been determined in *Bombyx* spermatocytes (Table 2). Chromosome interlockings (the entrapment of a chromosome between an unpaired interstitial segment of a partially paired bivalent) were present at a mean frequency of 1.6 per nucleus. Bivalent interlockings (the entrapment of a bivalent between an unpaired interstitial segment of another bivalent) had occurred with a mean frequency of 2.4 per nucleus. Occasionally, more than one chromosome or bivalent were entrapped by the same unpaired bivalent region (Figure 3), and up to nine interlockings have

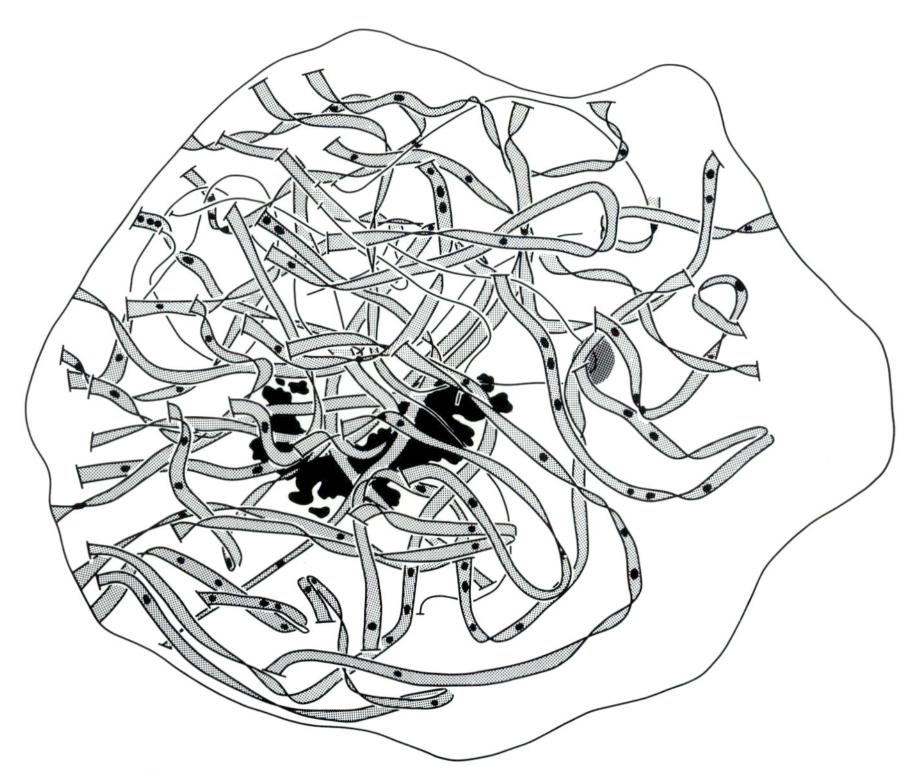

Figure 2. Reconstruction of synaptonemal complexes and lateral components of a *Bombyx* spermatocyte nucleus at late zygonema. Note the pronounced chromosome bouquet and the dense nucleolus. The dense bodies in the central region of the complexes are recombination nodules. (From Holm and Rasmussen, 1980. Reproduced with permission of the Carlsberg Laboratory.)

Table 2. Interlockings and Breaks in *Bombyx* Spermatocyte Nuclei at Late Zygonema

Nucleus number	Interlockings[a]		Breaks			Total number of aberrations
	Chromosome	Bivalent	Chromosome	Bivalent	Total	
1	0	6	0	2	4	10
2	0	2	0	0	0	2
3	0	1	—	—	—	1[b]
4	2	3	1	3	7	12
5	2	1	1	0	1	4
6	3	1	3	1	5	9
7	1	1	4	3	10	12
8	5	4	4	0	4	13
Mean	1.6	2.4	1.6	1.1	3.8	7.9

[a] The values include resolving interlockings.
[b] Several breaks in 5 pairs of homologues.

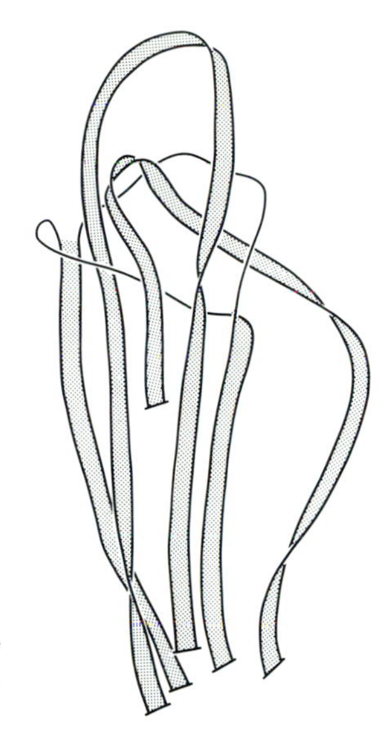

Figure 3. A double bivalent interlocking from a *Bombyx* spermatocyte nucleus at late zygonema. (From Holm and Rasmussen, 1980. Reproduced with permission of the Carlsberg Laboratory.)

been observed within one nucleus. Often a substantial part of the complement was intermingled, forming a large knot, held together by multiple interlockings (Figure 4).

3.1.3. Resolution of Interlockings

By mid pachynema in both *Bombyx* oocytes and spermatocytes, all interlockings have disappeared. Hence, mechanisms must exist that are capable of resolving interlockings during the transition from zygonema to pachynema. At late zygonema, major discontinuities of LCs and SCs were found in *Bombyx* spermatocytes at a mean frequency of 1.6 and 1.1 per nucleus respectively (Table 2). Broken chromosomes and bivalents were particularly frequent within the knotlike configuration of multiple interlocked chromosomes and bivalents, often in regions of presumptive interlockings (Figure 4).

In contrast, chromosome breaks were entirely absent and bivalent breaks were rare by the mid or late pachytene stage. Taken together, these observations show that interlockings in *Bombyx* spermatocytes are resolved by breakage of chromosomes or bivalents followed by reunion and repair of the broken ends.

The break–reunion–repair mechanism for resolution of interlockings was first revealed in human spermatocytes (Rasmussen and Holm, 1978) and has recently been found also in *Bombyx* oocyte nuclei (Rasmussen and Holm, unpublished observations). A preliminary analysis of late zygotene *Bombyx* females revealed, as in the male, knotlike configurations of interlocked chromosomes and bivalents and furthermore showed that both chromosome and bivalent breaks were frequent in the knot (Figure 5).

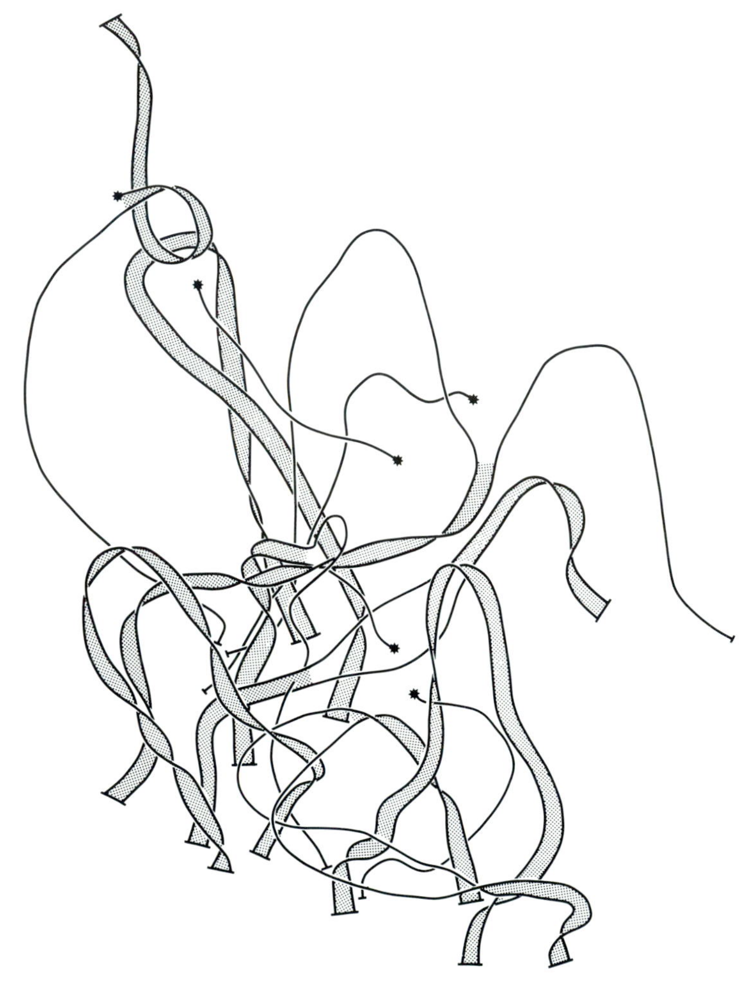

Figure 4. A complex interlocking involving eight bivalents at late zygonema in a *Bombyx* spermatocyte nucleus. Three chromosome breaks have given rise to six free chromosome ends (marked by stars). (From Holm and Rasmussen, 1980. Reproduced with permission of the Carlsberg Laboratory.)

The molecular mechanisms responsible for the breakage–reunion–repair events are yet unknown. The very low frequency of chromosome and bivalent fragments at pachynema and the virtual absence of chromosomal rearrangements do, however, indicate that these events are precisely controlled processes fundamentally different from, for example, the breakage–fusion–bridge cycle in maize described by McClintock (1941) where the breakage within a paracentric inversion is caused by physical tension. A controlled enzymatic breakage may be a necessary prerequisite for the precise and specific repair mechanism capable of joining the free ends of broken chromosomes and bivalents. The former may be a more simple process, as the intact homologue, by specific pairing with the broken one, facilitates the juxtaposition of the free ends. The latter

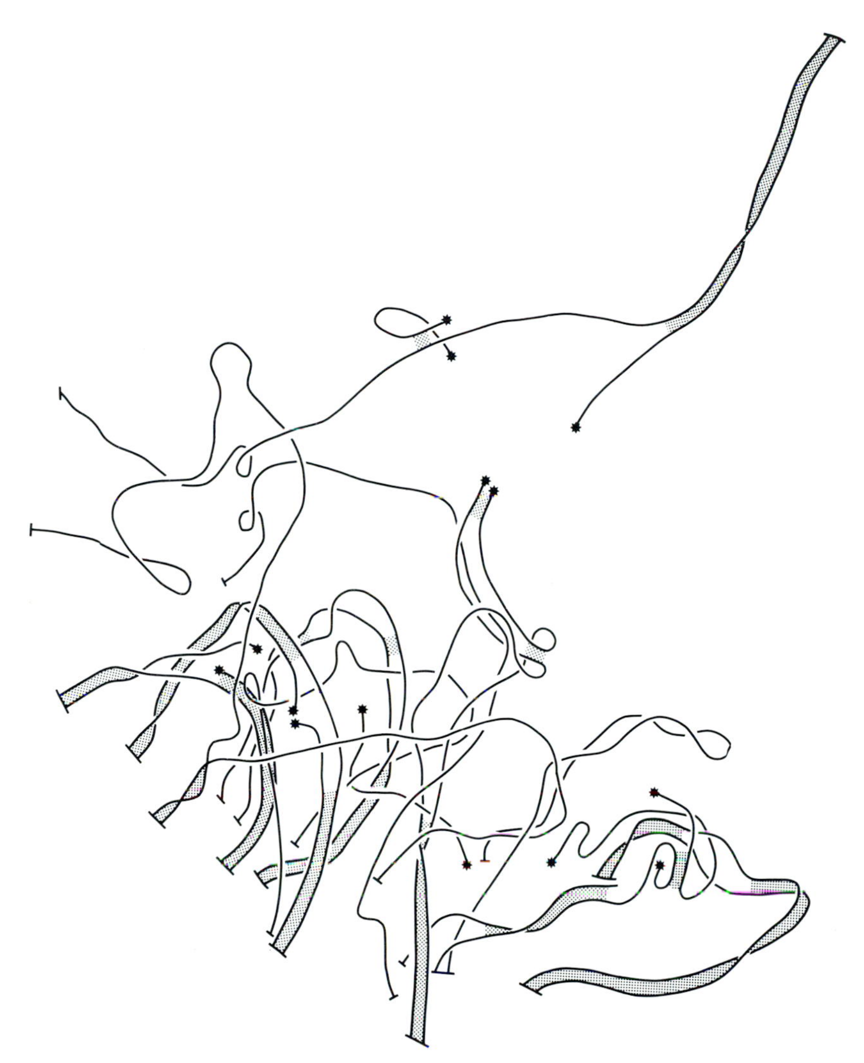

Figure 5. A complex interlocking involving eight bivalents and three univalents in a *Bombyx* oocyte nucleus in mid to late zygonema. The free chromosome ends are denoted by stars. The entire nucleus contained four chromosome interlockings and nine chromosome breaks.

type of break imposes a more difficult problem, as, in this case, bivalent continuity is disrupted and the free ends are often separated by distances up to several microns.

The observation that interlocking is a regular feature of zygotene chromosome pairing in *Bombyx* and human spermatocytes is at variance with the general contention from light microscopic observations that interlockings are rare events in normal diploid organisms (Bèlař, 1928; Darlington, 1937; Barber, 1942). These observations have been considered as evidence in favor of a presynaptic alignment of homologues, believed to occur at the last mitotic division.

The condensed state of the chromatin at this stage has been claimed to minimize the risk of entanglement during the succeeding pairing at zygonema (Buss and Henderson, 1971a,b). Evidently, this argument is no longer tenable. The contention that interlocking is a rare phenomenon stems rather from the difficulty in resolving the pairing zygotene chromosomes and from the fact that most cytological studies have been performed on squashes of cells at late diplonema or metaphase I. The preservation up to diakinesis of the few interlockings that escape resolution during the zygotene–pachytene transition furthermore requires the presence of chiasmata in the segments on either side of the interlocking.

3.1.4. The Pachytene Nucleus

At the ultrastructural level, the pachytene stage is characterized by a complete pairing of all homologues into bivalents, with continuous SCs throughout their entire length (Figure 6). At this stage, the SC attains its most distinct appearance, with two dense LCs flanking the chromatids and held in register by the central region. In both *Bombyx* oocytes and spermatocytes, the central region is scalariform (Figures 7a,b 8a). In spermatocytes it is 100–120 nm wide (King and Akai, 1971), as in most other organisms (Westergaard and von Wettstein, 1972), whereas the width of the central region in

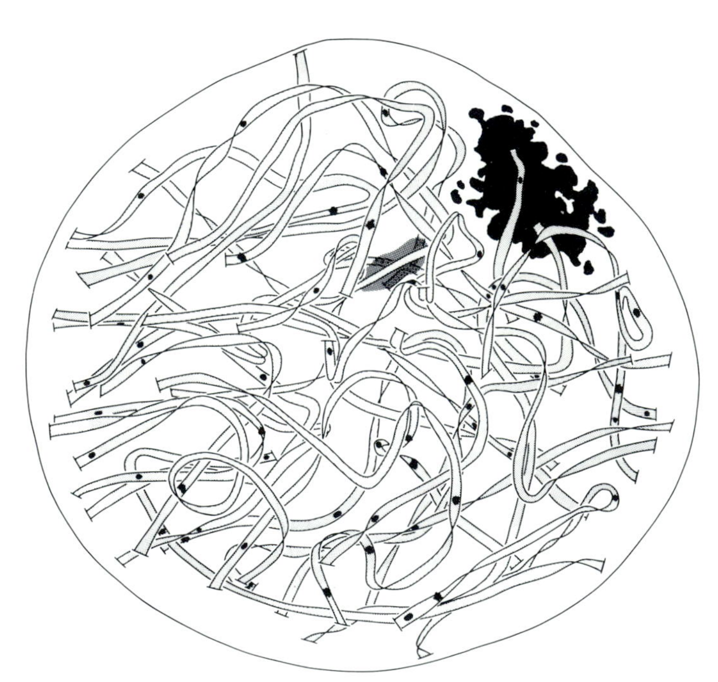

Figure 6. A complete reconstruction of a mid pachytene nucleus from a *Bombyx* spermatocyte. Recombination nodules and chromatin nodules in the central region of the synaptonemal complexes are denoted by ellipsoid and irregular black dots respectively. The hatched body is the knob on bivalent 1, and a dense nucleolus is attached to bivalent 2. (From Holm and Rasmussen, 1980. Reproduced with permission of the Carlsberg Laboratory.)

oocytes is only 70–80 nm (Rasmussen, 1976). Apart from this difference in width, the fine structure of the central region is very similar in both sexes. It is tempting to relate the difference in SC width between the male and female to the absence of crossing over in the female sex (Rasmussen and Holm, 1979). In both sexes, the chromosome bouquet disappears during early mid pachynema as the attachment sites of the telomeres disperse over the entire surface of the nuclear envelope. At mid pachynema, one, or more rarely both, LCs become longitudinally divided in some bivalent segments in spermatocytes. The splitting of the LCs becomes more pronounced by late pachynema and the central component of the SC appears more distinctly scalariform (compare Figures 7a and b).

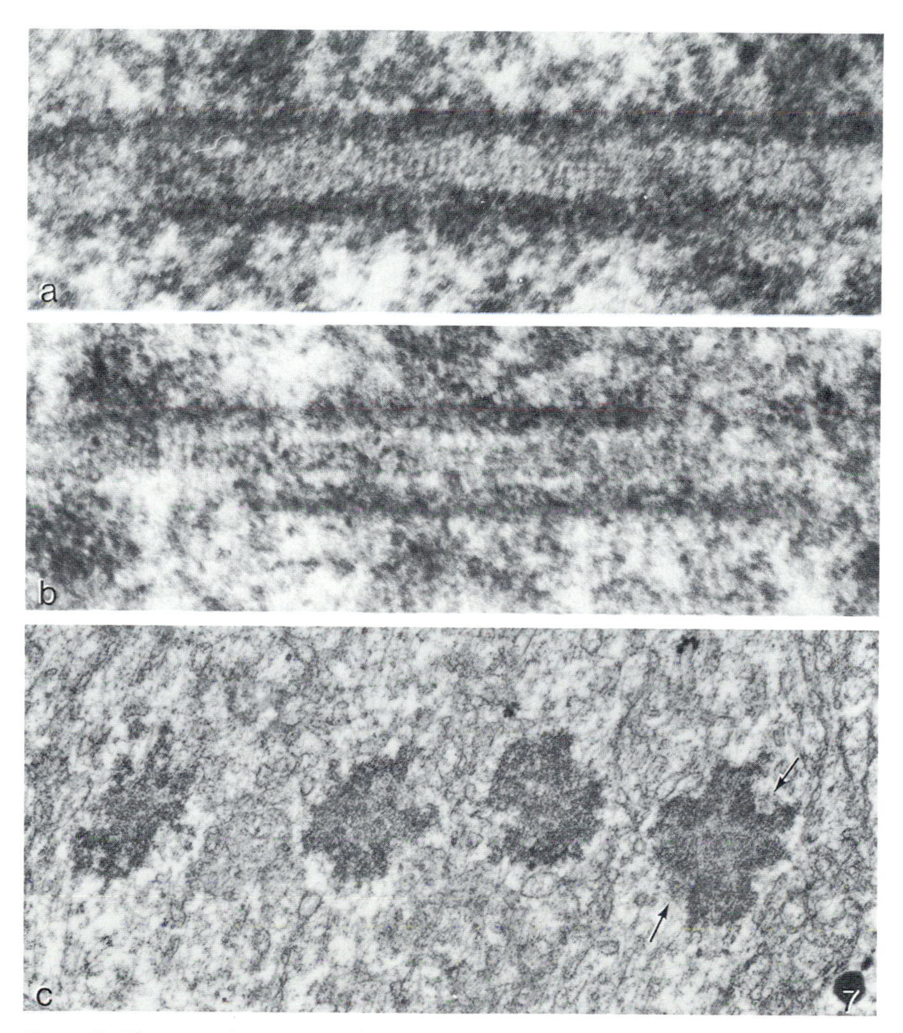

Figure 7. Electron micrographs of synaptonemal complexes and metaphase I bivalents of *Bombyx* spermatocytes. (a) From a mid pachytene nucleus. (b) From a nucleus at late pachynema. (c) Four bivalents at metaphase I (arrows denote centromere regions). (Figures 7a–b × 100,000. Figure 7c × 12,500.)

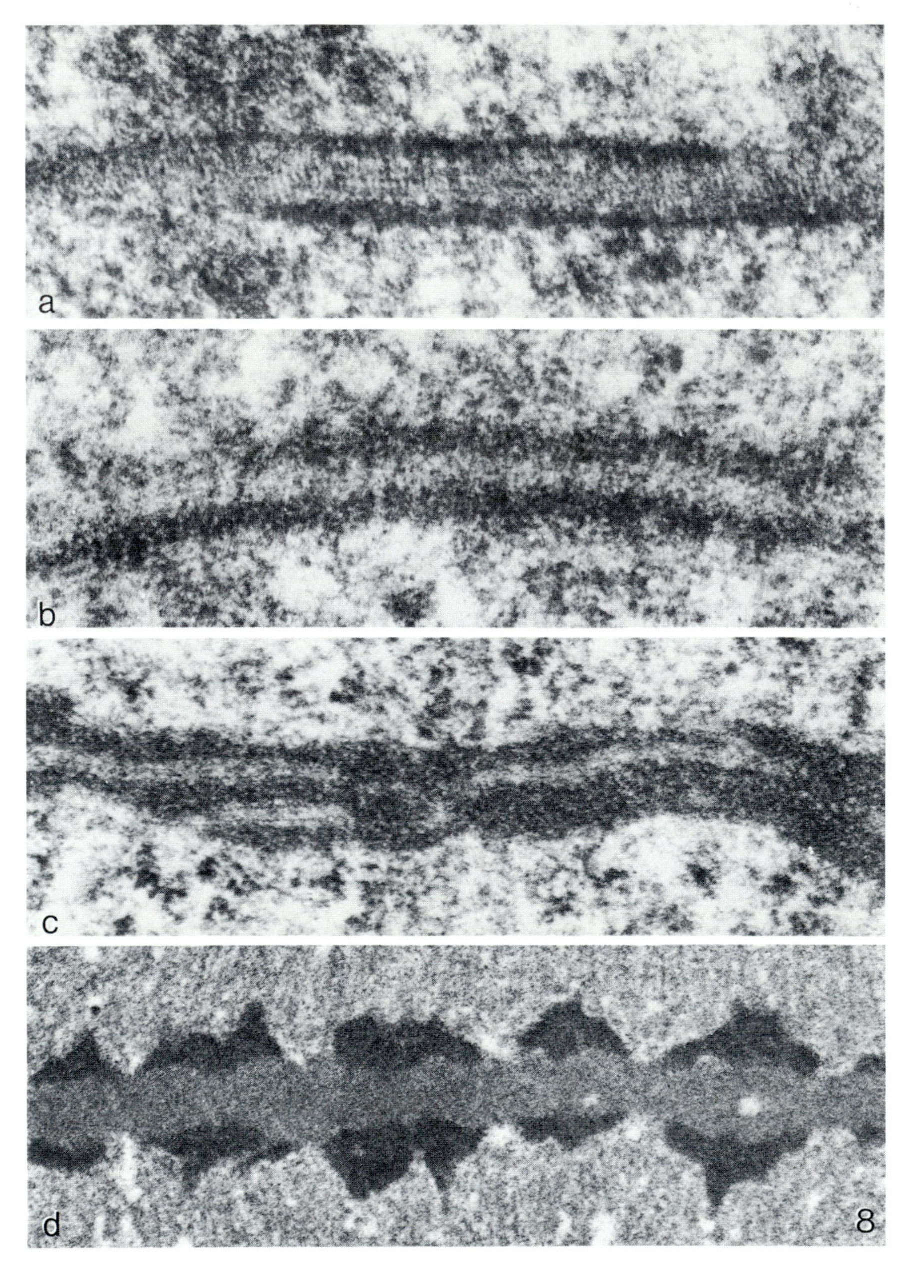

Figure 8. Electron micrographs of synaptonemal complexes and metaphase I bivalents of *Bombyx* oocytes. (a) Mid pachynema. (b) The initiation of synaptonemal complex modification. (c) A later stage in the modification. (d) A cross section through a metaphase plate showing that the modified synaptonemal complexes have fused into a continuous sheet between the dense chromatin of the homologues. (Figure 8a–c × 100,000. Figure 8d × 12,500.)

Similar ultrastructural modifications of the SC do not occur in females. The structure of the complex remains unaltered until the end of pachynema. Thereafter, both the central region and the LCs undergo a drastic modification. By the end of pachynema, the structural similarity between the male and the female meiosis terminates, as discussed in sections 3.2 and 3.3.

3.2. The Pachytene–Metaphase I Period in Males

3.2.1. Differentiation of Spermatocytes

The spermatocytes pass through a series of characteristic stages from diplonema to metaphase I, marked by distinctive changes in the chromatin condensation, elimination of the SC, centriolar behavior and evolution of chiasmata. Following a transient decondensation of most of the chromatin during diplonema, the chromatin condenses at early diakinesis, eventually resulting in the formation of compacted diakinesis–metaphase I bivalents (Kawaguchi, 1928; Holm and Rasmussen, 1980).

The degradation of the SC, initiated at the pachytene–diplotene transition, proceeds, and by early diplonema bivalent continuity is no longer recognizable (Figure 9). The central region and the LCs appear to be eliminated by a gradual degradation of the SC constituents and do not form polycomplexes

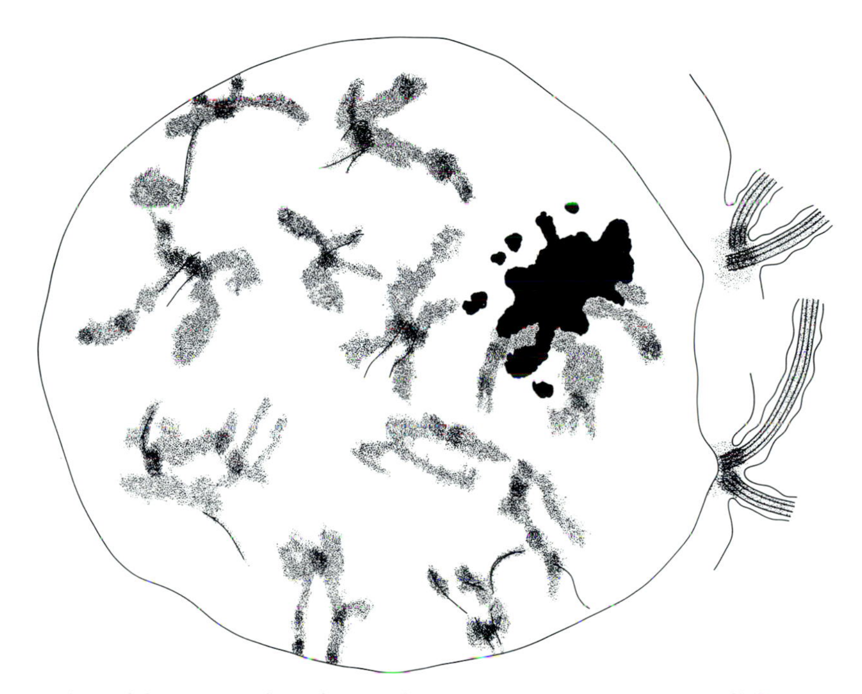

Figure 9. A partial reconstruction of a *Bombyx* spermatocyte nucleus at late diplonema. Note the condensed chromatin of the chiasmata with the associated remnants of synaptonemal complexes, the partially fragmented nucleolus, and the two pairs of flagellae. (From Holm and Rasmussen, 1980. Reproduced with permission of the Carlsberg Laboratory.)

as is the case in female nurse cells (cf. section 3.3). Only a few remnants of intact complexes are present at mid diplonema, and before diakinesis the LCs and the central region disappear.

Changes in centriolar morphology and location correlate well with the morphological changes of the nucleus, and thus serve as additional characteristics of the individual stages. The two pairs of centrioles establish contact with the cell membrane at mid pachynema, as the growth of the flagellae is initiated. By mid diplonema, the proximal part of the centrioles comes in contact with the nuclear envelope. The centrioles detach from the nuclear envelope at early diakinesis, and invaginations of the envelope evolve at the former attachment sites. The positions of these invaginations change in parallel with the migration of the two centriole pairs toward opposite poles at late diakinesis–metaphase I. (Friedländer and Wahrman, 1970; Holm and Rasmussen, 1980).

3.2.2. The Evolution of Chiasmata

As described in the introduction, available evidence suggests that each site where crossing over has taken place (or will take place) is marked by a dense, often ellipsoid structure, the recombination nodule (RN), which is associated with the central region of the SC. This hypothesis has gained additional support by the observation that RNs are absent in *Bombyx* oocytes but readily identified in *Bombyx* spermatocytes (Rasmussen and Holm, 1979; Holm and Rasmussen, 1980).

In Bombyx spermatocytes, RNs first appear at early zygonema simultaneously with the initiation of SC formation. During zygonema, the number of RNs increases, and by late zygonema (Figures 2, 10a) the mean number of nodules amounts to 91 per nucleus. If one assumes that unpaired bivalent regions upon pairing will have the same probability for receiving a nodule as already paired regions, 103 RNs per nucleus are expected. Between late zygonema and early pachynema, this number decreases by 41% (Table 3). Comparable decreases in the total number of RNs between late zygonema and early pachynema have been described in human spermatocytes (Rasmussen and Holm, 1978) and in the fungus *Schizophyllum* (Carmi *et al.*, 1978), and it has been suggested that this drastic reduction in the number of RNs somehow relates to the occurrence of reciprocal and nonreciprocal recombination (Rasmussen and Holm, 1978). Elimination of a given RN after zygonema may signify termination of a nonreciprocal recombination event, whereas reciprocal exchanges may require that the RNs remain attached to the central region of the SC at the site of a crossover.

At early and mid pachynema, several of the nodules appear larger, more electron dense, and more elongated than those present at late zygonema (Figures 10a,b), and the position of these nodules is always lateral to the central region of the SC. By mid pachynema, an increasing fraction of these dense nodules becomes larger and more irregular in shape by association of chromatin to the nodules (Figure 10c). These recombination nodule–chromatin aggregates, termed chromatin nodules, increase in size as well as in number

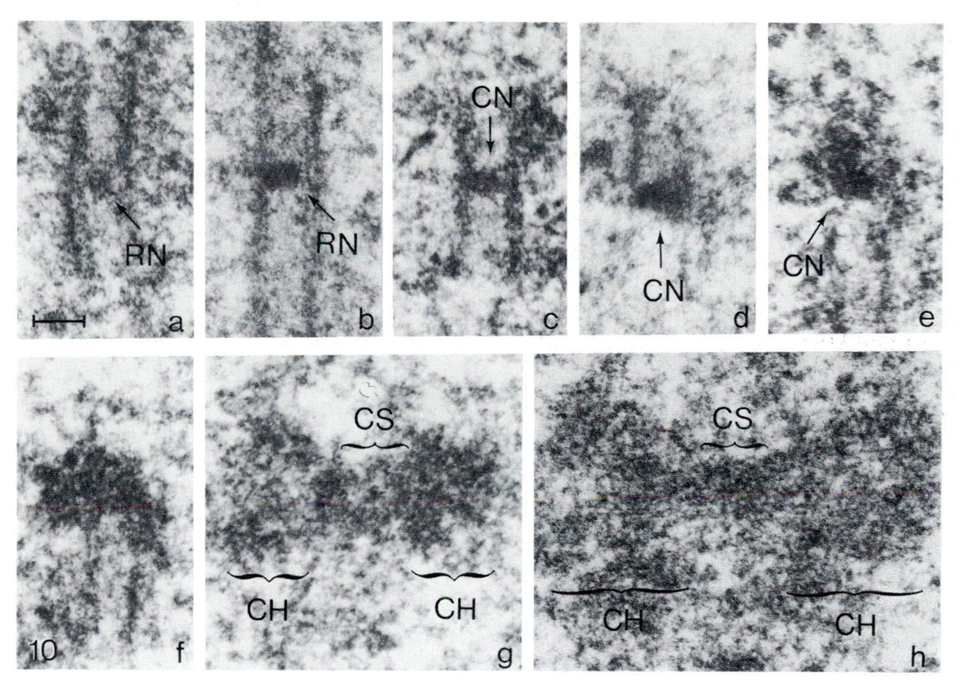

Figure 10. Electron micrographs showing the development of recombination nodules into chiasmata in *Bombyx* spermatocytes. (a) Recombination nodule (RN) at late zygonema. (b) Recombination nodule at mid pachynema. (c) Chromatin nodule (CN) at mid pachynema. (d) Chromatin nodule at late pachynema. (e) Chromatin nodule surrounded by condensed chromatin. (f) Chiasma at mid diplonema. Note the remnants of the synaptonemal complex. (g) Chiasma at late diplonema. Note the circular structure (CS) with a dense core between the homologues (CH). (h) Chiasma at mid diakinesis; part of the circular structure is present in the chromatin bridge between the two homologues (CH). (Scale bar = 0.1 μm.) (From Holm and Rasmussen, 1981. Reproduced with permission of Springer-Verlag.)

during the latter part of pachynema (Table 3), and at early diplonema nearly all nodules are of this type (Figure 10d).

A comparison with a random distribution of recombination nodules among the twenty-eight bivalents of a nucleus constructed by a computer simulation experiment (Holm and Rasmussen, 1980), showed that the observed

Table 3. Number of Recombination Nodules (RN) and Chromatin Nodules (CN) in *Bombyx* Spermatocytes

Stage	Number of nuclei	Number of RN	Number of CN	Total nodules
Late zygonema	8	91	0	103[a]
Early pachynema	16	58	2	61
Mid pachynema	8	37	18	55
Late pachynema	6	26	44	70

[a]The number expected upon completion of synaptonemal complex formation.

distribution deviates very little from a random distribution both at late zygonema and at early pachynema. From early to late pachynema, the random distribution changes by association of nodules preferentially to bivalents without nodules but also to some extent to bivalents with one and two nodules.

These observations demonstrate that the distribution of crossovers and chiasmata among bivalents are the result of an initially random distribution which is then slightly modified in order to ensure a minimum of one chiasma per bivalent, which is a prerequisite for a regular disjunction. The transition from pachynema to diplonema is marked by the initiation of SC degradation and by decondensation of the bulk of the chromatin. Maximum decondensation is reached at mid diplonema where only about sixty major chromatin condensations and a few minor condensations remain. The major condensations are in most cases associated with remnants of the SC (see Figures 9, 10e,f). Frequently, these SC remnants contain a dense core reminiscent of an RN. At late diplonema, the major chromatin condensations consist of two domains of condensed chromatin bridged by a circular structure, 120–160 nm in diameter (Figure 10g). The fine structure of these circular components remains unchanged until mid diakinesis. From the diplotene-diakinesis transition, the circular structures are frequently surrounded by a chromatin bridge that combines the two flanking regions of condensed chromatin (Figure 10h). Complete reconstructions of mid diakinesis nuclei reveal that these chromatin bridges together with their circular component constitute the chiasmata. At late diakinesis, the number of circular components decreases, and at metaphase I, very few remain. The bivalents condense at these stages into more or less spherical bodies, and chiasmata can no longer be identified.

The number of chiasmata is fairly constant from mid diplonema to mid diakinesis, and the mean number of fifty-eight per nucleus is very similar to the mean number of chromatin nodules at the transition from pachynema to diplonema and at early diplonema (Tables 3, 4). Furthermore, this value is approximately the same as the number of chiasmata at early and mid diakinesis reported by Maeda (1939). Finally, the distribution of chiasmata among

Table 4. Number of Recombination Nodules (RN), Chromatin Nodules (CN) and Chiasmata during Late Meiotic Prophase in *Bombyx* Spermatocytes

Stage	Number of nuclei	Number of RN	Number of CN and chiasmata
Pachynema–diplonema	3	16	59
Early diplonema	4	2	60
Mid diplonema	6	0	62
Late diplonema	2	0	46
Diplonema–diakinesis	4	0	55
Early diakinesis	1	0	58
Mid diakinesis	3	0	62
Late diakinesis	3	0	28 bivalents
Metaphase I	4	0	28 bivalents

bivalents closely resembles the distribution of RNs and chromatin nodules during the latter part of pachynema, as well as the distribution of chiasmata as determined by light microscopy (Maeda, 1939). These observations are consistent with a continuous morphological evolution of chiasmata in bivalent regions that at pachynema contain recombination nodules.

Altogether, these results demonstrate that in the male silkworm, RNs, in addition to their role in crossing over together with retained pieces of the SC, also function in the establishment and maintenance of stable chromatin bridges—chiasmata. Comparable data are at present not available for organisms other than *Bombyx*, and it remains to be seen whether this sequence of events represents a universal feature of meiosis or is unique to *Bombyx* spermatocytes.

Observations of retained pieces of SC after pachynema in the ascomycete *Neottiella* (Westergaard and von Wettstein, 1968; 1970) and in mouse spermatocytes (Solari, 1970) have suggested that the early chiasma in these organisms consists of a retained fragment of the SC, which is later replaced by a chromatin bridge. In the ascomycete *Sordaria*, Zickler (1977) found that the number of fragments was about the same as the number of chiasmata counted in the light microscope and that most of the SC fragments contained a distinct RN in their central region. The presence of RNs in retained fragments of SC at diplonema has also been reported in *Ascaris* (Kundu and Bogdanov, 1979) and in *Neurospora* (Gillies, 1979), altogether suggesting that both RNs and SC fragments are essential for the organization of chiasmata.

3.2.3. The Disjunction of the Homologues

The analysis of *Bombyx* spermatocytes at metaphase I in serial sections (Holm and Rasmussen, 1980) revealed in each bivalent four localized centromeres, two by two, facing opposite spindle poles (Figure 7c). In agreement with this, Gassner and Klemetson (1974) reported the presence of localized centromeres at metaphase I in an electron microscopic analysis of two species of Lepidoptera and the hemipteran, *Oncopeltus fasciatus*. Finally, Danilova (1973) reported that metaphase II bivalents of *Bombyx* spermatocytes also appear to contain localized centromeres. These observations are at variance with the general concept that *Bombyx*, like other lepidopteran species, has nonlocalized centromeres. This discrepancy may find its explanation in the differences in centromere behavior at mitosis and meiosis described by Hughes-Schrader and Schrader (1961). In three species of Hemiptera, the centromeres appeared nonlocalized at mitotic metaphase and anaphase, as evidenced by the orientation of the chromosomes parallel to the equatorial plane both at metaphase and during separation of the chromatids at anaphase. In contrast, the centromere activity in bivalents at metaphase I of meiosis was restricted to one telomere region per chromosome, resulting in an orientation and separation of the homologues parallel to the pole-to-pole axis. Similar differences in chromosome orientation have been observed at mitosis and meiosis in *Bombyx* by Murakami and Imai (1974) and Maeki (1980).

As pointed out by Hughes-Schrader and Schrader (1961), such meiosis-specific modifications of the centromere activity, leaving only one active centromere region per chromatid, possibly represent an evolutionary adaptation of the diffuse centromere to the problem of meiotic segregation. By restricting centromere activity to only one region per chromatid, a regular disjunction is ensured also in cases of incomplete chiasma elimination prior to spindle formation.

3.3. The Pachytene–Metaphase I Period in Females

In *Bombyx* females, the oocyte develops as a member of a cluster of eight interconnected cells (Miya *et al.*, 1970). The oocyte cannot be distinguished from the seven nurse cells, either by cellular morphology or by the ultrastructure of the SC, until the end of pachynema. Thereafter, the developmental sequences begin to diverge. In nurse cells, the chromatin condenses and the SCs are shed from the bivalents. The released SC constituents often fuse into various aggregates, some resembling the synaptonemal polycomplexes described in many insects (see, for example, Fiil and Moens, 1972; Rasmussen, 1975a), while others are more irregular assemblies of SC subunits. At first, these aggregates reside inside the condensed chromatin of the bivalents, but later the chromatin unravels and the SC aggregates disappear (Rasmussen, 1976).

In the oocytes, the LCs of the SC grow in width as well as in length (Figure 8b and Table 1), ultimately forming a dense, almost solid rod combining the dispersed chromatin of the homologues (Figure 8c). Subsequently, the chromatin and the modified SCs of the bivalents condense, and often two or more bivalents fuse end-to-end, forming short chains of bivalents. By metaphase I, all bivalents congregate into the metaphase plate, and the modified SCs fuse into an almost continuous sheet (Figure 8d). This material is eliminated from the bivalents and left behind at the position of the metaphase plate as the homologues move to opposite poles during anaphase I (Rasmussen, 1977a).

These fused, modified SCs have previously been described by Seiler (1914), Sorsa and Suomalainen (1975), Vereiskaya (1975), and others as "elimination chromatin." In *Bombyx* oocytes, and probably in other lepidoptera species as well, the modified SCs function as a substitute for chiasmata in retaining the association of the homologous chromosomes up to anaphase I, and thereby ensuring a regular disjunction in the absence of crossing over.

4. Meiosis in Polyploid Bombyx Females

The normal course of chromosome pairing in diploid females of *Bombyx* has been described in section 3. However, a more detailed analysis of chromosome pairing is possible when the pairing occurs under competitive conditions, as is the case in polyploid organisms. The lack of crossing over and chiasma formation in the female sex of *Bombyx* furthermore provides the opportunity to study pairing and SC formation in the absence of nonsister chromatid exchange.

Table 5. Chromosome Configurations in Triploid *Bombyx* Oocytes

Stage	Univalents	Bivalents	Trivalents	Nonhomologous associations
Zygotene-pachytene transition[a]	22.0	22.0	6.0	0
Early pachynema[b]	23.5	26.0	1.0	2.5
Mid late pachynema[c]	18.6	26.7	0	6.4

[a]Mean of 2 nuclei.
[b]Mean of 2 nuclei.
[c]Mean of 7 nuclei.

4.1. Homologous and Nonhomologous Chromosome Pairing in Triploids

At the zygotene-pachytene transition in triploid *Bombyx* oocytes (Rasmussen, 1977b), the majority of chromosomes are paired in sets of three, two of the homologues being intimately paired with an SC and the third homologue lying parallel to the bivalent at some distance. Occasionally, exchange of a pairing partner results in trivalent formation (see Table 5).

In contrast, trivalents are absent at mid late pachynema and the nuclei contain the maximum number of homologously paired bivalents. In addition, a number of nonhomologous associations have appeared in the form of fold-back pairing of univalents, associations of two chromosomes of unequal length, associations of three chromosomes completely paired with continuous SCs in the three arms, and associations of four chromosomes (Table 5).

This remarkable decrease in the number of trivalents during the zygotene-pachytene transition and the accompanying increase in the number of nonhomologous associations, show that pairing in the form of trivalents is unstable at late zygonema and that stability, in the absence of crossing over and chiasma formation, is achieved at this stage by the shift of a pairing partner whereby a bivalent and a univalent are formed at the expense of a trivalent. Furthermore, at early mid pachynema pairing and SC formation are no longer restricted to homologous chromosomes or chromosome segments but can also occur between nonhomologous chromosomes. The entire pairing sequence is illustrated schematically in Figure 11.

4.2. Correction of Chromosome Pairing in Tetraploids

The two phases of chromosome pairing revealed in triploids are further illustrated in autotetraploid *Bombyx* oocytes. (Rasmussen and Holm, 1979). As in the triploids, the pairing and SC formation occur under competitive conditions, but in tetraploids complete pairing in the form of bivalents is theoretically possible.

By the end of the homologous pairing phase in zygonema, a mean of 8.4 quadrivalents is present per nucleus (Table 6), in addition to 2 univalents and 0.7 trivalents, while the rest of the complement is paired into bivalents. This

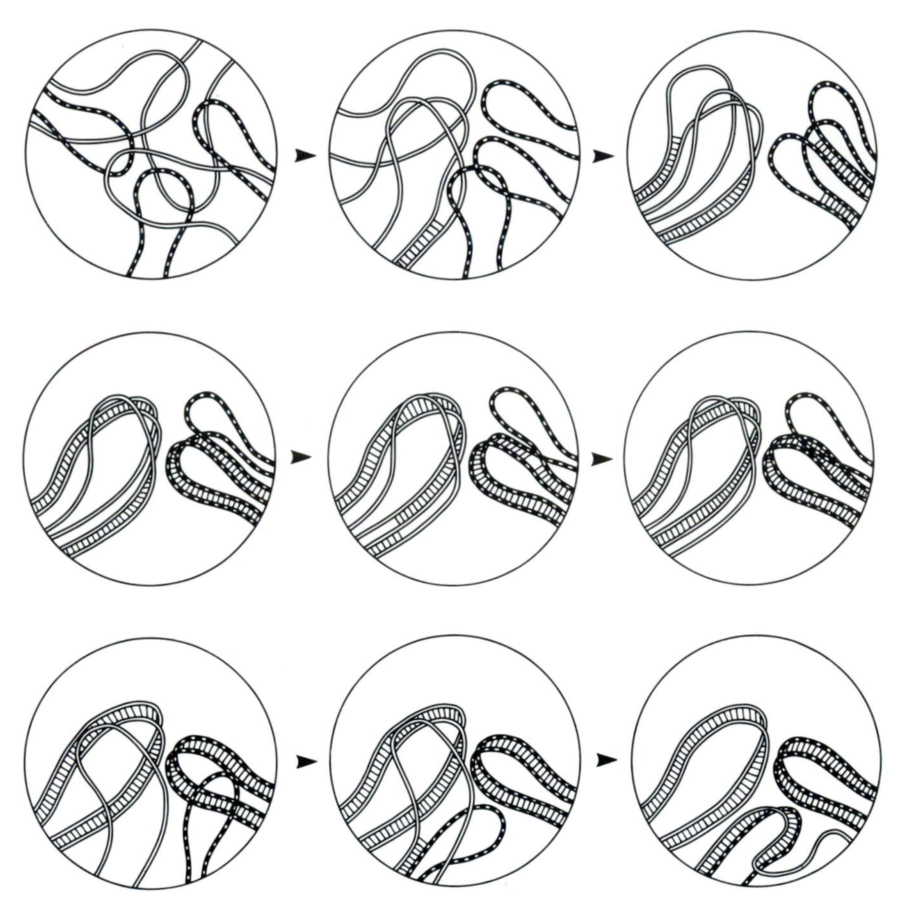

Figure 11. Schematic drawings illustrating the two phases of chromosome pairing in triploid *Bombyx* oocytes. Upper row, the homologous pairing phase at zygonema. Middle row, the transformation of trivalents into bivalents and univalents. Lower row, second round of synaptonemal complex formation. (From Rasmussen and Holm, 1980. Reproduced with permission of Mendelska Selskabet.)

Table 6. Chromosome Configurations in Tetraploid *Bombyx* Oocytes and Spermatocytes

Stage	Bivalents	Quadrivalents	Chromosomes not in bivalents and quadrivalents
Early pachynema, oocytes[a]	36.7	8.4	4.1
Mid late pachynema, oocytes[b]	51.9	0.9	3.4
Metaphase I, spermatocytes[c]	42.2	6.7	0.5

[a]Mean of 7 nuclei.
[b]Mean of 11 nuclei
[c]Data from Kawaguchi (1938).

picture is drastically altered by mid late pachynema, when nearly all chromosomes are paired into bivalents. The transition of quadrivalents into bivalents is illustrated schematically in Figure 12. At metaphase I, only bivalents are seen in the light microscope, and the resultant eggs are fully balanced and produce viable offspring (Astaurov, 1967; Vereiskaya and Vinnik, 1973).

It is thus clear that the pairing of homologous into quadrivalents, in the absence of crossing over and chiasma formation, also represents an unstable situation, quadrivalent formation being followed by a turnover of SCs, which tends to optimize pairing in the form of bivalents. As the number of univalents remains low throughout pachynema in tetraploid oocytes, the dissolution of the central region of the complex appears to be followed immediately by formation of a new central region between a different combination of LCs.

Comparable data for chromosome pairing in triploid and tetraploid spermatocytes are not available at present. However, in tetraploid *Bombyx* spermatocytes, light microscopical observations have revealed approximately the same number of bivalents and quadrivalents at metaphase I (Kawaguchi, 1938) as have been observed in tetraploid oocytes at early pachynema (Rasmussen and Holm, 1979). Since crossing over and chiasma formation take place in spermatocytes, the preservation of quadrivalents up to metaphase I strongly indicates that the correction of chromosome pairing by dissolution and reformation of the central region of the SC is prevented in regions where crossing over has occurred.

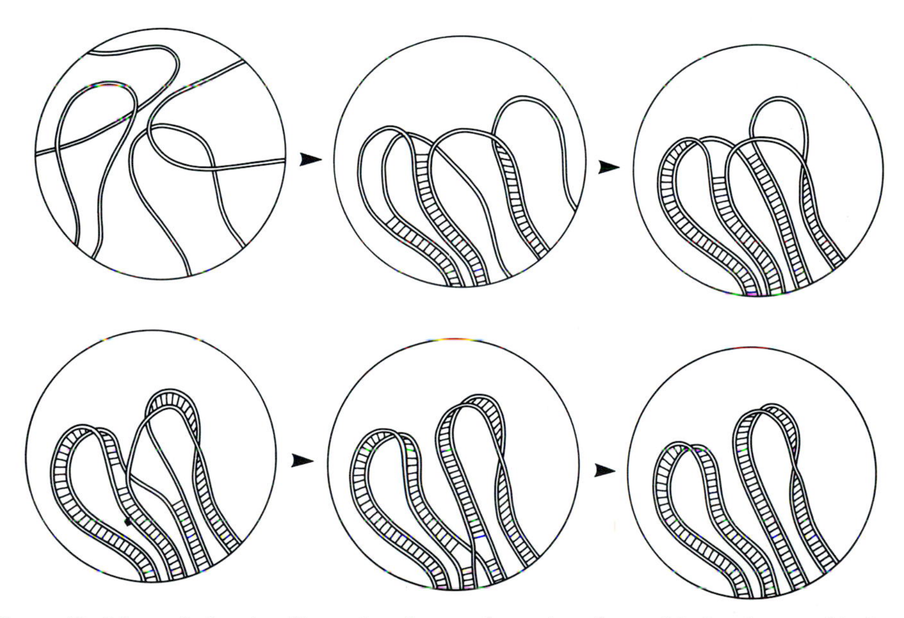

Figure 12. Schematic drawing illustrating the transformation of a quadrivalent into two bivalents in tetraploid *Bombyx* oocytes. (From Rasmussen and Holm, 1980. Reproduced with permission of Mendelska Selskabet.)

5. Conclusions

The present description of meiosis, although based primarily on observations in males and females of the silkworm, *Bombyx mori*, is consistent with a large number of reports on the normal course of meiosis in other organisms, such as fungi and insects, higher plants and animals, but also provides detailed insight into some of the individual processes ultimately resulting in a regular disjunction of homologous chromosomes.

Among these are the correction of chromosome interlocking during late zygonema and the transformation of multivalents into bivalents during pachynema. Despite the fact that the two phases of chromosome pairing were revealed in triploid and tetraploid *Bombyx* oocytes without crossing over, the mechanism may also operate in other eukaryotes, correcting illegitimate pairing of nonhomologous chromosomes or chromosome segments in regions where crossing over has not occurred. In this way, formation of bivalents with continuous SCs from telomere to telomere is optimized. Evidence for a temporal separation of homologous and nonhomologous pairing has been obtained in mice, heterozygous for duplications and inversions (Moses, 1977), and in a human translocation heterozygote (Holm and Rasmussen, 1978). In these cases, rearranged segments often pair nonhomologously at pachynema after the phase of homologous pairing and SC formation at zygonema. The second round of nonspecific pairing can also account for the formation of SCs between nonhomologous regions in hybrids (Menzel and Price, 1966), haploids (Gillies, 1974) and other chromosomal rearrangements (Gillies, 1973; Fletcher and Hewitt, 1978; Gillies, 1981). However, the preservation of multivalents up to metaphase I in autotetraploid *Bombyx* spermatocytes does indicate that the occurrence of crossing over in the involved chromosome segment effectively prevents such secondary modification of chromosome pairing.

Following the bivalent formation during zygonema and pachynema, crossing over ensures the reassortment of linked genes. In addition to its function in providing for genetic variability, the exchange of nonsister chromatids ensures that the bivalent condition is maintained up to metaphase I. The analysis of *Bombyx* spermatocytes has shown that recombination nodules are somehow involved in the establishment and organization of stable chiasmata.

In the absence of crossing over and chiasma formation, compensatory mechanisms are required for the maintenance of bivalents from pachynema to metaphase I as exemplified by the retention and modification of SCs in *Bombyx* females. The retention of SCs until metaphase I in the achiasmatic meiosis of *Bolbe nigra* spermatocytes (Gassner, 1969) may be another example of a mechanism compensating for the lack of chiasmata.

ACKNOWLEDGMENTS

The materials, on which most of this work has been performed were kindly provided by the late Professor B. L. Astaurov, by Dr. V. N. Vereiskaya, and by Dr. H. Akai. The work was supported by grant 202-76-1 BIO DK from the Commission of the European Communities.

References

Astaurov, B. L., 1967, Experimental alterations of the developmental cytogenetic mechanisms in mulberry silkworms: Artificial parthenogenesis, polyploidy, gynogenesis and androgenesis, *Adv. Morphogen.* **6**:199–257.

Barber, H. N., 1942, The experimental control of chromosome pairing in *Fritillaria*, *J. Genet.* **43**:359–374.

Bělař, K., 1928, Die Cytologische Grundlagen der Vererbung. In *Handbuch der Vererbungwissenschaft*, vol. 1, edited by E. Baur and M. Hartmann, Verlag Gebrüder Borntraeger, Berlin.

Buss, M. E., and Henderson, S. A., 1971a, Induced bivalent interlocking and the course of chromosome synapsis, *Nature New Biol.* **234**:243–246.

Buss, M. E., and Henderson, S. A., 1971b, The induction of orientational instability and bivalent interlocking at meiosis, *Chromosoma* **35**:153–183.

Carmi, P., Holm, P. B., Rasmussen, S. W., Sage, J., and Zickler, D., 1978, The pachytene karyotype of *Schizophyllum commune* analyzed by three dimensional reconstructions of synaptonemal complexes, *Carlsberg Res. Commun.* **43**:117–132.

Carpenter, A. T. C., 1975, Electron microscopy of meiosis in *Drosophila melanogaster*. II. The recombination nodule—a recombination associated structure at pachytene? *Proc. Nat. Acad. Sci. USA* **72**:3186–3189.

Carpenter, A. T. C., 1979a, Synaptonemal complex and recombination nodules in wild type *Drosophila melanogaster* females, *Genetics* **92**:511–541.

Carpenter, A. T. C., 1979b, Synaptonemal complex and recombination nodules in recombination-defective mutants of *Drosophila melanogaster*, *Chromosoma* **75**:259–292.

Danilova, L. V., 1973, An electron microscopic study of meiosis in diploid males of the silkworm, *Ontogenez* **4**:40–48.

Darlington, C. D., 1937, *Recent Advances in Cytology*, 2nd ed., Churchill, London.

Fawcett, D. W., 1956, The fine structure of chromosomes in the meiotic prophase of vertebrate spermatocytes, *J. Biophys. Biochem. Cytol.* **2**:403–406.

Fiil, A., and Moens, P. B., 1972, The development, structure and function of modified synaptonemal complexes in mosquito oocytes, *Chromosoma* **36**:119–130.

Fletcher, H. L., and Hewitt, G. M., 1978, Non-homologous synaptonemal complex formation in a heteromorphic bivalent in *Keyacris scurra* (Morabinae, Orthoptera), *Chromosoma* **65**: 271–281.

Friedländer, M., and Wahrman, J., 1970, The spindle as a basal body distributor: A study in the meiosis of the male silkworm moth, *Bombyx mori*, *J. Cell Sci.* **7**:65–89.

Gassner, G., 1969, Synaptonemal complexes in the achiasmatic spermatogenesis of *Bolbe nigra* Giglio-Tos (Mantoidea), *Chromosoma* **26**:22–43.

Gassner, G., and Klemetson, D. J., 1974, A transmission electron microscope examination of hemipteran and lepidopteran gonial centromeres, *Can. J. Genet. Cytol.* **16**:457–464.

Gillies, C. B., 1973, Ultrastructural analysis of maize pachytene karyotypes by three dimensional reconstruction of the synaptonemal complexes, *Chromosoma* **43**:145–176.

Gillies, C. B., 1974, The nature and extent of synaptonemal complex formation in haploid barley, *Chromosoma* **48**:441–453.

Gillies, C. B., 1975, Synaptonemal complex and chromosome structure, *Annu. Rev. Genet.* **9**:91–109.

Gillies, C. B., 1979, The relationship between synaptonemal complexes, recombination nodules and crossing over in *Neurospora crassa* bivalents and translocation quadrivalents, *Genetics* **91**:1–17.

Gillies, C. B., 1981, Spreading of maize pachytene synaptonemal complexes for electron microscopy, *Chromosoma* **83**:575–591.

Gowen, J. W., 1933, Meiosis as a genetic character in Drosophila melanogaster, *J. Exp. Zool.* **65**: 83–106.

Holm, P. B., and Rasmussen, S. W., 1978, Human meiosis. III. Electron microscopical analysis of chromosome pairing in an individual with a balanced translocation 46,XY,t(5p–;22p+), *Carlsberg Res. Commun.* **43**:329–350.

Holm, P. B., and Rasmussen, S. W., 1980, Chromosome pairing, recombination nodules and chiasma formation in diploid *Bombyx* males, *Carlsberg Res. Commun.* **45**:483–458.

Holm, P. B., and Rasmussen, S. W., 1981, Chromosome pairing, crossing over, chiasma formation and disjunction as revealed by three dimensional reconstructions. In *International Cell Biology 1980–1981*, edited by G. Schweiger, pp. 194–204, Springer Verlag, Berlin.

Hughes-Schrader, S., and Schrader, F., 1961, The kinetochores of the Hemiptera, *Chromosoma* **12:**327–350.

Kawaguchi, E., 1928, Zytologische Untersuchungen am Seidenspinner und seinen Verwanten. I. Gametogenese von *Bombyx mori* L. und *Bombyx mandarina* M. und ihrer bastarde, *Z. Zellforsch.* **7:**519–552.

Kawaguchi, E., 1938, Der Einfluss der Eierbehandlung mit Zentrifugierung auf die Vererbung bei dem Seidenspinner. II. Zytologische Untersuchung bei den polyploiden Seidenspinnern, *Cytologia* **9:**38–54.

King, R. C., 1970, The meiotic behavior of the *Drosophila* oocyte, *Int. Rev. Cytol.* **28:**125–167.

King, R. C., and Akai, H., 1971, Spermatogenesis in *Bombyx mori*. II. The ultrastructure of synapsed bivalents, *J. Morphol.* **134:**181–194.

Kundu, S. C., and Bogdanov, Yu. F., 1979, Ultrastructural studies of late meiotic prophase nuclei of spermatocytes of *Ascaris suum, Chromosoma* **70:**375–384.

Maeda, T., 1939, Chiasma studies in the silkworm *Bombyx mori* L., *Jpn. J. Genet.* **15:**118–127.

Maeki, K., 1980, The kinetochore of the Lepidoptera. I. Chromosomal features and behavior in mitotic and meiotic-I cells, *Proc. Jpn. Acad. B.* **56:**152–156.

McClintock, B., 1941, The stability of broken ends of chromosomes of *Zea mays. Genetics* **26:**243–282.

Menzel, M. Y., and Price, J. M., 1966, Fine structure of synapsed chromosomes in F_1 *Lycopersicon esculentum–Solanum lycopersicoides* and its parents, *Am. J. Bot.* **53:**1079–1086.

Meyer, G. F., 1964, A possible correlation between the submicroscopic structure of meiotic chromosomes and crossing over, *Proc. 3rd Europ. Conf. Electron Microscopy*, vol. B, pp. 461–462, Czechoslovak Acad. Sci., Prague.

Miya, K., Kurihara, M., and Tanimura, I., 1970, Electron microscope studies on the oogenesis of the silkworm *Bombyx mori* L. III. Fine structure of ovary in the early stage of fifth instar larva, *J. Fac. Agric. Iwate Univ.* **10:**59–83.

Moens, P. B., 1978, Ultrastructural studies of chiasma distribution, *Annu. Rev. Genet.* **12:**433–450.

Moses, M. J., 1956, Chromosomal structures in crayfish spermatocytes, *J. Biophys. Biochem. Cytol.* **2:**215–218.

Moses, M. J., 1968, Synaptinemal complex. *Annu. Rev. Genet.* **2:**363–412.

Moses, M. J., 1977, Microspreading and the synaptonemal complex in cytogenetic studies. In *Chromosomes Today*, vol. 6, edited by A. de la Chapelle, and M. Sorsa, pp. 71–82, Elsevier/North-Holland Biomedical Press, Amsterdam.

Murakami, A., and Imai, H. T., 1974, Cytological evidence for holocentric chromosomes of the silkworm, *Bombyx mori* and *B. mandarina* (Bombycidae, Lepidoptera), *Chromosoma* **47:**167–178.

Murakami, A., Ohnuma, A., and Imai, H. T., 1978, Cytological identification of the X-chromosome in *Bombyx mori, Annu. Rep. Nat. Inst. Genet.* (*Mishima*) **28:**68.

Rasmussen, S. W., 1973, Ultrastructural studies of spermatogenesis in *Drosophila melanogaster* Meigen, *Z. Zellforsch.* **140:**125–144.

Rasmussen, S. W., 1975a, Synaptonemal polycomplexes in *Drosophila melanogaster*, *Chromosoma* **49:**321–331.

Rasmussen, S. W., 1975b, Ultrastructural studies of meiosis in males and females of the $c(3)G^{17}$ mutant of *Drosophila melanogaster* Meigen, *C. R. Trav. Lab. Carlsberg* **40:**163–173.

Rasmussen, S. W., 1976, The meiotic prophase in *Bombyx mori* females analyzed by three dimensional reconstructions of synaptonemal complexes, *Chromosoma* **54:**245–293.

Rasmussen, S. W., 1977a, The transformation of the synaptonemal complex into the "elimination chromatin" in *Bombyx mori* oocytes, *Chromosoma* **60:**205–221.

Rasmussen S. W., 1977b, Chromosome pairing in triploid females of *Bombyx mori* analyzed by three dimensional reconstructions of synaptonemal complexes, *Carlsberg Res. Commun.* **42:**163–197.

Rasmussen, S. W., and Holm, P. B., 1978, Human meiosis. II. Chromosome pairing and recombination nodules in human spermatocytes, *Carlsberg Res. Comm.* **43:**275–327.

Rasmussen, S. W., and Holm, P. B., 1979, Chromosome pairing in autotetraploid *Bombyx* females: Mechanism for exclusive bivalent formation, *Carlsberg Res. Commun.* **44:**101–125.

Rasmussen, S. W., and Holm, P. B., 1980, Mechanics of meiosis, *Hereditas* **97**:187–225.

Seiler, J., 1914, Das Verhalten der Geschlechtschromosomes bei Lepidopteren, *Arch. Zellforsch.* **23**:159–269.

Smith, P. A., and King, R. C., 1968, Genetic control of synaptinemal complexes in *Drosophila melanogaster*, *Genetics* **60**:335–351.

Solari, A. J., 1970, The behaviour of chromosomal axes during diplotene in mouse spermatocytes, *Chromosoma* **31**:217–230.

Sorsa, M., and Suomalainen, E., 1975, Electron microscopy of chromatin elimination in *Cidaria* (Lepidoptera), *Hereditas* **80**:35–40.

Sotelo, R. J., 1969, Ultrastructure of the chromosomes at meiosis. In *Handbook of Molecular Cytology* edited by A. Lima-de-Faria, pp. 412–434, North-Holland Publishing Company, Amsterdam.

Sturtevant, A. H., 1915, No crossing over in the female of the silkworm moth, *Amer. Natur.* **49**: 42–44.

Tanaka, Y., 1953, Genetics of the silkworm, *Adv. Genet.* **5**:240–313.

Tazima, Y., 1964, *The Genetics of the Silkworm*, Logos Press and Academic Press, London.

Tazima, Y., 1978, *The Silkworm, an Important Laboratory Tool*, Kodansha Ltd., Tokyo.

Tazima, Y., Doria, H., and Akai, H., 1975, The domesticated silkmoth, *Bombyx mori*. In *Handbook of Genetics*, edited by R. C. King, pp. 63–124, Plenum Press, New York.

Vereiskaya, V. N., 1975, A cytochemical study of the elimination chromatin in the silkworm (*Bombyx mori*) meiosis, *Tsitologiya* **17**:603–606.

Vereiskaya, V. N., and Vinnik, T. A., 1973, Meiosis in thermoactivated oocytes of *Bombyx mori*, *Ontogenez* **4**:139–144.

Westergaard, M., and von Wettstein, D., 1968, The meiotic cycle in an ascomycete. In *Effect of Radiation on Meiotic Systems*, pp. 113–121, International Atomic Energy Agency, Vienna.

Westergaard, M., and von Wettstein, D., 1970, Studies on the mechanism of crossing-over. IV. The molecular organization of the synaptinemal complex in *Neottiella* (Cooke) Saccardo (Ascomycetes), *C. R. Trav. Lab. Carlsberg* **37**:239–268.

Westergaard, M., and von Wettstein, D., 1972, The synaptinemal complex, *Annu. Rev. Genet.* **6**:71–110.

Zickler, D., 1977, Development of the synaptonemal complex and the "recombination nodules" during meiotic prophase in the seven bivalents of the fungus *Sordaira macrospora* Auersw., *Chromosoma* **61**:289–316.

4

Morphological Manifestations of Ribosomal DNA Amplification during Insect Oogenesis

M. DONALD CAVE

1. Introduction

Genetic expression during the early stages of embryogenesis is largely under the control of the maternal genome rather than the embryo genome and appears to be a manifestation of genetic activity realized during oogenesis (reviewed by Davidson, 1976). In many organisms, ribosomes assembled during oogenesis provide the protein synthetic machinery of early embryogenesis. In such organisms, large amounts of ribosomal RNA are synthesized and accumulated during oogenesis (Harris and Forrest, 1967). In order to accomplish this, the oocytes may amplify the genes coding for ribosomal RNA (see Gall, 1969, and Tobler, 1975, for reviews).

Reflecting the large amount of information available on the structure of the ribosomal RNA genes of *Xenopus laevis* and their amplification during oogenesis, previous reviews have concentrated on ribosomal gene amplification in oocytes of amphibians (Macgregor, 1972; Tobler, 1975).

Considerable variation exists throughout the animal kingdom in the development of the oocyte (see Raven, 1967, for review). The oocytes of insects show extensive diversity in the mechanisms that contribute to the development of the oocyte. Over the past ten years, a large amount of information has been gathered about the structure of insect ribosomal RNA genes (particularly those of *Drosophila melanogaster*) and the means by which the ovaries of various insect species enhance the production and accumulation of large quantities of

M. DONALD CAVE • Department of Anatomy, University of Arkansas for Medical Sciences, Little Rock, Arkansas 72205, USA.

ribosomal RNA. It is therefore the purpose of this review to examine these diverse mechanisms. According to the theme of this volume and of this chapter in particular, references to studies on insects are cited in preference to studies on other organisms.

2. *The Nucleolus and Its Role in the Synthesis of Ribosomal RNA*

The nucleolus is the site of ribosomal RNA (rRNA) synthesis (Penman *et al.*, 1966; Edström and Daneholt, 1967). The genes coding for 18S RNA and 28S RNA*, i.e., the large RNA components of the ribosome, are localized in the chromosomal nucleolus organizer(s) (Ritossa and Spiegelman, 1965; Pardue *et al.*, 1970). Although the repetition frequency (redundancy) of these genes varies among different species and exhibits polymorphism in a given population (Ritossa and Scala, 1969), in the insect species that have been examined to date the average values range from 50 to 300 copies per haploid genome (see review by Long and Dawid, 1980).

Within the nucleolus organizer region of the chromosome, the genes coding for rRNA are arranged in tandem repeating units. The overall structure of the repeating unit is similar for most organisms, but shows some species variation. The DNA coding for rRNA (rDNA) of the fruitfly *Drosophila melanogaster* (Diptera: Drosophilidae), is one of the best characterized, and its description follows (Figure 1).

Each repeating unit consists of a DNA sequence that is transcribed into the 38S rRNA precursor (pre-rRNA) and a DNA sequence that does not appear to be transcribed *in vivo* and is referred to as nontranscribed spacer (NTS). The NTS DNA is to be distinguished from the transcribed spacer (TS), which is DNA that is transcribed into pre-rRNA but is degraded in the subsequent formation of 18S and 28S RNA.

The order of sequences proceeds from the 5′ end of the pre-rRNA as follows: A stretch of external TS precedes the sequences coding for 18S rRNA. DNA coding for 18S rRNA is followed by internal TS, which contains sequences coding for the 5.8S RNA of the large ribosomal subunit. This is followed by sequences coding for 28S RNA. Insect 28S RNA consists of two fragments of similar size that are held together by hydrogen bonding. The break in the 28S RNA reflects the fact that the DNA coding for 28S RNA is interrupted by a short sequence (about 140 base pairs) that is removed from the RNA transcript during processing, and the two resulting fragments are not ligated but are held together by hydrogen bonds between complementary base pairs. The NTS separates the sequences coding for the 28S RNA of one rRNA gene from sequences in the adjacent rRNA gene, which are transcribed into the 5′ end of another pre-rRNA molecule.

The sequences coding for 28S RNA of some insects may be interrupted by sequences that are not represented in the final RNA transcript. These "inter-

*In insects that have been examined to date, the sedimentation values of the major RNA species of the small and large ribosomal subunits are actually 17S and 27S respectively (Edström and Daneholt, 1967).

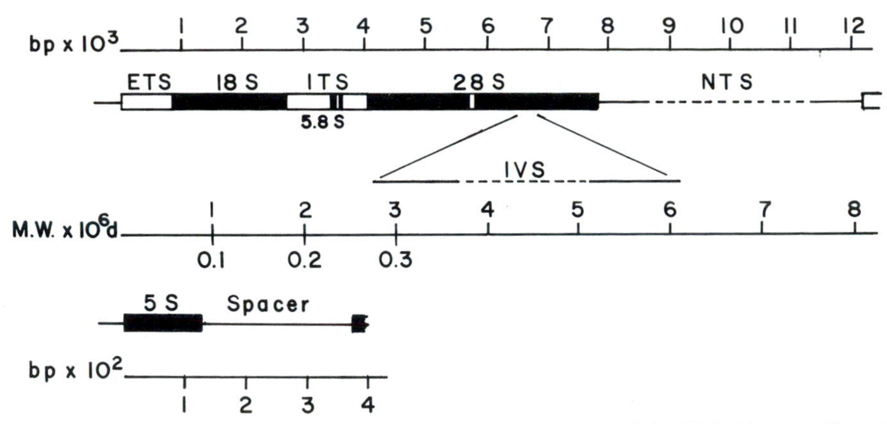

Figure 1. Maps of the repeat units of the rRNA genes (upper) and the 5S RNA genes (lower) of *Drosophila melanogaster*. Coding regions indicated by solid bars; transcribed spacers, by open bars; nontranscribed spacers by lines. The transcribed spacers are distinguished as external (ETS) and internal (ITS). The ITS include the regions coding for 5.8S RNA. The site of introns in some of the 28S rRNA genes is indicated (IVS). Nontranscribed spacers (NTS) and introns occur in various lengths as indicated by the broken line. MW, molecular weight; bp, base pairs. The map of the ribosomal RNA genes follows that of Long and Dawid (1980). The map of the 5S RNA genes follows that of Tschudi and Pirrotta (1980).

vening sequences" or "introns" are probably transcribed but are removed from the final RNA product. Therefore, "Processing of the primary transcript with . . . ligation may be what distinguishes this type of intervening sequence . . ." (Federoff, 1979). Introns have been identified in the rDNA of several dipterans (Wellauer and Dawid, 1977; Barnett and Rae, 1979). A large portion of the 28S rRNA genes of *D. melanogaster* and *D. virilis* is interrupted by such an intron. The introns, which vary in length, separate the gene into two parts. However, transcription of the rRNA genes containing introns does not contribute significantly to the synthesis of 28S RNA (Glatzer, 1979; Long and Dawid, 1979). The bulk of transcription occurs on rRNA genes lacking introns. Except for the presence of such introns, the DNA coding for rRNA is remarkably homogeneous within a given species. The NTS, however, may display extensive length heterogeneity.

Thus, the transcription of *D. melanogaster* rDNA results in the formation of a large 38S precursor molecule. In addition to the sequences conserved in 18S, 5.8S, and 28S rRNA molecules, the pre-rRNA contains sequences that are lost in processing or maturation of the final products.

The nucleoli of different types of cells present a variety of appearances in tissue sectioned for transmission electron microscopy. Two major components are, however, represented in virtually all cells. The granular component (*pars granulosa*) is rich in dark granules that measure 15–20 nm in diameter and are morphologically similar to cytoplasmic ribosomes. The fibrillar component (*pars fibrosa*) consists of extremely fine (4–8 nm thick) electron-dense filaments packed tightly together. In the nucleoli of some organisms, closely packed fibrils are arranged in aggregates around cores of less dense fibrillar centers (Recher *et al.*, 1969). The fact that the fibrillar centers are sensitive to

DNAse suggests that they contain the DNA of the nucleolus organizer (Chouinard, 1970). In some cells the fibrillar and granular components may be extensively intermingled. In others they may be separated. In the polytene chromosomes within the larval salivary gland cells of *Chironomus thummi* (Diptera: Chironomidae), a fibrillar zone surrounds the nucleolus organizer. The fibrillar area is surrounded by a peripheral granular zone (Stevens, 1964). Results of electron microscope autoradiographic studies on salivary gland cells of *C. thummi* indicate that transcription of rDNA occurs in the fibrillar component of the nucleolus, and that migration to the granular component coincides with steps in the processing of mature rRNA (von Gaudecker, 1967). Conceptually, the fibrillar centers represent the nucleolus organizer; the fibrillar component, the pre-rRNA precursor; and the granular component, ribonucleoproteins being processed to form the mature cytoplasmic ribosomal subunits (reviewed by Fakan and Puvion, 1980).

In the insects for which their location has been ascertained, the genes coding for 5S RNA are localized outside of the nucleolus organizer. Like the rRNA genes, 5S RNA genes are redundant, and the length of the repeating unit exceeds the length of the transcribed unit (i.e., each repeating unit includes spacer) (Figure 1). In *D. melanogaster* there are about 160 copies of the 5S RNA genes located at band 56 EF on the right arm of chromosome 2 (Wimber and Steffenson, 1970). The transcribed portion of the 5S RNA genes contain approximately 120 nucleotides, while the entire repeating unit is approximately three times that length.

Ribosomal proteins are synthesized on cytoplasmic polyribosomes, transported to the nucleolus, and associate with the rRNA precursor and 5S RNA to form ribonucleoprotein particles. Recent experiments with *D. melanogaster* nurse cell nucleoli, spread and treated with tagged antibodies to specific ribosomal proteins, indicate a sequential addition of ribosomal proteins to nascent pre-rRNA (Leiby and Chooi, 1979).

3. Mechanisms Which Can Alter the Reiteration Frequency of Genes

There are several mechanisms by which individual cells can alter the redundancy level of a gene, a group of genes, or an entire genome. These include polyploidization, polytenization, magnification, underreplication, chromatin diminution, chromosome elimination, and amplification. Various species of insects utilize one or several of these during the course of cell differentiation and development.

"Gene amplification denotes a process by which specific genes are synthesized preferentially in relation to others, thus leading to a quantitative increase of one or some particular gene sequences" (Tobler, 1975). Amplification differs from polyploidization in that the former results in extra copies of a limited portion of the genome, while the latter results in extra copies of the entire genome. As such, the term has been used to describe the process by which virus DNA is multiplied in a host cell (Griffin and Fried, 1975) or to describe the process by which multiple copies of plasmids are formed in bacteria (Clewell, 1972). In eukaryotes, amplification of the genes coding for rRNA occurs during macronuclear formation in many protozoans (Yao and Gall, 1977) and in the oocytes of many species of multicellular organisms (Tobler, 1975).

Amplification of the genes coding for rRNA, first discovered in the oocytes of amphibians (Brown and Dawid, 1968; Gall, 1968), also occurs in the oocytes of several species of insects. Conceptually, rDNA amplification results from repeated replication of the nucleolus organizer in the absence of replication of the remainder of the genome. This results in an excess of copies of the rRNA genes in the oocyte relative to their copy number in somatic cells; however, the genes comprising the remainder of the oocyte genome are present in the same copy number as they are in diploid somatic cells. The entire repeat unit including the NTS is amplified in the oocyte. The apparent importance of rDNA amplification is to provide additional DNA templates for the production and accumulation of a large number of ribosomes in oocytes. These ribosomes provide the protein synthetic machinery of early embryogenesis.

Although amplification of rDNA occurs in the oocytes of several species of insects, it is apparently not a universal phenomenon (i.e., it does not occur in the ooyctes of all organisms or for that matter all insects). Apparently other mechanisms can provide sufficient ribosomes for the early stages of development.

4. Amplification of rDNA in Insect Oocytes

4.1. Extrachromosomal DNA and the Occurrence of rDNA Amplification in Insect Oocytes

Giardina (1901) described a chromatin-containing body in the oocytes of the water beetle, *Dytiscus marginalis* (Coleoptera: Dytiscidae). This body, which subsequently was named for him (Giardina's mass), was believed to be an oocyte determinant, since it was passed on to the cells that were destined to become oocytes. Bauer, as early as 1933, demonstrated Giardina's mass to be Feulgen-positive and therefore rich in DNA. Microphotometric determinations indicate that Giardina's mass in oocytes of *D. marginalis* contains 20–30 times the amount of DNA present in a diploid cell (Gall *et al.*, 1969). That Giardina's mass contains amplified copies of the genes coding for rRNA has been demonstrated by RNA–DNA hybridization analysis. Oocytes of the water beetle *D. marginalis* and *Colymbetes fuscus* (Coleoptera: Dytiscidae) are rich in a satellite DNA that has a high guanine and cytosine content and hybridizes with rRNA (Gall *et al.*, 1969). In somatic cells the satellite represents only a small fraction of the total cellular DNA.

Examination of representatives of four subfamilies of the Family Dytiscidae indicates that Giardina's mass is present in some species but not in others (Urbani, 1969). It is not even found in all members of a subfamily. Therefore, its presence or absence shows no apparent relationship to the systematic position of the different forms.

A similar extrachromosomal DNA body is found in oocytes of the crane fly *Tipula oleracea* (Diptera: Tipulidae) (Bauer, 1932). Microphotometric determinations on Feulgen stained oocyte nuclei, indicate that the DNA body contains an amount of DNA equivalent to four times the diploid DNA content of somatic cells (Lima-de-Faria, 1962). A similar DNA body has been described in the oocytes of ten other species belonging to the subfamily Tipulidae (Bayreuther, 1956).

A large extrachromosomal DNA-containing body is found in oocytes of crickets (Orthoptera: Gryllidae). Buchner in 1909 discovered this mass of extrachromosomal chromatin in oocytes of *Gryllus campestris.* That the extrachromosomal body is rich in DNA was first demonstrated by the fact that the DNA body in oocytes of *Nemobius fasciatus* was intensely Feulgen-positive (Johnson, 1938).

More recent studies have been carried out on oocytes of the house cricket *Acheta domesticus.* A large extrachromosomal mass of Feulgen-positive material was first demonstrated in oocytes of *A. domesticus* by Nilsson (1966; 1968). That the DNA body is present in addition to the chromosomes is demonstrated by routine chromosome counts, which indicate that there are twenty-one and twenty-two chromosomes in diploid male and female tissues respectively. The fact that an XX-XO sex determining mechanism is operative in this species accounts for the sexual dimorphism. That the DNA within the body is indeed present in addition to chromosomal DNA is indicated by microphotometric determinations made on the nuclei of oocytes. The DNA body accounts for approximately 25% of the DNA in the oocyte nucleus (Cave and Allen, 1969b). This represents approximately 2 pg of DNA present in addition to the G_2 amount of chromosomal DNA (Cave and Allen, 1969b; Lima-de-Faria *et al.*, 1973b).

Like the oocytes of *D. marginalis,* the oocytes of *A. domesticus* are enriched with a high-density satellite DNA that hybridizes with rRNA (Lima-de-Faria *et al.*, 1969). Oocytes contain a much greater proportion of DNA coding for rRNA than do somatic tissues (Cave, 1972). Hybridization of *A. domesticus* oocytes with labeled 18S and 28S RNA *in situ* demonstrates that the extrachromosomal DNA body contains hundreds of copies of the genes coding for rRNA (Figure 2) (Cave, 1972; 1973).

By hybridizing oocytes *in situ* with labeled RNA that is complementary to the entire rDNA satellite (including the NTS as well as the coding regions) it is demonstrated that the rDNA of *A. domesticus* is localized in the extrachromosomal DNA body as well as in a single chromosomal locus (which can also be demonstrated in spermatocytes) (Figure 2) (Cave, 1973; Ullman *et al.*, 1973). Moreover, DNA complementary to 18S and 28S RNA comprises only a small portion of the rDNA satellite (<10%), all of which is amplified in the oocyte.

Lima-de-Faria *et al.* (1973a,b,d) report that there are five amplification sites in the oocyte genome of *A. domesticus.* These produce five "major chromomeres," or DNA bodies. Three of these disappear early in prophase and

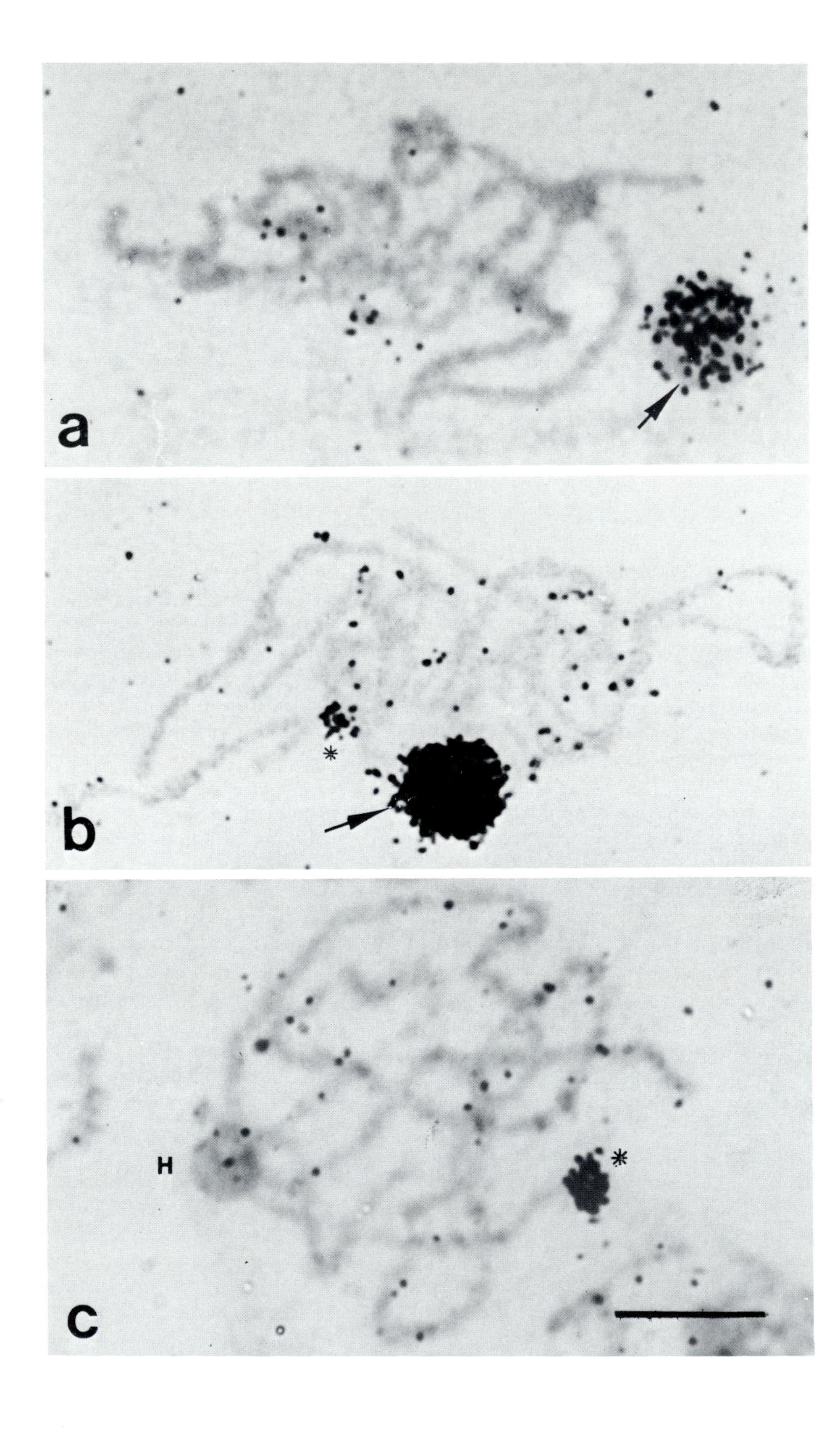
a
b
c
H

cannot be localized to specific chromosomes (Lima-de-Faria, 1973). The other two are localized on chromosomes 6 and 11, the one on chromosome 6 being the much larger of the two. Hybridization of oocytes with ^{3}H-labeled RNA that is complementary to rDNA *in situ* indicates that most of the hybridization is associated with the DNA body on chromosome 6 (Ullman *et al.*, 1973), the other four DNA bodies demonstrating little if any label. Indeed, in the clearest micrographs of these workers, the DNA body associated with chromosome 6 is the only DNA body visible. Alternatively, this may be accounted for by the fact that the DNA body associated with chromosome 11 disappears during the pachytene stage (Lima-de-Faria *et al.*, 1973c). Studies in our laboratory indicate that most oocytes of *A. domesticus* contain but a single extrachromosomal DNA body. The fact that a single chromosomal locus of oocytes or spermatocytes hybridizes with RNA complementary to rDNA supports this observation.

A survey of the oocytes of other species belonging to the family Gryllidae (true crickets) indicates that the extrachromosomal DNA body is a common feature of cricket oogenesis. An extrachromosomal DNA body similar to that found in *A. domesticus* has been observed in early prophase oocytes of all species examined to date. These include ten species belonging to the subfamily Gryllinae (field crickets, the subfamily to which *A. domesticus* belongs), seven species belonging to subfamily Nemobiinae (ground crickets), two species belonging to subfamily Oecanthinae (tree crickets), and one species belonging to the subfamily Gryllotalpinae (mole crickets) (Allen and Cave, 1972; Cave and Allen, 1974).

In addition to the species described above, extrachromosomal DNA, which probably represents amplified copies of the genes coding for rRNA, occurs in the oocytes of a large number of other insect species (see Table I).

4.2. *The Genesis of Amplified Extrachromosomal rDNA in the Oocyte*

Giardina's mass is first observed in the nuclei of oogonia. During the course of the four cystocyte divisions that give rise to the nurse cells and the oocyte, it is passed to only one daughter cell at each division (Giardina, 1901). Thus, at the first differential mitosis one cell receives Giardina's mass. The daughter cells undergo a second, third, and fourth differential mitosis in

←

Figure 2. Autoradiographs of early prophase oocytes and spermatocytes of *Acheta domesticus* hybridized *in situ* with labeled RNA. (a) An oocyte hybridized with ^{3}H-labeled 18S and 28S RNA. (b) An oocyte hybridized with ^{3}H-labeled rDNA-satellite cRNA. (c) A spermatocyte hybridized with ^{3}H-labeled rDNA-satellite cRNA. ^{3}H-labeled rDNA-satellite cRNA was prepared by transcribing isolated *A. domesticus* rDNA with *E. coli* RNA polymerase and ^{3}H-labeled nucleotide triphosphates. The rRNA and rDNA satellite cRNA hybridize with the DNA body of the oocytes. In addition to the DNA body, a single chromosomal locus of the oocytes (*) can be detected when hybridized with rDNA-satellite cRNA (b). A similar chromosomal locus (*) can be detected in spermatocytes (c). Labeling of the extrachromosomal DNA body is accounted for by the fact that it contains hundreds of amplified copies of the genes coding for rRNA. The chromosomal labeling demonstrates the site of the chromosomal nucleolus organizer. The arrows indicate the extrachromosomal DNA bodies of oocytes. H, the heteropyknotic X chromosome of spermatocytes. (Scale bar = 10 μm.)

Table 1. Occurrence of Extrachromosomal DNA and rDNA Amplification in Insect Oocytes

Species	Evidence[a]	Reference
Orthoptera		
Acheta domesticus	2	Nilsson (1966)
	2,3	Bier, *et al.* (1967), Heinonen and Halkka (1967), Kunz, (1967a), Lima-de-Faria *et al.* (1968), Cave and Allen (1969a), Kunz (1969), Allen and Cave (1972)
	4	Hansen-Delkeskamp (1969a)
	4,5	Lima-de-Faria, *et al.*, (1969), Cave (1972), Pero *et al.* (1973)
	6	Cave (1972, 1973), Ullman *et al.* (1973)
Acheta desertus	2	Nilsson *et al.* (1973)
Gryllus campestris	1	Buchner (1909)
Gryllus bimaculatus, G. mitratus	1	Favard-Sereno (1968)
Gryllus pennsylvanicus, G. veletis, G. rubens, G. assimilis, G. fultonii, Gryllodes sigillatus, Scapsipedis marginatus, Teleogryllus commodus, T. oceanicus.	2	Allen and Cave (1972)
Allonemobius fasciatus	2	Johnson (1938), Allen and Cave (1972)
Allonemobius maculatus, A. allardi, Neonemobius cubensis, N. mormonius, Eunemobius carolinus, Pictonemobius ambitiosus.	2	Allen and Cave (1972)
Oecanthus nigricornis, O. celerinictus, O. quadripunctatus	2	Cave and Allen (1974)
Scapteriscus acletus	2,6	Cave and Allen (1974)
Locusta migratoria	7	Kunz (1967a)
Neuroptera		
Chrysopa perla	2,6	Gruzova *et al.* (1972), Gaginskaya and Gruzova (1975)
Chrysopa carnea, C. vittata	2	Gruzova *et al.* (1972)
Coleoptera		
Laccophilus hyalinus	2	Urbani (1950)
Colymbetes fuscus	1	Debaisieux (1909)
	4,5	Gall *et al.* (1969)
Rhantus consimilis, Agabus lutosus, A. disintegratus	2	Kato (1968)

(Continued)

Table 1. *(Continued)*

Species	Evidence[a]	Reference
Rhantus punctatus, Agabus bipustulatus, A. nebulosus, A. brunneus. A. dydimus	2	Urbani (1950)
Dytiscus marginalis	1	Giardina (1901)
	2	Bauer (1952)
	2,3	Urbani and Russo-Caia (1964), Bier *et al.* (1967), Kato (1968)
	4,5	Gall *et al.* (1969)
Hydraticus leander	2	Urbani (1950)
Acilius semisulcatus abbreviatus, Dytiscus marginicollis	2	Kato (1968)
Dineutes nigrior	1	Gunthert (1910), Hegner and Russell (1916)
Gyrinus natator	2	Matuszewski and Hoser (1975)
Creophilus maxillosus	2	Matuszewski and Kloc (1976)
	3	Kloc (1976)
Diptera		
Tipula paludosa	2	Bauer (1932), Bayreuther (1956)
Tipula oleracea	2	Bayreuther (1956), Lima-de-Faria (1962)
Tipula caesia, T. lateralis, T. pruinosa. T. marginata, Pales crocata, P. scurra, P. quadristriata, P. pratensis	2	Bayreuther (1956)
Anopheles maculipennis, Stegomyia fasciata	2	Bauer (1952)
Siphonaptera		
Nosopsylla fasciatus, Xenopsylla cheopis	2	Bayreuther (1957)

[a]Number indicates analytical basis for evidence: 1, Morphological analysis; 2, Feulgen stain; 3, ^{3}H-thymidine autoradiography; 4, Enrichment of rDNA satellite by isopycnic banding; 5, RNA-DNA hybridization analysis; 6, *In situ* hybridization; 7, Other types of analysis.

which the mass is only included in one daughter cell. The one cell that finally receives the mass becomes the oocyte; the other fifteen cells become nurse cells. The fact that ^{3}H-thymidine is incorporated into Giardina's mass of interphase cystocytes indicates that synthesis of extrachromosomal DNA begins in the oogonia and continues in the cystocytes, which after four divisions give rise to the oocyte–nurse cell complex (Urbani and Russo-Caia, 1964; Kato, 1968).

In *T. oleraeca,* an extrachromosomal DNA body appears in the two nuclei that result from the first oogonial mitosis (Bayreuther, 1956). From then on, these are passed on to one daughter cell at each of the next three pre-oocyte divisions. Therefore, of the sixteen cells formed from the original oogonium, only two contain the DNA body. One of these becomes the oocyte; the other becomes a nurse cell. The fourteen cells without the body also become nurse cells. The fact that the DNA body of the cystocyte incorporates ^{3}H-thymidine indicates that DNA synthesis occurs during the four interphases that precede meiosis (Lima-de-Faria, 1962).

The extrachromosomal DNA body of *A. domesticus* is first observed as a dense mass of chromatin localized within the nucleolus of oogonia (Cave and Allen, 1969a). Microphotometric determinations on Feulgen-stained nuclei of early prophase cells demonstrate that the amount of DNA within the extrachromosomal body approximately doubles as the cells pass from the zygotene to the pachytene stage of meiotic prophase I (Cave and Allen, 1969b). Furthermore, in autoradiographs of oocytes exposed to ^{3}H-thymidine for very short intervals (10–30 minutes) silver grains are localized over the extrachromosomal DNA body in cells at the lepotene, zygotene, and pachytene stages. The chromosomes of cells exposed to ^{3}H-thymidine for 10–30 min are unlabeled. Label does not appear over the chromosomes until 3–4 hr after isotope administration. The fact that the 30 minutes is too brief an interval to enable the cells to complete S, pass through G_2 and the leptotene, zygotene, and pachytene stages indicates that extrachromosomal DNA is synthesized during prophase of meiosis rather than during S phase, when chromosomal DNA synthesis occurs (Cave and Allen, 1969b). Silver grains do not appear over the DNA body of diplotene oocytes until approximately two weeks after the administration of thymidine, indicating that synthesis of extrachromosomal DNA ceases in late pachytene cells. The combined morphological, microphotometric, and ^{3}H-thymidine studies indicate that amplification of rDNA begins within the nucleolus of premeiotic oogonia and continues into the pachytene stage of meiotic prophase I.

Combined autoradiographic and biochemical studies (Cave, 1973) indicate that at a time in the life cycle of *A. domesticus* when a large proportion of the oocytes incorporate ^{3}H-thymidine into the extrachromosomal DNA body, ovarian synthesis takes place of a guanine- and cytosine-rich satellite DNA that contains DNA sequences complementary to rDNA. Synthesis of this DNA satellite is not detectable in cells that do not contain an extrachromosomal DNA body or in ovaries of older animals in which most of the cells are in diplonema. During the time when the DNA within the extrachromosomal body is being synthesized, the DNA body is closely associated with one of the chromosome bivalents (Cave and Allen, 1969b). In electron micrographs of pachytene stage cells, the extrachromosomal DNA body is seen to be in contact with the synaptonemal complex representing a single chromosomal bivalent (Figure 3). A small amount of nucleolar material is associated with the DNA body at this time.

Figure 3. Electron micrograph of a pachytene oocyte of *Acheta domesticus*. The DNA body (D) is in contact with a prominent synaptonemal complex (s). A small amount of nucleolar material (n) is associated with the DNA body. (Scale bar = 0.5 μm.) *Inset:* A light micrograph of a pachytene oocyte stained according to the Feulgen procedure. The arrow indicates the DNA body. (Inset scale bar = 10 μm.)

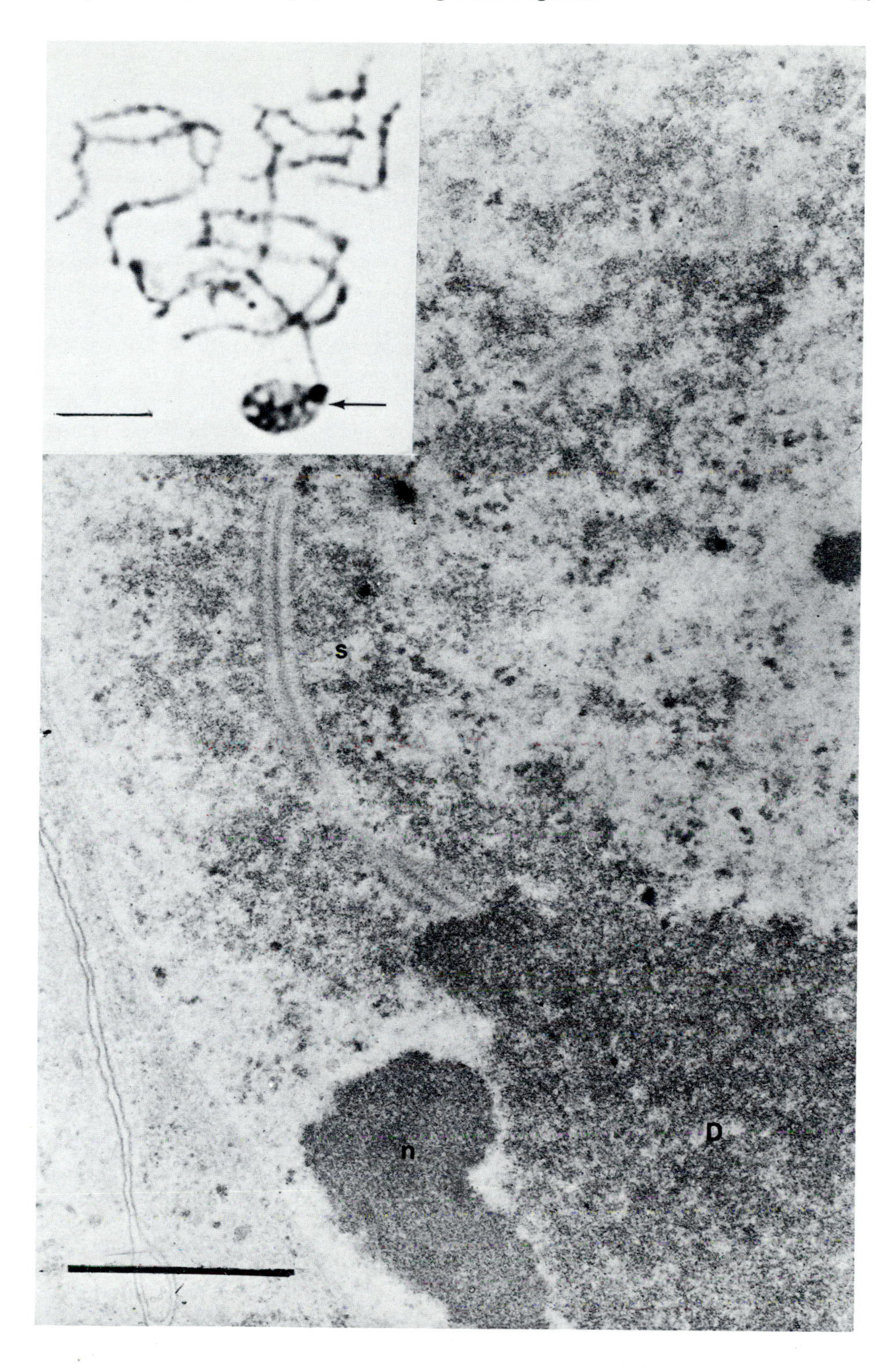
s
n
D

4.3. Transcription of Amplified rDNA

In *D. marginalis* the extrachromosomal DNA body begins to break up as the cells enter the first meiotic prophase. This disintegration is correlated with the onset of a burst of amplified rDNA transcription. This can be demonstrated histochemically by staining with methyl green-pyronin, azure B, or acridine orange and by autoradiographic detection of increased uptake of ^{3}H-uridine into the RNA associated with the DNA body (Urbani and Russo-Caia, 1964; Kato, 1968). Electron micrographs of oocytes show many small masses of nucleolar material dispersed throughout Giardina's mass (Kato, 1968). As the oocyte proceeds through diplonema, the size and the number of nucleolar masses increase as Giardina's mass disintegrates (Bier *et al.*, 1967; Kato, 1968; Trendelenberg *et al.*, 1977).

In oocytes of *T. oleracea*, nucleoli first appear within the extrachromosomal DNA body during early prophase of meiosis. These nucleoli, which develop extensive nucleolonemata, lie within the structure of the extrachromosomal DNA body of the oocyte. Nucleolar material also forms a thin shell that covers the extrachromosomal DNA body (Lima-de-Faria and Moses, 1966). The extrachromosomal DNA body disappears rather suddenly in late diplotene oocytes (Lima-de-Faria, 1962). The disposition of the nucleoli that were previously associated with the extrachromosomal DNA body has not been investigated.

As stated previously, a very small amount of nucleolar material is associated with the extrachromosomal DNA of *A. domesticus* during the pachytene stage of meiotic prophase I. Scanning electron microscopy demonstrates that the extrachromosomal DNA body has a smooth surface at this time (Lima-de-Faria, 1974). As the cells pass into the diplotene stage of meiotic prophase I, the amount of nucleolar material associated with the DNA body increases. In scanning electron microscope preparations one can see that indentations divide the surface of the DNA body into polygonal plates (Lima-de-Faria, 1974). Correspondingly, little if any RNA synthesis can be detected in association with the DNA body by means of ^{3}H-uridine autoradiography during the pachytene stage of meiotic prophase I. As the cells proceed into diplonema, there is a marked increase in RNA synthetic activity of the DNA body (Cave and Allen, 1971). Associated with this increased RNA synthetic activity of the extrachromosomal DNA body at diplonema, there is a change in the histochemically detectable (fast green, pH 8.5) histone associated with the body. The DNA bodies of pachytene cells are rich in fast green-staining histones, while those of diplotene cells show a very weak histone staining (Cave and Allen, 1971).

On the electron microscope level, nucleoli begin to accumulate at the periphery of the DNA body during the early diplotene stage (Figure 4) (Cave

→

Figure 4. Electron micrograph of an extrachromosomal DNA body in an early diplotene oocyte of *Acheta domesticus*. A large number of nucleoli (n) have begun to accumulate at the periphery of the DNA body. The arrow indicates the secondary nucleolus. (Scale bar = 5 μm.) *Insets:* Light micrographs of early (upper) and late (lower) diplotene oocytes stained with azure B. The DNA body (arrow), which is a prominent structure within the nucleolar complex of early diplotene oocytes, is no longer apparent in late diplotene oocytes.

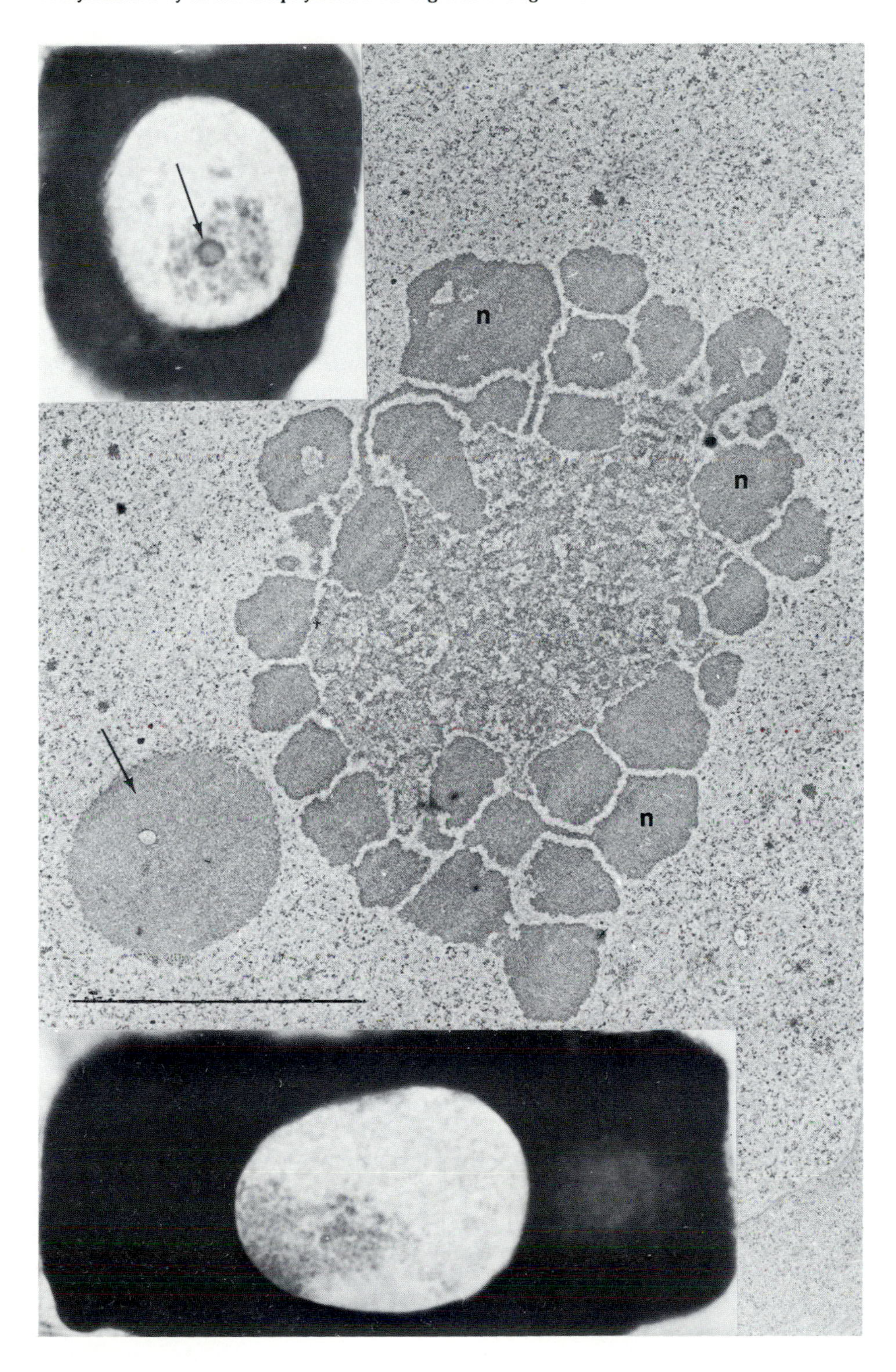
n
n
n

and Allen, 1969a). As the cells proceed through diplonema, the number of nucleoli associated with the DNA body increases. That most of the RNA synthesized in association with the extrachromosomal DNA body is rRNA is indicated by combined biochemical and autoradiographic experiments. Concentrations of actinomycin D, which selectively inhibit the incorporation of ^{3}H-uridine into 18S and 28S RNA but not into the 4–7S RNAs, selectively inhibit the incorporation of ^{3}H-uridine into the RNA associated with the extrachromosomal DNA body of diplotene oocytes. Microphotometric measurements on azure B-stained RNA indicate that as the cells proceed through diplonema there is an apparent 200-fold increase in the amount of RNA associated with the extrachromosomal body (Cave and Allen, 1969b). The extrachromosomal DNA body is no longer visible in late diplotene stage oocytes, and the nucleoli that previously surrounded the DNA body have dispersed to one end of the nucleus (Figure 4). The gradual disappearance of the DNA body and the concurrent increase in the number of nucleoli suggest that the amplified copies of the rRNA genes become incorporated into the multiple nucleoli that form around the DNA body.

The fact that individual nucleoli contain DNA that serves as a template for RNA synthesis was implied from the actinomycin-D-sensitive rapid uptake of ^{3}H-uridine by these nucleoli. The individual ring-shaped nucleoli that can be seen by phase contrast microscopy are completely disrupted by desoxyribonuclease (Kunz, 1967b). Direct evidence for the presence of rDNA is provided by examination of nucleoli from *A. domesticus* oocytes spread according to the method of Miller and Beatty (1969) (Figure 5). The nucleoli of cricket oocytes are composed of a 10-nm-thick ring of deoxyribonucleoprotein that demonstrates a repeating pattern of fibril-covered (the portion coding for rRNA precursor) and fibril-free (nontranscribed spacer) areas along its long axis (Trendelenberg *et al.*, 1973; 1976). Therefore, a single repeat unit (made up of a fibril-covered matrix region, and a fibril-free nonmatrix region) comprises one rRNA gene. The matrix covered areas are composed of ribonucleoprotein molecules oriented at right angles to the main deoxyribonucleoprotein fiber axis. There is a linear gradient of ribonucleoprotein fibers, the longest ones (at the presumptive termination site of transcription) measuring approximately 0.5 μm. Measurements of the length of a matrix unit allow one to estimate that a 5.5×10^{6} d rRNA precursor molecule is transcribed on the gene. This is 2.5 times greater than the combined molecular weights of 18S and 28S rRNA (2.2×10^{6} d). In spread nucleoli from the toad *Xenopus laevis*, the length of the transcribed portion of rDNA is approximately that which would be expected to code for the precursor to rRNA (Miller and Beatty, 1969). Estimates from polyacrylamide gel electrophoresis measurements demonstrate that in *A. domesticus* the precursor-rRNA molecule has a weight of 2.8×10^{6} d (Trendelenberg *et al.*, 1973). However, according to its length the matrix unit codes for twice as much RNA as is in the precursor. This discrepancy may be accounted for by assuming that the 2.8×10^{6} d RNA is an intermediate precursor. The average length of the matrix-free units (NTS), also suggests that the repeat unit of *A. domesticus* is very large. Electron microscope analysis of Kleinschmidt-spread (1968) *A. domesticus* rDNA indicates that the repeat unit

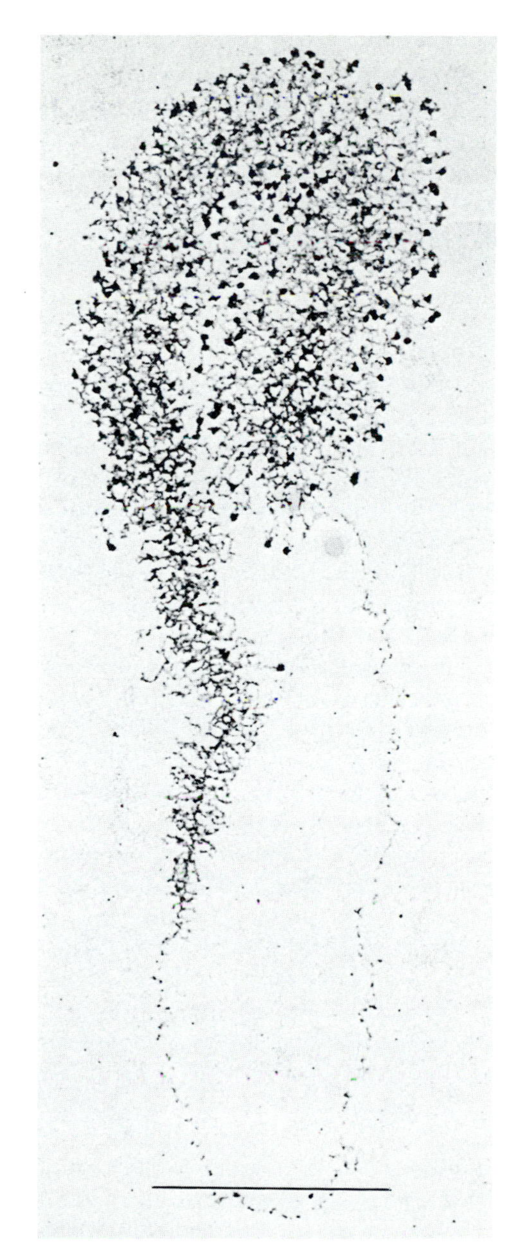

Figure 5. Electron micrograph of transcriptionally-active, nucleolar chromatin isolated from the ovary of *Acheta domesticus* and spread according to the technique of Miller and Beatty (1969). This circle contains one repeat unit of amplified rDNA. The central deoxyribonucleoprotein fiber axis is divided into a fibril-covered segment (the portion coding for rRNA precursor) and a fibril-free segment (nontranscribed spacer). In the fibril-covered area, ribonucleoprotein molecules are oriented at right angles to the axis of the ring of deoxyribonucleoprotein. There is a linear gradient of ribonucleoprotein molecules; the shortest ones are opposite the initiation site of transcription, and the longest ones border the termination site. (Scale bar = 1 μm.) (Courtesy of Dr. Michael Trendelenberg, German Cancer Research Center, Heidelberg, Federal Republic of Germany, with permission from Trendelenberg *et al.*, Journal of Molecular Biology, 108). Copyright by Academic Press, Inc. (London) Ltd.

measures about 37×10^6 d (Cave, 1979). Restriction endonuclease analysis of the rDNA of *A. domesticus* indicates that the repeat unit measures about 35×10^6 d, and that the transcribed region measures about 9×10^6 d (Sharp and Cave, 1980).

Similar studies on oocytes of *D. marginalis* provide evidence that the multiple nucleoli contain amplified copies of rDNA that were previously localized in Giardina's mass. Spread nucleoli from oocytes of *D. marginalis* also exhibit a 10-nm-thick ring of deoxyribonucleoprotein with alternating fibril-covered matrix areas and fibril-free nontranscribed spacer areas (Trendelenberg, 1974; Trendelenberg *et al.*, 1976). According to the average length of the matrix-covered unit, it should produce an RNA molecule with a weight of 3.7×10^6 d. However, polyacrylamide electrophoresis measurements indicate that the RNA has a molecular weight of 2.8×10^6 d. As in *A. domesticus,* the discrepancy between the length of the transcribed area and the weight of the precursor rRNA molecule suggests that the 2.8×10^6 d RNA is an intermediate precursor.

4.4. The Mechanism of rDNA Amplification

During the amplification process, the rDNA of *X. laevis* takes on a circular configuration (Hourcade *et al.*, 1973). The contour length of the smallest rDNA circles corresponds to the expected length of one rDNA repeat unit, which for *X. laevis* measures 8×10^6 d. The lengths of larger circles are integral multiples of the basic repeating unit. Circles with attached tails are also observed and are probably the replicative form of rDNA (Rochaix *et al.*, 1974). Amplification, therefore, is believed to occur by a rolling-circle mechanism reminiscent of the manner in which the genome of certain viruses is replicated (Gilbert and Dressler, 1968).

Light microscopic observations on the nucleoli of *A. domesticus* oocytes indicate the presence of circular nucleoli, the continuity of which can be disrupted by DNAase (Kunz, 1967b). The circles from nucleoli of *A. domesticus* and *D. marginalis* spread according to the Miller and Beatty technique (1969) contain between one and six matrix units (Trendelenberg *et al.*, 1973; 1976; 1977). Each of the matrix units on the circles are separated from adjacent matrix units by NTS. Electron micrographs of Kleinschmidt-spread *A. domesticus* and *D. marginalis* amplified rDNA (Figure 6) provide direct evidence for the circularity of rDNA during the amplification process (Gall and Rochaix, 1974; Trendelenberg *et al.*, 1976; Cave, 1979). Like the amplified circular rDNA of *X. laevis,* the contour lengths of amplified circular rDNA molecules of *A. domesticus* and *D. marginalis* fall into size classes the lengths of which are integers of the smallest molecules. The mean molecular weights of the smallest circles are 8×10^6 d for *X. laevis* (Hourcade *et al.*, 1973), 19×10^6 d for *D. marginalis* (Gall and Rochaix, 1974), and 37×10^6 d for *A. domesticus* (Cave, 1979). In *X. laevis,* the entire repeating unit, which weighs 8×10^6 d, includes DNA that codes for 18S and 28S RNA (4.4×10^6 d) as well as spacer DNA (3.6×10^6 d) (Hourcade *et al.*, 1973). Assuming the molecular weight of DNA coding for 18S and 28S rRNA to be 4.4×10^6 d, the data indicate that 55% of the *X. laevis* rDNA, 23% of the *D. marginalis* rDNA, and 12% of the *A. domesticus* rDNA code for 18S and 28S RNA. These values are similar to the

→

Figure 6. An electron micrograph of a circular rDNA molecule from oocytes of *Acheta domesticus* spread according to the Kleinschmidt procedure. The molecular weight of this molecule is approximately 34×10^6. The small circles are ϕX-174 bacteriophage DNA, which serve as a molecular weight standard (mol. wt. = 3.4×10^6 d). The average molecular weight of 100 amplified circular rDNA molecules was found to be 37.4 ± 4.2 (S.D.) $\times 10^6$ d (Cave, 1979). (Scale bar = 1 μm.)

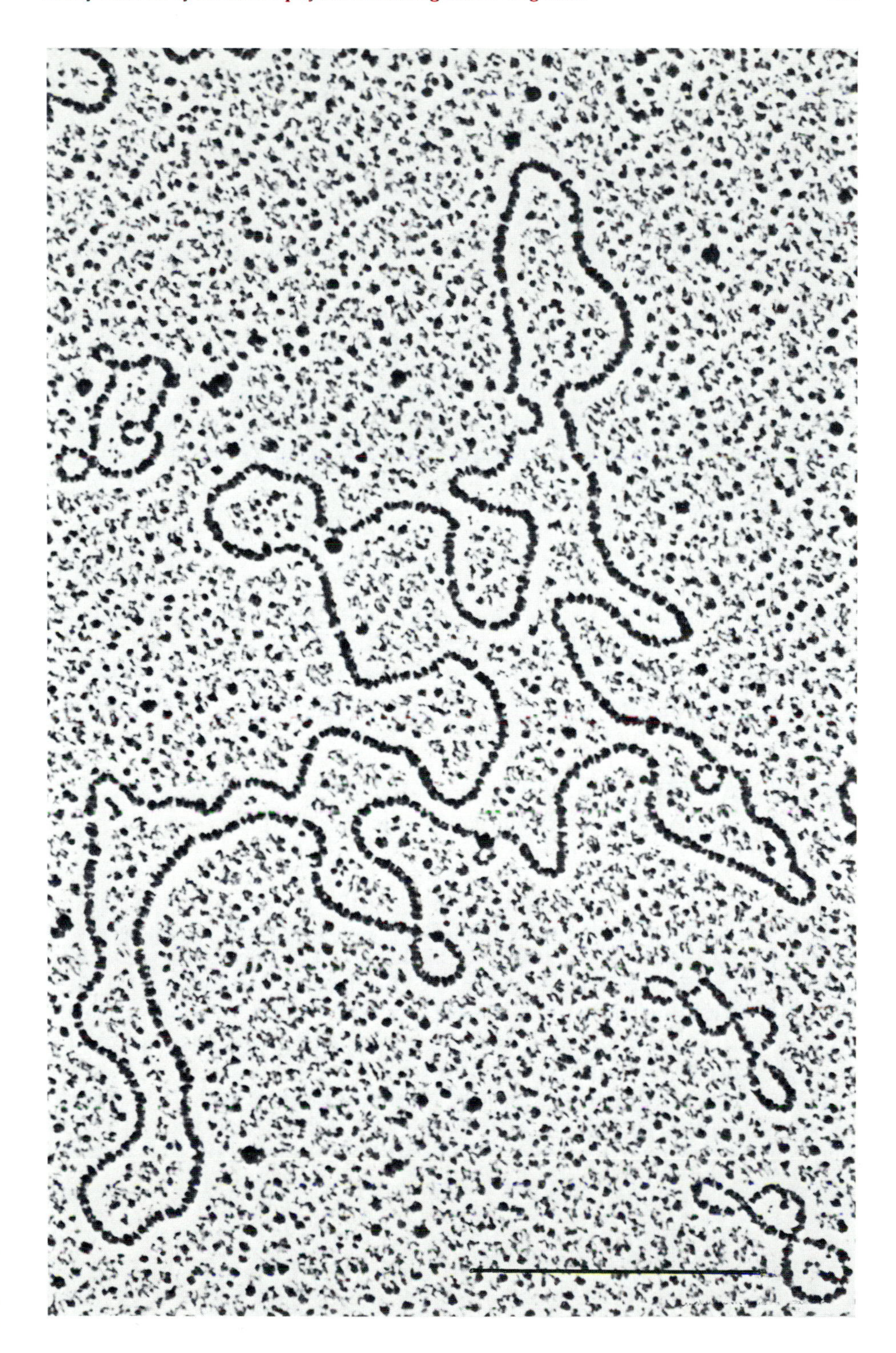

values for two of the three species estimated from RNA–DNA hybridization analysis [50% for *X. laevis* (Wensink and Brown, 1971), 8% for *A. domesticus* (Pero *et al.*, 1973)].

4.5. The Role of the Products of Amplified rDNA in Early Embryogenesis

In *A. domesticus*, 80–90% of the RNA that is synthesized during oogenesis and accumulates in the egg is rRNA (Hansen-Delkeskamp, 1969b; Cave and Allen, 1971). The rRNA synthesized during oogenesis is stable through the yolk-cleavage stage of development (Hansen-Delkeskamp, 1969b), and there is little if any rRNA synthesis during the early stages of embryogenesis. Synthesis of rRNA does not begin until the cellular blastoderm stage, 27 hours (23°C) after egg laying (Hansen-Delkeskamp *et al.*, 1967). In spite of this, synthesis of 4S RNA is detected at earlier stages of development. The onset of rRNA synthesis coincides with a lengthening of interphase and the appearance of nucleoli in the nuclei of the embryo.

Although there is little or no rRNA synthesis during early embryogenesis, extensive synthesis of protein can be readily detected by the incorporation of ^{14}C-labeled amino acids as early as 5 hr after egg laying (Hansen-Delkeskamp, 1969b). The fact that protein synthesis during these early stages proceeds in the absence of rRNA synthesis indicates that these proteins are synthesized on ribosomes that were stored during oogenesis. That is, ribosomes synthesized during oogenesis serve as the protein-synthetic machinery of early embryogenesis.

4.6. The Relationship of rDNA Amplification to the Organization of the Oocyte Nucleolar Apparatus

Amplification of the genes coding for rRNA appears to be correlated with the formation of multiple nucleoli in the diplotene oocyte nucleus. In oocytes of *A. domesticus*, multiple nucleoli are formed in close association with the DNA body during the early diplotene stage of meiotic prophase I. As the cells proceed through diplonema, the number of nucleoli associated with the DNA body increases as the extrachromosomal DNA body gradually diminishes in size. Finally, the extrachromosomal DNA body disappears. The multiple nucleoli that originally surrounded the DNA body are dispersed throughout the nucleus. These data are consistent with the idea that amplified copies of the genes coding for rRNA that were originally localized within the DNA body become incorporated into the developing nucleoli. Several further lines of evidence support this concept.

(1) Autoradiographs of sectioned early diplotene oocytes hybridized *in situ* with ^{3}H-labeled 18S and 28S RNA reveal silver grains over the developing nucleolar mass as well as over the extrachromosomal DNA body. Autoradiographs of late diplotene oocytes in which the extrachromosomal

DNA body is no longer apparent reveal silver grains associated with the dispersed nucleoli (Cave, 1972). Therefore, the individual nucleoli contain rDNA that was originally localized in the DNA body.

(2) The multiple nucleoli of *A. domesticus* oocytes contain a DNA core. This was first indicated by the fact that treatment of the nucleoli with DNAase leads to their disruption (Kunz, 1967b). Direct observation of spread oocyte nucleoli of *A. domesticus* demonstrate a circular DNA core on which transcription of rDNA can be seen (Trendelenberg *et al.*, 1976).

Examination of the oocytes of various species of crickets belonging to the family Gryllidae and representing at least five of twelve subfamilies reveals that, although the extrachromosomal DNA body is present in the pachytene oocytes of all species examined to date, the structure of the nucleolar apparatus during the diplotene stage is quite variable (Allen and Cave, 1972; Cave and Allen, 1974). The late diplotene oocytes of some species have thousands of small (1-2 μm) nucleoli, while those of other species have several (10-20) large (20-30 μm) nucleoli. Such large nucleoli frequently contain elaborate nucleolonemata. To explain these differences, Trendelenberg *et al.* (1977) have proposed that amplified rDNA may (1) remain in close association with the chromosomal nucleolus organizer(s) or (2) may immediately or later detach from the chromosomal nucleolus organizer to lie free in the nucleoplasm as (a) isolated single units or as (b) extrachromosomal aggregates of amplified rDNA. The isolated single units would give rise to thousands of small nucleoli, the aggregates to large and complex nucleoli. Of course, intermediates between these possibilities would be expected (i.e., situations in which there are both large and small nucleoli in the diplotene oocyte).

In *A. domesticus*, the extrachromosomal DNA body is formed in association with the chromosomal nucleolus organizer during the zygotene and pachytene stages. At this time, the DNA body is in intimate association with one of the chromosomal bivalents. Several lines of evidence indicate that during diplonema the DNA body detaches from the nucleolus organizer to lie free in the nucleoplasm as a large aggregate of amplified rDNA. Subsequently, individual units detach from this aggregate to give rise to the small, individual nucleoli of the late diplotene oocyte. Detachment of the DNA body from the nucleolus organizer is indicated by the appearance of another structure in the nucleus. This structure has been referred to as the "Binnenkörper," "secondary nucleolus," "secondary nucleolar component," or "dense nucleolar body" (Bier *et al.*, 1967; Allen and Cave, 1969; Jaworska and Lima-de-Faria, 1973; Trendelenberg *et al.*, 1977). Although various workers disagree on the significance of this structure, the timing of its appearance suggests that it may represent the expression of ribosomal genes in the chromosomal nucleolus organizer that had been suppressed until this time by their association with the amplified rDNA.

In *A. domesticus*, the development of the complex containing the extrachromosomal DNA body and the multiple nucleoli proceeds in a stepwise fashion:

(1) The amplified rDNA remains attached to the chromosomal nucleolus organizer during the period of rDNA amplification.
(2) During early diplonema, when amplification is completed, the body detaches from the chromosomal nucleolus organizer as a single aggregate.
(3) As the cells proceed through diplonema, individual units (containing several rDNA repeat units) detach from the aggregate forming thousands of small nucleoli.

Studies on several species belonging to the subfamily Oecanthinae indicate that although Step 1 is the same as in *A. domesticus,* Steps 2 and 3 differ in that during early diplonema several rather large nucleoli are formed, and the extrachromosomal DNA body is not observed in any of these (Cave and Allen, 1974). Studies on oocytes of various species of Gryllinae (the subfamily to which *A. domesticus* belongs) suggest that, although the extrachromosomal DNA body is a characteristic of zygotene and pachytene oocytes of all species examined and appears to be associated with the chromosomal nucleolus organizer, the organization of the nucleoli in diplotene oocytes is quite variable. In some species, it forms a single aggregate during early diplonema, gradually discharging individual units to form small individual nucleoli of late-diplotene oocytes. In other species, several large nucleoli are formed, a conspicuous DNA body is not visible in any of these, and the nucleolar number remains approximately the same throughout diplonema. In still other species, several large nucleoli form during early diplonema, some of which are dispersed further during late diplonema.

Therefore, there appear to be differences in regard to the degree of dispersal of the individual amplified rDNA units. The possible forces involved in the association or dissociation of the units of amplified rDNA, although not elucidated, appear to be responsible for these species differences.

Multiple nucleoli are prominent structures within the diplotene oocyte nucleus of many other species of Orthopterans. Although various types of evidence indicate that amplification of the genes coding for rRNA occurs in these oocytes, a large extrachromosomal DNA body similar to that observed in *A. domesticus* and the pachytene oocytes of all of the Gryllid crickets examined to date has not been visualized in these oocytes (Kunz, 1967a,b). This absence of an extrachromosomal DNA body during the pachytene stage of meiotic prophase may reflect the fact that the amplified copies of the rDNA do not remain in close association with the nucleolus organizer, but immediately detach from the nucleolus organizer to lie free in the nucleoplasm. Differences in the organization of nucleoli in the diplotene stage of meiosis in these widely differing species of Orthopterans would again reflect the aggregation or dispersal of individual rDNA units in different species. A notable case is that of several species of Acrididae in which the multiple nucleoli resemble "pearl necklaces" (Kunz, 1967a,b). These are very long threads that extend throughout the nucleus and on which the nucleoli are arranged. Treatment with DNAase breaks the threads, dispersing the nucleoli.

Multiple nucleoli appear to be a characteristic feature of oogenesis in

Orthopterans or, for that matter, of oogenesis in panoistic ovaries in general. Notable exceptions to this rule are oocytes of the cockroaches (Family: Blattidae), which have been known for many years to be uninucleolate (Wilson, 1928). This view has been challenged by Bier *et al.* (1967), who interpret the nucleolar apparatus of the cockroach oocyte to be multinucleolate on the basis of its ultrastructural organization. However, combined light and electron-microscope studies on oocytes of *Blattella germanica* demonstrate that a single nucleolus appears in the oocytes at early diplonema and although the nucleolus increases in size, it remains single throughout oogenesis (Cave, 1976).

Routine cytological analysis of spermatocytes and oocytes of *B. germanica* indicate chromosome numbers of 21 (male) and 22 (female), the difference being accounted for by an XX–XO sex determining mechanism. Although both oocytes and spermatocytes display a heteropyknotic knob on the X chromosome, an extrachromosomal DNA body similar to that observed in the Gryllid crickets is not seen. Detailed cytological analyses including *in situ* hydridization with ^{3}H-labeled 18S and 28S RNA show that in *B. germanica* the nucleolus organizer is located on the X chromosome, and therefore female cells have twice the number of ribosomal genes as male cells. Furthermore, RNA–DNA hybridization analysis demonstrates that the genes coding for rRNA are not amplified during oogenesis, oocytes having the same number of ribosomal RNA genes as do female somatic cells (Cave, 1976). The fact that amplification of rRNA occurs in the multinucleolate oocytes of many orthopteran species, but is absent from species the oocytes of which are uninucleolate, indicates that amplification of rDNA is associated with the development of multiple nucleoli in the oocyte. This conclusion is in agreement with the findings of other investigators on the uninucleolate oocytes of several invertebrates belonging to phyla other than Arthropoda (Vincent *et al.*, 1969).

5. *Modification of the Nurse Cell Genome Associated with rRNA Synthesis*

In the panoistic ovary, lampbrushlike chromosome structures and multiple nucleoli develop in the germinal vesicles of oocytes during the diplotene stage of oogenesis (Kunz, 1967a). The chromosomes and nucleoli of the oocyte are actively engaged in RNA synthesis (Bier *et al.*, 1967). In contrast to this, the chromosomes of the germinal vesicle in oocytes within the typical meroistic ovary are condensed in a karyosphere and do not synthesize RNA. As a general rule, active nucleoli are not observed within the germinal vesicle. The RNA synthetic activity of the oocyte nucleus is taken over by the nurse cells, which synthesize RNA and transport this RNA to the cytoplasm of the oocyte (Bier, 1963). In the telotrophic type of meroistic ovary, the nurse cells are clustered at the apical end of each ovariole and through nutritive cords establish continuity with the oocyte. In the polytrophic type of meroistic ovary, an ovariole consists of a string of egg chambers each containing an oocyte and a nest of nurse cells. The cells within a chamber are interconnected by a system of canals (see Chapter

1, this volume). Electron microscopic studies demonstrate that ribosomes synthesized by nurse cells are a substantial component of the nurse cell's contribution to the oocyte (Koch *et al.*, 1967; Huebner and Anderson, 1972).

Oocytes of certain Chrysopidae (Neuroptera); Dytiscidae, Staphylinidae, and Gyrinidae (Coleoptera); and Tipulidae (Diptera); are exceptional for insects having meroistic ovaries in that they are known to amplify rDNA (Table 1). Autoradiographic studies indicate that RNA is synthesized by oocyte nuclei within the polytrophic ovaries of *D. marginalis* (Urbani and Russo-Caia, 1964), *Creophilis maxillosus* (Coleoptera: polyphaga) (Kloc, 1976), *Gyrinus natator* (Coleoptera: adephaga) (Matuszewski and Hoser, 1975), *T. oleracea* (Lima-de-Faria and Moses, 1966), and *Chrysopa perla* (Chrysopidae: Neuroptera) (Gruzova *et al.*, 1972). Moreover, the oocyte nuclei of these species develop many well-defined nucleoli, which actively snythesize RNA. The association of multiple nucleoli with extrachromosomal DNA bodies in oocytes of these species lends further support to the hypothesis that amplification of rDNA is associated with the development of the multinucleolate oocyte. The fact that the nuclei of the nurse cells are also active in the synthesis of RNA and develop many active nucleoli indicates that the rRNA that accumulates in the cytoplasm of oocytes of these species represents contributions of both the nurse cell and the oocyte.

Cytological observations suggest that in addition to the increased transcription capacity provided for by polyploidization of the nurse cell chromosomes, DNA within the nucleolar organizer of the nurse cell may also be amplified. It has already been noted that in the polytrophic ovary of *T. oleracea* one of the nurse cells in each nest contains a DNA body, while the other six do not (Lima-de-Faria and Moses, 1966).

The fact that DNA from the polytrophic ovaries of the fungus gnat, *Rhynchosciara angelae* (Diptera: Sciaridae), contains approximately twice as much DNA complementary to rRNA as does DNA extracted from salivary glands, led to the suggestion that rDNA is amplified in the polytrophic ovary of *R. angelae* (Gambarini and Meneghini, 1972). Since DNA complementary to rRNA is underreplicated in salivary glands of most Dipterans (Hennig and Meer, 1971), it was not clear whether differences in the amount of rDNA reflect amplification of the somatic cell value or underreplication in salivary gland cells. In contrast to the data on *R. angelae,* the percent of DNA hybridizing with rRNA in the polytrophic ovary of *D. melanogaster* is the same as that of the total DNA extracted from adult flies (Mohan and Ritossa, 1970; Mohan, 1976). Further studies on polytene tissues of several *Rhyncosciara* species (Gambarini and Lara, 1974) indicate that tissues show low polyteny (such as fat body) have a greater proportion of DNA hybridizable to rRNA than do tissues that show high polyteny (such as salivary gland). The fact that cells of the ovary, which are a mixture of polytene and polyploid cells, carcass (which are also a mixture of cells), and fat body (the cells of which show low polyteny) have the same percentage of DNA that hybridizes to rRNA indicates that differences in the amount of rDNA in ovary and salivary gland are accounted for by underreplication of rDNA cistrons in the salivary gland rather than their amplification in the ovary (Gambarini and Lara, 1974).

Large numbers of nucleoli develop within the nurse cell nuclei of the blowfly, *Calliphora erythrocephala* (Diptera: Calliphoridae) (Ribbert and Bier, 1969). In specific strains that show polytenization of nurse cell chromosomes, the nucleoli are seen to lie free of the chromosomes. In other polytene tissues (e.g., trichogen cells) of the same animals, there is but a single nucleolus, which is attached to a specific chromosomal locus. The free nucleoli in which DNA particles can be visualized by histochemical techniques are known to incorporate ^{3}H-uridine. DNA extracted from ovaries of *C. erythrocephala* contains on a proportional basis approximately 35% more DNA hybridizing with rRNA than does DNA from somatic cell nuclei (Renkawitz and Kunz, 1975). Whether or not the increase in the proportion of rDNA related to total DNA is due to amplification, as is suggested by the cytological studies (Ribbert and Bier, 1969), to underreplication of other repetitious DNA sequences, which has been demonstrated in ovaries of *D. virilis* (Renkawitz-Pohl and Kunz, 1975), or simply to independent replication of rDNA has not been ascertained.

Mermod *et al.*, (1977) report in *D. melanogaster* that the rate of rRNA and 5S RNA synthesis increases during oogenesis and is paralleled by an increase in rDNA. Further studies indicated that the ribosomal genes were replicated in proportion to the total DNA. However, the technique for estimating "ribosomal gene number" in these studies did not enable the authors to determine whether or not there was "preferential ribosomal gene replication." Therefore, although the number of rRNA genes in each egg chamber was shown to increase, it was not known whether this merely reflects an increase in total DNA that also occurs in the egg chamber, amplification, or underreplication of other sequences. More detailed study (Jacobs-Lorena, 1980) has led to the conclusion that as the egg chamber of *D. melanogaster* proceeds through oogenesis, there is a slight underreplication of rDNA, while 5S DNA is replicated to the same extent as the rest of the genome DNA.

Underreplication of rDNA has been detected in ovaries of *D. virilis* and *Sarcophaga barbata* (Diptera: Sarcophagidae) (Endow and Gall, 1975; Renkawitz and Kunz, 1975). Renkawitz and Kunz (1975) have conclusively demonstrated an underreplication of rDNA in nurse cells of *D. hydei*. In that study the ribosomal DNA content of isolated nurse cells was 55% lower than that of diploid brain cells.

Polytenization of chromosomes, although not exclusively a dipteran phenomenon, is found in few organisms outside of that order. This probably reflects the incidence of somatic chromosome pairing which is almost unique to the Diptera. In most other species, polyploidization rather than polytenization increases the DNA content of the cell. The phenomena of underreplication, overreplication, and independent replication, which are frequently associated with polytenization, have not been observed in cells with polyploid nuclei. Since these phenomena tend to obscure the detection of rDNA amplification, studies in my laboratory have been directed toward species the nurse cells of which are polyploid rather than polytene. The polytrophic ovaries of the Lepidoptera fulfill these criteria.

In the oriental silkmoth, *Bombyx mori* (Lepidoptera: Bombyciidae), there is extensive diversity in the degree of polyploidy expressed by cells of different tissues. Silk glands of prepupae are believed to contain up to one million times the amount of DNA found in the gamete (Rasch, 1974). Most of the nuclei from carcasses of the same animals are diploid (Gage, 1974a). Several lines of evidence indicate that uniform replication of the entire genome without any specific loss or gain due to underreplication or overreplication of any portion of its accounts for the increased DNA content of polyploid cells.

(1) The relative proportions of highly repetitive, middle repetitive, and nonrepetitive sequences in DNA are the same in tissues composed of diploid cells and tissues composed of highly polyploid cells (Gage, 1974b).
(2) The gene coding for silk fibroin, a nonrepetitive DNA sequence, comprises the same relative proportion of the total DNA from diploid cells as it does in the polyploid cells of the silk gland (Suzuki *et al.*, 1972). If independent replication (overreplication or underreplication) of any portion of the genome had occurred during polytenization, this would be expected to be reflected in an increase or decrease in the proportion of DNA coding for fibroin.
(3) The proportion of the genome coding for rRNA is the same for polyploid cells, as it is for diploid and haploid cells (Gage, 1974a; Cave and Sixbey, 1976).

Nurse cells of *B. mori* display a high level of polyploidy (Cave and Sixbey, 1976). The fact that the proportion of the genome coding for rRNA is the same for silk gland, the cells of which are highly polyploid, for carcass, most of the cells of which are diploid, for testis, which contains diploid and haploid cells, and for isolated ovarioles of prepupae and 10-day-old pupae, indicates that rDNA is not amplified in the nurse cells (Cave and Sixbey, 1976).

The nurse cell cysts and the oocytes of the giant silkmoth, *Antheraea pernyi* (Lepidoptera: Saturniidae), are large enough to enable their isolation by microdissection. The highly polyploid nurse cells and the dechorionated diploid oocytes contain the same proportion of rDNA as spermatogenic cells, which are haploid and diploid, as somatic cells from brain, which are diploid, and as somatic cells from Malpighian tubules, which are polyploid (Cave, 1978). The fact that the oocytes contain the same proportion of rDNA as diploid somatic cells indicates that this relatively quiescent oocyte does not amplify its rDNA. The fact that the nurse cells of *B. mori* and *A. pernyi* contain the same relative proportion of DNA hybridizing with rRNA as do spermatogenic or somatic cells indicates that amplification in excess of the level of ploidy expressed by the trophocytes does not occur. In the typical polytrophic ovary, polyploidization of the entire nurse cell genome without any further amplification of a small part of it appears to account for the increased amount of rDNA available for transcription.

Nurse cells in the telotrophic ovary of the large milkweed bug, *Oncopeltus fasciatus* (Hemiptera: Lygaeidae), are highly polyploid (Bonhag, 1955; Cave, 1975). The morphological separation of the nurse cells and oocytes in each ovariole enables isolation of the nurse cells by microdissection. The fact that the percentage of DNA hybridizing with rRNA is the same for isolated nurse cells, spermatogenic cells, and various somatic cells indicates that amplification of rDNA does not occur in nurse cells of *O. fasciatus* (Cave, 1975). As in polytrophic ovaries, polyploidization appears to increase the rDNA content of the nurse cells by providing increased DNA templates for the massive accumulation of rRNA that accumulates in the oocyte.

How effective is this polyploidization of the nurse cell genome in increasing rRNA synthetic activity? Endoreduplication of the nurse cell genome multiplies the number of rRNA genes, the products of which are transported to the oocyte. In *D. melanogaster* the increased rate of rRNA synthesis per egg chamber parallels the number of rRNA genes (Mermod *et al.*, 1977). The oocytes accumulate large amounts of rRNA (King, 1970). The fact that fertilized eggs of *D. melanogaster* that lack sex chromosomes (and therefore lack ribosomal genes) can undergo 10–12 mitotic divisions before dying suggests that the protein synthesized during early embryogenesis occurs on ribosomes provided by nurse cells during oogenesis (von Borstel and Rekemeyer, 1958).

In *O. fasciatus*, the newly fertilized egg contains approximately 180 ng of ribosomal RNA, most of which was synthesized by maternal nurse cells (Harris and Forrest, 1967). This is several orders of magnitude more rRNA than that found in a diploid somatic cell. The synthesis of rRNA does not begin in embryos of *O. fasciatus* until approximately 44 hr after fertilization, when the germ band undergoes gastrulation (Harris and Forrest, 1967). Therefore, rRNA synthesized by the nurse cells probably serves the developing embryo in a manner similar to the RNA product of amplified oocyte rDNA, i.e., it provides the protein synthesizing machinery of early embryogenesis. Hence, polyploidization of the genome in the nurse cell apparently performs the same function with respect to rRNA synthesis as amplification of the genes coding for rRNA in the oocyte.

6. Summary

Amplification of the genes coding for rRNA in insect oocytes is correlated with the development of multiple nucleoli. It occurs commonly in the oocyte nuclei of insects with panoistic ovaries and rarely in those with polytrophic meroistic ovaries.

In many species of insects amplification of rDNA is correlated with the appearance of large, extrachromosomal DNA bodies, which are readily observed in the nucleus of the oocyte. Multiple nucleoli develop in close association with these extrachromosomal DNA bodies. Amplified rDNA originally localized within such an extrachromosomal DNA body becomes incorporated into the forming multiple nucleoli.

Just as there are species differences in the organization of the ribosomal RNA genes, there are morphological differences in the organization of amplified extrachromosomal rDNA. Such differences probably reflect the fact that amplified copies may remain in association with the chromosomal nucleolus organizer or may directly detach and come to lie in the nucleoplasm as isolated single units or as extrachromosomal aggregates.

Amplification of rDNA begins during the oogonial stages and proceeds into prophase of the first meiotic division, long after chromosomal DNA synthesis is complete.

Amplification of rDNA in insect oocytes occurs via a rolling-circle mechanism reminiscent of the manner in which the DNA of certain bacteriophage viruses is replicated.

Amplification of rDNA provides increased DNA template for the massive synthesis and accumulation of rRNA that occurs in the insect oocyte. rRNA is not synthesized by the embryo during the early stages of development. Ribosomes synthesized during oogenesis provide the protein synthetic machinery of early embryogenesis.

Amplification of rDNA is not a universal feature of insect oogenesis. Amplification of rDNA occurs in multinucleolate oocytes. In uninucleolate oocytes that have been examined to date, amplification of the genes coding for rRNA does not occur or proceeds at a very low level.

In the typical polytrophic meroistic ovary the oocyte nucleus is quiescent in regard to RNA synthesis; the RNA synthetic functions of the oocyte are taken over by nurse cells, which actively synthesize rRNA and transport this RNA to the oocyte through ring canals. Endoreduplication of the entire nurse cell genome rather than amplification of a small part of it appears to provide the increased amount of DNA template necessary for the synthesis and accumulation of large amounts of rRNA by the oocyte. Ribosomes synthesized by the trophocytes provide the protein synthesizing machinery for early development.

References

Allen, E. R., and Cave, M. D., 1969, Cytochemical and ultrastructural studies of ribonucleoprotein containing structures in oocytes of *Acheta domesticus*, *Z. Zellforsch.* **101:**63–71.

Allen, E. R., and Cave, M. D., 1972, Nucleolar organization in oocytes of Gryllid crickets: Subfamilies Gryllinae and Nemobiinae, *J. Morphol.* **137:**433–447.

Barnett, T., and Rae, P. M. M., 1979, A 9.6 kb intervening sequence in *D. virilis* rDNA, and sequence homology in rDNA interruptions of diverse species of *Drosophila* and other Diptera, *Cell* **16:**763–775.

Bauer, H., 1932, Die Histologie des Ovars von *Tipula paludosa* Meig, *Z. wiss. Zool.* **143:**53–76.

Bauer, H., 1952, Die Chromosomen im Soma der Metazoen, *Ver. Dtsch. Zool. Ges.*, pp. 252–268.

Bauer, H., 1933, Die wachsenden Oocytekerne einiger Insekten in ihrem Verhalten zur Nuklealfärbung, *Z. Zellforsch. v. mikroskop. Anat.* **18:**254–298.

Bayreuther, K., 1956, Die Oögenese der Tipuliden, *Chromosoma* **7:**508–540.

Bayreuther, K., 1957, Extrachromosomales DNS-haltiges Material in der Oögenese der Flohe, *Z. Naturforsch.* **12b:**458–461.

Bier, K., 1963, Synthese, Interzellulärer Transport und Abbau von Ribonukleinsäuer im Ovar der Stubenfliege *Musca domestica*, *J. Cell Biol.* **16:**436–440.

Bier, K., 1967, Oögenese, das Wachstum von Riesenzellen, *Naturwis.* **54:**189–195.

Bier, K., Kunz, W., and Ribbert, D., 1967, Struktur and Funktion der Oöcyten Chromosomen und Nukleolen sowie der Extra DNS während der Oögenese panoistischer und meroistischer Insekten, *Chromosoma* **23:**214–254.

Bonhag, P. F., 1955, Histochemical studies of ovarian nurse tissues and oocytes of the milkweed bug, *Oncopeltus fasciatus (Dallas)*. I. Cytology, nucleic acids, and carbohydrates, *J. Morphol.* **96:**381–439.

Brown, D. D., and Dawid, I. B., 1968, Specific gene amplification in oocytes, *Science* **160:**272–280.

Buchner, P., 1909, Das accessorischen Chromosom in Spermatogenese und Oögenese der Orthopteren, zugleich ein Beitrag zur Kenntnis der Reduktion, *Arch. Zellforsch.* **3**:355–430.

Cave, M. D., 1972, Localization of ribosomal DNA within oocytes of the house cricket, *Acheta domesticus* (Orthoptera: Gryllidae), *J. Cell Biol.* **55**:310–321.

Cave, M. D., 1973, Synthesis and characterization of amplified DNA in oocytes of the house cricket, *Acheta domesticus* (Orthoptera: Gryllidae), *Chromosoma* **42**:1–22.

Cave, M. D., 1975, Absence of ribosomal DNA amplification in the meroistic (telotrophic) ovary of the large milkweed bug *Oncopeltus fasciatus* (Dallas) (Hemiptera: Lygaeidae), *J. Cell Biol.* **66**:461–469.

Cave, M. D., 1976, Absence of rDNA amplification in the uninucleolate oocyte of the cockroach *Blatella germanica* (Orthoptera: Blattidae), *J. Cell Biol.* **71**:49–58.

Cave, M. D., 1978, Absence of amplification of ribosomal DNA in the polytrophic meroistic ovary of the giant silkworm moth, *Antheraea pernyi* (Lepidoptera: Saturniidae), *Wilhelm Roux's Arch. Dev. Biol.* **184**:135–142.

Cave, M. D., 1979, Length heterogeneity of amplified circular rDNA molecules in oocytes of the house cricket *Acheta domesticus* (Orthoptera: Gryllidae), *Chromosoma* **71**:15–27.

Cave, M. D., and Allen, E. R., 1969a, Extra-chromosomal DNA in early stages of oogenesis in *Acheta domesticus, J. Cell Sci.* **4**:593–609.

Cave, M. D., and Allen, E. R., 1969b, Synthesis of nucleic acids associated with a DNA-containing body of oocytes of *Acheta, Exp. Cell Res.* **58**:201–212.

Cave, M. D., and Allen, E. R., 1971, Synthesis of ribonucleic acid in oocytes of the house cricket (*Acheta domesticus*), *Z. Zellforsch.* **120**:309–320.

Cave, M. D., and Allen, E. R., 1974, Nucleolar DNA in oocytes of crickets: Representatives of the subfamilies Oecanthinae and Gryllotalpinae (Orthoptera: Gryllidae), *J. Morphol.* **142**: 379–394.

Cave, M. D., and Sixbey, J., 1976, Absence of ribosomal DNA amplification in a meriostic polytrophic ovary, *Exp. Cell Res.* **101**:23–30.

Chouinard, L. A., 1970, Localization of intranucleolar DNA in root meristem cells of *Allium cepa, J. Cell Sci.* **6**:73–86.

Clewell, D. B., 1972, Nature of Col E_1 plasmid replication in *Escherichia coli* in the presence of chloramphenicol, *J. Bact.* **110**:667–676.

Davidson, E. H., 1976, *Gene Activity in Early Development*, Academic Press, New York.

Debaisieux, P., 1909, Les débuts de l'ovogénèse dans le *Dytiscus marginalis, La Cellule*, **25**:207–237.

Edström J. E., and Daneholt, B., 1967, Sedimentation properties of the newly synthesized RNA from isolated nuclear components of *Chironomus tentans* salivary gland cells, *J. Mol. Biol.* **28**:331–343.

Endow, S. A., and Gall, J. G., 1975, Differential replication of satellite DNA in polyploid tissues of *Drosophila virilis, Chromosoma* **50**:175–192.

Fakan, S., and Puvion, E., 1980, The ultrastructural visualization of nucleolar and extranucleolar RNA synthesis and distribution, *Int. Rev. Cytol.* **65**:255–299.

Favard-Séréno, C., 1968, Évolution des structures nucléolaires au cours de la phase d'accroissement cytoplasmique chez le grillon (Insecte: Orthoptere), *J. Microsc.* (Paris) **7**:205–230.

Federoff, N. V., 1979, On spacers, *Cell* **16**:697–710.

Gage, L. P., 1974a, Polyploidization of the silk gland of *Bombyx mori, J. Mol. Biol.* **86**:97–108.

Gage, L. P., 1974b, The *Bombyx mori* genome: Analysis by DNA reassociation kinetics, *Chromosoma* **45**:27–42.

Gaginskaya, E. R., and Gruzova, M. N., 1975, Detection of the amplified rDNA in ovarial cells of some insects and birds by hybridization of nucleic acids *in situ, Tsitologiya* **17**:1132–1136.

Gall, J. G., 1968, Differential synthesis of the genes for ribosomal RNA during amphibian oögenesis, *Proc. Nat. Acad. Sci. USA* **60**:553–560.

Gall, J. G., 1969, The genes for ribosomal RNA during oögenesis, *Genetics* 61, Suppl. **1**:121–132.

Gall, J. G., and Rochaix, J. D., 1974, The amplified ribosomal DNA of Dytiscid beetles, *Proc. Nat. Acad. Sci. USA* **71**:1819–1823.

Gall, J. G., Macgregor, H. C., and Kidston, M. E., 1969, Gene amplification in the oocytes of Dytiscid water beetles, *Chromosoma* **26**:169–187.

Gambarini, A. G., and Lara, F. J. S., 1974, Under-replication of ribosomal cistrons in polytene chromosomes of *Rhynchosciara, J. Cell Biol.* **62**:215–222.

Gambarini, A. G., and Meneghini, R., 1972, Ribosomal RNA genes in salivary gland and ovary of *Rhynchosciara angelae*, *J. Cell Biol.* **54**:421–426.

Giardina, A., 1901, Origine dell'oocite e delle cellule nutrici nei *Dytiscus*, *Int. Mschr. Anat. Physiol.* **18**:417–484.

Gilbert, W., and Dressler, D., 1968, DNA replication: The rolling circle model, *Cold Spring Harbor Symp.* **33**:473–484.

Glatzer, K. H., 1979, Lengths of transcribed rDNA repeating units in spermatocytes of *Drosophila hydei:* Only genes without an intervening sequence are expressed, *Chromosoma* **75**:161–175.

Griffin, B. E., and Fried, M., 1975, Amplification of a specific region of the polyoma virus genome, *Nature* **256**:175–178.

Gruzova, M. N., Zaichikova, Z. P., and Sokolov, I. I., 1972, Functional organization of the nucleus in the oogenesis of *Chrysopa perla* L. (Insecta: Neuroptera), *Chromosoma* **37**:353–386.

Gunthert, T., 1910, Die Eibildung der Dytisciden, *Zool. Jb. Abt. Anat. Ontog.* **30**:301–345.

Hansen-Delkeskamp, E., 1969a, Satelliten-Desoxyribonucleinsäure in Gonaden und somatischen Gewebe von *Acheta domestica* L., *Z. Naturforsch.* **24b**:1331–1335.

Hansen-Delkeskamp, E., 1969b, Synthese von RNS und Protein während der Oögenese und frühen Embryogenese von *Acheta domestica*, *Arch. Enw. Mech. Org.* **162**:114–120.

Hansen-Delkeskamp, E., Sauer, H. W., and Duspiva, F., 1967, Ribonucleinsäuren in der Embryogenese von *Acheta domestica* L., *Z. Naturforsch.* **22b**:540–545.

Harris, S. E., and Forrest, H. S., 1967, RNA and DNA synthesis in developing eggs of the milkweed bug *Oncopeltus fasciatus* (Dallas), *Science* **156**:1613–1615.

Heinonen, L., and Halkka, O., 1967, Early stages of oögenesis and metabolic DNA in the oöcytes of the house cricket, *Acheta domesticus* (L.), *Ann. Med. Exp. Biol. Fenn.* **45**:101–109.

Hegner, R. W., and Russell, C. P., 1916, Differential mitosis in the germ-cell cycle of *Dineutes nigrior*, *Proc. Nat. Acad. Sci. USA* **2**:356–367.

Hennig, W., and Meer, B., 1971, Reduced polyteny of ribosomal RNA cistrons in giant chromosomes of *Drosophila hydei*, *Nature New Biol.* **233**:70–72.

Hourcade, D., Dressler, D., and Wolfson, J., 1973, The amplification of ribosomal RNA genes involves a rolling circle intermediate, *Proc. Nat. Acad. Sci. USA* **70**:2926–2930.

Huebner, E., and Anderson, E., 1972, A cytological study of the ovary of *Rhodnius prolixis*. II. Oocyte differentiation, *J. Morphol.* **137**:385–415.

Jacobs-Lorena, M., 1980, Dosage of 5S and ribosomal genes during oogenesis of *Drosophila melanogaster*, *Develop. Biol.* **80**:134–145.

Jaworska, H., and Lima-de-Faria, A., 1973, Amplification of ribosomal DNA in *Acheta*. VI. Ultrastructure of two types of nucleolar components associated with ribosomal DNA, *Hereditas* **74**:169–186.

Johnson, M. W., 1938, A study of the nucleoli of certain insects and the crayfish, *J. Morphol.* **62**:113–139.

Kato, K., 1968, Cytochemistry and fine structure of elimination chromatin in Dytiscidae, *Exp. Cell Res.* **52**:507–522.

King, R. C., 1970, *Ovarian Development in Drosophila melanogaster*, Academic Press, New York.

Kleinschmidt, A. K., 1968, Monolayer techniques in electron microscopy of nucleic acid molecules. In *Methods in Enzymology*, vol. XII, edited by L. Grossman and K. Moldave, pp. 361–379. Academic Press, New York.

Kloc, M., 1976, Extrachromosomal DNA and RNA synthesis in oocytes of *Creophilus maxillosus* (Staphylinidae, Coleoptera, Polyphaga), *Experientia* **32**:375–379.

Koch, E. A., Smith, P. A., and King. R. C., 1967, The division and differentiation of *Drosophila* cystocytes, *J. Morphol* **121**:55–70.

Kunz, W., 1967a, Funktionsstrukturen im Oocytenkern von *Locusta migratoria*, *Chromosoma* **20**:332–370.

Kunz, W., 1967b, Lampenbürsten Chromosomen und multiple Nukleolen bei Orthopteren, *Chromosoma* **21**:446–462.

Kunz, W., 1969, Die Entstehung multipler Oöcyten Nukleolen aus akzessorischen DNS-Körpern bei *Gryllus domesticus*, *Chromosoma* **26**:41–75.

Leiby, K. R., and Chooi, Y. W., 1979, Group fractionation and template level assembly of ribosomal proteins in *Drosophila melanogaster*, *J. Cell Biol.* **83**:169a.

Lima-de-Faria, A., 1962, Metabolic DNA in *Tipula oleracea, Chromosoma* **13:**47–59.
Lima-de-Faria, A., 1973, The molecular organization of the chromomeres of *Acheta* involved in ribosomal DNA amplification, *Cold Spring Harbor Symp.* **38:**559–571.
Lima-de-Faria, A., 1974, Amplification of ribosomal DNA in *Acheta.* IX. The isolated ribosomal DNA-RNA complex studied in the scanning electron microscope, *Hereditas* **78:**225–264.
Lima-de-Faria, A., and Moses, M. J., 1966, Ultrastructure and cytochemistry of metabolic DNA in *Tipula, J. Cell Biol.* **30:**177–192.
Lima-de-Faria, A., Nilsson, B., Cave, D., Puga, A. and Jaworska, H. 1968, Tritium labelling and cytochemistry of extra DNA in *Acheta, Chromosoma* **25:**1–20.
Lima-de-Faria, A., Birnstiel, M., and Jaworska, H., 1969, Amplification of ribosomal cistrons in the heterochromatin of *Acheta, Genetics* 61, Suppl. **1:**145–159.
Lima-de-Faria, A., Daskaloff, S., and Enell, A., 1973a, Amplification of ribosomal DNA in *Acheta.* I. The number of chromosomes involved in the amplification process, *Hereditas* **73:** 99–118.
Lima-de-Faria, A., Gustafsson, T., and Jaworska, H., 1973b, Amplification of ribosomal DNA in *Acheta.* II. The number of nucleotide pairs of the chromosomes and chromomeres involved in amplification, *Hereditas* **73:**119–142.
Lima-de-Faria, A., Jaworska, H., Gustafsson, T., and Daskaloff, S., 1973c, Amplification of ribosomal DNA in *Acheta.* III. The release of DNA copies from chromomeres, *Hereditas* **73:** 163–184.
Lima-de-Faria, A., Jaworska, H., and Gustafsson, R., 1973d, Release of amplified ribosomal DNA from the chromomeres of *Acheta, Proc. Nat. Acad. Sci. USA* **70:**80–83.
Long, E. O., and Dawid, I. B., 1979, Expression of ribosomal DNA insertions in *Drosophila melanogaster, Cell* **18:**1185–1196.
Long, E. O., and Dawid, I. B., 1980, Repeated genes in eukaryotes, *Annu. Rev. Biochem.* **49:**727–764.
Macgregor, H. C., 1972, The nucleolus and its genes in amphibian oogenesis, *Biol. Rev. Cambridge Phil. Soc.* **47:**177–210.
Matuszewski, B., and Hoser, P., 1975, Gene amplification and its effects on the structure and function of the oocyte nucleus in the Whirligig beetle *Gyrinus natator* (Gyrinidae, Coleoptera-Adephaga), *Experientia* **31:**431–432.
Matuszewski, B., and Kloc, M., 1976, Gene amplification in oocytes of the rove beetle, *Creophilus maxillosus* (Staphylinidae, Coleoptera, Polyphaga), *Experientia* **32:**34–38.
Mermod, J. J., Jacobs-Lorena, M., and Crippa, M., 1977, Changes in rate of RNA synthesis and ribosomal gene number during oogenesis of *Drosophila melanogaster, Develop. Biol.* **47:** 393–403.
Miller, O. L., Jr., and Beatty, B. R., 1969, Visualization of nucleolar genes, *Science* **164:**955–957.
Mohan, J., 1976, Ribosomal DNA and its expression in *Drosophila melanogaster* during growth and development, *Mol. Gen. Genet.* **147:**217–223.
Mohan, J., and Ritossa, F. M., 1970, Regulation of ribosomal RNA synthesis and its bearing on the *bobbed* phenotype in *Drosophila melanogaster, Develop. Biol.* **22:**495–512.
Nilsson, B., 1966, DNA-bodies in the germ line of *Acheta domesticus* (Orthoptera), *Hereditas* **56:** 396–398.
Nilsson, B., 1968, *Acheta domesticus* (Orthoptera) some cytological observations, *Lantbrukshogsk. Ann.* **34:**437–464.
Nilsson, B., Larsson, M., Hofsten, A., 1973, Chromosomes and DNA bodies of *Acheta desertus* (Orthoptera) with parallels to *Acheta domesticus, Hereditas* **75:**251–258.
Pardue, M. L., Gerbi, S. A., Eckhardt, R. A., and Gall, J. G., 1970, Cytological localization of DNA complementary to ribosomal RNA in polytene chromosomes of Diptera, *Chromosoma* **29:** 268–290.
Penman, S., Smith, I., and Holtzman, E., 1966, Ribosomal RNA synthesis and processing in a particulate site in the HeLa cell nucleus, *Science* **154:**786–789.
Pero, R., Lima-de-Faria, A., Ståhle, U., Granstrom, H., and Ghatnekar, R., 1973, Amplification of ribosomal DNA in *Acheta.* IV. The number of cistrons for 28S and 18S RNA, *Hereditas* **73:**195–210.
Rasch, E. M., 1974, The DNA content of sperm and hemocyte nuclei of the silkworm *Bombyx mori* L., *Chromosoma* **45:**1–26.

Raven, C. P., 1967, *Oogenesis: The Storage of Developmental Information*, Pergamon Press, London.

Recher, L., Whitescarver, J., and Briggs, L., 1969, The fine structure of a nucleolar constituent, *J. Ultrastruct. Res.* **29**:1–14.

Renkawitz, R., and Kunz, W., 1975, Independent replication of the ribosomal RNA genes in the polytrophic meroistic ovaries of *Calliphora erythrocephala, Drosophila hydei,* and *Sarcophaga barbata, Chromosoma* **53**:131–140.

Renkawitz-Pohl, R., and Kunz, W., 1975, Underreplication of satellite DNAs in polyploid ovarian tissue of *Drosophila virilis, Chromosoma* **49**:375–382.

Ribbert, D., and Bier, K., 1969, Multiple nucleoli and enhanced nucleolar activity in the nurse cells of the insect ovary, *Chromosoma* **27**:178–197.

Ritossa, F. M., and Scala, G., 1969, Equilibrium variations in the redundancy of rDNA in *Drosophila melanogaster, Genetics* 61, Suppl. **1**:305–317.

Ritossa, F. M., and Spiegelman, S., 1965, Localization of DNA complementary to ribosomal RNA in the nucleolus organizer region of *Drosophila melanogaster, Proc. Nat. Acad. Sci. USA* **53**:737–745.

Rochaix, J. D., Bird, A., and Bakken, A., 1974, Ribosomal RNA gene amplification by rolling circles *J. Mol. Biol.* **87**:473–487.

Sharp, Z. D., and Cave, M. D., 1980, Restriction endonuclease analysis of the ribosomal genes of *Acheta domesticus, J.* Cell Biol. **87**:109a.

Stevens, B. J., 1964, The effect of antinomycin D on nucleolar and nuclear fine structure in the salivary gland cell of *Chironomus thummi, J. Ultrastruct. Res.* **11**:329–335.

Suzuki, Y., Gage, L. P., and Brown, D. D., 1972, The genes for silk fibroin in *Bombyx mori, J. Mol. Biol.* **70**:637–699.

Tobler, H., 1975, Occurrence and developmental significance of gene amplification. In *The Biochemistry of Animal Development,* vol. 3., *Molecular Aspects of Animal Development,* edited by R. Weber, pp. 91–143, Academic Press, New York.

Trendelenburg, M. F., 1974, Morphology of ribosomal RNA cistrons in oocytes of the water beetle, *Dytiscus marginalis* L., *Chromosoma* **48**:119–135.

Trendelenburg, M. F., Scheer, U., and Franke, W., 1973, Structural organization of the transcription of ribosomal DNA in oocytes of the house cricket, *Nature New Biol.* **245**:167–170.

Trendelenburg, M. F., Scheer, U., Zentgraf, H., and Franke, W. W., 1976, Heterogeneity of spacer lengths in circles of amplified ribosomal DNA of two insect species, *Dytiscus marginalis* and *Acheta domesticus, J. Mol. Biol.* **108**:453–470.

Trendelenburg, M. F., Franke, W. W., and Scheer, U., 1977, Frequencies of circular units of nucleolar DNA in oocytes of two insects *Acheta domesticus* and *Dytiscus marginalis,* and changes of nucleolar morphology during oogenesis, *Differentiation* **7**:133–158.

Tschudi, C., and Pirrotta, V., 1980, Sequence and length heterogeneity in the 5S RNA gene cluster of *Drosophila melanogaster, Nuc. Acid Res.* **8**:441–451.

Ullman, J. S., Lima-de-Faria, A., Jaworska, H., and Bryngelsson, T., 1973, Amplification of ribosomal DNA in *Acheta.* V. Hybridization of RNA complementary to ribosomal DNA with pachytene chromosomes, *Hereditas* **74**:13–24.

Urbani, E., 1950, Studio comparativo della massa di Giardina nei Dytiscidae, *Rend. Acc. Naz. Lincei* **9**:384–395.

Urbani, E., 1969, Cytochemical and ultrastructural studies of oogenesis in the Dytiscidae, *Monit. Ital.* **3**:55–87.

Urbani, E., and Russo-Caia, S., 1964, Osservazioni citochimiche e autoradiografiche sul metabolismo degli acidi nucleici nella oogenesi di *"Dytiscus marginalis"* L., *Rend. Ist. Sci. Univ. Camerino* **5**:19–50.

Vincent, W. S., Halvorson, H. O., Chen, H. R., and Shin, D., 1969, A comparison of ribosomal gene amplification in uni- and multi-nucleolate oocytes, *Exp. Cell Res.* **57**:240–250.

von Borstel. R. C., and Rekemeyer, M. L., 1958, Division of a nucleus lacking a nucleolus, *Nature* **181**:1597–1598.

von Gaudecker, B., 1967, RNA synthesis in the nucleolus of *Chironomus thummi,* as studied by high resolution autoradiography, *Z. Zellforsch.* **82**:536–557.

Wellauer, P. K., and Dawid, I. B., 1977, The structural organization of ribosomal DNA in *Drosophila melanogaster*, *Cell* **10**:193–212.

Wensink, P., and Brown, D. D., 1971, Denaturation map of the ribosomal DNA of *Xenopus laevis*, *J. Mol. Biol.* **60**:235–247.

Wilson, E. B., 1928, *The Cell in Development and Heredity*, Macmillan, Inc., New York.

Wimber, D. E., and Steffensen, D. M., 1970 Localization of 5S RNA genes on *Drosophila* chromosomes by RNA–DNA hybridization, *Science* **170**:639–641.

Yao, M-C., and Gall, J. G., 1977, A single integrated gene for ribosomal RNA in a eucaryote, *Tetrahymena pyriformis*, *Cell* **12**:121–132.

5

The Cell Biology of Vitellogenic Follicles in *Hyalophora* and *Rhodnius*

WILLIAM H. TELFER, ERWIN HUEBNER, AND
D. SPENCER SMITH

1. The Roles of the Oocyte and of the Follicle Cells

Insect oocytes assemble their protein yolk bodies by receptor-mediated endocytosis of specific extracellular proteins (Telfer, 1961; Anderson, 1964; Roth and Porter, 1964; other papers reviewed by Telfer, 1965; Engelmann, 1979; Hagedorn and Kunkel, 1979). The major constituent incorporated is the sex-limited hemolymph protein, vitellogenin, but smaller amounts of lipophorin and many other hemolymph proteins are also internalized (Telfer, 1960; Telfer *et al.*, 1981a). In special cases, several apparent modifications of this general pattern have been described: In *Drosophila melanogaster* and related species a fraction of the total vitellogenin accumulated is apparently synthesized within the ovary (Bownes, 1980); in *Hyalophora cecropia* a secretory product of the follicle cells, now termed paravitellogenin, supplements the hemolymph-derived proteins (Anderson and Telfer, 1969; Bast and Telfer, 1976; Rubenstein, 1979); the fine structure of *Glossina austeni* suggests that yolk in this species may be assembled primarily from follicle cell products (Huebner *et al.*, 1975); and in some apterygotes fine structure implies that some yolk may be synthe-

WILLIAM H. TELFER • Department of Biology, University of Pennsylvania, Philadelphia, Pennsylvania, 19104, USA.
ERWIN HUEBNER • Department of Zoology, University of Manitoba, Winnipeg, Manitoba R3T 2N2, Canada.
D. SPENCER SMITH • Department of Zoology and the Hope Entomological Collection, Oxford University, Oxford, England.

sized within the oocyte itself (Bilinski, 1977). While yolk formation in insects has thus evolved a certain amount of diversity, it is nevertheless clear that in most species vitellogenin synthesized by the fat body and transported through the hemolymph is the primary source of the protein that is amassed in the oocyte (reviewed by Engelmann, 1979; Hagedorn and Kunkel, 1979).

The model proposed over twenty years ago to explain vitellogenin incorporation remains essentially intact. Extracellular channels facilitate protein diffusion through the ovarian sheaths and between the follicle cells (Figures 1, 7, and 8), and yolk precursors that have traversed these routes bind to putative receptor molecules on the surface of the oocyte. Endocytosis of bound proteins is effected by clathrin-coated micropinocytotic vesicles (Figure 2), which, after appropriate modification in the cortex of the oocyte, deliver their load of extracellular protein to the growing yolk spheres by membrane fusion.

In a general way, the respective roles of the follicle cells and of the oocyte in this sequence are clear. The epithelium formed by the follicle cells can either facilitate or obstruct the transmission of yolk precursors from the hemolymph, and in fact both roles may be played at appropriate stages of development. Facilitation occurs during vitellogenesis, when intercellular spaces are generated by any of several active processes (see section 2). At the end of vitellogenesis, permeation is forcibly terminated in at least one group by the genesis of occlusion zones between the follicle cells (Rubenstein, 1979).

In addition to their permissive role in vitellogenesis, the follicle cells may contribute directly to both the yolk and the cytoplasm of the growing oocyte. We have already noted the synthesis and secretion of paravitellogenin by the follicle cells of *Hyalophora*. This 70,000-d protein is secreted into the intercellular spaces where it apparently mixes with hemolymph proteins, and the mixture is then endocytosed by the oocyte. It is still uncertain how widespread this phenomenon may be; identical methods failed to reveal a paravitellogenin in *Oncopeltus fasciatus* (Kelly and Telfer, 1979), but in *Locusta migratoria* there is evidence of follicle cell products that are specific to vitellogenesis (Glass and Emmerich, 1981).

A second class of contributions by the follicle cells is made possible by the recently discovered high permeability junctional complexes between the oocyte and follicle cells (Woodruff, 1979; Huebner, 1981). The follicle cells have in effect two routes for contributing synthetic products to the vitellogenic oocyte, one for large molecules *via* the intercellular spaces and receptor-mediated endocytosis, and the second for small hydrophilic solutes *via* junctional complexes directly into the ooplasm.

The oocyte, for its part, must synthesize clathrin, receptor molecules, and other membrane components and be able to assemble these at appropriate loci in the oolemma. When the receptors have been loaded with extracellular ligands, the oocyte must then be able to form these constituents into coated vesicles and transport them to the underlying stratum of yolk spheres. The rate and magnitude of these processes can be truly exceptional. As an example, it takes 3 hr to make a new stratum of 20-μm diameter yolk spheres in *Hyalophora* (Melius and Telfer, 1969), and volume considerations indicate that for each yolk sphere this would require over 10^6 coated vesicles (assuming a conservatively large diameter of 0.2 μm and that neither structure gains or loses water). The yolk spheres are close packed in this species, and each must therefore be generated by an approximately circular patch of oocyte surface whose diameter is also about 20 μm. The 10^6 coated vesicles have a total surface area of 1.2×10^5 μm^2, while the patch of oocyte surface producing them approximates 100 π, or 314 μm^2. It therefore follows that the vitellogenic surfaces of the oocyte must generate over 300 times their area in vesicle membranes every 3 hr. Because the oocyte surface of *Hyalophora* is a folded structure (King and Aggarwal, 1965; Stay, 1965), the area of cell membrane in each patch is several times greater than 314 μm^2, but even with this correction it is clear that endocytosis would consume all available cell membrane in less than a minute if membrane replacement were not equally rapid.

Similarly, the small number of coated vesicles in the cortex at any time, and the limited numbers of tubular and noncoated vesicular elements that we interpret as fusion-related configurations (section 4) argue, in the face of the measurably prodigious rate of endocytosis, for a rapid turnover of these elements as well. It therefore seems reasonable to predict that total recycling times for the individual molecules of the endocytotic mechanism may well prove to be in the range of 1 to 10 min. By this reasoning, if a single molecule of vitellogenin receptor or clathrin should persist in the cortex for as long as one day it could participate in several hundred to a thousand rounds of endocytosis, and would thus contribute a considerable saving in the synthetic effort of the oocyte. Without turnover, the oocyte could as economically synthesize its own yolk protein as become such an extraordinary specialist in endocytosis.

We compare in this review the vitellogenic follicles of *Hyalophora cecropia* and *Rhodnius prolixus*, the two insects in which the cell biology of yolk formation has been most thoroughly studied. Included are the mechanisms of formation of the intercellular spaces, the structure and possible functions of the high permeability junctional complexes between the oocyte and the follicle cells, and the fine structure of endocytosis in the oocytes. The first two topics have been recently studied and have not been broadly reviewed, while the third, which is much reviewed, has not been seriously studied for over ten years. There are many recent advances in the understanding of receptor-mediated endocytosis in other systems, however, as well as a growing awareness that insect vitellogenesis is a developmentally and hormonally controlled process of general interest. A reiteration in current terms of the ultrastructural basis of endocytosis is therefore timely.

2. *The Formation of Intercellular Spaces*

The follicle cells have recently been shown in two species to play active roles in maintaining the patency of the intercellular spaces, but their mechanisms of doing so appear at this point to be entirely different. In *Rhodnius*, an essentially mechanical system involving cell shrinkage and cytoskeletal changes has been proposed (Pratt and Davey, 1972; Davey and Huebner, 1974;

Abu-Hakima and Davey, 1977a; Huebner and Injeyan, 1980), while in *Hyalophora*, the secretion of a proteoglycan matrix with affinity for hemolymph proteins appears to play a dominant role (Anderson and Telfer, 1970a; Telfer, 1979). It seems plausible to expect that many other insects producing yolk from hemolymph vitellogenins will be proven to rely on one or the other or a combination of these two mechanisms. It is fortunate for our perspective of the problem that the first two systems to be analyzed should have been proven to work on such different principles.

2.1. *Hyalophora* and the Secretion of Proteoglycans

In transmission electron micrographs of vitellogenic *Hyalophora* follicles that have been fixed directly in the animal, the meshwork of spaces that separate the follicle cells from one another, as well as the perioocytic space between the epithelium and the oocyte and between the folds in the oocyte surface, appear to be filled with a densely staining matrix (Figures 1, 2). If the follicle is soaked in physiological saline prior to fixation, however, the spaces are empty, except for a sparse population of fibers that appear to attach to the surfaces of the follicle cells (Figure 3). Analysis of the washings from soaked follicles indicated that they contain suprisingly large amounts of hemolymph proteins (Anderson and Telfer, 1970a), and it is assumed that these account for the high density of the spaces in the freshly fixed follicles. When the amounts of the eluted proteins were compared to the intercellular space volume measured by ^{14}C-inulin elution, their concentrations were estimated to be up to five times greater in the spaces than in the hemolymph. It was therefore proposed that the intercellular spaces contain an insoluable binding agent that can reversibly hold hemolymph proteins in place. The sites of hemolymph protein absorption presumably include the cell surfaces and the fibers that are attached to them, although we cannot preclude the possibility that they are all or in part lost during fixation.

That the intercellular spaces contain a proteoglycan matrix was shown in follicles labeled with ^{35}S-sulfate or ^{3}H-glucosamine (Telfer, 1979), Autoradiograms of such follicles that have been labeled either *in situ* or *in vitro* indicate extremely heavy concentrations of label in all of the intercellular spaces of the epithelium, as well as in the basement membrane (Figure 5). Even when soaked in the physiological saline that elutes bound hemolymph proteins the follicles did not lose most of this label. Furthermore, very little label appeared in the yolk spheres, even after a 24-hr incubation. Thus, sulfate and glucosamine are precursors for a structural component of the spaces, and not a diffusible substance or a yolk precursor.

The following observations indicated that the extracellular label was in the form of very large, sulfated proteoglycans. (1) Prolonged TCA-extraction of intact follicles failed to extract the label; on the other hand, when treated with pronase under conditions that were shown not to damage the permeability of the cells, the follicles released large quantities of a nondialyzable, TCA-soluble label. A high molecular weight, sulfated conjugate that is normally linked or entrapped by extracellular protein was implied by this result.

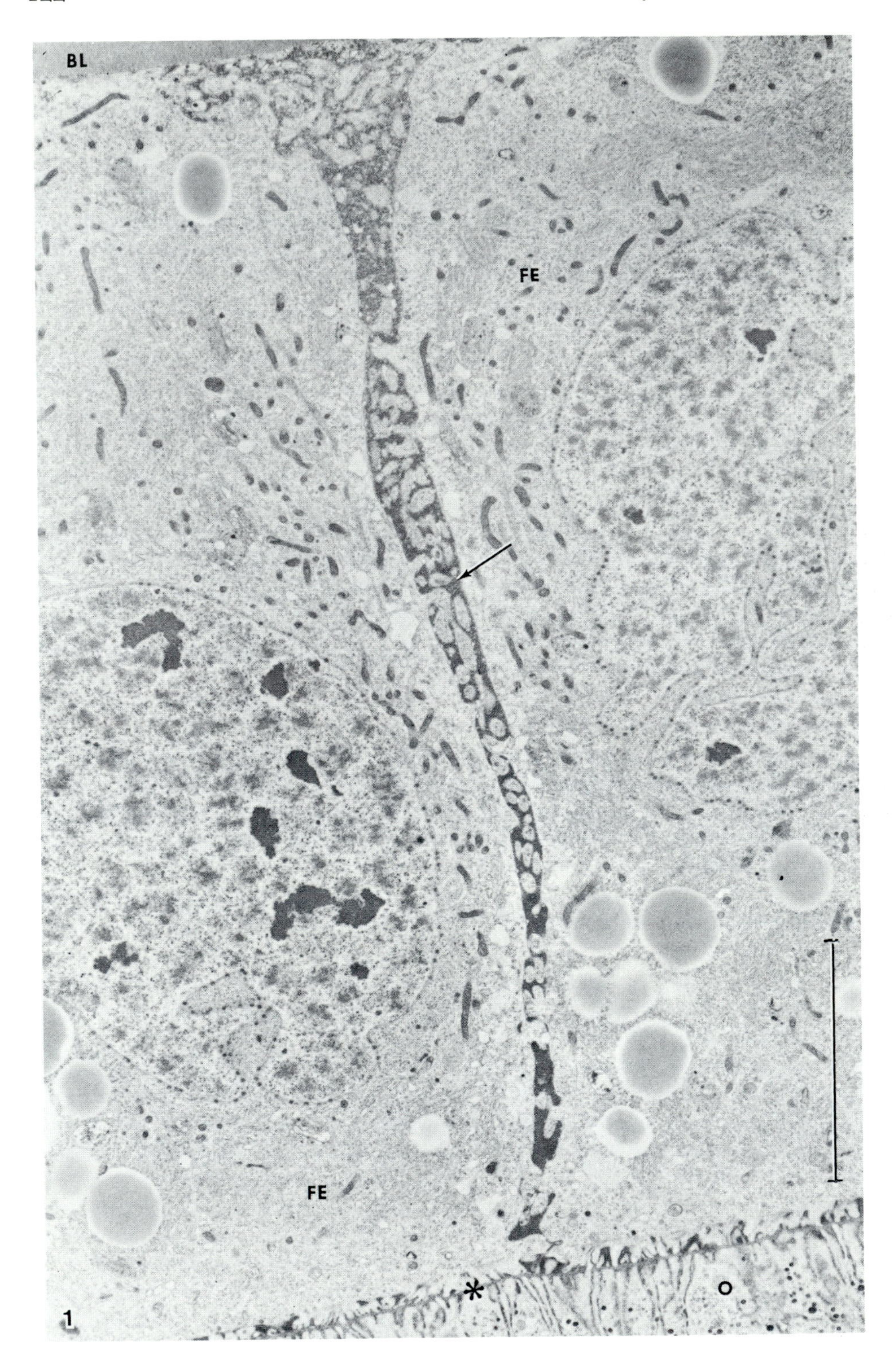
BL
FE
FE
*
O
1

(2) In a manner characteristic of many large polysaccharides, the sulfate-labeled material exhibited a wide range of molecular weights in agarose gel filtration columns. (3) ^{3}H-glucosamine and ^{35}S-sulfate both labeled pronase-insensitive materials with similar ranges of molecular weights. (4) Both glucosamine- and sulfate-labeled pronase shavings were degraded to a low molecular weight (in the oligosaccharide range) by bovine testicular hyaluronidase.

High molecular weight, glucosamine-labeling, and hyaluronidase sensitivity combine to suggest that the protease-insensitive conjugate is a hyaluronic acid or chondroitin sulfate-like glycosaminoglycan. These properties explain well the affinity of the intercellular spaces for hemolymph proteins, for sulfated glycans are known to have a divalent cation-mediated affinity for lipoproteins in particular (Comper and Laurent, 1978).

When vitellogenesis in *Hyalophora* terminates, occlusion zones are produced between the follicle cells near their apical ends (Rubenstein, 1979). The perioocytic spaces between the occlusion zones and the oocyte continue to label, and this suggests that the sulfated proteoglycan is secreted apically to the occlusion zones. The spaces between the follicle cells basal to these zones (with a total length of about 50 μm) lose their intercellular matrix, however (Telfer and Anderson, 1968) (Figure 6), and can no longer be labeled with sulfate and glucosamine (Telfer 1979).

Disassembly suggested that labeled matrix fragments might be detectable in the hemolymph of females containing follicles that are nearing maturity, and this in fact has been proven to be the case (Telfer and Tsiongas, in preparation). Gel filtration chromatography of the hemolymph of females at late stages of metamorphosis revealed easily detected quantities of a sulfate- and glucose-labeling macromolecule that can be degraded by testicular hyaluronidase. An unexpected result was that the material is present not only in females with postvitellogenic follicles, but also at an earlier stage when yolk formation is reaching its peak. *In vitro* studies have confirmed that vitellogenic follicles release heterogeneously sized sulfated glycans.

The sulfated proteoglycan matrix of the intercellular spaces appears from this information to be in a dynamic state, with continuous deposition and gradual degradation going on simultaneously throughout vitellogenesis. Furthermore, since secretion appears to be primarily from the apical surfaces of the follicle cells, while release to the hemolymph is from the basal end of the spaces, there must be a gradual apicobasal creeping of the matrix as it is degraded. Electron micrographs confirm this concept by showing that the matrix is much denser near the apex of the spaces than it is near the base (Figure 1).

←

Figure 1. Two follicle cells (FE) of an early vitellogenic follicle of *Hyalophora cecropia*. The cells are separated by a space containing a densely staining intercellular matrix (arrow) that is penetrated by many poorly organized cytoplasmic processes. The matrix appears more compact and darkly staining near the oocyte (O) than near the basement lamina (BL). Matrix material is also visible between the follicle cells and the oocyte (*). Nuclei are large, folded, and, reflecting their endopolyploidy, contain much chromosomal material. The cytoplasm is highly basophilic at this stage and is organized for protein synthesis and secretion. The follicle cells are large, measuring 45 μm in depth and 30 μm in diameter. Scale bar, 10 μm.

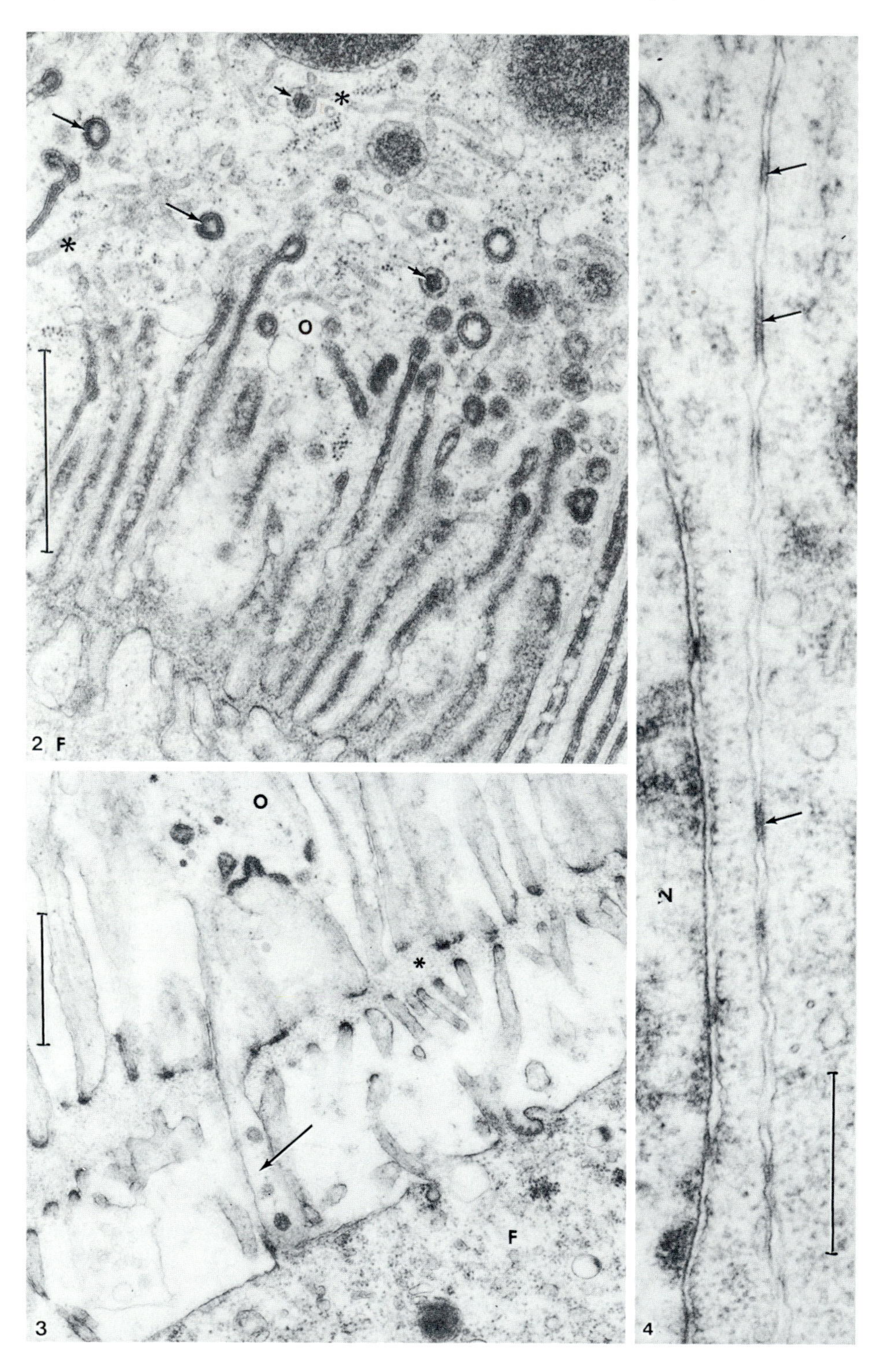
*
*
O
2 F
O
*
F
3
N
4

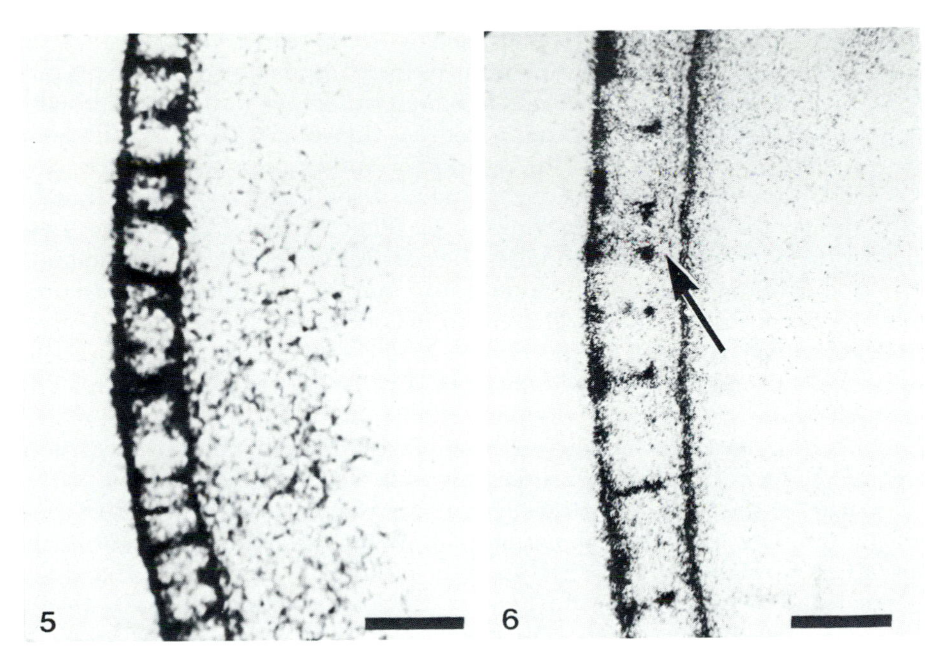

Figures 5 and 6. *Hyalophora cecropia.* Autoradiograms of follicles labelled for 24 hr *in situ* with ^{35}S-sulfate. In Figure 5, the follicle was fixed while still vitellogenic, and the label is concentrated around the follicle cells. In Figure 6, the follicle had terminated vitellogenesis several hours before fixation. Labelled matrix has initiated disassembly between the follicle cells, but remains intact in the perioocytic spaces that are enclosed by the apical occlusion zones (arrows). Space bars, 100 μm.

Figures 2–4. *Hyalophora cecropia.*
Figure 2. The cortex of a vitellogenic oocyte (O) with the edge of a follicle cell (F) shown at the lower left. The oocyte membrane contains 1- to 2-μm folds, and the spaces between the folds contain heavily staining extracellular matrix material. Coated pinocytotic pits are seen primarily where the folds have penetrated most deeply into the cortex. Two kinds of vesicles are seen in the cortex between the folded area and the yolk spheres (the dark structures at the top of the picture). Coated vesicles (long arrows) have clathrin coats on their cytoplasmic surfaces, and a layer of densely staining material on their luminal surfaces. A clear lumen is also visible. The second class of vesicles (short arrows) are about the same size but have lost both their clathrin and their luminal coats. This region of the cortex also contains high concentrations of tubular elements (*), which are interpreted as vesicle fusion residues. Scale bar, 1 μm.
Figure 3. The interface between an oocyte (O) and a follicle cell (F) in a vitellogenic follicle that, before fixation, has been soaked 30 min in physiological saline in order to elute soluble hemolymph proteins from the intercellular spaces. After this treatment, the spaces are seen to retain a filamentous material, which is thus implicated as the sulfated proteoglycan component of the extracellular matrix. Also clear after this treatment are the vitelline envelope (*), and the hemidesmosomes, by which the tips of the oocyte folds and the follicle cell microvilli attach to it. A cytoplasmic process (arrow) is seen to cross the vitelline envelope from the oocyte. In the absence of hemolymph proteins, the extracellular material attached to the oocyte membrane in the coated pits is presumed to be primarily paravitellogenin. Scale bar, 1 μm.
Figure 4. The space between two follicle cells in a previtellogenic follicle. The extracellular matrix seen in Figure 1 is absent at this stage, and the cells are tacked together by large numbers of densely staining junctional complexes (arrows). A nucleus (N) lies close to the membrane of the cell on the left. Scale bar, 0.5 μm.

When this information is combined with the earlier finding that the matrix has a relatively nonspecific affinity for a number of hemolymph proteins, an interesting model develops in which sulfated glycan-bound proteins are conveyed centrifugally against a counter-flux of centripetally diffusing, free proteins. Furthermore, since the oocyte surface represents an effective sink for yolk precursor porteins, the bound proteins being carried centrifugally would be relatively enriched in those species that are not selectively accumulated by the oocyte. This in turn raises the possibility that one of the functions of matrix binding and centrifugal creeping may be to reduce the concentration of adventitious proteins adjacent to the oocyte where they would otherwise be passively trapped in the lumina of the endocytotic pits and vesicles (section 4). In a qualitative way, this model fits the facts that are known about the intercellular matrix of *Hyalophora* follicles; but whether the rate of matrix secretion is in fact fast enough to counter significantly the influx of nonvitellogenic hemolymph proteins remains to be determined.

Whatever the function of turnover may prove to be, its degradative component provides a ready explanation of how the matrix is destroyed at the end of yolk formation, for all that would be required to cause the matrix to disappear would be to stop its deposition. In the spaces between the follicle cells, where degradation first occurs, this interference is provided by the occlusion zones, which would presumably block the centrifugal movements of apically-secreted matrix as well as the influx of vitellogenic proteins.

Less is known about the onset of matrix and space formation at the start of vitellogenesis, because onset is a gradual process with no clear marker to identify the follicles that have begun to form yolk. It is clear, however, that previtellogenic follicles have neither the dense matrix nor the well-developed intercellular spaces that characterize vitellogenesis (Figure 4). The follicle cells at this stage are tacked to each other and to the oocyte by large numbers of densely staining junctional complexes.

2.2. *Rhodnius* and the Mechanical Genesis of Intercellular Spaces

In *Rhodnius prolixus,* transmission electron microscopy of vitellogenic follicles indicates that the perioocytic space contains interdigitating processes from the oocyte and the follicle cells, and a loose, flocculent material, which presumably contains primarily hemolymph proteins (Figure 7). There is no indication of the dense matrix seen in the intercellular spaces of the *Hyalophora* follicle (Huebner and Anderson, 1972a), and it seems clear that a very different basis for space formation must be entailed. Thick sections (Figure 8) indicate that the apical follicle cell surfaces form bulbous projections, and this is confirmed by scanning electron micrographs of the apical surface of epithelia from which the oocyte has been removed (Figure 9). In these preparations, eight to ten bulbous processes are seen to protrude from the surface of each follicle cell, with smaller microvilli arising in turn from these. Each follicle cell is surrounded laterally by broad spaces that average in width almost half the diameters of the cells. At the mid to basal surface of the epithelium, by

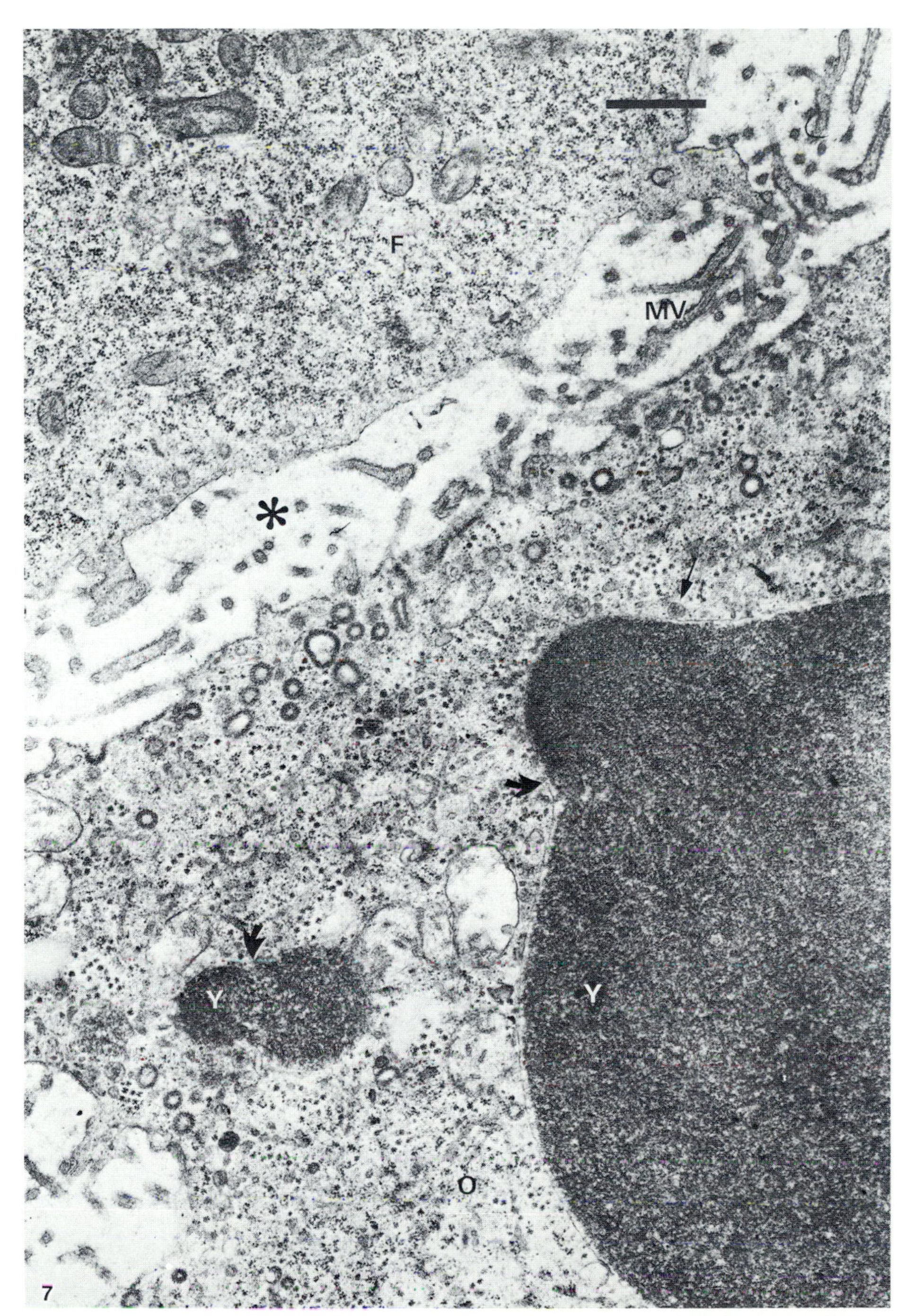

Figure 7. *Rhodnius prolixus.* The extracellular space between the oocyte (O) and the follicle cells (F) contains only a faint flocculent material (*). The oocyte cortex has numerous coated and uncoated vesicles and shows the vesicle fusions (arrows) that lead to progressively larger yolk spheres. (Y). MV, microvilli. Scale bar, 1 μm.

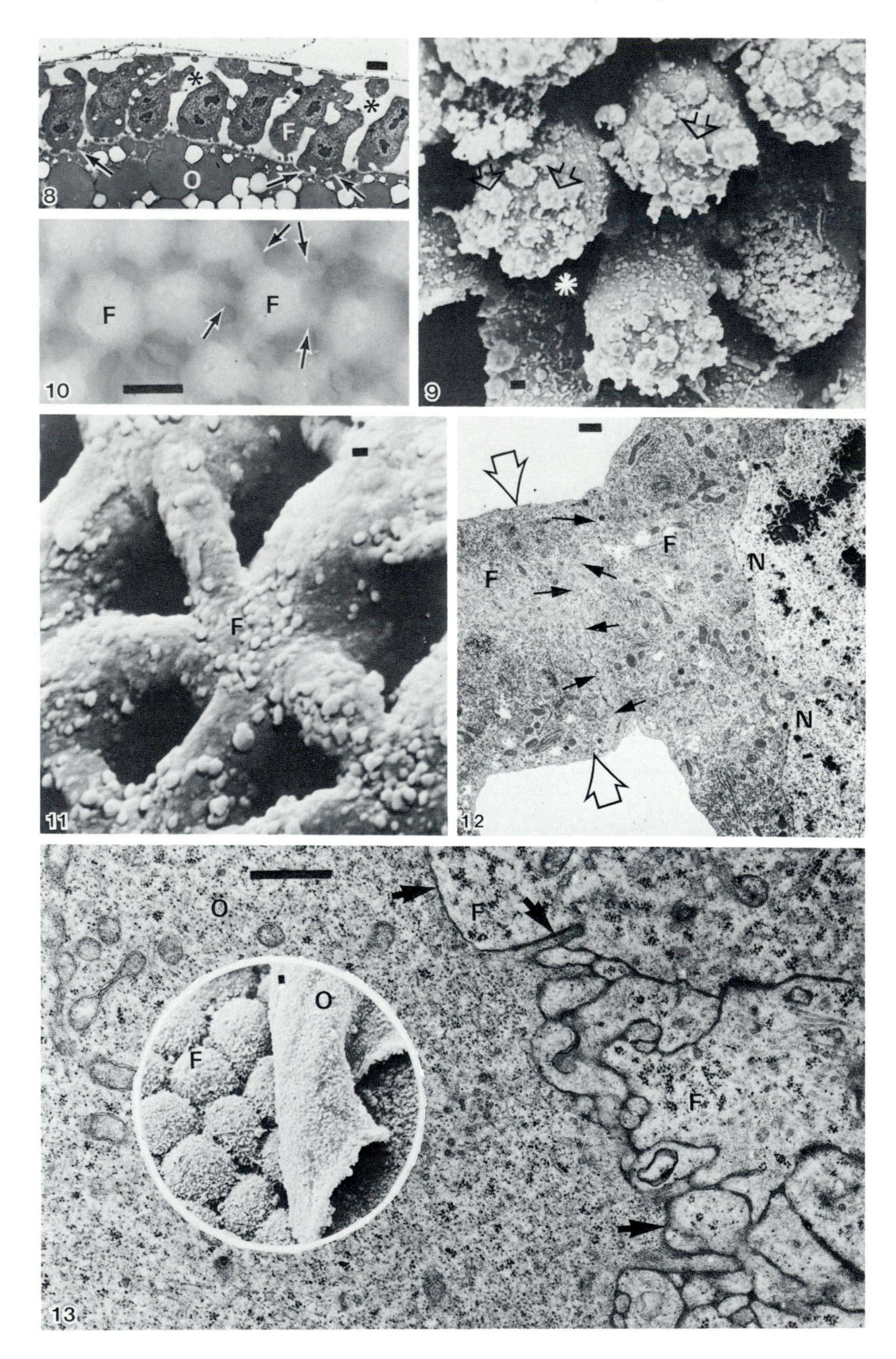
8
*
F
*
O
10
F
F
9
*
11
F
12
F
F
N
N
13
O
F
F
F
O

contrast, each cell maintains contact with its neighbors with a broad extension of cytoplasm ranging 5–10 μm in diameter and 3–8 μm in length (Figures 8, 10–12). Transmission electron microscopy suggests that this highly characteristic shape of the vitellogenic follicle cell is maintained by a complex cytoskeleton of microtubules and microfilaments (Abu-Hakima and Davey, 1977c; Huebner and Injeyan, 1981). Morphology thus suggests a cell that is corseted into a self-determined shape by its own cytoskeleton, rather than being constrained, as in *Hyalophora*, by its neighbors and an intercellular matrix.

In previtellogenic follicles, intercellular spaces are virtually absent, and the membranes of follicle cells and the oocyte are closely apposed to each other (Figure 13 and inset). After a blood meal and juvenile hormone secretion, spaces first appear at the corners between adjacent follicle cells, and these widen progressively as the follicle increases in size; the struts of cytoplasm elongate, and the intercellular spaces become wider (Figure 10) in a way that correlates directly with the length of the follicle (Huebner and Injeyan, 1980).

Studies on the mechanism of space formation became practical when it was found that allatectomy inhibits the development of patency (Pratt and Davey, 1972) and that juvenile hormone (JH) promotes patency in follicles incubated in a physiological salt solution (Davey and Huebner, 1974). Removal of vitellogenic follicles from their normal JH-containing environment causes a partial relaxation of the follicle cells, so that patency is somewhat reduced. After incubation under these conditions, the addition of 10^{-8} M JH stimulates a rapid increase in the size of the spaces, at the expense of the volume of the cells, and therefore appears to involve a loss of cellular water, as well as cytoskeletal changes (Abu-Hakima and Davey, 1977a,c; Huebner and Injeyan, 1981). The response requires neither RNA nor protein synthesis, but is inhibited by metabolic poisons (Abu-Hakima and Davey, 1977b). Ouabain also

Figures 8–13. *Rhodnius prolixus.*

Figure 8. A toluidine blue-stained, 1-μm-thick section of a vitellogenic follicle. Note the extracellular spaces (*) and the short, bulbous projections of the follicle cells (arrows) abutting the oocyte. Scale bar, 10 μm. (From Huebner and Anderson, 1972a, courtesy of the Wistar Institute Press.)

Figure 9. This scanning electron micrograph reveals the apical surface of a follicular epithelium from which the oocyte has been dissected. Note the numerous bulbous projections (arrows), and the large intercellular spaces (*). Scale bar, 1 μm (Huebner and Injeyan, unpublished).

Figures 10 and 11. These micrographs depict the upper or basal surface of a vitellogenic follicle in a living, Evans blue-stained preparation, and a scanning electron microscope preparation, respectively. Neighboring follicle cells (F) retain contact with each other across the intercellular spaces by extensions (arrows) 3–6 μm in diameter that form intercellular struts. Scale bars, 10 μm and 1 μm (Huebner and Injeyan, 1981, unpublished).

Figure 12. Thin section of a cytoplasmic strut (open arrows) between adjacent follicle cells (F). Solid arrows indicate the cell membranes at the point of contact. N, nucleus. Scale bar, 1 μm.

Figure 13 and inset. Thin section of a previtellogenic follicle. Note the close apposition of the follicle cells (F) to each other and to the oocyte (O) (arrows). The inset shows a scanning electron microscope preparation of a previtellogenic follicle with the oocyte membrane and cortex (O) peeled away from the tightly packed follicle cells (F). Scale bars, 1 μm. (From Huebner and Anderson, 1972a, courtesy of Wistar Institute Press, and Huebner and Injeyan, unpublished.)

inhibits the response, which suggests the participation of a Na^+/K^+-sensitive ATPase in the steps leading to water loss (Abu-Hakima and Davey, 1979; Ilenchuk, 1980). Finally, the integrity of the cytoskeleton is also required, for the response is inhibited also by colchicine and by cytochalasin B (Huebner, 1976; Abu-Hakima and Davey, 1977c).

The influence of JH on patency can be imitated by a variety of pharmacological agents, though only at concentrations 10^4- to 10^6-times greater than the effective JH concentration required for this purpose (Huebner 1976; and Huebner and Injeyan, 1981). These include theophylline (10^{-3} M), dibutyryl cAMP (10^{-4} M), ethanol (10^{-3} M), procaine (10^{-3} M), and a variety of vertebrate control factors, such as vasopressin, norepinephrine, and prostaglandins. The mechanism of action of 10^{-8} M JH in producing its *in vitro* effects thus seems to have a physiological basis differing fundamentally from the transcriptional controls that have been proposed for its action on vitellogenin synthesis in the fat body (reviewed by Engelmann, 1979; Hagedorn and Kunkel, 1979). For our purposes, however, the important message from these studies is that cell shape changes entailing volume reduction and cytoskeletal elements can generate intercellular spaces in a manner quite different from the proteoglycan matrix secretion seen in *Hyalophora*.

The contrasts between *Rhodnius* and *Hyalophora* are evident not only in the density of the intercellular matrix, but also in the cytoplasmic processes that connect neighboring cells. In *Hyalophora*, these appear to be relatively numerous, small, and structureless relative to the broad struts of *Rhodnius* (compare Figure 1 with Figures 11 and 12). Despite these differences in appearance and behavior, the two species could conceivably have features in common. Thus, matrix secrection is not entirely ruled out in *Rhodnius*, since even in JH-free media the intercellular spaces of vitellogenic follicles do not disappear entirely (Huebner and Injeyan, 1980). And on intuitive grounds it seems unlikely that the follicle cells of *Hyalophora* are as flaccid as our model for this species would imply. For any other species whose intercellular space formation should be investigated in the future it will be judicious to assume at the outset that both models may apply to some degree.

3. Gap Junctions

3.1. Physiological Studies

High-permeability junctional complexes between epithelial cells have been recognized in insects since 1964. They have been characterized functionally by applying electrical coupling and fluorescent dye injection techniques, particularly to salivary glands in dipteran larvae (Loewenstein and Kanno, 1964), but also to hemipteran and coleopteran epidermal cells (Warner and Lawrence, 1973; Caveney and Podgorsky, 1975; Caveney and Berdan, 1982). The cells in these tissues are apparently connected by low-resistance channels that are permeable to substances with molecular weights no larger than 1500 (Loewenstein, 1978). More recently is has become apparent that high-permeability junctions can also form between cells of differing types, and in particular between follicle cells and oocytes.

The earliest suggestion of this for insects was the finding by Wollberg *et al.* (1976) that oocytes in adjacent follicles in the ovarioles of *Locusta migratoria* are electrically coupled. The oocytes in this panoistic ovary, having no association with nurse cells, are completely surrounded by a follicular epithelium; adjacent follicles are separated by a plug of modified follicle cells that form an interfollicular connective. While the authors did not comment on this issue, the morphology of their system requires that ion-permeable junctions exist between oocytes and follicle cells, as well as between the cells of the connective and the follicles on either side. The measurements also indicated that the equilibrium potential of the oocyte becomes more negative as it matures, and that rectification occurs between adjacent follicles, so that current would tend to flow more readily toward the mature end of the ovariole.

The second paper dealing with this problem in insect ovaries concerned the polytrophic system of *Hyalophora cecropia* (Woodruff, 1979). Here also the oocytes of neighboring follicles were shown to be electrically coupled, though there were differences in detail from *Locusta*. In particular, there were no indications of rectification in the moth—hyperpolarizing and depolarizing currents yielded identical coupling ratios, indicating that ions moved with equal facility in the two directions between adjacent follicles. It is not clear whether this discrepancy resulted from technical difficulties or from differences in developmental patterns. The *Locusta* measurements were made between the single vitellogenic follicle in each ovariole and the adjacent previtellogenic follicle; in the moth, by contrast, 30–40 vitellogenic follicles are simultaneously present in an ovariole, and coupling was thus studied between follicles nearly identical in developmental stage.

The moth study also made use of fluorescein microinjections (Woodruff, 1979). When this dye was administered to vitellogenic oocytes, it moved into all other cells in the follicle within 15 min, and crossed the interfollicular connective to adjacent follicles, both apically and basally, within 30 min. Fluorescein-labeled protein, by contrast, either remained confined to the injected oocyte, or, at the most, crossed the intercellular bridges to the nurse cells, depending on their electrical charge. In no case was there any indication of protein movement from the oocyte or the nurse cells to the follicle cells, even when labeled lysozyme, a small protein with a molecular weight of only 14,400 d, was injected (Woodruff and Telfer, 1980). The permeability characteristics of these heterologous junctional complexes are in these regards similar to those described earlier for the homologous complexes of insect salivary glands and epidermis.

A third case of high permeability junctional complexes is in the telotrophic ovary of *Rhodnius prolixus* (Huebner, 1981). Here also coupling occurs between oocytes, but apparently mediated by very different routes from those of panoistic and polytrophic ovaries. The oocytes are connected by strands of cytoplasm, the trophic cords, to the tropharium at the apex of the adult ovariole. The tropharium is a syncytium that is formed by incomplete cytokinesis, and by the fusion of nurse cells during metamorphosis (Huebner and Anderson, 1972b; Huebner, 1981; Lutz and Huebner, 1981), and all oocytes attached to it are necessarily coupled to each other *via* this route. To demonstrate coupling between the oocytes and the follicle cells, therefore, it was necessary to study the spread of injected fluorescent markers. Here, as in the moth, tracers like fluorescein and lucifer yellow injected into the oocyte moved into the surrounding follicle cells within a few minutes (Huebner, 1981; Telfer *et al.*, 1981b), while fluorescent proteins failed to do so even after several hours of incubation (Figures 15, 16). Equally striking, and in contrast to the moth, was the inability of fluorescein to move from the epithelium of the injected follicle into either the more basal or the more distal follicles. In this ovary, with its oocytes coupled *via* the tropharium, high-permeability junctional complexes between the follicles are absent.

3.2. Structural Studies

Of the several kinds of junctional complexes that are recognized by electron microscopy, only gap junctions have now been correlated with the trait of intercellular diffusion (Revel *et al.*, 1971; Gilula *et al.*, 1972). If the issue were still in doubt, the same answer would be provided by the oocyte–follicle cell example, for gap junctions are the only complexes that have thus far been identified between these two cells. In insect epithelia, by contrast, these junctions often occur as plaques embedded in a field of pleated, septate desmosomes (Noirot-Timothée and Noirot, 1980).

In transverse sections, gap junctions appear as areas where the membranes of adjacent cells are strictly parallel to each other and separated by a clear space that is 20–40 Å wide (Revel and Karnovsky, 1967). The space can be penetrated by lanthanum ions and, when viewed *en face,* such junctions appear as areas containing concentrated arrays of negatively stained, 100 Å particles (Figure 14 and inset). The particles can also be viewed in freeze-fracture preparations where, in arthropod material, they are associated with the outer leaflet of the cell membrane (reviewed by Satir and Gilula, 1973; Gilula, 1980). They contain

Figures 14–18. *Rhodnius prolixus.*

Figure 14 and inset. A lanthanum-impregnated preparation of an oocyte–follicle cell interface during late previtellogenesis. There are numerous *en face* views of gap junctional arrays (open arrows). The higher magnification inset shows that the arrays are clusters of nonstaining particles (dark arrows), each containing a small central pore. Scale bars, 0.1 μm. (From Huebner, 1981, courtesy of Academic Press.)

Figure 15. Fluorescence micrograph of a living ovariole. The oocyte (O) was microinjected with lucifer yellow, and the dye had spread up the trophic cord (TC) and into the follicluar epithelium (F, arrow). Incubation time, 30 min. Scale bar, 5 μm.

Figure 16. Fluorescence micrograph of an oocyte microinjected with fluorescein-labelled lysozyme, a small protein. The conjugate was too large to cross gap junctions into the epithelium (F, arrow) though it was able to move into the trophic cord. Incubation time, 30 min. Scale bar, 50 μm.

Figure 17. Lanthanum-impregnated, early vitellogenic follicle showing a gap junction in transverse section (arrow) between follicle cell and oocyte processes. Scale bar, 0.1 μm.

Figure 18. A freeze-fracture preparation of a vitellogenic follicle showing a gap junctional plaque (arrow) on a bulbous follicle cell process (FP). Microvilli (MV) extend into the intercellular space from the oocyte at the bottom of the picture. Scale bar, 1 μm.

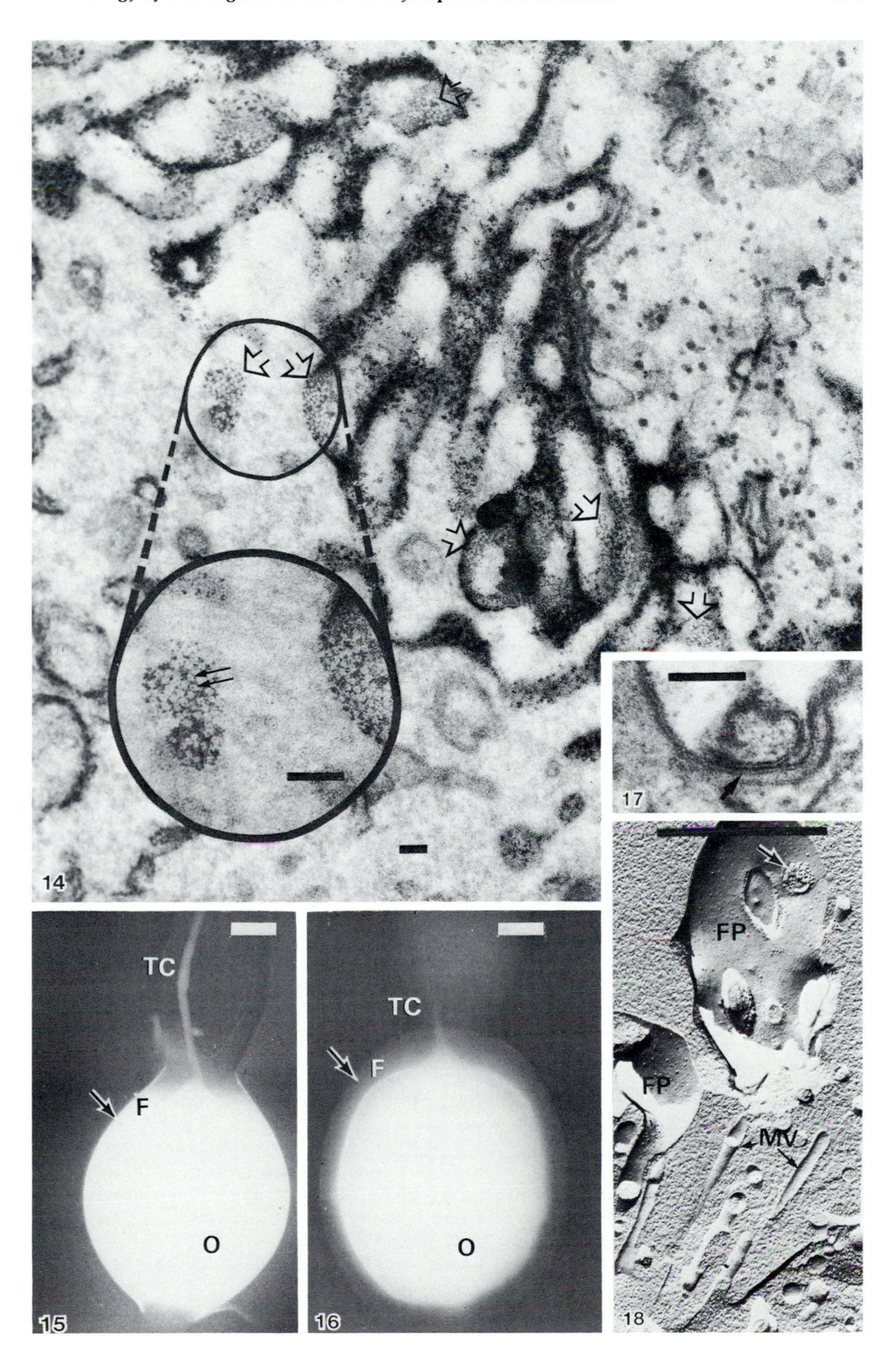
14
TC
F
O
15
TC
F
O
16
17
FP
FP
MV
18

a central hole 10–20 Å in diameter, which has been interpreted as the pore through which small molecules may pass (Unwin and Zampighi, 1980).

These features have now been seen at the interface between the follicle cells and the oocyte in *Rhodnius*. As shown in Figures 7 and 14, these two cells send interdigitating processes into the perioocytic space, and it is between these processes that gap junctions are formed. Transverse and *en face* views of lanthanum-permeated complexes are illustrated in Figures 14 and 17, and freeze-fracture images of junctional particles are illustrated in Figure 18. These configurations are seen in both late previtellogenic and vitellogenic follicles of *Rhodnius*, stages at which microinjected fluorescein has been found to move from the oocyte to the follicle cells, and it is therefore highly probable that they are in fact the intercellular complexes responsible for the exchange.

Of particular significance in this regard is the finding that gap junctions are the only junctional complexes that have been found between the oocyte and the follicle cells. This is in marked contrast to follicle cell–follicle cell interfaces of *Rhodnius*, which include not only gap junctions (Figures 20 and 21), but also prominent arrays of pleated septate desmosomes (Figures 19 and inset). and occasional plaque desmosomes. During the initiation of vitellogenesis these complexes, which earlier are generally distributed, become restricted to the intercellular struts, apparently so that the epithelium can maintain mechanical and physiological integrity even as it generates the system of intercellular spaces (Huebner and Injeyan, 1981).

In *Hyalophora*, the interface between the oocyte and follicle cells has a different configuration. Here the cells are separated by a thin vitelline envelope that prevents their contacting each other over most of the interface (Figures 3 and 22). The oocyte surface is folded, rather than microvillar, and the tips of the folds attach to the envelope with hemidesmosomes. Follicle cell microvilli attach in the same manner to the outer surface of the vitelline envelope, and the two cells thus remain largely separated from each other throughout vitellogenesis. The only direct contact between the two cells is provided by occasional microvilli that arise from among the folds of the oocyte surface (Woodruff, 1979), penetrate the vitelline envelope (Figure 3), and terminate against the main body of a follicle cell (Figures 22, 23). The termini are characteristically hemispherical domes with diameters of about 0.3 μm, and they fit into a correspondingly sized depression in the follicle cell surface. Since these interfaces are the only points of contact between the two cells, they necessarily contain the junctions responsible for electrical coupling and fluorescein exchange. The width of the zone between the two cell membranes is consistent

→

Figures 19–21. *Rhodnius prolixus.*

Figure 19 and inset. Lanthanum-treated follicle revealing the extensive junctional complexes between follicle cells. Note the regular pleated septate junctions in various planes (arrows). The inset show these septate junctions in freeze-fracture. Scale bars, 0.1 μm and 1 μm. (Inset from Huebner and Injeyan, 1981, courtesy of Academic Press.)

Figure 20. Transverse section of a gap junction (arrow) between follicle cells. Scale bar 0.1 μm. (From Huebner and Injeyan, 1981, courtesy of Academic Press.)

Figure 21. Follicle cell–follicle cell gap junction (arrow) as seen in a freeze-fracture preparation. Scale bar, 0.1 μm. (From Huebner and Injeyan, 1981, courtesy of Academic Press.)

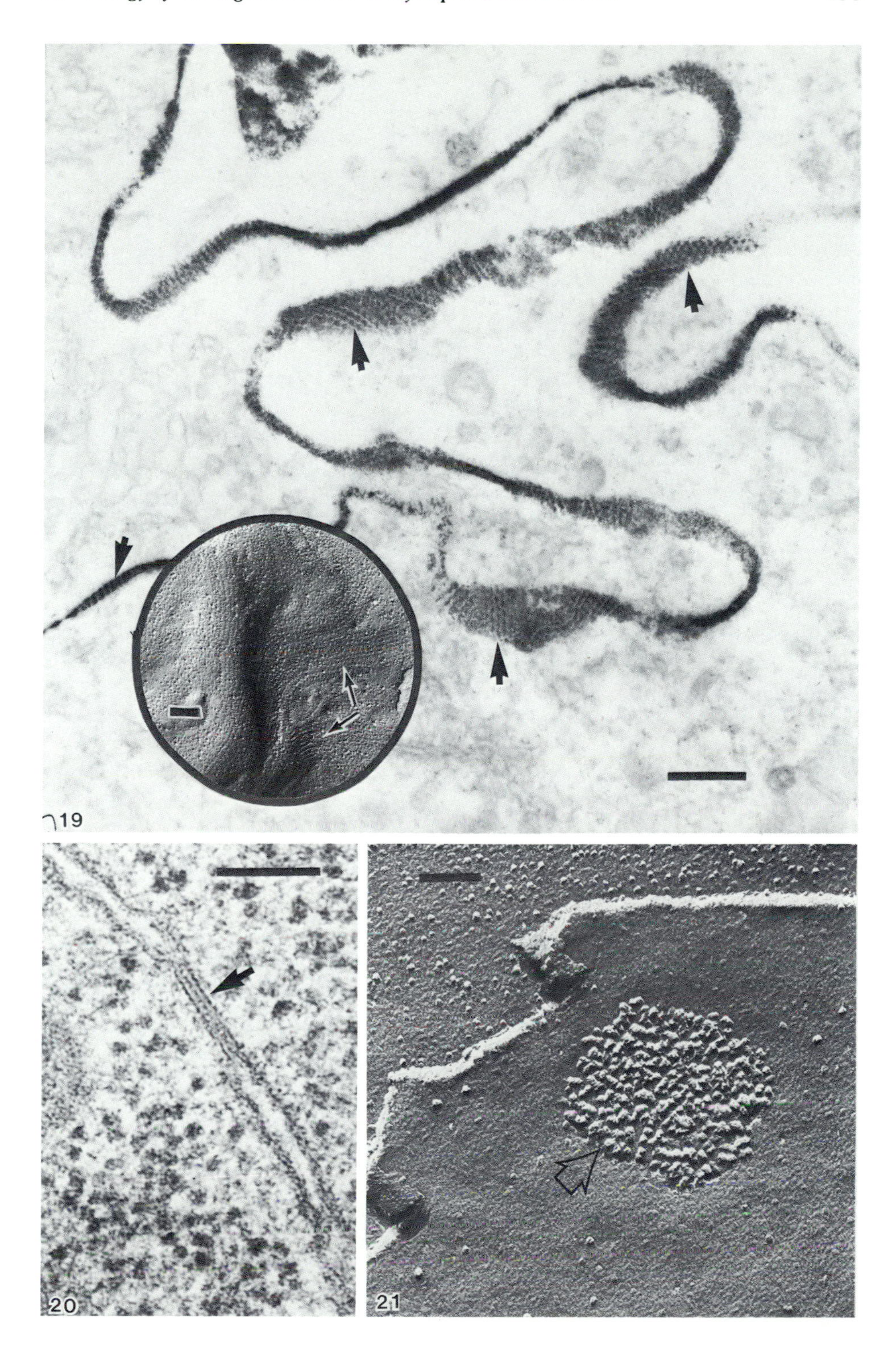
19
20
21

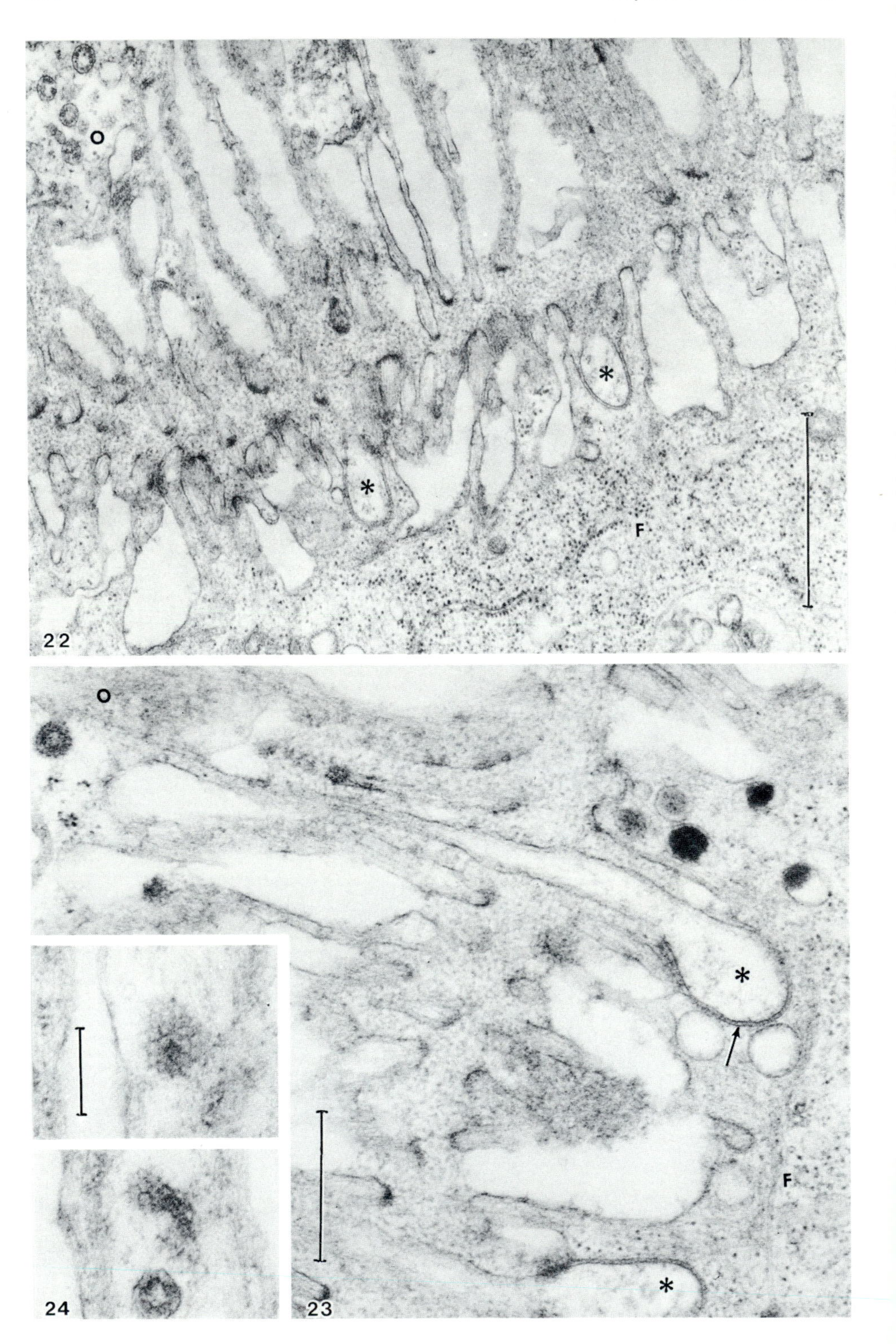
O
*
*
*
F
22
O
*
*
F
24
23

with this interpretation (Figure 23), and though they have not yet been seen in lanthanum-stained or freeze-fracture preparations, it is possible that the domes will prove to be conventional gap junctions.

Domes may well be a general feature of germ cell–somatic cell contact among Lepidoptera, for Szöllösi and Marcaillou (1980) reported an identical configuration between spermatocytes and sheath cells in the spermatic cysts of *Anagasta kuehniella*. In this case, the fine structural attributes of gap junctions could be decisively visualized, and this adds support to the assumption that the domes will prove to contain gap junctions in *Hyalophora* follicles.

Except for the Lepidoptera, most insects display interdigitating cell processes like those of *Rhodnius* during vitellogenesis, and it may safely be predicted that high permeability junctional complexes will prove to be a general feature of insect follicles at this stage. These relations make possible a kind of cellular interaction that was not imagined several years ago. First clues to the functional significance of these junctions have already come from recent work on mouse ovaries, in which follicle cells and oocytes can be cultured separately or together. In this system, it has been shown that follicle cells pick up nucleotides from the medium, phosphorylate them, and release them to the oocyte (Heller and Schultz, 1980). Since transfer occurs only at stages when gap junctions are present, it is presumed that these are the routes of transport. In short, gap junctions have made much more concrete the older concept, temporarily abandoned in insects with the discovery of vitellogenin, that follicle cells materially abet the growth of the vitellogenic oocyte.

4. Receptor-Mediated Endocytosis

4.1. Early Background

A combination of molecular and ultrastructural studies had made clear by 1965 that receptor-mediated endocytosis accounts for yolk deposition in many insects. Endocytosis was first implicated in saturniid moths by the finding that foreign proteins injected into the hemocele were deposited in membrane-limited protein yolk spheres in the cytoplasm of the oocytes (Telfer, 1960). Selectivity was demonstrated by the finding that vitellogenin in particular was deposited in the yolk at least 300 times more readily than the foreign proteins. In these experiments, the endogenous synthesis of vitellogenin was ruled out by injecting *Hyalophora* protein into the hemolymph of *Antheraea polyphemus*

←

Figures 22–24. *Hyalophora cecropia.*
Figures 22 and 23. The interface between the oocyte (O) and a follicle cell (F) in follicles fixed after removal of the extracellular matrix. As in Figure 3, the vitelline envelope separates the two cells, which attach to it by hemidesmosomes. Oocyte processes that have penetrated the vitelline envelope indent the follicle cell surface to form 0.3-μm-diameter termini (*). Adjacent membranes in the termini are consistently separated by a 10- to 20-nm-wide space often seen to contain a periodically arranged material. Scale bar, 1 μm for Figure 22 and a 0.5 μm for Figure 23.
Figure 24. Two examples of the clathrin lattice seen in tangential sections of coated vesicles. Scale bar, 0.2 μm.

and measuring with species-specific antibodies to vitellogenin the amounts deposited in the yolk of the recipient.

Shortly thereafter, ultrastructural studies on a wide variety of insects showed that the oocyte surface generates coated pinocytotic vesicles during yolk formation (Kessel and Beams, 1963; Anderson, 1964; Bier and Ramamurty, 1964; Favard-Séreno, 1964; Roth and Porter, 1964; King and Aggarwal, 1965; Stay, 1965; Hopkins and King, 1966), and that these vesicles incorporate exogenously provided ferritin or horseradish peroxidase (Stay, 1965; Anderson, 1969; Mahowald, 1972; Giorgi and Jacob, 1977).

While there has been much recent and essential progress on the isolation and biochemistry of vitellogenin (Mundall and Law, 1979; Kunkel *et al.*, 1980; Chinzei *et al.*, 1981) and of lipophorin, the second most prominent constituent of the yolk (Chino *et al.*, 1969; Pattnaik *et al.*, 1979), the endocytotic process itself has been relatively neglected for the past ten years. In the field of receptor-mediated endocytosis, the baton has passed instead to a number of simplified vertebrate models, and our review will attempt to correlate the results of these studies with the ultrastructure of hemolymph protein transfer to the yolk spheres in oocytes.

4.2. Clathrin

Coated vesicles were first isolated by Pearse (1975; 1976) from pig brain homogenates, and the principle component of their cytoplasmic coat was shown to be protein with a subunit molecular weight of 180,000 d. Bowers (1964) had first shown with insect pericardial cells that, in tangential view, the structural element of the cytoplasmic coat is a hexagonal lattice; Pearse confirmed this configuration in negatively stained, isolated vesicles, and named the principle protein clathrin. Examples of the clathrin lattice are shown in tangential view in Figure 24 for *Hyalophora* and in Figure 32 for *Rhodnius*.

Transformations of the clathrin lattice during vesicle formation were examined by Heuser and Evans (1980) by deep freeze etching of cultured vertebrate fibroblasts. They found that patches of clathrin deposited on a planar area of cell surface have a predominantly hexagonal configuration, but that the hexagons are interspersed with pentagonal elements in coated pits and vesicles. Whether the hexagon-pentagon transformation is a force-generating process responsible for vesicle formation, as suggested by Kanesaki and Kadota (1969), is not as yet clear. (There have also been suggestions that coated pits may be associated with microfilaments, and this is confirmed in Figures 29 and 30, in which tannic-acid-stained cortical filaments appear to be associated with the coated pits of *Rhodnius*). Woodward and Roth (1978) were able to form clathrin baskets *in vitro* in the absence of cell membranes by lowering the pH of soluble clathrin obtained from isolated coated vesicles.

4.3. The Clathrin–Receptor Association

When a patch of clathrin appears on the inner surface of the oolemma, extracellular protein is already attached to the corresponding outer membrane surface, while under many conditions the adjacent membrane surfaces may be relatively clean (Figure 29). This configuration, which was initially described by Roth and Porter (1964), is especially clear in *Hyalophora* follicles that have been soaked in physiological saline prior to fixation in order to remove unabsorbed hemolymph proteins from the intercellular spaces (Figures 3, 33). The oocyte continues to pinocytose the follicle cell secretion, paravitellogenin, under these conditions (Anderson and Telfer, 1969), and the material adsorbed to the extracellular surfaces of the oocyte presumably reflects the distribution of this protein and of its receptor. Thus, either clathrin is itself a transmembrane protein with receptor sites on its extracellular moiety, or else the receptors are transmembrane, bifunctional protein with an affinity for clathrin at one end and for their extracellular ligand on the other end (Pearse, 1980).

Receptor characterization has thus far been reported only in vertebrate systems (reviewed by Kaplan, 1981). As an example, in the uptake of low density lipoprotein by human fibroblasts in tissue culture, the connection between high affinity binding and subsequent endocytosis has been especially clearly established (Anderson *et al.*, 1977; 1978). The binding constant is in this case 2.5×10^{-8} M (Brown and Goldstein, 1974), so the affinity of the receptor for its ligand is extremely high. The receptor is rapidly destroyed by extracellular proteases; it has recently been solubilized from adrenal tissue while still functionally intact (Mello *et al.*, 1980), so that isolation and characterization can be expected to follow shortly. In this case, the endocytosed receptor–ligand complex is delivered by membrane fusion to lysosomes, which destroy the protein, instead of to storage vesicles.

Perhaps more pertinent to insect vitellogenesis is the study by Woods and Roth (1979) of phosvitin receptors in isolated membrane preparations from hen's eggs. In this case also, high affinity has been demonstrated, with a binding constant of 10^{-6} M. Receptor–ligand complexes have been solubilized with detergent, so there is hope that characterization is iminent here also.

The transfer of precursor proteins to the yolk spheres occurs by vesicle fusion, and many configurations suggestive of this process are visible in the oocyte cortex (Figures 7, 25). Prior to this event, however, the clathrin–receptor–ligand complex undergoes the first stages of disassembly. As shown in Figures 2 and 26, the oocyte cortex contains two populations of vesicles, one that retains its clathrin coat and its lining of receptor-bound ligands, and a second in which the clathrin has disappeared and the internalized vitellogenic precursors are free in the lumen of the vesicle. This is the first step in a putative recycling process that should ultimately make clathrin, the receptors, and the membrane constituents of the vesicles, available for reincorporation into the cell surface (Figure 27). [The existence of recycling is strongly supported by the fact that vitellogenin, lipophorin, and other proteins derived from outside the oocyte, are the only proteins thus far recognized in significant quantities within the yolk spheres (Telfer *et al.*, 1981a). In particular, no yolk constituents have been found that are sufficiently abundant to have served as 1 : 1 receptors for vitellogenin.] One of the important challenges for the future will be to determine the conditions that are responsible for dissociation of the clathrin–receptor–ligand complex.

*
MV
V
Y
OL
Y
Y
25
OL
Y
26
28
*
*
29
30
27
*
31
32

Viewed in these terms, the fine structure of insect vitellogenesis raises the following molecular questions that may prove to be answerable in the near future. Are there separate receptors for vitellogenin, lipophorin, paravitellogenin, and other yolk precursor proteins, or does a single receptor react with a binding site that all precursors share in common? Does dissociation of the receptor–ligand complexes occur because of an ionic, pH, or other change within the vesicle, or because of something that happens to the clathrin on the outside? And finally, are the clathrin, receptors, and other membrane constituents that fail to appear in the yolk in fact reassembled at the oocyte surface for new rounds of endocytosis? With the recent studies on coated vesicles of vertebrates to serve as models, it can be hoped that these and other questions may prove to be answerable in insect oocytes with their extraordinary specializations for endocytosis.

4.4. The Tubular Elements

In addition to coated and uncoated vesicles, the cortex of vitellogenic oocytes characteristically contains large numbers of tubular elements (Figures 2, 36). These were first described by Favard-Sereno (1964) in *Acheta domestica*, in which they appear as branching canaliculi that are continuous with the membranes of the incipient yolk spheres. They appeared in this material to

---

Figures 25–32. *Rhodnius prolixus.*

Figure 25 and insets. Freeze-fracture preparation of the oocyte cortex during midvitellogenesis. Microvilli (MV) extend into the perioocytic space (*). The cortex contains many vesicles (V), and nascent yolk spheres (Y) of varying sizes. Note the fusion of small vesicles with the larger yolk sphere (large arrows). Scale bar, 1 μm. The insets illustrate various stages of vesicles either pinching off or fusing with the oolemma (OL). Scale bars, 0.1 μm. (From Huebner, 1981, courtesy of Academic Press.)

Figure 26. This thin section of the oocyte cortex illustrates that both coated (solid arrows) and uncoated vesicles (open arrows) are present. The uncoated vesicles vary in size and are able to fuse to form yolk spheres. Scale bar, 0.1 μm.

Figures 27–32. These thin sections illustrate various features of the coated pinocytotic vesicles of the material fixed in tannic acid, glutaraldehyde, and paraformaldehyde (Batten *et al.*, 1980), in order to preserve cytoplasmic filaments. (From Huebner, Batten, and Anderson, unpublished.) In all cases the scale bars represent 0.1 μm.

Figure 27. An uncoated vesicle either pinching off or fusing with the oolemma. This configuration, which might represent the recycling of cell membrane to the oolemma in *Rhodnius*, has not been reported in other insects.

Figure 28. An oblique section of the oolemma showing the clathrin lattice on a coated pit.

Figure 29. This section, which is perpendicular to the oolemma, shows two coated pits in an early stage of formation (*), and illustrates that the dense patches of extracellular material are coincident with the clathrin coats. Arrows indicate cytoplasmic filaments that are associated with the coated plaques.

Figure 30. A coated vesicle with cytoplasmic filaments attached at various points (arrows).

Figure 31. A large pinocytotic pit with an attenuated attachment to the oocyte surface (*). The classical bristlelike appearance of the clathrin basket (arrows) in transverse section is particularly clear.

Figure 32. Coated vesicles sectioned at various levels revealing both the bristlelike and lattice (arrows) aspects of the clathrin coat.

be the elements with which coated vesicles fuse in transmitting their exogenous proteins to the yolk spheres. Similar elements were seen by Bowers (1964) in *Myzus persicae* pericardial cells, and later in a number of additional insect oocytes, including *Hyalophora* (Stay, 1965; Telfer and Smith, 1970), *Periplaneta americana* (Anderson, 1969), and *Drosophila* (Mahowald, 1972). In *Hyalophora*, they are 300–450 Å in diameter and appear to consist of a unit membrane lined by a poorly staining material similar in appearance to the most intimate lining of the membranes of the coated vesicles (Figures 2, 36). In *Drosophila*, their diameter is around 500 Å, with an 85-Å unit membrane lined by, in this case, a more clearly defined material with a periodicity of 125 Å along the length of the tubule. In all cases there is a narrow lumen that opens directly into the yolk sphere to which the tubule is attached. Since the tubular elements are attached only to small or immature yolk spheres, they are best interpreted as the membrane residues of vesicles that have delivered their contents to the yolk and have not yet been shaved off for recycling to the oocyte surface. The necessity for such a mechanism is indicated by surface–volume considerations: the 10^6 coated vesicles required to deliver vitellogenin to a single yolk sphere in *Hyalophora*, for instance, would have about 100 times too much membrane to cover a single yolk sphere with the smooth, spherical shell that is characteristic of this system.

4.5. Experimental Modification

It is a curious fact that so few efforts have been made to explore the function of the tubular elements or to look for morphological evidences of later stages in the process of recycling. As a result of its early abandonment, the potential contributions of the insect vitellogenic cortex to cell biology have been only minimally exploited. While the newly aroused interest in clathrin and vitellogenin receptors may well reverse this neglect at the molecular level, other approaches are suggested by the startling degree to which the morphology of the vitellogenic cortex can be modified by altering the environment of the follicle. We have already seen an example in the effects of blood protein elution

Figures 33–36. *Hyalophora cecropia.*

Figure 33. A coated pit (arrow) of a follicle that had been soaked in physiological saline for 30 min to remove hemolymph proteins from the extracellular spaces. Preferential binding of extracellular protein, in this case most likely paravitellogenin, to regions transmembrane from the clathrin coat is illustrated by this figure. Scale bar, 0.2 μm.

Figure 34. A greatly elongated coated pit or vesicle produced during the reversal of trypan blue inhibition. The animal had been injected with trypan blue (5 mg/g live weight), held for 12 hr, and then injected with female pupal hemolymph (0.067 ml/g live weight) to reverse the effects of the dye. Fixation was 4 hr after the hemolymph injection. Scale bar, 0.5 μm.

Figure 35. Branching, membrane-limited masses interpreted as uncoated derivatives of vesicles such as that shown in Figure 34. These are seen only in the cortex of oocytes recovering from *in situ* trypan blue inhibition. Scale bar, 0.5 μm.

Figure 36. Tubular elements in the vitellogenic cortex appear as branching structures (long arrows), and are often attached to the membrane of a growing yolk sphere (*). The short arrow indicates a cross section. Scale bar, 0.5 μm.

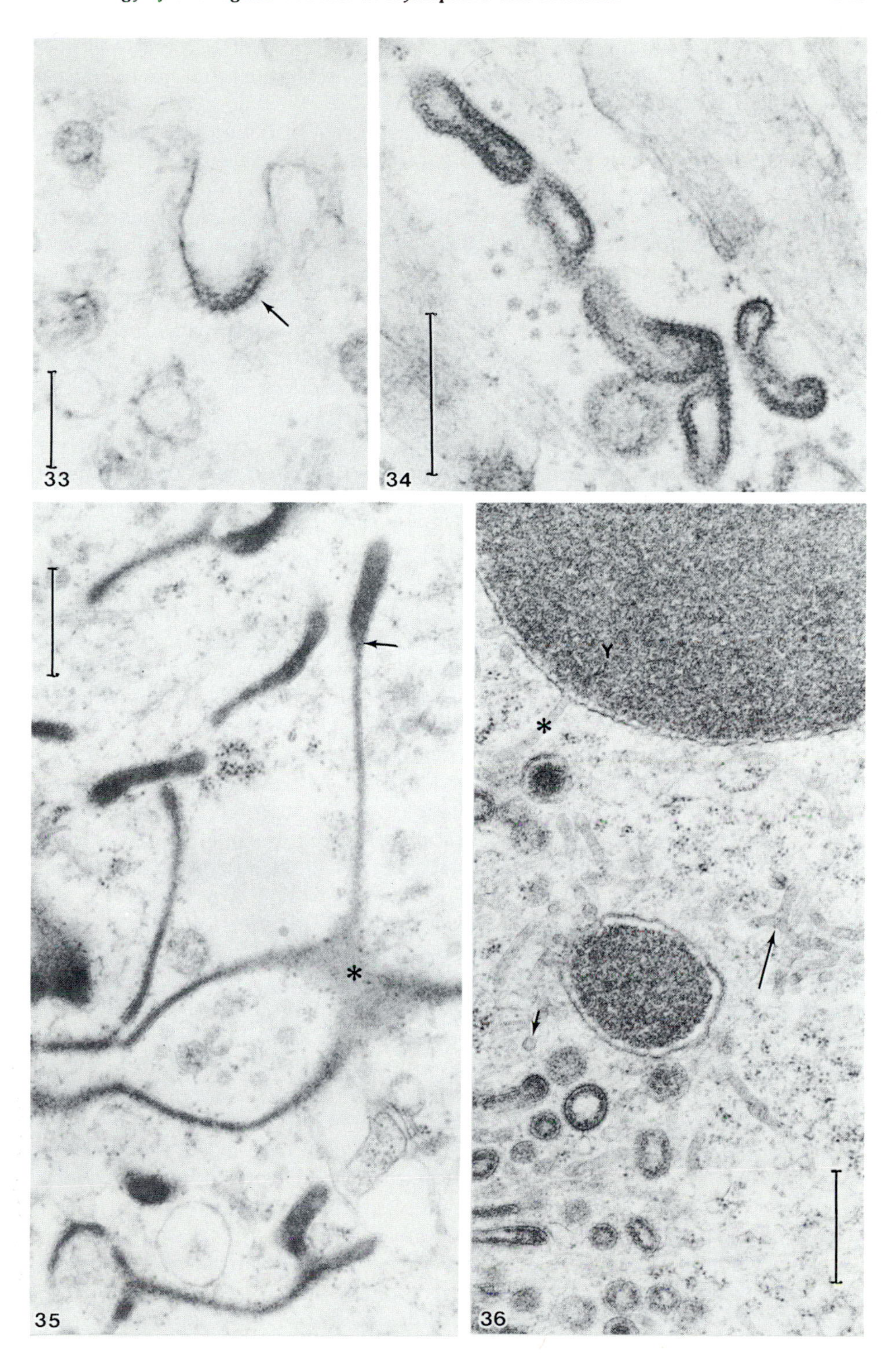
33
34
35
36
*
*
Y

on the intercellular matrix of the follicle. A second example is presented here in two micrographs showing the reorganization of the cortex that can be elicited by treating vitellogenic follicles with trypan blue.

This polyionic dye has long been recognized as having an affinity for endocytotic mechanisms, and it was this rationale that led Ramamurty (1964) to test its ability to be incorporated into yolk in *Panorpa communis*. Trypan blue proved indeed to be avidly accumulated by the vitellogenin uptake mechanism (Anderson and Telfer, 1970b; Wojc *et al.*, 1977), but at the same time it caused in *Hyalophora* a substantial reduction in the amount of yolk produced. Mid-vitellogenic follicles normally lay down a 100-μm-deep stratum of 20-μm-diameter yolk spheres every 24 hr is this species. But when trypan blue was injected into the hemocoele (5mg/g live weight), 5-μm-diameter yolk spheres were accrued, which formed a stratum well under 50 μm deep. Even more striking was the finding that this inhibition could be entirely repaired by injecting the animal 12–20 hr later with dialyzed samples of pupal hemolymph proteins (0.1 ml/g live weight). Within several hours after the injection, the miniature yolk spheres that contained the dye had grown to a nearly normal size and had expanded to form the usual 100 μm depth of yolk spheres that is formed in a 24 hr period. Either male or female blood had the same effect. Unlike normal yolk spheres, which stop growing within 3 hr, the inhibited yolk spheres were thus still within the endocytotic zone of the cortex and in a postion to incorporate new protein 12–20 hr after their formation. The response suggested that unutilized endocytotic elements had accumulated at the cell surface in extraordinary amounts, and that, when the system was provided with fresh proteins, they could be incorporated at a greatly accelerated rate.

Transmission electron microscopy revealed a cortical anatomy that agreed with this interpretation. The inhibited follicle had an abnormally diffuse intercellular matrix (suggesting that trypan blue had interfered either with protein binding or with fixation of the matrix), and at the oocyte surface there were very few pinocytotic pits. The yolk spheres were less dense than normal, so that they appeared to be deficient in content as well as in volume.

In follicles fixed several hours after hemolymph injection, two remarkable configurations were seen. Coated pits and vesicles now took the form of long, wormlike structures that were many times the size of normal endocytotic configurations (Figure 34). When the clathrin coat had been lost, they appeared as densely staining and branching structures lying deep in the cortex of the oocyte (Figure 35). It was as though the internalized vitellogenin concentrate had not been successfully transferred entirely to the yolk spheres and had remained in part lodged in swollen tubular elements. The normal processes of vesicle formation and fusion are thus grossly distorted when release from inhibition occurs, and it becomes clear that the normal shape of the endocytotic vesicles and tubular elements are to some degree the product of a particular balance between ligand and receptor availabilities.

There are many practical questions raised by observations such as these. Where are the backlog of clathrin and vitellogenin receptors localized during inhibition? How does trypan blue inhibit endocytosis? And how does the inhibited yolk sphere manage to remain receptive to new additions for as long as 24 hr, when normally growing yolk spheres are completed within 3 hr? These and other questions demonstrate that experimental interference with the normal process of vitellogenin uptake can contribute much to the understanding of the endocytotic processes that make the oocyte such an extraordinary specialist in this mode of growth.

5. Summary and Developmental Implications

The special differentiations of the vitellogenic follicle treated in this review include: (1) its ability to generate and regulate a meshwork of intercellular spaces through which protein yolk precursors are made available to the oocyte surface; (2) the production of heterologous gap junctions whose existence raises the possibility of a direct contribution of the follicle cells to the growth of the ooplasm by the uptake, metabolic conversion, and intercytoplasmic transfer of small molecules; and (3) the extraordinary specialization of the oocyte membrane for receptor-mediated endocytosis, with its implied synthesis, assembly, and turnover of vitellogenin–receptor–clathrin complexes and other membrane components.

In addition to their utility in the study of intercellular transport mechanisms, these three phenomena provide an essential physiological context for a number of problems in insect development. Three examples seem particularly appropriate for mention at this time:

In all insects, a discrete period of previtellogenic development must be completed before a follicle initiates blood protein uptake, and there is an equally well defined termination of the process. Questions concerning how vitellogenesis is activated and terminated must in the future be phrased in terms of the cellular mechanisms that we have described. Contrasts will be particularly interesting between the two insects treated in this review. In unfed *Rhodnius* adults, the largest follicle in each ovariole has completed its previtellogenic development and remains in a state of arrest until protein uptake is activated by a blood meal and the subsequent secretion of JH. In *Hyalophora,* by contrast, the transformations into and out of vitellogenesis occur spontaneously in each follicle as it reaches the appropriate stages of development. What is the defect that stops development in the unfed *Rhodnius?* Does JH activate the synthesis of clathrin, vitellogenin receptors, and the other molecular tools of protein uptake, or simply their insertion into functional sites? And how does *Hyalophora* manage to bypass this deficit in the absence of JH?

A second developmental issue concerns the origins and deposition of the conjugated ecdysteroids that accumulate in extraordinary amounts in the eggs of many insects. There is preliminary evidence from *Locusta migratoria* that these substances, which appear to be converted to active molting hormone during embryonic development, may be synthesized by the follicle cells (Lageaux *et al.*, 1977). Conjugated ecdysteroids, with their peculiarly hydrophilic characteristics, should be insoluble in cell membranes, and thus should not readily move directly from the follicle cells to the oocyte without special transport provisions. One possibility suggested by the information in this review is that they are transmitted directly into the oocyte *via* the gap junctions, in which case they might be expected to be stored in the ooplasm. Alternatively, they could be secreted by the follicle cells into the intercellular spaces, and taken in by endocytosis, a route that would most likely lead to storage in the yolk spheres. In this event, precautions to prevent their escape to the hemolymph would be expected, and for this purpose an oocyte surface receptor equivalent to that for vitellogenin might be expected.

Finally, the vitellogenic follicle plays a special role in producing the substratum for embryogenesis, and this lends an added dimension of significance to the anaylsis of its ultrastructure. But, while the oocyte cortex in particular will at a later time become the site of blastoderm and germ band formation, it is difficult to conceive of a way in which the localization of factors responsible for orienting and organizing the embryo could be preserved in the cortex during the physical turmoil of vitellogenesis. It will instead be more plausible to focus the search for the origins of embryonic localizations in insect eggs on the period of cortical reorganization that immediately follows vitellogenesis. Substantial changes occur in the structure of the cortex during the terminal growth phase of the oocyte in *Hyalophora* (Telfer and Anderson, 1968; Telfer and Smith, 1970), and this supports the proposal that the switch from vitellogenesis to embryogenesis entails a fundamental retooling of the cortex.

References

Abu-Hakima, R., and Davey, K. G., 1977a, The action of juvenile hormone on the follicle cells of *Rhodnius prolixus*: The importance of volume change, *J. Exp. Biol.* **69:**33–44.

Abu-Hakima, R., and Davey, K. G., 1977b, The effects of hormones and inhibitors of macromolecular synthesis on the follicle cells of *Rhodnius, J. Insect Physiol.* **23:**913–917.

Abu-Hakima, R., and Davey, K. G., 1977c, The action of juvenile hormone on the follicle cells of *Rhodnius prolixus*: The effect of colchicine and cytochalasin B, *Gen. Comp. Endocrinol.* **32:**360–370.

Abu-Hakima, R., and Davey, K. G., 1979, A possible relationship between ouabain-sensitive (Na^+-K^+) dependent ATPase and the effects of juvenile hormone on the follicle cells of *Rhodnius prolixus, Insect Biochem.* **9:**195–198.

Anderson, E., 1964, Oocyte differentiation and vitellogenesis in the roach *Periplaneta americana, J. Cell Biol.* **20:**131–155.

Anderson, E., 1969, Oogenesis in the cockroach *Periplaneta americana*, with special reference to the specialization of the oolema and the fate of coated vesicles, *J. Microsc.* (Oxford) **8:**721–738.

Anderson, L. M., and Telfer, W. H., 1969, A follicle cell contribution to the yolk spheres of moth oocytes, *Tissue Cell* **1:**633–644.

Anderson, L. M., and Telfer, W. H., 1970a, Extracellular concentrating of proteins in the Cecropia moth follicle, *J. Cell Physiol.* **76:**37–53.

Anderson, L. M., and Telfer, W. H., 1970b, Trypan blue inhibition of yolk deposition–a clue to follicle cell function in the Cecropia moth, *J. Embryol. Exp. Morphol.* **23:**35–52.

Anderson, R. G. W., Brown, M. S., and Goldstein, J. L., 1977, Role of the coated endocytotic vesicle in the uptake of receptor-bound low density lipoprotein in human fibroblasts, *Cell* **10:**351–364.

Anderson, R. G. W., Vasile, E., Mello, R. J., Brown, M. S., and Goldstein, J. L., 1978, Immunocytochemical visualization of coated pits and vesicles in human fibroblasts: Relation to low density lipoprotein receptor distribution, *Cell* **15:**919–933.

Bast, R. E., and Telfer, W. H., 1976, Follicle cell protein synthesis and its contribution to the yolk of the Cecropia moth oocyte, *Develop. Biol.* **52**:83–97.

Batten, B. E., Aalberg, J. J., and Anderson, E., 1980, The cytoplasmic filamentous network in cultured ovarian granulosa cells, *Cell* **21**:885–895.

Bier, K., and Ramamurty, P. S., 1964, Elektronoptische Unterzuchungen zur Einlagerung der Dotterproteine in die Oocyte, *Naturwis.* **51**:223–224.

Bilinski, S., 1977, Oogenesis in *Acerentomen gallicium* Jonescu (Protura). *Cell Tissue Res.* **179**: 401–412.

Bowers, B., 1964, Coated vesicles in the pericardial cells of the aphid (*Myzus persicae* Sulz), *Protoplasma* **59**:351–367.

Bownes, M., 1980, The use of yolk protein variations in *Drosophila* species to analyse the control of vitellogenesis, *Differentiation* **16**:109–116.

Brown, M. S., and Goldstein, J. L., 1974, Familial hypercholesterolemia: Defective binding of lipoproteins to cultured fibroblasts associated with impaired regulation of 3-hydroxy-3-methyl-glutaryl coenzyme A reductase activity, *Proc. Nat. Acad. Sci. USA* **71**:788–792.

Caveney, S., and Berdan, R., 1982, Selectivity in junctional coupling between cells of insect tissues In this volume.

Caveney, S., and Podgorsky, C., 1975, Intercellular communication in a positional field: Ultrastructural correlates and tracer analysis of communication between insect epidermal cells, *Tissue Cell* **7**:559–574.

Chino, H., Murokami, S., and Harashima, K., 1969, Diglyceride-carrying proteins in insect hemolymph, *Biochim. Biophys. Acta* **176**:1–26.

Chinzei, Y., Chino, H., and Wyatt, G. R., 1981, Purification and properties of vitellogenin and vitellin from *Locusta migratoria. Insect Biochem.* **11**:1–8.

Comper, W. D., and Laurent, T. C., 1978, Physiological function of connective tissue polysaccharides, *Physiol. Rev.* **58**:255–315.

Davey, K. G., and Huebner, E., 1974, The response of the follicle cells of *Rhodnius prolixus* to juvenile hormone and antigonodotropin in vitro, *Can. J. Zool.* **52**:1407–1412.

Engelmann, F., 1979, Insect vitellogenin: Identification, biosynthesis, and role in vitellogenesis, *Adv. Insect Physiol.* **14**:49–108.

Favard-Séréno, C., 1964, Phénomone de pinocytose au cours de la vitellogenese proteique chez le Grillon (Orthoptere), *J. Miscrosc.* (Paris) **3**:323–338.

Gilula, N. B., 1980, Cell-to-cell communication and development, *Symp. Soc. Develop. Biol.* **38**: 23–44.

Gilula, N. B., Reeves, O. R., and Steinbach, A., 1972, Metabolic coupling, ionic coupling, and cell contacts, *Nature* (London) **235**:262–265.

Giorgi, F., and Jacob, J., 1977, Recent findings on oogenesis in *Drosophila melanogaster.* II. Further evidence on the origins of yolk platelets, *J. Embryol. Morphol.* **38**:125–137.

Glass, H., and Emmerich, H., 1981, Properties of two follicle proteins and their possible role for vitellogenesis in the African locust, *Wilhelm Roux's Arch. Dev. Biol.* **190**:22–26.

Hagedorn, H. H., and Kunkel, J. G., 1979, Vitellogenin and vitellin in insects, *Annu. Rev. Entomol.* **24**:475–505.

Heller, D. T., and Schultz, R. M., 1980, Ribonucleoside metabolism by mouse oocytes: Metabolic cooperativity between fully-grown oocyte and cumulus cells, *J. Exp. Zool.* **214**:355–364.

Heuser, J., and Evans, L., 1980, Three-dimensional visualization of coated vesicle formation in fibroblasts, *J. Cell Biol.* **84**:560–583.

Hopkins, C. R., and King, P. E., 1966, An electron microscopical and histochemical study of the oocyte periphery in *Bombus terrestris* during vitellogenesis, *J. Cell Sci.* **1**:201–216.

Huebner, E., 1976, Experimental modulation of the follicular epithelium in *Rhodnius* oocytes by juvenile hormone and other agents, *J. Cell Biol.* **70**:251A.

Huebner, E., 1981, Oocyte-follicle cell interaction during normal oogenesis and atresia in an insect, *J. Ultrastruct. Res.* **74**:95–104.

Huebner, E., and Anderson, E., 1972a, A cytological study of the ovary of *Rhodnius prolixus.* I. The ontogeny of the follicular epithelium, *J. Morphol.* **136**:459–493.

Huebner, E., and Anderson, E., 1972b, A cytological study of the ovary of *Rhodnius prolixus* III. Cytoarchitecture and development of the trophic chamber, *J. Morphol.* **138**:1–40.

Huebner, E., and Injeyan, H. S., 1980, Patency of the follicular epithelium in *Rhodnius prolixus:* A re-examination of the hormone response and technique refinement, *Can. J. Zool.* **58:** 1617–1625.

Huebner, E., and Injeyan, H., 1981, Follicular modulation during oocyte development in an insect: Formation and modification of septate and gap junctions, *Develop. Biol.* **83:**101–113.

Huebner, E., Tobe, S. S., and Davey, K. G., 1975, Structural and functional dynamics of oogenesis in *Glossina austeni* vitellogenesis with special reference to the follicular epithelium, *Tissue Cell* **7:**535–558.

Ilenchuk, T. T., 1980, Juvenile hormone stimulates (Na^+-K^+) ATPase in vitellogenic follicle cells of *Rhodnius prolixus, Amer. Zool.* **20:**900.

Kanesaki, T., and Kadota, K., 1969, The "vesicle in a basket." A morphological study of the coated vesicle isolated from the nerve endings of the guinea pig brain with special reference to the mechanism of membrane movement, *J. Cell Biol.* **42:**202–220.

Kaplan, J., 1981, Polypeptide-binding membrane receptors: Analysis and classification, *Science* **212:**14–20.

Kelly, T. J., and Telfer, W. H., 1979. The function of the follicular epithelium in vitellogenic *Oncopeltus* follicles, *Tissue Cell* **11:**663–672.

Kessel, R. G., and Beams, H. W., 1963, Micropinocytosis and yolk formation in oocytes, *Exp. Cell Res.* **30:**440–443.

King, R. C., and Aggarwal, S. K., 1965, Oogenesis in *Hyalophora cecropia, Growth* **29:**17–83.

Kunkel, J. G., Shepard, G. L., McCarthy, R. A., Ethier, D. B., and Nordin, J. H., 1980, Concanavalin A reactivity and carbohydrate structure of *Blattella germanica* vitellin, *Insect Biochim.* **10:**703–714.

Lageaux, M., Hirn, M., and Hoffman, J., 1977, Ecdysone during ovarian development in *Locusta migratoria, J. Insect Physiol.* **23:**109–119.

Loewenstein, W., 1978, Junctional intercellular communication and the control of growth, *Biochim. Biophys. Acta* **560:**1–65.

Loewenstein, W., and Kanno, Y., 1964, Studies on epithelial (gland) cell junction. I. Modification of surface membrane permeability, *J. Cell Biol.* **22:**565–586.

Lutz, D., and Huebner, E., 1981, Development of nurse cell–oocyte interactions in the insect telotrophic ovary of *Rhodnius, Tissue Cell.* **13:**321–335.

Mahowald, A. P., 1972, Ultrastructural observations on oogenesis in *Drosophila, J. Morphol.* **137:**29–48.

Melius, M. E., and Telfer, W. H., 1969, An autoradiographic analysis of yolk deposition in the cortex of the Cecropia moth oocyte, *J. Morphol.* **129:**1–16.

Mello, R. J., Brown, M. S., Goldstein, J. L., and Anderson, R. G. W., 1980, LDL receptors in coated vesicles isolated from bovine adrenal cortex: Binding sites unmasked by detergent treatment, *Cell* **20:**829–837.

Mundall, E. C., and Law, J. H. 1979, Physical and chemical characterization of vitellogenin from the hemolymph and eggs of the tobacco hornworm, *Comp. Biochem. Physiol.* **63B:**459–465.

Noirot-Timothee, C., and Noirot, C., 1980, Septate and scalariform junctions in arthropods, *Int. Rev. Cytol.* **63:**97–140.

Pattnaik, N. M., Mundall, E. C., Trambust, B. C., Law, J. H., and Kezdy, F. J., 1979, Isolation and characterization of a larval lipoprotein from the hemolymph of *Manduca sexta, Comp. Biochem. Physiol.* **63B:**469–476.

Pearse, B. M. F., 1975, Coated vesicles from pig brain: Purification and biochemical characterization, *J. Mol. Biol.* **97:**93–98.

Pearse, B. M. F., 1976, Clathrin: A unique protein associated with intercellular transfer of membrane by coated vesicles, *Proc. Nat. Acad. Sci. USA* **73:**1255–1259.

Pearse, B. M. F., 1980, Coated vesicles, *Trends Biochem. Sci.* **5:**131–134.

Pratt, G., and Davey, K. G., 1972, The corpus allatum and oogenesis in *Rhodnius prolixus* (Stal), *J. Exp. Biol.* **56:**201–214.

Ramamurty, P. S., 1964, On the contribution of the follicle epithelium to the deposition of yolk in the oocyte of *Panorpa communis, Exp. Cell Res.* **33:**601–604.

Revel, J. P., and Karnovsky, M. J., 1967, Hexagonal array of subunits in intercellular junctions of the mouse heart and liver, *J. Cell Biol.* **33:**C7–C12.

Revel, J. P., Yee, A. G., and Hudspeth, A. J., 1971, Gap junctions between electrotonically coupled cells in tissue culture and in brown fat, *Proc. Nat. Acad. Sci. USA* **68:**2924–2927.

Roth, T. F., and Porter, K. R., 1964, Yolk protein uptake in the mosquito *Aedes egypti* L., *J. Cell Biol.* **20:**313–332.

Rubenstein, E. C., 1979, The role of an epithelial occlusion zone in the termination of vitellogenesis in *Hyalophora cecropia* ovarian follicles. *Develop. Biol.* **71:**115–127.

Satir, P., and Gilula, N. B., 1973, The fine structure of membranes and intercellular communication in insects, *Annu. Rev. Entomol.* **18:**143–166.

Stay, B., 1965, Protein uptake in the oocytes of the Cecropia moth, *J. Cell Biol.* **26:**49–62.

Szöllösi, A., and Marcaillou, C., 1980, Gap junctions between germ cells and somatic cells in the testis of the moth *Anagasta keuhniella* (Insecta: Lepidoptera), *Cell Tissue Res.* **213:**137–147.

Telfer, W. H., 1960, The selective accumulation of blood proteins by the oocytes of saturniid moths, *Biol. Bull.* **118:**338–351.

Telfer, W. H., 1961, The route of entry and localization of blood proteins in the oocytes of saturniid moths, *J. Biophys. Biochem. Cytol.* **9:**747–759.

Telfer, W. H., 1965, The mechanism and control of yolk formation, *Annu. Rev. Entomol.* **10:** 161–184.

Telfer, W. H., 1979, Sulfate and glucosamine labelling of the intercellular matrix in vitellogenic follicles of a moth, *Wilhelm Roux's Arch. Dev. Biol.* **185:**347–362.

Telfer, W. H., and Anderson, L. M., 1968, Functional transformation accompanying the initiation of a terminal growth phase in the Cecropia moth oocyte, *Develop. Biol.* **17:**512–535.

Telfer, W. H., and Smith, D. S., 1970, Aspects of egg formation. In *Insect Ultrastructure*, edited by C. Neville, pp. 117–133, Blackwell Scientific Publications, Oxford and Edinburgh.

Telfer, W. H., Rubenstein, E. C., and Pan, M. L., 1981a, How the ovary makes yolk in *Hyalophora*. In *Regulation of Insect Development and Behavior*, edited by F. Sehnal, A. Zabza, J. J. Menn, and B. Cymborowski, pp. 637–654, Wroclaw Technical Univ., Wroclaw.

Telfer, W. H., Woodruff, R. I., and Huebner, E., 1981b, Electrical polarity and cellular differentiation in meroistic ovaries, *Amer. Zool.* **21:**675–686.

Unwin, P. N. T., and Zampighi, G., 1980, Structure of the junction between communicating cells, *Nature* (London) **283:**545–549.

Warner, A. E., and Lawrence, P. A., 1973, Electrical coupling across developmental boundaries in insect epidermis, *Nature* (London) **245:**47–48.

Wojc, E., Bakker-Grunwald, T., and Applebaum, S. W., 1977, Binding and uptake of trypan blue by developing oocytes of *Locusta migratoria migratorioides*, *J. Embryol. Exp. Morphol.* **37:**1–11.

Wollberg, Z., Cohen, E., and Kalina, M., 1976, Electrical properties of developing oocytes of the migratory locust, *Locusta migratoria*, *J. Cell Physiol.* **88:**145–158.

Woodruff, R. I., 1979, Electrotonic junctions in Cecropia moth ovaries, *Develop. Biol.* **69:**281–295.

Woodruff, R. I., and Telfer, W. H., 1980, Electrophoresis of proteins in intercellular bridges, *Nature* (London) **286:**84–86.

Woods, J. W., and Roth, T. F., 1979, Selective protein transport: Characterization and solubilization of the phosvitin receptor from chicken oocytes, *J. Supramol. Struct.* **12:**491–504.

Woodward, M. P., and Roth, T. F., 1978, Coated vesicles: Characterization, selective dissociation and reassembly, *Proc. Nat. Acad. Sci. USA* **75:**4394–4398.

6

Order and Defects in the Silkmoth Chorion, A Biological Analogue of a Cholesteric Liquid Crystal

GRACE DANE MAZUR, JEROME C. REGIER, AND FOTIS C. KAFATOS

". . . all life is perhaps a knot, a tangle, a blemish in the eternal smoothness."

E. M. Forster, *A Room With A View* (1908)

1. Introduction

1.1. Analogies between Chorion and Cuticle

The silkmoth chorion or eggshell is a strong, extracellular protective covering that exhibits complex physiology, chemistry, and morphology. Analogous, in many ways, to the cuticle, the chorion protects the egg and developing embryo from physical assaults of predators, as well as from the more passive

GRACE DANE MAZUR AND FOTIS C. KAFATOS • Department of Cellular and Developmental Biology, The Biological Laboratories, Harvard University, 16 Divinity Avenue, Cambridge, Massachusetts 02138, USA.
JEROME C. REGIER • Department of Biochemistry and Molecular and Cellular Biology, Northwestern University, 2153 Sheridan Road, Evanston, Illinois 60201, USA.

dangers of dessication and drowning, while permitting air exchange and the penetration of sperm. The chorion is produced by the monolayer of follicular epithelial cells, as the cuticle is produced by the monolayer of epidermis.

While the literature on cuticle is rich and of long standing (see Richards, 1951; Neville, 1975; Hepburn, 1976, for reviews), chorion has more recently become the focus of active investigations (Kafatos *et al.*, 1977; Kafatos, 1980). Both chorion and cuticle are intriguing for their high degree of order and for the striking patterns of defects that punctuate it. Both the ordered structure and the faults are presumed to have active physiological functions. The structure and morphogenesis of chorion and cuticle are best studied with reference to the pioneering work of Bouligand, who showed that their basic architecture is that of a cholesteric liquid crystal, and that geometric analysis of liquid crystals is a useful and necessary frame of reference for an understanding of biological materials built on this plan (Bouligand, 1965; 1969; 1972a; 1975). As Bouligand's work is not sufficiently known in the biological literature, we will review it briefly to lay the foundation for our discussion of chorion.

Both cuticle and silkmoth chorion consist of a large number of proteins, synthesized according to temporal and spatial developmental programs (Neville, 1975; Hackman and Goldberg, 1976; Kafatos *et al.*, 1977; Mazur *et al.*, 1980; Regier *et al.*, 1980; Willis *et al.*, 1981). In the chorion, 186 distinct polypeptides have been resolved by two-dimensional gel electrophoresis. Sequence analysis of nineteen purified chorion proteins and seventeen distinct chorion-encoding cDNA and chromosomal clones has established that distinct chorion proteins are evolutionarily related and are encoded by at least three district multigene families (Regier *et al.*, 1978a; Rodakis, 1978; Jones *et al.*, 1979; Tsitilou *et al.*, 1980; Hamodrakas, *et al.*, 1981). The various chorion proteins are synthesized over different periods of choriogenesis, and at widely different rates (Paul and Kafatos, 1975; Kafatos *et al.*, 1977). Major differences in the temporal and spatial program of protein synthesis appear to correspond to different modes of growth of the chorion (Mazur *et al.*, 1980; Regier *et al.*, 1980; Mazur *et al.*, 1982), while minor differences may be reflected in the subtly different structure of the circumferential layers of the chorion (see below).

The silkmoth cuticle, like the chorion, is biochemically complex (Hackman, 1974; Regier *et al.*, 1978b; Willis *et al.*, 1981). Of fifty polypeptides resolved by gel isoelectric focusing, twelve have been isolated and analyzed; their compositional similarities suggest that cuticle proteins are also evolutionarily related and encoded by one or more multigene families (Willis *et al.*, 1981). The morphological differences in horizontal layers are interpreted as reflecting differences in timing of synthesis of the various cuticle proteins (Neville, 1975; Willis *et al.*, 1981). Distinct spatial programs of synthesis are reflected in the consistent differences in amino acid composition between various anatomical regions of the cuticle (Willis *et al.*, 1981).

In both cuticle and chorion, the bulk of the structure consists of fibrils in lamellar arrangement (Bouligand, 1965; Telfer and Smith, 1970; Weis-Fogh, 1970; Regier *et al.*, 1982). Some of these fibrils are deposited by apposition, layer upon layer. In addition, however, later-synthesized components permeate the structure. In part, this penetration occurs by means of specific conduits:

wax canals, pore canals, and dermal glands in cuticle (Neville, 1975) and aeropyles in chorion (Hinton, 1969; 1981; Kafatos *et al.*, 1977). In part, however, percolation of new components occurs between already formed elements. Late permeation leads to obscuring of the fibrils in both cuticle (Neville, 1975) and chorion (Smith *et al.*, 1971; Kafatos *et al.*, 1977; Mazur *et al.*, 1982; Regier *et al.*, 1982). In crustacean cuticle, late permeation also leads to calcification (Bouligand, 1970; Giraud, 1977).

Despite these analogies between chorion and cuticle, important differences also exist. Silkmoth chorion contains no chitin, and is almost entirely composed of protein (Smith *et al.*, 1971), whereas cuticle contains lipids and large amounts of chitin.

Although both structures are secreted by the apical surface of a monolayer of epithelial cells, that surface is directed inwards in the follicle and outwards in the cuticular epidermis. Thus, chorion is secreted inwards, around the oocyte, and when the follicular epithelium is shed at ovulation the chorion surface that is revealed is the most recently formed surface. By contrast, the outer layer of the cuticle is that which was deposited first. That outer layer (epicuticle) is structurally distinct and has been implicated in the control of water loss (Neville, 1975). The analogous, early-secreted vitelline membrane (Telfer and Smith, 1970) of the follicle is internal, interposed between oocyte and chorion.

1.2. Chorion Morphology

The surface of the silkmoth chorion is richly ornamented with a polygonal network corresponding to the boundaries of secretory cells. In *Antheraea polyphemus*, the species on which we shall focus here, four different regions of the mature chorion are distinguished by differences in cell imprints (Figure 1a,b) (Regier *et al.*, 1980). The flattened sides of the eggshell show a simple polygonal network of ridges punctuated by holes at points corresponding to three-cell junctions. These holes, which are called aeropyles, pierce the chorion and serve in its system of air exchange. A second region, along the longest meridian of the egg, is the "aeropyle crown" region, where each aeropyle opening is surrounded by a tall, many-pronged crown. The aeropyle crown region is bisected by the third region, the meridional "stripe," which contains neither aeropyles nor crowns. At the anterior pole, the stripe leads to the fourth region, the micropyle, which is also devoid of aeropyles and crowns, but contains distinctive cell imprints and the sole point of entry for sperm.

The chorion surface patterns show interspecies variation. While *A. polyphemus* has only a restricted region of aeropyle crowns, *A. pernyi* has crowns at all three-cell junctions (except in the micropyle region), and *Hyalophora cecropia* and *Bombyx mori* have no crowns at all (Figure 1g) (Sakaguchi *et al.*, 1973; Kafatos *et al.*, 1977; Hinton, 1981). The presence and shape of the ridges, and the size and position of the aeropyle channels, also vary between species (cf. Figures 1b,g) (Hinton, 1969; 1981).

The interior architecture of all silkmoth chorions examined to date is basically very similar. All contain a low colonnade (trabecular layer) covered by a very much thicker lamellar roof, although in detail these structures vary widely (cf. Figures 1d,h) (Furneaux and Mackay, 1972). In *A. polyphemus,* the innermost, nonlamellar trabecular layer is very thin (0.4 μm), and consists of columns surrounded by air space, standing on a very thin floor that is directly apposed to the vitelline membrane (Figures 1c,d, 7a,b,e) On top of the trabecular layer lie around sixty lamellae that can be classified into four circumferential layers on the basis of thickness, angle, and patterns formed by holes and lamellar faults. Adjacent to the trabecular layer are twenty-six thin (0.33 μm) lamellae of the *inner lamellar layer,* then eleven thick disrupted lamellae of the *holey layer,* six lamellae (0.5 μm) of the *outer lamellar layer,* and finally, jutting out on an oblique angle to the chorion surface, the sixteen thick lamellae of the *oblique layer.*

In oblique sections of immature chorion, the lamellae appear as stacked rows of parallel arcs (Figure 1e), while in perpendicular sections the arcs disappear, leaving alternating bands of lines and dots (Figure 1f). Similar arced patterns have been seen in oblique sections of many different arthropod cuticles, as well as of other totally unrelated biological materials such as dinoflagellate chromosomes (Bouligand, 1965; 1972a; 1978a). As explained by Bouligand's "twisted model", the arcs result not from curvature of the component fibrils, but from their helicoidal arrangement. In any given horizontal plane, the fibrils are straight and lie parallel. In a neighboring plane, the fibrils are again straight and parallel, but their direction has rotated by an angle that is proportional to the distance separating the two planes (Figure 2e). The structure is thus that of a continuous helical plywood; it has been called a 'universal plywood' (Neville, 1975), to indicate that the grain rotates progressively in successive layers through all angles up to 360°, while in a conventional plywood the direction of the wood grain alternates successively between two orthogonal directions. Each row of arcs is termed a *lamella* and corresponds to a 180° rotation of the direction of the fibrils. Lamellae are conventionally defined as starting and finishing where the fibrils are parallel to the plane of section. Zones where fibrils have this orientation generally, though not invariably, appear darker in electron micrographs (e.g., Figures 1e,f, 3b, 5b). The Bouligand model and its consequences will be further described below.

1.3. Chorion Morphogenesis

Chorion morphogenesis has been studied in some detail (King and Aggarwal, 1965; Telfer and Smith, 1970; Smith *et al.,* 1971; Mazur *et al.,* 1980). We have described four different modes of growth that occur during formation of the lamellar chorion, following the construction of the trabecular layer (Mazur *et al.,* 1982; Regier *et al.,* 1982). (1) The first mode is *lamellogenesis,* or framework formation; a small number of helicoidally arranged fibril planes are laid down, largely by apposition, to form very thin lamellae, which attain their final number when less than 20% of the final dry weight of the chorion has been deposited. (2) Permeation then leads to *expansion* of this framework, with the insertion of new sheets of fibrils that thicken the individual lamellae throughout the chorion. (3) *Densification* of the entire structure follows, as

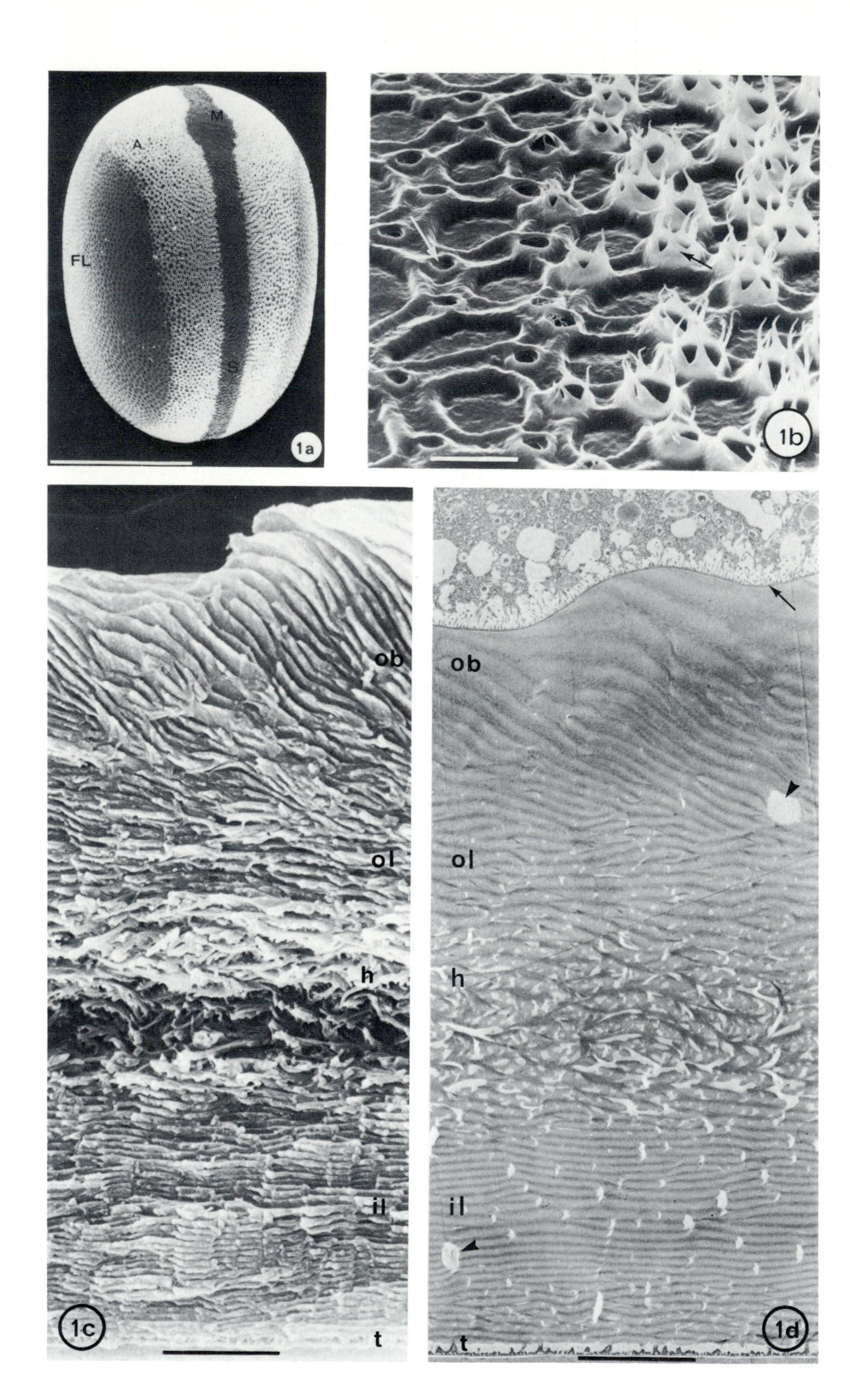

M
A
FL
S
1a
1b
ob
ol
h
il
t
1c
1d

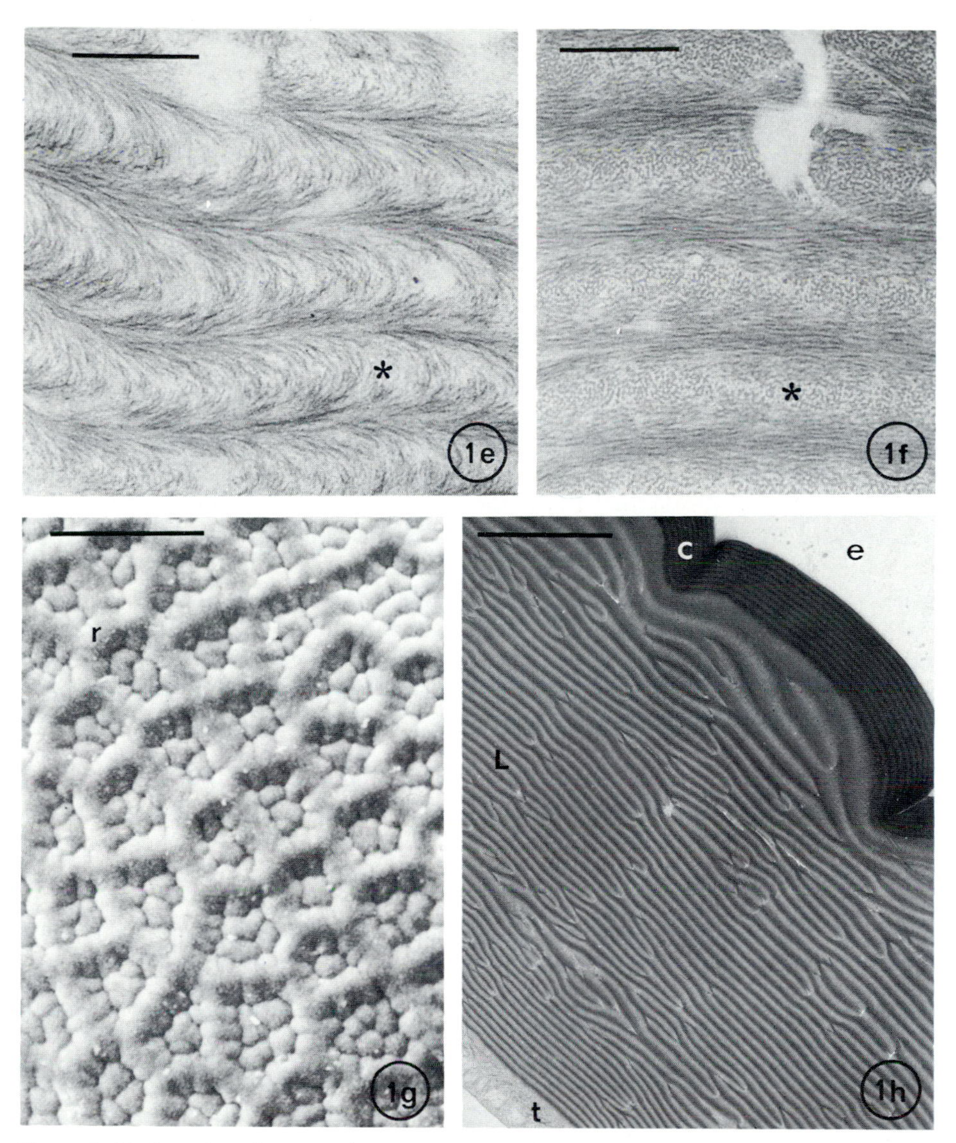

Figure 1. (a) Scanning electron micrograph (SEM) of the outer surface of a nearly mature (last pre-ovulatory) silkmoth eggshell from which the follicular epithelial cells have been removed by dissection. This, and all following micrographs, are of *Antheraea polyphemus* except where noted. Four surface regions are evident: flat (FL), aeropyle crown (A), stripe (S), and micropyle (M). (Scale marker = 1 mm.) (b) SEM. Detail of the border between flat and aeropyle crown regions. At left are cell imprint ridges with simple aeropyle openings (white arrow) characteristic of flat region. Tall many-pronged crowns (black arrow) surround these openings in the aeropyle crown region at right. (Scale marker = 20 μm.) (c) SEM. Ripped section through the flat region of an almost mature chorion, from which follicle cells have been removed. This and all other cross-sections are oriented with the outer chorion surface and adjacent follicular epithelial cells at the top and the inner chorion surface and oocyte at bottom. At bottom is the thin trabecular layer (t) consisting of short columns surrounded by air spaces. The bulk of the chorion is lamellar. Four

(*Caption continued on next page.*)

the fibrils thicken until their individual structure is obscured, although the lamellar banding remains obvious (cf. Figures 1e,f, 5b and 1f, 5c) (Smith *et al.*, 1971). (4) Finally, *regionalization* occurs as additional thin lamellae are deposited and sculpted into aeropyle crowns by the combined action of the follicle cells on the outer surface and of the nonlamellar filler on the inside (Mazur *et al.*, 1980). Each of these modes of growth has been correlated with restricted subsets of the chorion proteins (Mazur *et al.*, 1980; Regier *et al.*, 1982).

2. *The Structure of Liquid Crystals and Their Analogues*

2.1. *Molecular Structure of Liquid Crystals*

Liquid crystals are defined as ordered liquids, which are birefringent in the absence of bulk flow. These liquids represent a state of matter intermediate between, yet sharply distinct from, the amorphous liquid and the crystalline solid states (Friedel, 1922). Liquid crystals consist of long molecules that become oriented in either one, two, or three dimensions, though still fluid, within a well-defined range of temperatures and concentrations (Gray, 1962; Bouligand, 1972a). The synonyms for liquid crystal, mesophase and mesomorph, stress the intermediate, rather than the contradictory, nature of the state. Liquid crystals are never at the same time both very liquid and very crystalline, but rather show inversely varying degrees of both properties; when they have three-dimensional order, they exhibit very little fluid character (Frank, 1958).

Among biological materials there are many true liquid crystals: solutions of DNA (Robinson, 1961), tRNA (Wilkins, 1963), plasma membranes (Chap-

---

types of lamellae can be distinguished: thin and even in the inner lamellar layer (il), thick and distorted in the holey layer (h), thick and even in the outer lamellar layer (ol), and thick and at an oblique angle to the chorion surface in the oblique layer (ob). A cell-imprint ridge is evident at top right. (Scale marker = 5 μm.) (d) Composite transmission electron micrograph (TEM) of a section through the flat region of nearly mature chorion, as in Figure 1c. Above the trabecular layer (t) the stratified order of the lamellar chorion is apparent in the inner lamellar (il), holey (h), outer lamellar (ol), and oblique (ob) layers. Three types of holes are seen: large aeropyle channels cut transversely (arrowheads); horizontal (circumferenial) holes, elongated at lamellar boundaries and rounded in the middle of lamellae, in the holey layer; and vertical (radial) holes in the inner lamellar layer, appearing as linked crescents in this slightly oblique section. (Scale marker = 5 μm.) (e) TEM. Oblique section of lamellae of immature chorion. Lamellar boundaries, where fibrils are seen in longitudinal section, appear darker; because of the oblique angle of the section, lamellae appear as rows of fibrous arcs. (*) marks the middle of one lamella. (Scale marker = 0.5 μm.) (f) TEM. Vertical section of lamellae of immature chorion. Lamellae appear as rows of lines and dots, representing fibrils in longitudinal and transverse sections. (*) marks the middle of one lamella. (Scale marker = 0.5 μm.) (g) SEM. Surface view of the eggshell of *Bombyx mori* showing network of cell imprint ridges (r). Within each polygon formed by the ridges are several raised bumps. *Bombyx* chorion lacks crowns, and aeropyle openings are not evident at this magnification. (Scale marker = 50 μm.) (h) TEM. Cross-section of a chorion of *Bombyx mori*. Adjacent to the follicular epithelial cell (e) at upper right is the highly electron dense outer crust (c) containing very thin, even lamellae, then the middle lamellar chorion (L) with distortions in the outer part and translation defects throughout, and finally the trabecular layer, which in this species has a flat roof (t). (Scale marker = 5 μm.)

man, 1970), and a praying mantis oothecal protein (Neville and Luke, 1971). There are also many liquid crystal analogues, including cuticle and silkmoth chorion (see Bouligand, 1978a, for a recent review). The analogues are naturally occuring biopolymers, or combinations of polymers, fibrous in conformation, which, though often hardened by biochemical processes, are arranged in a structure geometrically similar to the structure of liquid crystals. Only a few molecular arrangements have been encountered in liquid crystals and their analogues, despite widely differing chemical compositions. Thus, the structures of liquid crystals represent a very basic set of natural forms.

Two fundamental types of molecular arrangement in liquid crystals were described by Friedel in 1922. They are *smectics* (from σμῆγμα = detergent; referring to certain soaps that show this phase) and *nematics* (from νῆμα = thread; referring to the threadlike appearance of defects in this phase). A third type, *discotics,* has been described more recently (see reviews in Bouligand, 1980a,b).

Both smectics and nematics are usually formed of rod-shaped molecules with a rigid center, often containing benzene rings, and flexible lateral extensions, consisting of paraffinic chains. In smectics, the molecules stand parallel, in stacked layers, their centers of gravity lying on equidistant parallel planes. Molecules diffuse freely within, but not between, layers; thus a smectic behaves as a liquid in the two dimensions of a layer, and as a solid in the direction perpendicular to the layer (Figure 2a). There are several types of smectic arrangements; in smectics A, molecules stand perpendicular to the layers, as shown. In smectics C, the molecules are tilted at a defined angle with respect to the layers. Smectics A and C can each have a variant form B in which there is local hexagonal close-packing of the molecules (Bouligand, 1980a).

Nematics represent a lower degree of order than smectics. There are no layers, only the directions of the molecules are defined. The rod-shaped molecules are still parallel, but their centers of gravity are arranged at random in all three dimensions (Figure 2b), in some ways like a school of small fish.

Unlike the rod-shaped molecules that make up smectics and nematics, the molecules of discotics are flat, generally having a rigid planar aggregate of benzene rings in the center and a lateral set of flexible paraffinic chains (Bouligand, 1980b). In this phase, the disc-shaped molecules are stacked in cylinders that stand parallel and may exhibit hexagonal packing, rather like stacks of coins, except that the distance between adjacent disc-shaped molecules in a cylinder is not constant, and the molecules of neighboring stacks are not in register (Figure 2c). The cylinders can bend in parallel and can glide with respect to one another (Bouligand, 1980a). Bouligand notes that in concentrated preparations, densely packed nucleosomes exhibit hexagonal discotic symmetries.

2.2. *The Cholesteric Phase*

The type of liquid crystal that is most relevant to silkmoth chorion is the helical variant of a nematic (Bouligand, 1972a). This is called *cholesteric,* for the simple historical reason that it was first seen in some esters of cholesterol (Gray, 1962). In this arrangement (Figure 2d), which corresponds to Bouligand's

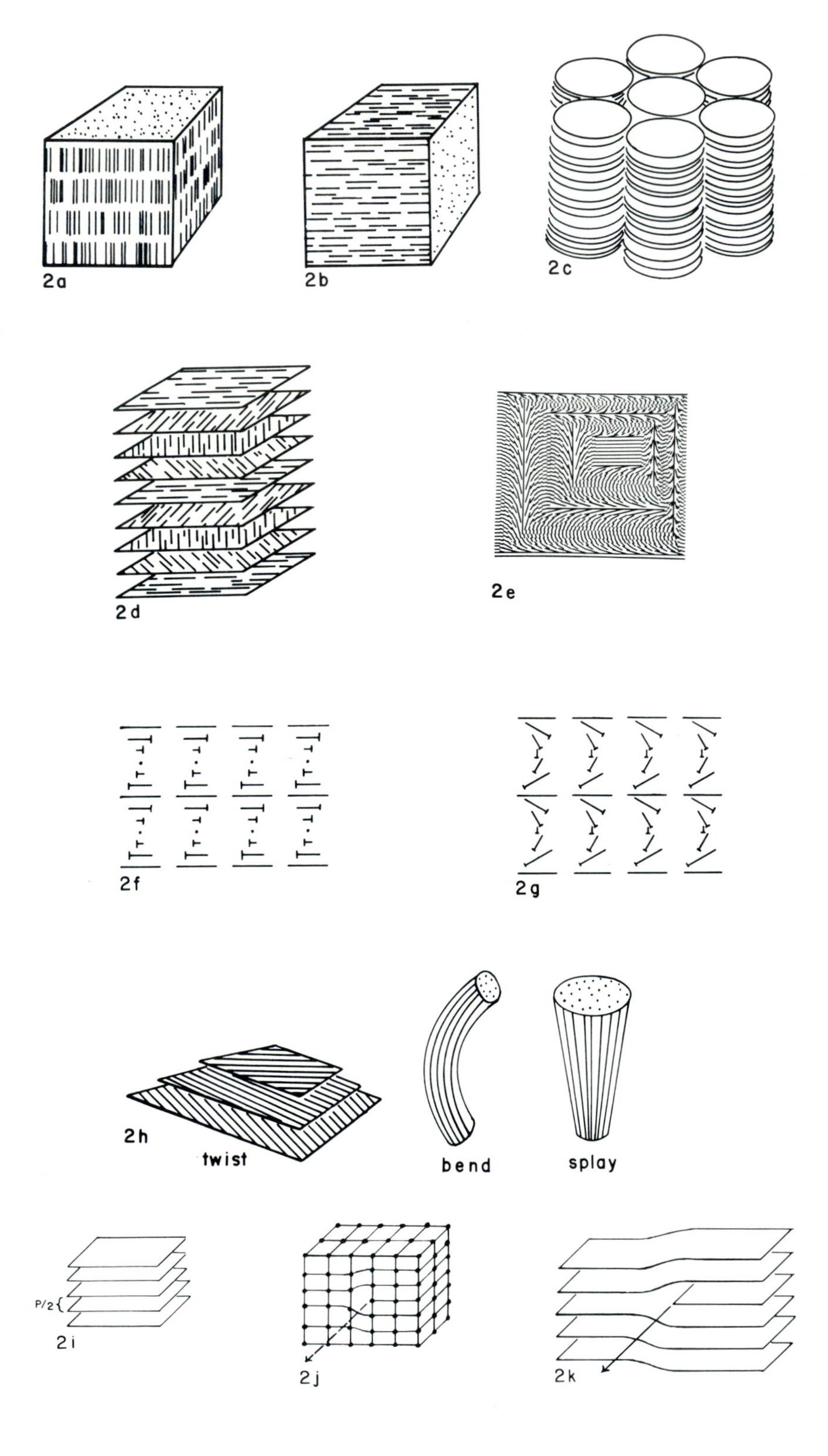
2a
2b
2c
2d
2e
2f
2g
2h
twist
bend
splay
P/2
2i
2j
2k

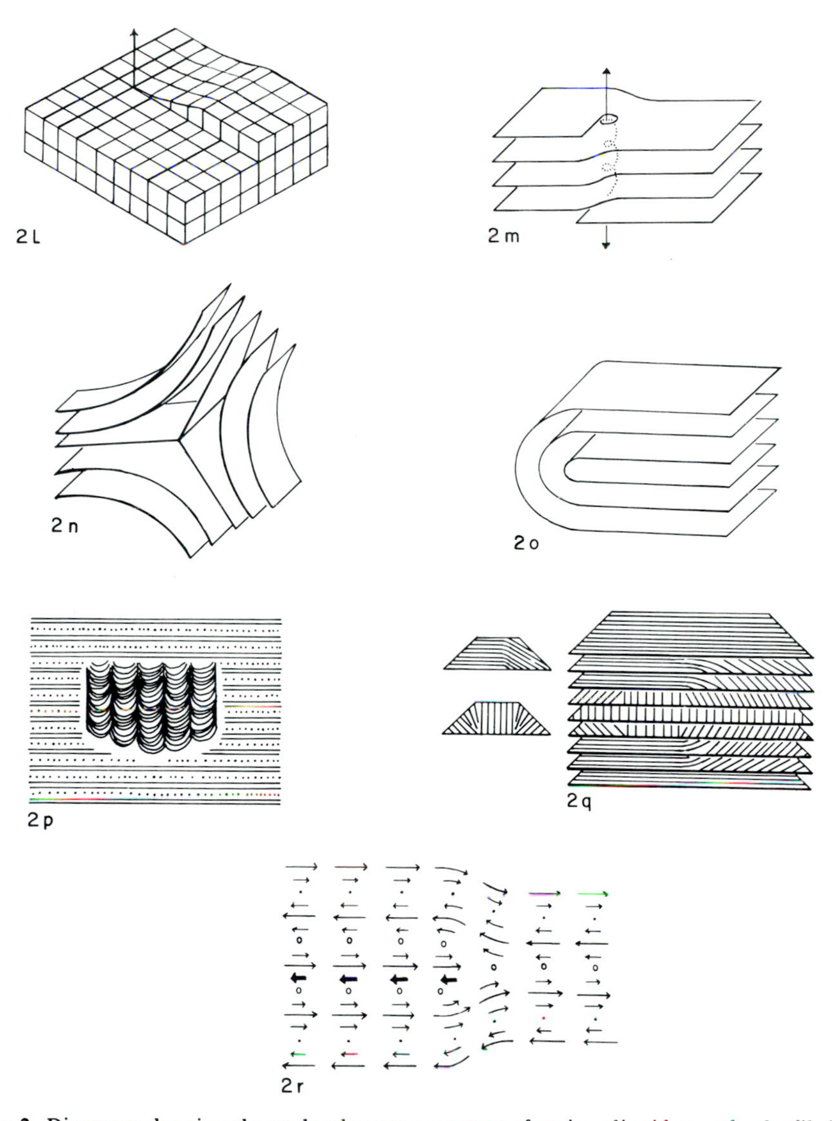

Figure 2. Diagrams showing the molecular arrangements of various liquid crystals, the fibrillar arrangements of their biological analogues, and the structure of various defects in solid and liquid crystals and their analogues. (a) The arrangement of molecules in a smectic liquid crystal (after Bouligand, 1972a). (b) The arrangement of molecules in a nematic liquid crystal (after Bouligand, 1972a). (c) The arrangement of molecules in a discotic liquid crystal (after Bouligand, 1980b). (d) The arrangement of molecules in a cholesteric liquid crystal, or of fibrils in a cholesteric liquid crystal analogue; 360° rotation of molecule or fibril direction corresponds to two lamellae (after Bouligand, 1972a). (e) Photograph of an adaptation of the "pyramid model" of Bouligand (1972a). The model consists of pieces of cardboard inscribed with straight parallel lines representing the average fibril direction in that plane. The angle between the directions in two adjacent planes is 15°. The cardboard pieces are stacked in a truncated pyramid, with the sides of the pyramid each corresponding to a different possible oblique section. 360° of rotation of fibril direction is shown. (f) Conventional notation for representing a cholesteric liquid crystal or biological analogue,

(*Caption continued on next page.*)

twisted model, the molecules lie parallel in a given plane, with their centers of gravity aligned at random. In a neighboring parallel plane, the molecules are again parallel to each other, but their direction has been rotated through an angle proportional to the distance separating the two planes. This rotation imparts a helicoidal twist to the cholesteric liquid crystal or analogue. The helix can be either right or left handed; most biological analogues so far examined have been shown to be left handed (reviewed in Neville, 1975). The pitch of a cholesteric is defined as the minimum distance corresponding to a

---

sectioned normal to the helicoidal layering. Here, dots (. .) represent molecules of fibrils perpendicular to the plane of the drawing, while lines (—) stand for molecules or fibrils parallel to this plane. Molecules or fibrils oblique to the plane of the drawing are shown by nails (⊢) whose points are directed towards the observer. Their length is proportional to their projection onto the drawing plane. It is important to note that the nail points refer only to that part of the molecule or fibril that happens to be closer to the observer, and make absolutely no reference whatsoever to one end or the other i.e., "head" or "tail," of the asymmetrical molecule or fibril in question (after Bouligand, 1974). (g) Oblique section of a cholesteric arrangement shown in the same notation as Figure 2f (after Bouligand, 1974). (h) The distortions bend, twist, and splay (after Livolant, 1977). (i) Schematic drawing of a cholesteric liquid crystal or its analogue showing equidistant planes with no defects. The planes represent levels in which the fibrils have the same direction in a twisted system. For the sake of clarity in this type of diagram, all the intermediate levels, where the fibril directions change through 180°, are left out. As the cholesteric pitch is defined as a 360° rotation of the fibril direction, the distance between any two adjacent levels is equal to ½ pitch (after Livolant, 1977). (j) Schematic diagram of an *edge dislocation* in a solid crystal. Each solid dot represents a subunit of the crystal. Arrow indicates the dislocation line (after Gordon, 1968). (k) Schematic diagram of an *edge dislocation* in a cholesteric liquid crystal or its analogue. As in Figure 2i, the planes represent levels where molecules or fibrils have the same direction in the twisted system. The dislocation line, indicated by the arrow, is parallel to the cholesteric layering (modified from Livolant, 1977). (l) Schematic diagram of a *screw dislocation* in a solid crystal. Each cube represents a subunit of the crystal. Arrow indicates the dislocation line (modified from Gordon, 1968). (m) Schematic diagram of a *screw dislocation* in a cholesteric liquid crystal or its analogue. The planes represent levels where molecules of fibrils have the same direction in the twisted system. The dislocation line, indicated by the arrow, is perpendicular to the cholesteric layering (modified from Livolant, 1977). (n) Disclination. Rotation defect in the form of a triple point corresponding to a rotation of lamellar direction after insertion of new lamellae. The same conventions apply here as in Figure 2i (after Livolant, 1977). (o) Disclination. Rotation defect in the form of cofocal parabolae corresponding to a rotation of lamellar direction after deletion of lamellae. The same conventions apply here as in Figure 2i (after Livolant, 1977). (p) Schematic representation of a block defect (cf. Figure 5). Lines and dots indicate lamellae in vertical section, rotated stacked arcs indicate that the block defect has been rotated about two orthogonal axes. (q) Schematic representation of the central region of a complex defect seen in Figure 6b. This region can be understood as a combination of distortions with some fiber planes having bent fibrils (as in upper fibril plane detail at left) and others having splayed fibrils (as in lower fibril plane detail at left). (r) Diagram of a simple translation defect showing its relation to fibril polarity. The fibrils are represented as lines with the "head" end of the fibril as an arrowhead and the "tail" end plain. The fibrils are stacked in helicoidal orientation. The fibrils pointing out of the plane of the page are shown as dots when their "heads" are pointing towards the observer, and as empty circles when their "tails" are pointing towards the observer.

From the diagram it can be seen that although the shape of the parabolic patterning of the lamellae repeat with each twist of 180°, the head-tail alignment repeats itself every 360°. Where an edge or screw dislocation introduces part of a lamella into the arrangement, however, then the head-to-tail parallel fibril orientation within a layer, and/or head-close-to-head fibril orientation between layers, is destroyed.

360° rotation of the direction of the molecules. Therefore, the molecules are in parallel alignment every half pitch, which is equivalent to one lamella. In cholesterics, the rod-shaped, asymmetric molecules are optically active (Neville, 1975).

The pitch of a cholesteric is not fixed, but varies with the temperature and can also be changed experimentally by varying the concentration of components of a cholesteric mixture (Gray, 1962). In certain mixtures, increasing the concentration of one of the components can cause such an increase in pitch that the cholesteric twist disappears entirely and the mixture becomes a simple nematic. Further addition of the same component causes the cholesteric twist to reappear, but in the opposite direction, the pitch now decreasing with increasing concentration as the helical arrangement becomes more tightly twisted (Robinson, 1961; reviewed in Neville, 1975). This type of dynamic transformation of lamellar structures should be kept in mind when one studies the morphogenesis of cholesteric bioanalogues such as chorion.

A few of the biological analogues of cholesterics, such as copepod cuticle (Bouligand, 1965; Gharagozlou-Van Ginneken and Bouligand, 1973) and coelocanth fish scales (Giraud *et al.*, 1978), show a relatively large, discrete or stepwise rotation between the fibril directions of adjacent layers, with little variation of fibril direction within a layer. Theoretically, the rotation in the cholesteric structure can be continuous, although for reasons of convenience it is generally drawn as a stack of discrete planes.

2.3. *Properties and Consequences of the Twisted Model*

Many properties of cholesteric analogues can be understood if the twisted model is drawn as a set of discrete planes, as in Figure 2e. This adaptation of the "pyramid" model of Bouligand is made of sheets of cardboard inscribed with straight parallel lines representing the average fibril direction in that plane. The angle between the directions in two adjacent sheets is 15°. The cardboard pieces are stacked in a truncated pyramid, with the sides of the pyramid each representing a different possible oblique section.

Stacked rows of nested parallel arcs appear on the sides of the pyramid. We can now see that the arcs are formed by the projection onto the viewing plane of the closely juxtaposed cut ends of the straight fibrils of successive layers. It is easy to see that the only twist needed to produce the arcs is the twist of fibril direction between adjacent layers.

Certain consequences follow from the model; many of these have in fact already been observed in biological material (Livolant *et al.*, 1978). First, as seen at opposite sides of the pyramid, sections of opposite obliqueness with respect to the vertical axis will show arcs pointing in opposite directions (Bouligand, 1978b). Similarly, vertical sections observed by the transmission electron microscope (TEM) will show arcs of opposite orientations when tilted on either side of a vertical axis by means of a goniometer stage (Livolant *et al.*, 1978).

Second, the apparent thickness of the lamellae and the shape of the arcs vary as the obliqueness of the angle of section is changed. As the sections approach the vertical, the same number of lamellae are present, but they are narrower and the arcs are less apparent. In fact, when the section becomes completely perpendicular to the layers, the arcs disappear, and we see bands of lines representing fibrils cut more or less longitudinally, alternating with bands of dots, representing head-on views of transversely cut fibrils. This is easily seen in chorion (Figure 1f).

Third, horizontal sections should obviously give fields of parallel lines. In chorion, however, this will be obscured in TEM because of the finite thickness of the thin sections used, even if the planes of fibrils are indeed discrete. Since typical thin sections are 0.06 μm–0.08 μm thick and lamellae of mature chorion (synthetic stage X) have an average thickness of 0.4 μm, a horizontal section should contain 15% to 20% of a lamella, fibrils whose directions diverge from each other by 27° to 36°. For immature chorions with thinner lamellae, the case is even worse: when lamellae first appear, they have an average thickness of 0.07 μm, and thus we would expect an angular range of 150° to 205° in fibril orientation within each horizontal section.

2.4. Departures from the Ideal Model

The Bouligand model represents a single statistical average of fibril directions in each plane; actually, in biological materials there is variation of fibril direction within any given plane (Bouligand *et al.*, 1968).

Two other ways in which cholesterics and their analogues depart from the ideal model are *distortions* and *defects* (Livolant 1977; Bouligand 1972a; 1975; 1980c,d). A distortion can come about by a continuous deformation of the ideal model; a defect, however, cannot be produced by a continuous operation, but must result either from insertion or excision of material. The mutual arrangement of fibrils in an organized system can be described by a combination of three fundamental parameters: *twist*, a measure of the angle between planes of fibrils; *bend*, a measure of the curvature of the fibrils; and *splay*, a measure of how far the fibrils depart from parallel (Figure 2h) (Frank, 1958; Livolant, 1977). In a purely parallel arrangement, twist, bend, and splay are all equal to zero. In Bouligand's ideal model there is twist, but as the fibrils are straight and parallel in any given plane, the bend and splay are again zero. In chromosomes (Livolant, 1977), cuticle (Barth, 1973; Gharagozlou-van Ginneken and Bouligand, 1973), and chorion (Figures 1d,h, 4d), distortions of bend and splay are introduced into the twisted arrangement.

Figures 2f,g, and i illustrate conventions that are convenient abstractions for describing the cholesteric arrangement, its distortions, and its topological defects. Here we will consider only those defects relevant to a study of the chorion; these are summarized in Table 1. For a more complete treatment of defects in liquid crystals and their analogues, the reader is directed to the works of Bouligand (1972b; 1974; 1975; 1980c,d).

Many defects are found both in solid and in liquid crystals and their analogues: *substitutions*, where a different type of subunit is found within the array; *vacancies*, where a position is left empty; and *dislocations*, or translation defects, which result from the translation on one part of the crystal with respect to another. There is an *edge* dislocation (Figures 2j,k), when adjacent planes of the crystalline network are locally wedged apart by the insertion of one or more additional planes; or when adjacent lamellae of a cholesteric liquid crystal are wedged apart by the insertion of one or more lamellae. The other main type of translation defect is the *screw* dislocation (Figure 21, 21m); this can be formed by cutting part way through a crystalline or lamellar array and sliding the cut surfaces relative to each other, by an integral number of crystalline subunits or lamellae, along the direction of the cut (Gordon, 1968; Bouligand, 1980d).

Dislocations can arise by accident during the formation of a crystal. Some, however, such as screw dislocations, can also facilitate crystalline growth (Gordon, 1968). The screw dislocation presents a step at the point where it emerges, offering sheltered corners for bonding first in two, then in three directions. This is most obvious in schematic drawings of solid crystalline screw dislocations (Figure 2l). In this way, as new subunits fall into place, the dislocation is preserved, its position within the crystal rotating about its axis as each new layer of the crystalline array is added (Gordon, 1968).

A diagnostic feature of the two types of translation defect is the direction of the line of dislocation. In an edge dislocation, the dislocation line runs along the edge of the extra layer of subunits, in a direction parallel to the layers (Figures 2j,k). In a screw dislocation, however, the dislocation line is perpendicular to the layers, and runs down the central axis of the screw (Figure 2l,m). In both cases, the region around the dislocation line is termed the *core*.

It is often difficult to determine from a section whether a given defect is an edge or a screw dislocation. One reason for this is that there are defects that are actually intermediate between edge and screw (Gordon, 1968; Livolant, 1977). But another, much more common reason for this difficulty is that if the section passes through a screw defect, but does not include the line of dislocation or the core around that line, then the screw dislocation is indistinguishable from a set of multiple alternating edge dislocations (Figure 3b). Conversely, if the cores of multiple edge dislocations fuse locally, they could generate a compound core indistinguishable in section from the core of a screw dislocation. Thus, serial sections may be necessary for distinguishing edge and screw dislocations. Often, however, this problem can be satisfactorily resolved by scanning electron microscopy of fractured material. As we shall see below, such fractures can allow us to see both the staggered alternation of lamellae caused by the defect and the direction of the core, which is visible as a hole, around the line of dislocation (Figures 3c,d).

While translation defects form in both solid and liquid crystals, rotation defects or *disclinations* are found only in liquid crystals (Frank, 1958; Harris, 1977) and their biological analogues. In the analogues, disclinations can be found either in the arrangement of fibrils within a layer, or in the array of the lamellae (Bouligand, 1975). They can be formed by the insertion or deletion of material, followed by the rotation of the remaining fibrils or lamellae to rectify the even spacing of lamellae and to fill any gaps formed by changing the amount of material (Figures 2n,o). The rotation can be around the direction of the molecules or around the perpendicular to that direction (Bouligand, 1972b).

Table 1. Some Defects and Distortions

Type	Description	Origin in liquid crystalline systems and biological analogues
Distortion 	Lamellae that appear bent or curved but without breaks	Mechanical deformation; local differences in chemical composition
Screw Dislocation 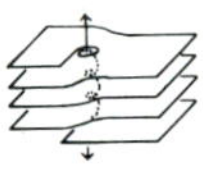	A disruption in the alignment of lamellae that appears as through several lamellae were cut transversely and slid relative to each other along the direction of the cut. This leads to a screw-shaped hole, or dislocation core, perpendicular to the lamellae	Inhomogeneous nucleation of lamellae; mechanical perturbations during lamellar formation
Edge Dislocation 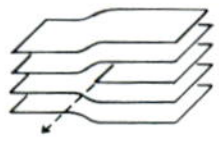	Insertion of one or more extra lamellae by wedging apart adjacent lamellae. Dislocation core is parallel to lamellar layering	As for screw dislocation
Vacancy 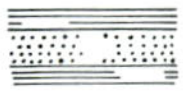	Missing fibrils	Not documented in liquid crystals; Presumed to be due to chemical differences in chorion
Substitution 	Disruptions in lamellae caused by presence of nonlamellar components; in biological analogues these components are often cytoplasmic extensions or nonlamellar fibrils	Chemical impurities or inhomogeneities In biological analogues: cell mediation; localized differences in secretion
Block Defect 	Rotation around one or two orthogonal axes of a small isolated block of lamellae	Not documented in liquid crystals In chorion: may be due to boundary conditions causing orthogonal nucleation or to mechanical perturbation

in Helicoidal Lamellar Systems

Examples in chorion		Examples in cuticle
In *A. polyphemus*:	Aeropyle crowns; oblique layer; holey layer	Soft cuticle of intersegmental membrane (Neville, 1975); opisthosoma of spider (Barth, 1973); copepod cuticle (Gharagozlou-van Ginneken and Bouligand, 1973)
In *Bombyx mori*:	In surface lamellae of ridges (under two cell junctions) and bumps (within cell imprints).	
In *A. polyphemus*:	Very prominent in inner lamellar layer; found to a lesser extent in the other lamellar layers.	
		Often seen under tubercles of crab cuticle (Bouligand, 1972a)
In other silkmoths:	Found in all silkmoth chorions so far examined (Smith *et al.*, 1971; Kafatos *et al.*, 1977)	
Not documented		Locust cuticle (Bouligand, 1972a)
In *A. polyphemus*:	Holey layer (Figures 1c,d)	Not documented
In *Hyalophora cecropia*:	Holey layer (Smith *et al.*, 1971)	
In *A. polyphemus*:	Aeropyle channels (Figures 1c,d; 4a–d)	Pore canals (Neville, 1975)
In *Bombyx mori*:	Aeropyle channels (Kafatos *et al.*, 1977)	
In *Hyalophora cecropia*:	Aeropyle channels (Smith *et al.*, 1971)	
In *A. polyphemus*:	In regions that crystallize into helicoidal arrangement with delayed timing	Not documented

(Continued on following pages.)

Table 1.

Type	Description	Origin in liquid crystalline systems and biological analogues
Disclination	Incomplete lamellae in which portions have been inserted or excised with rotation of remainder to accomodate added material or to fill in gaps. Insertion leads to formation of triple points; excision leads to cofocal parabolae	In liquid crystals: mechanical deformation, boundary condition, or chemical composition

Dislocations and disclinations are associated with a certain amount of strain energy. As material is added (edge dislocations, disclinations), subtracted (disclinations), or sheared (screw dislocations), the molecular bonds undergo changes in length and/or direction, which introduce strain. In dislocations, this strain increases as one approaches the dislocation line; at the line itself, the strain would be infinite (Harris, 1977). The structure of the core surrounding the dislocation line in liquid crystals is not well understood (Livolant, 1977), and the cores are diagramatically represented as hollow cylinders. Disclinations are defects of high strain energy (Harris, 1977). They occur mainly in the more fluid regions of a cholesteric, near the border between liquid crystal and amorphous (isotropic) liquid (Bouligand, 1975).

3. Distortions and Defects in Silkmoth Chorion

3.1. Distortions

While the twist between fiber sheets is responsible for the striking helicoidal geometry of the chorion, it is the differential distribution of abundant distortions and defects that gives each of the circumferential chorion layers its characteristic texture (Figures 1c,d, 3e).

Distortions are most prominent toward the outer chorion surface. They appear to be related to surface sculpturings rather than to the global curvature of the chorion, which is imperceptible on the scale of individual fibrils and maximal in the inner layer.

The strongest distortions are evident in the tall aeropyle crowns of *A. polyphemus,* where the thin lamellae are sharply bent, almost orthogonally to the rest of the chorion (Figure 1b). This orientation must be imposed by the combination of molding by the changing shape of the secretory cell surface (Figures 4c,d) and the inner, buttressing action of the filler (Mazur *et al.*, 1980). Bend and splay distortions are also evident in the outer lamellae of *Bombyx* (Figure 1h), which are sculpted into a series of rounded bumps within each polygonal cell imprint (Figure 1g). In *polyphemus,* distortions in the neighbor-

(*Continued*)

Examples in chorion		Examples in cuticle
In *A. polyphemus*:	In lamellae at the outer chorion surface in the aeropyle crown region (Figures 6a,b)	Fibrils in horizontal planes of a lamella, in beetle cuticle (Neville, 1975)
In *Hyalophora cecropia*:	Fibrils in horizontal planes of a lamella (Smith *et al.*, unpublished, cited in Bouligand, 1975)	

hood of cell imprint ridges are minimized by the obliqueness of the lamellae (Figure 1c).

3.2. Defects

There are six major types of defects or holes found in the chorion, not counting the micropyle.

3.2.1. Translation Defects

Edge dislocations have not been identified with certainty in *A. polyphemus* chorion. By contrast, abundant vertical screw dislocations penetrate the inner lamellar layer, with a periodicity of approximately 2 μm. Depending on the plane of section or rip, these defects can be seen to extend through at least eight to ten lamellae, sometimes twenty. In transmission electron microscopy (TEM), they appear as empty screw-shaped holes, when the plane of section passes vertically through the center of the core (Figure 3a); as a series of empty linked crescents, when the section cuts vertically through the edge of the core, and as a series of closely staggered lamellae alternating in zig-zag fashion, when the section cuts vertically but just misses the core (Figure 3b). In oblique sections, these screw defects appear as single or linked pairs of empty crescents (Figure 1d).

In scanning electron microscopy (SEM), vertical rips show the screw defect cores as shallow indentations running perpendicular to the lamellae (Figure 1c) or as deep screw-shaped vertical holes (Figure 3d). In horizontal rips, parallel or almost parallel to the lamellae, the cores appear as empty rounded holes; the lamellae often show a slight helical twist around the holes (Figures 3c,e).

Screw dislocations are more rare in the holey, outer, and oblique layers, where they lack prominent cores and have no obvious periodic spacing (Figures 1d, 3e).

It is particularly interesting to find defects of translation (i.e., edge and/or screw dislocations) in chorion, because their presence implies a lack of polarity

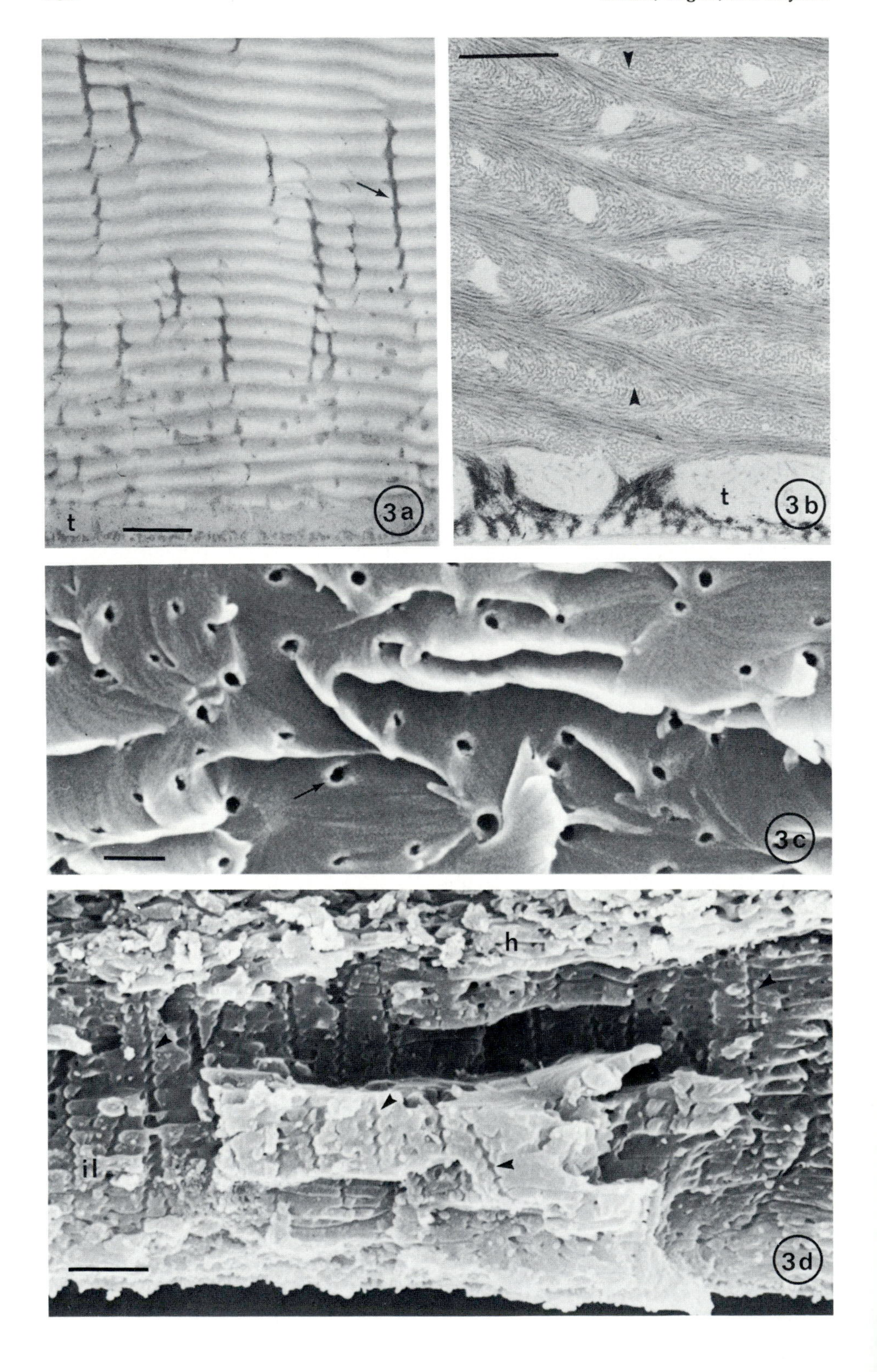
t
3a
t
3b
3c
h
il
3d

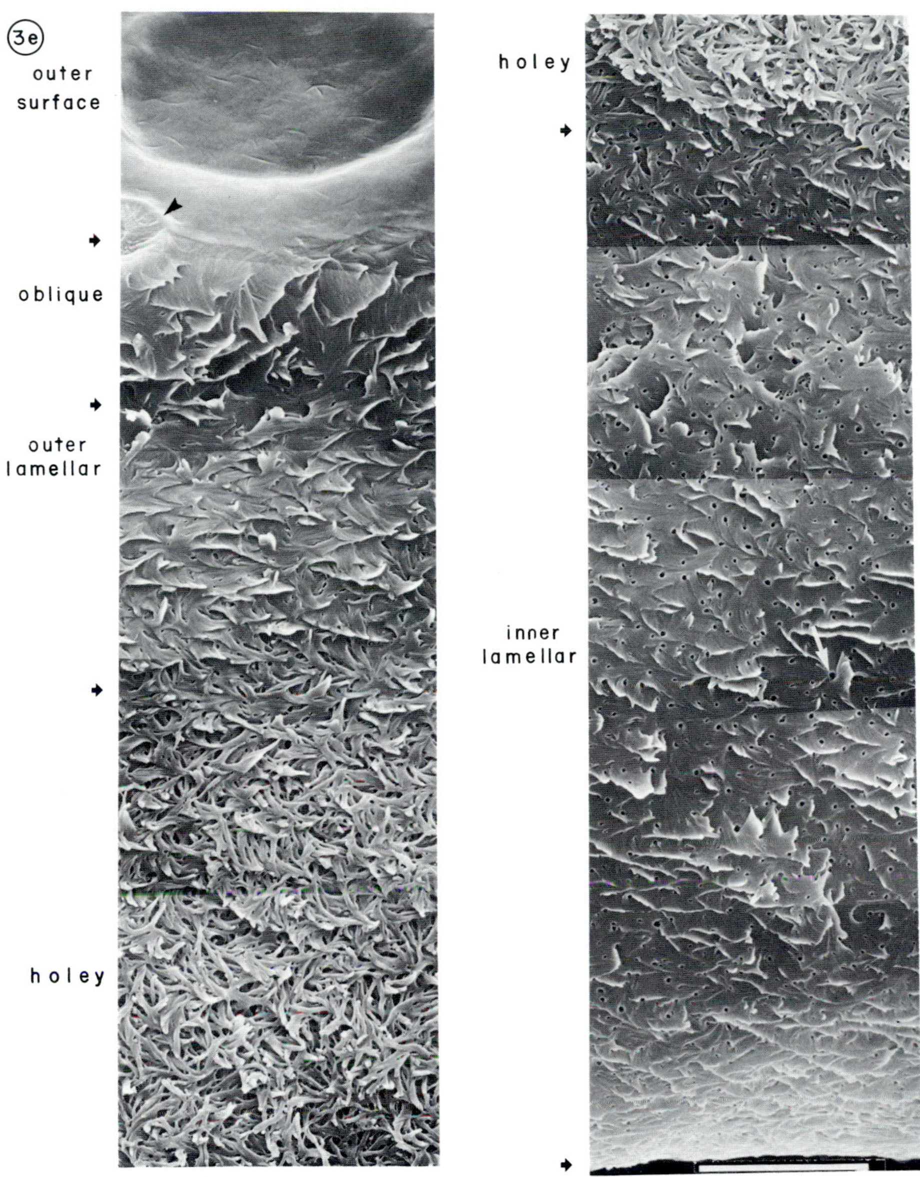

Figure 3. Screw defects. (a) TEM of a vertical section of the inner lamellar layer of mature chorion. The cores of several screw defects are seen as vertical screw shaped holes (arrow) passing through many lamellae. Lamellae are seen to originate directly from the tops of the trabeculae of the trabecular layer (t). (Scale marker = 1 μm.) (b) TEM of a detail of the inner lamellar layer of immature chorion. This vertical section passes through the edge of a screw defect, which is seen as a zig-zag alternation of lamellae. Arrowheads indicate the direction of the dislocation core. Zones where fibrils are cut in longitudinal section appear darker. (t) = trabecular layer. (Scale marker = 0.5 μm.) (c) SEM of a detail of an extremely shallow, almost horizontal rip through the inner lamellar layer of the chorion (the entire rip from which this detail is taken is seen in Figure 3e). Abundant screw defects appear as rounded empty holes (arrow), often with a slight helical component. (Scale marker = 1 μm.) (d) SEM of a vertical rip through the inner lamellar layer (il) and the lower part of the holey layer (h) of a mature chorion. Empty, screw shaped cores of screw defects

(Caption continued on next page.)

in fibril arrangements (Livolant, 1977), or at least a toleration of localized abnormal polarity; in such defects, the fibrils of one layer align either head-to-head or head-to-tail with the fibrils of the neighboring layers (Figure 2r).

3.2.2. Holey Layer–Axial Vacancies

As soon as they appear, the holey layer lamellae are significantly thicker than lamellae in the other chorion layers (Regier *et al.*, 1982). Their most striking defect is a set of large, apparently empty holes running parallel to the direction of the fibrils. Thus in sectioned material, the holey layer shows rounded holes in the middle of lamellae and elongated, often interconnected cavities at lamellar boundaries (Figures 1d, 7d).

Scanning electron micrographs of ripped material emphasize the porous nature of this layer and the frequency of interconnections between the screw dislocations in the inner lamellar layer, the holes of the holey layer, and the aeropyle channels prominent in the outer and oblique layers (Figures 1c, 3d, 4a).

3.2.3. Aeropyles

The aeropyle channels are another set of periodically occurring radial disruptions (Figures 1b, 4a), often causing both distortions (Figures 4c,d) and translation defects (Figure 4b). These channels are formed around groups of microvilli that extend from each of the cells of a three-cell junction and penetrate deeply into the chorion. They are spaced 10–30 μm apart, with cores approximately 1 μm in diameter. Although aeropyle channels penetrate the chorion from the outer surface to the trabecular layer (Smith *et al.*, 1971; unpublished observations), they are rarely in one plane and so are hardly ever visualized in their entirety in a single micrograph, even in material ripped for SEM. They usually penetrate the outermost chorion obliquely, parallel to the lamellae of the oblique layer (Figure 4a), and change direction at the holey layer. In the aeropyle crown region, however, they are occasionally perpendicular.

3.2.4. Block Defects

Small blocks of five to twelve lamellae are occasionally rotated at right angles to the rest, in what we call "block defects" (Figures 2p, 5a,b,c). The rotation may be along two orthogonal axes, accentuating the arced patterns

---

(arrowheads) penetrate the inner lamellar layer and extend into the holey layer. (Scale marker = 2 μm.) (e) Composite SEM of an extremely shallow, almost horizontal, rip through chorion. At top left is the outer chorion surface with part of one cell imprint and an aeropyle channel (arrowhead) at the corner of the imprint. The oblique, outer lamellar, holey, and inner lamellar layers are shown, with arrows indicating boundaries between layers. The cores of screw defects, seen as rounded holes, appear larger and with more regularity in the inner lamellar layer. The white arrow in the right-hand panel indicates the region from which the detail in Figure 3c is taken. (Scale marker = 10 μm.)

(Figures 5a,b). These blocks are found in both the aeropyle crown and the flat regions throughout choriogenesis, starting with the first chorion that shows more than a couple of lamellae; they are usually associated with aeropyle channels.

3.2.5. Rotation Defects: Disclinations

Chorion lamellae exhibit rotation defects particularly toward the outer edge of the aeropyle crown region (Figures 4d, 6a). In this area, the lamellae that form the crowns extend at right angles to the rest of the chorion, and the neighboring lamellae of the oblique layer often appear in a state of upheaval. Although the lamellar arrangement is often too complex to be analyzed from single thin sections, occasionally the triple points and cofocal parabolae characteristic of disclinations can be made out (Figures 2n,o, 6a,b). Disclinations in the arrangements of fibrils in a horizontal plane within a lamella have been documented by Smith *et al.*, (cited in Bouligand, 1975).

3.2.6. Complex Defects

More complex defects are also found, such as that seen in Figure 6b, in which the arcs appear to form a closed figure. The central part of this figure may be resolved into a combination of distortions (Figure 2q).

4. Genesis of Defects in the Chorion

In general, in *A. polyphemus* chorion, the defects are a primary phenomenon; that is, they appear as soon as the lamellae surrounding them are formed.

Screw dislocations appear very early in *polyphemus* choriogenesis, in fact, as soon as one sees the earliest framework of the inner lamellar layer (Regier *et al.*, 1982) these defects are present within it, both in sectioned material (Figure 7b) and in rips examined by SEM (Figure 7c).

The chorion layers are formed sequentially, starting with the construction of a trabecular colonnade on a thin floor apposed to the vitelline membrane of the oocyte. The inner lamellar layer forms adjacent to the trabeculae; then, on top of it the holey and outer layers are formed, and outermost, the oblique layer (Regier *et al.*, 1982). The first lamellae of the inner lamellar layer actually form at the tops of individual trabeculae, as though these columnar structures were serving as nucleation sites for the helicoidal crystallization (Figure 7a). These lamellae grow outwards from the tops of the trabeculae until they fuse laterally (Regier *et al.*, 1982). At the same time, however, it appears that new lamellar components are being added onto the material centered over the trabeculae. The result is that the lamellae right on top of trabeculae are often out of phase with the lamellae forming over the spaces between trabeculae. These areas of mismatch, primitive dislocations, can be seen when the chorion is only two or three lamellae thick (Figure 7b).

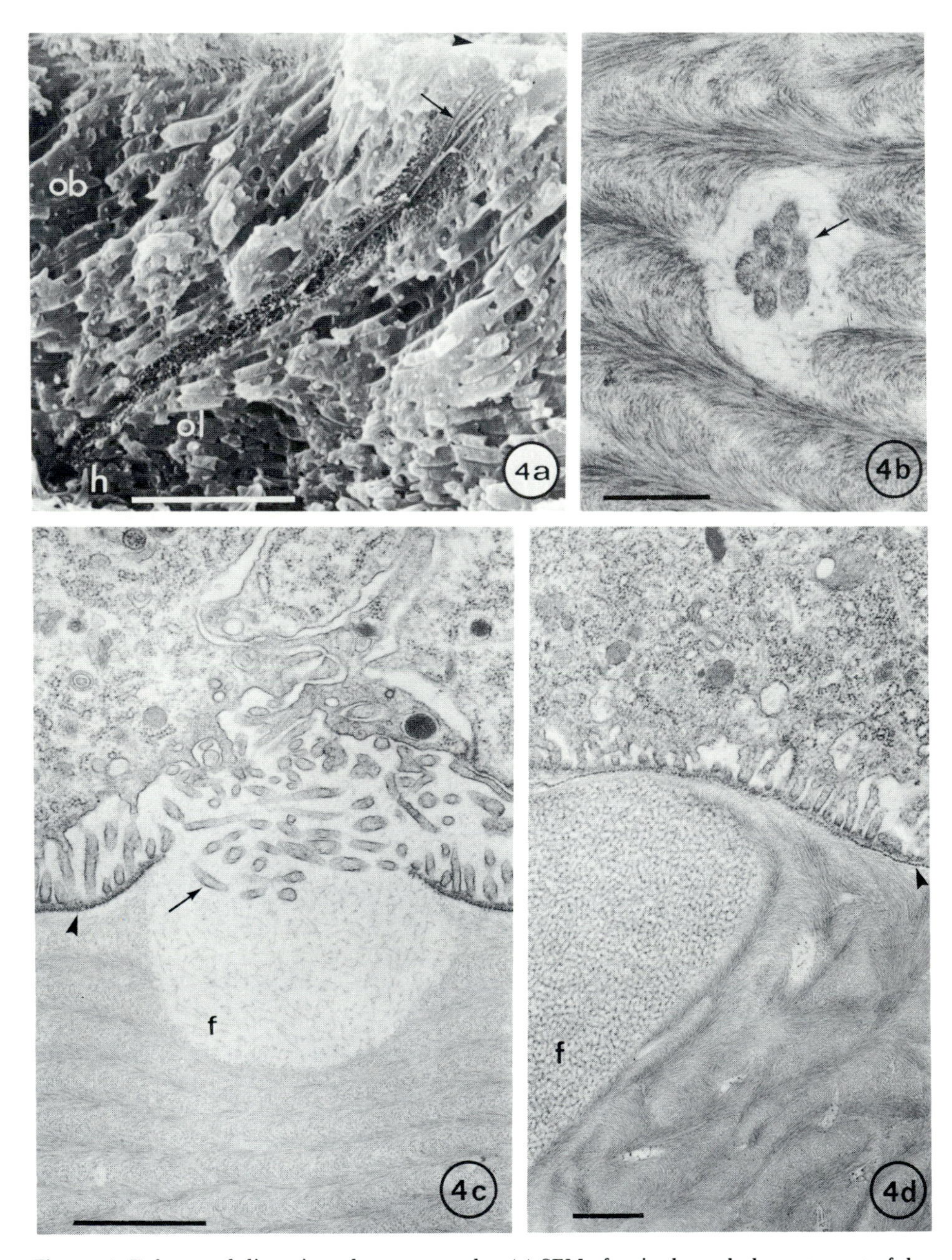

Figure 4. Defects and distortions due to aeropyles. (a) SEM of a rip through the outer part of the flat region. An aeropyle channel originating at the outer chorion surface (arrowhead) passes through the oblique layer (ob) parallel with lamellae, through the outer lamellar layer (ol) at an angle, and into the holey layer (h). The channel is filled with flocculent "filler" material and contains a group of microvilli (arrow) that have extended from an overlying three-cell junction. (Scale marker = 5 μm.) (b) TEM of an oblique section through an aeropyle channel that has penetrated to the inner lamellar layer of an immature chorion. The unequal number of lamellae on either side of the channel signals a translation dislocation. The aeropyle channel contains sparse fibrils

The holes of the holey layer are present and appear empty from the beginning of choriogenesis (Figure 7d) to the end (Figure 1d). They expand with the lamellar expansion, retaining their relative diameter of about one third the lamellar thickness.

The appearance of the aeropyle channels changes during choriogenesis. In early choriogenesis, the aeropyle channel is totally filled by the extended microvilli (Smith *et al.*, 1971; unpublished observations). In mid-choriogenesis, the channel is also filled with a loose network of the non-lamellar proteinaceous "filler" material we have described elsewhere (Mazur *et al.*, 1980; Regier *et al.*, 1980) (Figure 4c). By late choriogenesis, the aeropyle channels have increased their diameter by a factor of two to four; this is even more pronounced in the outermost lamellae of the aeropyle crown region. The filler material at this stage forms a dense, highly connected network (Figure 4d).

When the block defects are formed, they are surrounded by narrow spaces containing sparse disorganized fibrils (Figure 5a). Later, however, these spaces disappear. Occasionally, in older chorion, some of the vertical lamellae of the block defect appear confluent with surrounding horizontal lamellae (Figures 5b,c).

5. *Discussion*

5.1. Origin of Patterns in Helicoidal Structures

What is the origin of the different patterns of lamellae and their defects and distortions in the chorion? We can visualize two types of mechanisms, which are not mutually exclusive. The first is primarily mechanical: preexisting structures might impose boundary conditions on fibrillar direction or lamellar nucleation, and changing shapes of the cellular secretory surface or changing amounts of secreted nonlamellar material might also affect the geometry of the lamellar chorion. The second major cause of pattern variation is the nature and concentration of the chorion proteins present, which depend on the temporally and spatially changing patterns of chorion gene expression.

In cholesteric liquid crystals, the term "texture" has been defined as the pattern formed by defects (Bouligand, 1975). Mechanical disruptions, variations in the concentration of crystal components or chemical solvents, and changes in temperature can transform one texture into another, and can also cause changes in the cholesteric pitch (Bouligand, 1973). Boundary conditions

---

of "filler" material and a group of seven microvilli (arrow) sectioned transversely. (Scale marker = 0.5 μm.) (c) TEM of an aeropyle channel in an immature chorion and an overlying three-cell junction. Interposed between the lamellar chorion and the microvillar border of the follicular epithelial cell is the electron-dense sieve layer (arrowhead). Where the sieve is absent, microvilli (arrow) penetrate the aeropyle channel, which also contains sparse fibrils of filler (f). (Scale marker = 1 μm.) (d) TEM of a forming aeropyle crown from an almost mature chorion. A prong of the aeropyle crown one to two lamellae in thickness separates the electron-dense sieve layer (arrowhead) from the dense network of filler (f) in the aeropyle channel. Chorion lamellae adjacent to the growing prong show multiple defects and distortions. (Scale marker = 0.5 μm.)

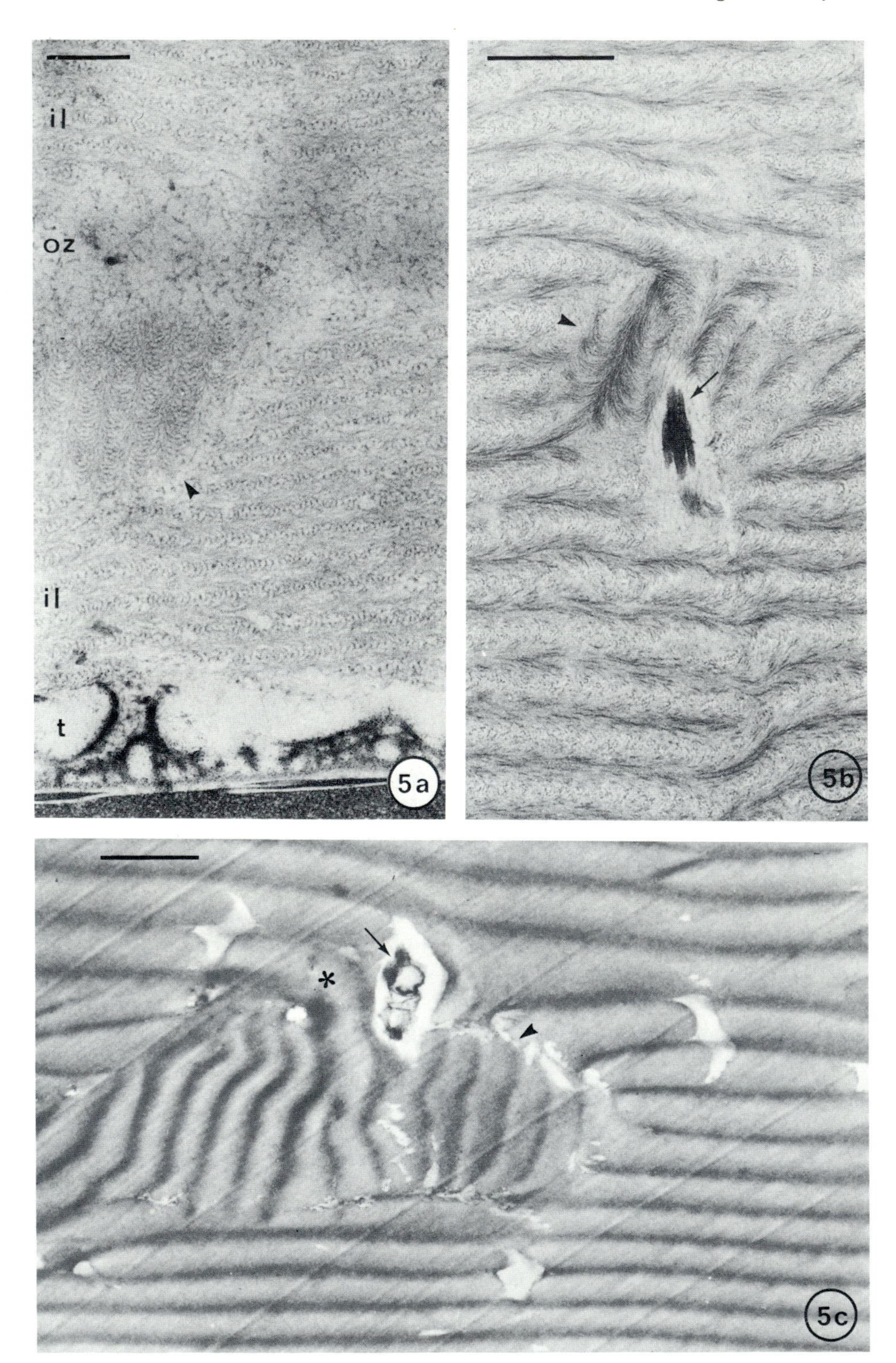
il
oz
il
t
5a
5b
*
5c

are also very important in determining texture; the type, spacing, and frequency of defects are a function of the surface (and its perturbations) on which the liquid crystal grows (Bouligand, 1975). For example, the limiting surfaces between which liquid crystals grow are often glass slides and coverslips that have been etched with fine parallel lines; defects arise due to the stringent anchoring conditions imposed on molecular direction by these highly oriented boundaries. Textures are also known to vary as a function of distance from the boundary between liquid crystal and isotrophic liquid; textures involving defects of higher strain energy are formed nearest the border (Bouligand, 1975).

5.2. The Morphogenetic Role of Changing Protein Concentrations

Changing protein concentrations appear to be the main source of different types of order within the chorion. We suspect that the prominent variations in pitch between the inner, holey, outer, and oblique layers of the chorion result from the different nature and relative concentrations of their early component proteins. It is known that the five major classes of chorion proteins (A–E) are produced according to a stable program of synthesis and secretion (Paul *et al.*, 1972; Paul and Kafatos, 1975). We have proposed that the multiple C-class proteins that are synthesized only briefly, early in choriogenesis, are responsible for the initial helicoidal framework of sixty thin lamellae (Regier *et al.*, 1982). Various temporal subclasses of C proteins can be distinguished. Thus, as the four circumferential chorion layers are laid down in sequence, they probably come to contain different amounts of "C" components. Relative variations in pitch between layers are evident as soon as the layers appear and may reflect these postulated differences in C protein compositions.

Crab and spider cuticles are known to show stable pitch variations, similar to those of the chorion layers. In crab, the thickest lamellae occur in the middle of the cuticle (Bouligand, 1971); while in spider exocuticle, lamellar width increases continuously from outside to inside (Barth, 1973).

The axial holes (vacancies) of the holey layer may also be due to its early protein composition; these holes are evident as soon as the holey layer lamellae are laid down, and are not obviously correlated with any mechanical constraint.

←

Figure 5. Transmission electron micrographs of block defects. (a) A very early chorion in which a block (arrowhead) of six or seven lamellae has been rotated around two axes. Arcs are prominent in the block defect lamellae, while the surrounding lamellae of the inner lamellar layer (il) show lines and dots indicative of vertically sectioned material. The defect is largely surrounded by the inner organizing zone (oz) containing sparse unoriented fibrils that have not yet crystallized into helicoidal arrangement. (t) = trabecular layer. (Scale marker = 0.25 μm.) (b) Inner lamellar region in early mid-choriogenesis. A block of four or five rotated lamellae (arrowhead) is seen in close apposition to an aeropyle channel containing a group of microvilli (arrow). (Scale marker = 1 μm.) (c) TEM of the inner lamellar region in very late choriogenesis (last pre-ovulatory follicle), showing a rotated block of at least twelve lamellae (arrowhead) in conjunction with what appears to be an aeropyle channel containing degenerating microvilli (arrow). Certain horizontal lamellae are confluent with rotated ones (*). (Scale marker = 1 μm.)

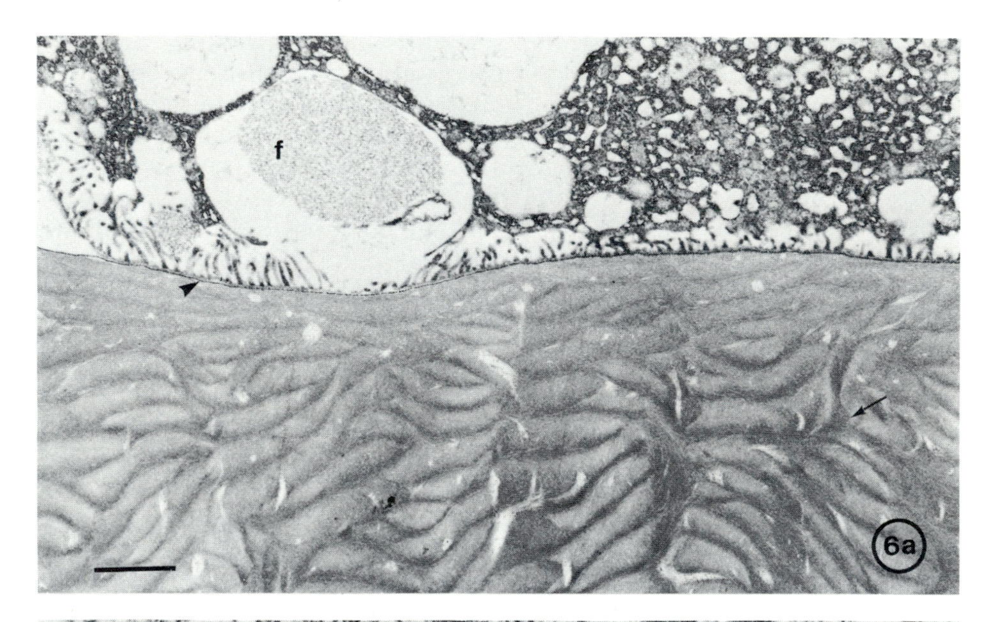
f
6a

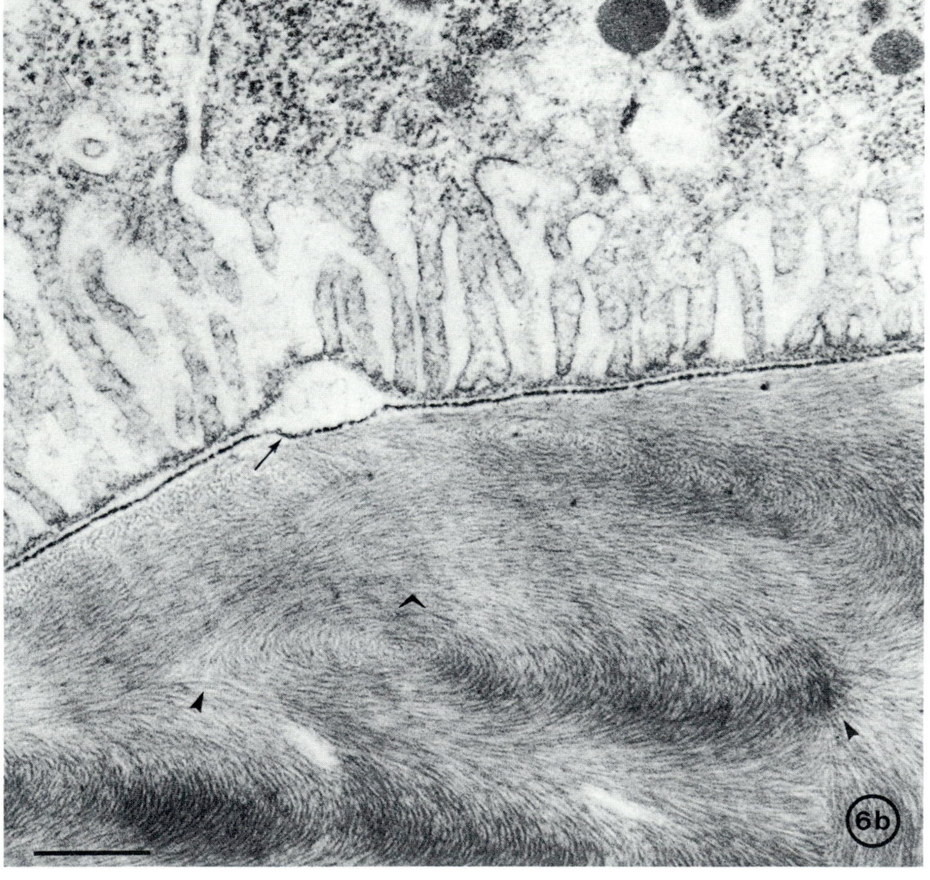
6b

New protein components appearing in middle and late choriogenesis modify the ordered structure of the early framework, as the insertion of new fibrils causes chorion expansion, and fibril thickening results in densification (Mazur *et al.*, 1982; Regier *et al.*, 1982). Very late proteins are responsible for the thin lamellar construction and projecting shapes of the aeropyle crowns formed on the chorion surface toward the end of choriogenesis (Mazur *et al.*, 1980; Regier *et al.*, 1980).

5.3. The Morphogenetic Role of Mechanical Constraints and Forces

While the changing protein concentrations are important for the orderly structure of the chorion and for its radial and regional variations, it appears likely that mechanical constraints are of major importance in the establishment of the various defects.

5.3.1. Screw Defects

The texture of defects in the inner lamellar layer may be dictated by the spacing of nucleation sites for the initial lamellae. Since lamellae are nucleated by the columns of the trabecular layer and spread laterally (Figures 7a,b), it is clear that mismatching of lamellae can easily occur as new lamellae are added on top of the columns while the first lamella is still forming between columns. According to this interpretation, the texture of the inner lamellar layer might be primarily due to periodic inhomogeneities of the surface on which it grows. It is noteworthy that in the chorion of *Bombyx*, which has fewer screw dislocations, the first few lamellae are deposited not on the tops of columns, but on a rather flat ceiling of the trabecular layer (Figure 1h); this ceiling does not exist in *A. polyphemus* (Figures 3b, 7e). Judging by one published micrograph (Smith *et al.*, 1971), the ceiling is also absent in *Hyalophora cecropia* and, correspondingly, screw defects are abundant.

Many cuticles (Neville and Caveney, 1969; Bouligand, 1971; Delachambre, 1971; Filshie and Smith, 1980) are notable for the extreme regularity of lamellae and the relative absence of translation defects. We suggest that the explanation may lie in the regularity of the surface of the epicuticle on which the lamellar cuticle forms. Apparently, the thin pore canals that characterize these cuticles are generally not sufficient to cause dislocations.

Figure 6. TEMs of disclinations and complex defects. (a) The outer edge of the aeropyle crown region in late choriogenesis The chorion lamellae contain abundant distortions and defects, including a triple point disclination (arrow). At the outer edge of the chorion, the sieve layer (arrowhead) separates the lamellae from the microvillar surface of the follicular epithelial cell in which a large bundle of filler (f) is undergoing exocytosis. (Scale marker = 1 μm.) (b) Outer edge of the aeropyle crown region in mid-choriogenesis. A complex defect in the form of a closed oblong is centered at ($\wedge$), and surrounded by disclinations (arrowheads). At the outer edge of the chorion beneath a two-cell junction, the sieve layer has parted, revealing its two components: a fuzzy layer directly apposed to microvilli, and a thinner highly electron-dense granular layer (arrow). (Scale marker = 0.5 μm.)

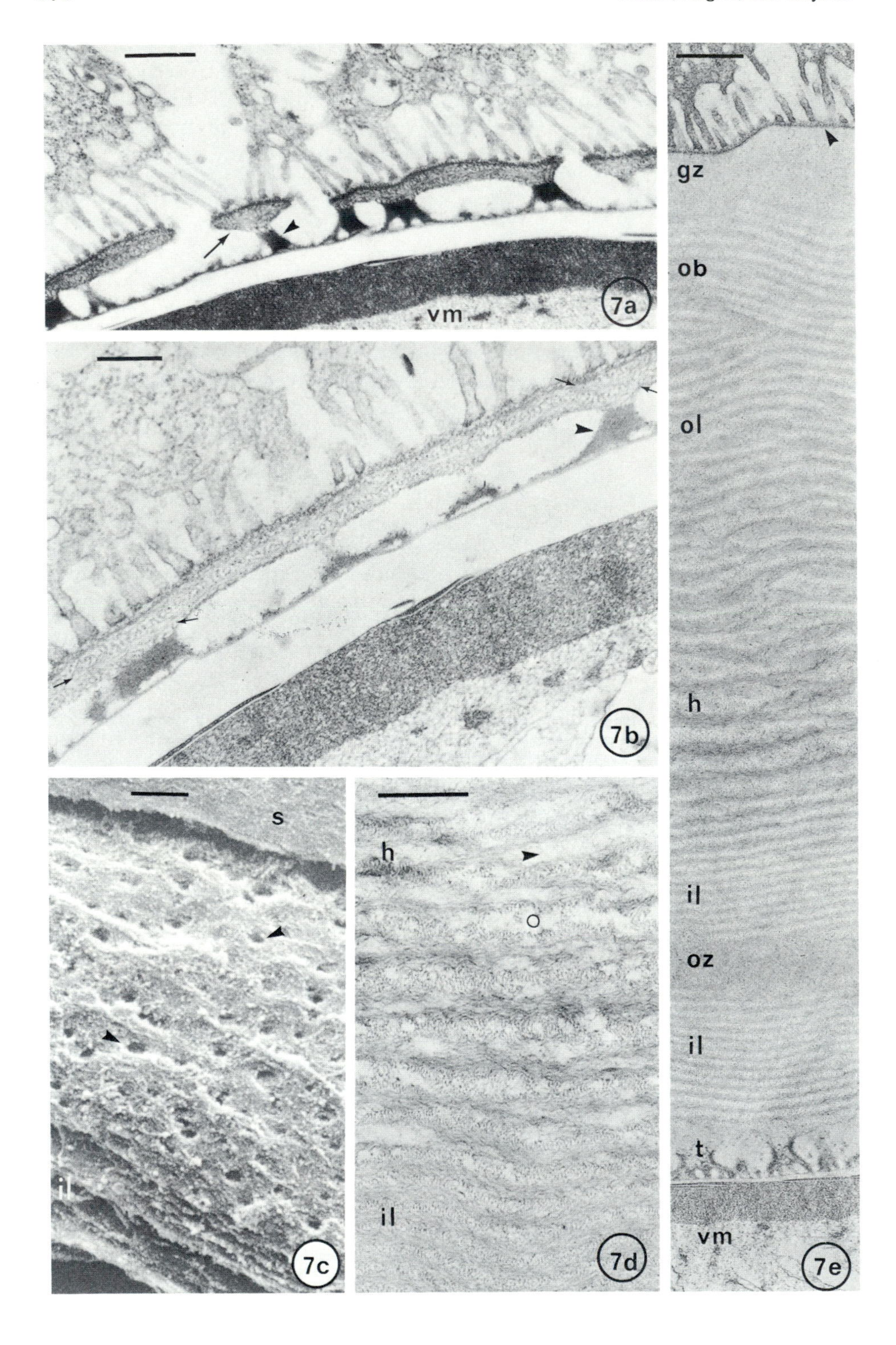
vm
7a
7b
s
il
7c
h
il
7d
gz
ob
ol
h
il
oz
il
t
vm
7e

Compared to the screw dislocations in the inner lamellar layer, those in the outer layer are not very prominent; they are fewer in number, and the cores are smaller in diameter and not evenly spaced (Figures 1d, 3e). It is possible that these dislocations are formed in response to the waviness of the outer surface of the holey layer, upon which the outer lamellar layer is formed.

5.3.2. Distortions and Disclinations

More than any other layer, the oblique layer is progressively distorted as choriogenesis proceeds, presumably because of changes in the two boundaries formed by the cell surface and by the architectural filler. In agreement with this interpretation, distortions and disclinations are maximal in the aeropyle crown region, while the lamellae of the flat region are rather smoothly bent and do not show many disclinations (Figure 1d). In the aeropyle crown region, filler synthesis and secretion increases sharply towards the end of choriogenesis, causing engorgement of the aeropyle channels (Mazur *et al.*, 1980). This enlargement is greatest in the outermost part of the chorion, and there distortions become most abundant (Figures 4d, 6a,b). Similarly, the newly forming lamellae of the aeropyle crowns are progressively sculpted by the combined forces of increasing amounts of filler on the inside of the crown and by the microvillar surface of the follicle cells on the outside (Figures 4c,d); this molding process results in prominent distortions and disclinations, mainly near the bases of the crowns (Figure 4d, 6a). It is interesting that disclinations of lamellae are found at the outermost part of the chorion; in cholesteric liquid crystals disclinations occur maily in the neighborhood of the border between liquid crystal and disordered isotropic liquid (Bouligand, 1975).

Figure 7. The origin of defects in very early chorion, (a) TEM. Initial crystallization of lamellae in the helicoidal arrangement (arrow) occurs at the top of the electron-dense trabeculae (arrowhead) of the trabecular layer. At this stage in choriogenesis (synthetic stage Ib; Mazur *et al.*, 1980), the chorion is about 4% of its final dry weight. (vm) = vitelline membrane. (Scale marker = 0.5 μm.) (b) TEM. The lamellae crystallize laterally between trabeculae (arrowhead) and continue to crystallize vertically above the trabeculae. Two sets of lamellar mismatch are indicated (pairs of small arrows). These probably represent primitive screw dislocations. At this stage (synthetic stage Ib+) the chorion is about 5% of its final dry weight. (Scale marker = 0.25 μm.) (c) SEM of a ripped section of very early chorion. The inner surface of the chorion (s) is uppermost, and the trabecular layer has been removed, revealing the early inner lamellar layer (il) with abundant screw dislocations seen as rounded holes (arrowheads). At this stage (synthetic stage Ic) the chorion is about 6% of its final dry weight. (Scale marker = 1 μm.) (d) TEM of the holey layer (h) and part of the inner lamellar layer (il) of very early chorion. Holes parallel to the direction of fibrils are seen in longitudinal section at lamellar boundaries (arrowhead) and in transverse section in the middle of lamellae (o). At this stage (synthetic stage Id) 8% of final dry weight of chorion has been deposited. (Scale marker = 0.5 μm.) (e) Transmission electron micrograph of a cross section through very early chorion. Adjacent to the sieve layer (arrowhead), which covers the microvillar surface of the follicular epithelial cell, is the outer growth zone (gz), a zone of apposition that contains unoriented, presumably newly secreted fibrils. The oblique (ob), outer lamellar (ol), holey (h), inner lamellar (il), and trabecular (t) layers are all seen. The transient inner organizing zone (oz) is slower to crystallize into helicoidal arrangement than the rest of the surrounding inner lamellar layer. This chorion is from the same stage as that in Figure 7d. (Scale marker = 0.5 μm.)

5.3.3. Fluidity and the Chorion

One of the biologically most interesting properties of liquid crystals is the possibility of movement within an ordered array. By this means, external materials can be translated to the interior of a crystal without unusual energy requirements, and with no global disruption of the pre-existing order. Dislocation cores can also serve as conduits for bulk flow of materials, if they remain empty as a consequence of their high strain. Obviously, screw dislocations can act as channels in a direction perpendicular to the layers, and edge dislocations in a direction parallel to the layers. In liquid crystals and fluid biological systems, such as dinoflagellate chromosomes, distortions as well as defects are probably continually changing (Livolant, 1977). In the chorion, the distortions and defects that appear during choriogenesis are frozen into place as the structure becomes progressively chemically bonded. The permanence of the resulting channels and the extensive interconnections between them presumably have important physiological consequences.

In the biological analogues of liquid crystals, such as chorion, the rate of hardening of the various components must be an important factor in morphogenesis. Newly secreted chorion components are highly soluble and can penetrate to the inner layers of the chorion, as autoradiographic studies have shown (Blau and Kafatos, 1979). Extraction studies imply that the secreted components are progressively bonded noncovalently, presumably after they reach their destinations within the chorion structure, and are massively cross-linked by disulfide bonds only at ovulation (Blau and Kafatos 1979; Regier and Wong, unpublished observations). It is intriguing that lamellar expansion by insertion of new fibrils and densification by fibril thickening both start from the innermost chorion lamellae and proceed outward, i.e., toward, rather than away from the follicle cells that secrete the components that cause expansion and densification (Smith *et al.*, 1971; Mazur *et al.*, 1982).

It appears that different parts of the chorion may have different rates of crystallization. Two zones of the chorion are free of lamellae during the first 10% of choriogenesis, even though the regions adjacent to them are already in helicoidal arrangement (Figure 7e) (Regier *et al.*, 1982). One of these zones (the outer growth zone) occurs at the outer edge of the chorion and may simply represent the most recently secreted lamellar precursors; the second zone, however, (the inner organizing zone) is in the middle of the inner lamellar layer, and cannot be explained in this manner.

Block defects are generally found in or at the edges of these zones, especially of the inner organizing zone (Figure 5a). These blocks first appear as crystallized domains in a disorganized surrounding. Later, as the organizing zone is replaced by the helicoidal lamellae, the spaces around the blocks disappear, and are filled with fibrils that align themselves helicoidally, in phase with either the bulk of the lamellae, or with the rotated block (Figures 5b,c).

The elongated microvilli occurring in aeropyle channels are often associated with block defects, and may promote their formation, either by physically rotating an already crystallized block in its more liquid surroundings, or by introducing new boundary conditions for the lamellae of the block to nucleate at right angles to the bulk of the lamellar stacking. According to this interpretation, the block defects would be caused by the interplay between mechanical constraints of the microvilli and the temporal program of protein synthesis, which may be responsible for the special composition of the inner organizing zone, causing it to lag behind the adjacent chorion in crystallization.

5.4. Possible Functions of Chorion Defects

5.4.1. Stress Dispersal

Analogies with other mechanical systems, both biological and nonbiological, might be useful in considering the biological and evolutionary significance of the observed patterns of defects in the chorion. It is known that notches, holes, and other flaws in an otherwise perfect material act as concentrators of stress (when ships break apart at sea, the crack that wrecks them often starts from a hatchway; Gordon, 1968). In bone it has been shown that the major discontinuities are aligned to minimize the stress concentrations that arise from them (Currey, 1962). Furthermore, the minor discontinuities in bone (for the most part, osteocyte lacunae) may compensate for their deleterious stress concentrating effects by acting as crack stoppers (Currey, 1962; Gordon, 1968).

The largest holes in the chorion are the aeropyle channels, 1 to 8 μm in diameter. They occur 10 to 20 μm apart, and penetrate throughout the thickness of the chorion. In the flat region, however, these channels curve in one plane through the oblique and outer lamellar layers, and change direction at the holey layer. In this fashion, by curving out of the linear direction, they may avoid acting as crack conductors.

In the holey layer, the abundant circumferential holes are about 0.2 μm in diameter, and at least 2 μm long. This layer may well act as a plastic sponge, deforming easily to relieve applied stress. It is also ideally placed to stop radially traveling cracks, in a manner analogous to that proposed for lacunae in bone. When the chorion is experimentally ripped, the rip will occasionally travel perpendicularly through the oblique and outer lamellar layers, and then change direction at the holey layer to run horizontally.

The finely dispersed radial screw dislocations of the inner lamellar layer are ideally placed to stop circumferential or oblique cracks. It may be that the screw shape of these dislocations confers added strength on the chorion structure by causing the helical interleaving of fibrils from one lamella to another around the core of the defect.

5.4.2. Transport Across the Chorion

In arthropod cuticle, the major system of conduits is the pore canals, which, like the chorion aeropyles, are formed by microvilli extending from the apical surface of the secretory cells (see review in Neville, 1975). Unlike the aeropyles of saturniids, pore canals are not spatially restricted to three-cell junctions; single cells in beetles, for example, may form as many as thirty pore canals (Neville, 1975).

Pore canals are narrow in relation to the lamellar thickness, and generally do not cause distortions or defects; they are twisted into a helicoidal ribbon shape by the cholesteric arrangement of the chitin–protein system. They can function as conduits to transport wax and impregnating materials such as polyphenols bound to proteins (Neville, 1975). There is some evidence that the microvilli in pore canals are capable of secreting chitin fibrils (Neville, 1975).

It is not known whether the microvilli in aeropyles secrete lamellar or other components, or even whether the aeropyles serve as extracellular conduits for components coming from the rest of the follicle cell. That possibility should be examined in future work, especially in connection with densification or other secondary modifications of the developing chorion.

5.4.3. Ventilation

The contradictory demands of mechanical strength, water conservation, and permeability to gases for respiration of the developing embryo are met by the chorion through four very different ventilation systems. The abundant radial pipes formed by the screw dislocations of the inner lamellar layer result in extensive but controlled communication between the cavities of the trabecular layer and the spongy holey layer (Figure 3d). Leading out from the holey layer are the wide flues or chimneys of the aeropyle channels, curving towards the chorion surface (Figure 4a). Each of these interconnecting systems runs predominantly in a different direction through the lamellae, minimizing both the possibilities for cracks and the rate of water loss.

Each of these systems may have a different origin: the aeropyles are cell-mediated channels, the holey layer may depend solely on its chemical components, and the radial screw dislocations may arise from spatial constraints on the initial lamellar assembly. While the holes in the holey layer and the cores of the screw dislocations generally appear empty throughout choriogenesis, the aeropyle channels are filled at first only with microvilli, then with a sparse distribution of filler as well (Figure 4b,c), and finally with a dense network of filler (Figure 4d) (Regier *et al.*, 1982). When the egg is laid, much of the filler disappears (unpublished observations), leaving the aeropyle channels free to function in air exchange.

An extensive system of wide, cylindrical, radial holes filled with hexagonally packed fibrils has been documented in the developing chorion of *H. cecropia* (Smith *et al.*, 1971). We do not find these channels in *Bombyx* or *polyphemus*, although both *cecropia* and *polyphemus* chorions have a similar set of aeropyle channels containing microvilli and filler, as well as holey layers of similar appearance. *Bombyx* chorion has very few cavities except for aeropyle channels; the holey layer is almost nonexistent, and the cores of randomly occuring screw dislocations do not appear to form interconnecting channels.

These differences may be related to different physiological needs of the three species. *Bombyx* diapauses at the egg stage, and the sparsity of air spaces may be a way of avoiding excessive water loss. While *cecropia* and *polyphemus* diapause as pupae, *cecropia* chorion is twice as thick (60 μm) as *polyphemus* chorion (30 μm), and so may require more air spaces for adequate ventilation. As we have noted elsewhere, the crowns of *polyphemus* appear to meet the necessary conditions for setting up convection currents within the eggshell by a Bernouli effect (Vogel and Bretz, 1972). Lacking aeropyle crowns, the *cecropia* chorion may have evolved its large, radial air spaces instead.

ACKNOWLEDGMENTS

It is a pleasure to acknowledge our great debt to Yves Bouligand for many illuminating conversations. We thank Ed Seling of the Museum of Comparative Zoology SEM Facility, Harvard University, for expert technical assistance with the scanning electron microscopy, Bianca Klumper for help with the figures, and Susann Foy for secretarial assistance. This work was supported by grants from NSF and NIH to F.C.K.

References

Barth, Friedrich G., 1973, Microfiber reinforcement of an arthropod cuticle: Laminated composite material in biology, *Z. Zellforsch.* **144**:409–433.

Blau, H. M., and Kafatos, F. C., 1979, Morphogenesis of the silkmoth chorion: Patterns of distribution and insolubilization of the structural proteins, *Develop. Biol.* **72**:211–225.

Bouligand, Y., 1965, Sur une architecture torsadée répandue dans de nombreuses cuticules d'arthropodes, *C. R. Acad. Sci. Paris* **261**:3665–3668.

Bouligand, Y., 1969, Sur l'existence de "pseudomorphoses cholestériques" chez divers organismes vivants, *J. Physique*, Colloque C4 **30**:C4-90.

Bouligand, Y., 1970, Aspects ultrastructuraux de la calcification chez les crabes, *Septième Congr. Intern. Micr. Grenoble* **3**:105–106.

Bouligand, Y., 1971, Les orientations fibrillaires dans le squelette des arthropodes, *J. Microsc.* (Paris) **11**:441–472.

Bouligand, Y., 1972a, Twisted fibrous arrangements in biological materials and cholesteric mesophases, *Tissue Cell* **4**:189–217.

Bouligand, Y., 1972b, Recherches sur les textures des états mésomorphes 2. Les champs polygonaux dan les cholestériques, *J. Physique* **33**:715–736.

Bouligand, Y., 1973, Recherches sur les textures des états mésomorphes. 4. La texture à plans et la morphogenèse des principales textures dans les cholestériques, *J. Physique* **34**:1011–1020.

Bouligand, Y., 1974, Recherches sur les textures des états mésomorphes. 5. Noyaux, fils et rubans de moebius dans les nématiques et les cholestériques peu torsadés, *J. Physique* **35**:215–236.

Bouligand, Y., 1975, Defects and textures in cholesteric analogues given by some biological systems, *J. Physique*, Colloque Cl **36**:Cl-331.

Bouligand, Y., 1978a, Cholesteric order in biopolymers. In *Mesomorphic Order in Polymers and Polymerization in Liquid Crystalline Media*, edited by A. Blumstein, American Chemical Society, Washington, D.C.

Bouligand, Y., 1978b, Liquid Crystalline Order in Biological Materials. In *Liquid Crystalline Order in Polymers*, edited by A. Blumstein, Academic Press, New York.

Bouligand, Y., 1980a, Geometry of (non smectic) hexagonal mesophases, *J. Physique* **41**:1297–1306.

Bouligand, Y., 1980b, Defects and textures of hexagonal discotics, *J. Physique* **41**:1307–1315.

Bouligand, Y., 1980c, Defects in ordered biological materials: Bibliographical notes. Lecture notes for Les Houches, Summer School 1980, North Holland Publishing Co., New York.

Bouligand, Y., 1980d, Geometry and topology of defects in liquid crystals: Bibliographical notes. Lecture notes for Les Houches, Summer School 1980, North Holland Publishing Co., New York.

Bouligand, Y., Soyer, M. O., and Puiseux-Dao, 1968, La structure fibrillaire et l'orientation des chromosomes chez les dinoflagellés, *Chromosoma* **24**:251–287.

Chapman, D., 1970, The chemical and physical characteristics of biological membranes. In *Membranes and Ion Transport,* edited by E. E. Bittar, vol. 1, Wiley-Interscience, London.

Currey, J. D., 1962, Stress concentrations in bone, *Quart. J. Microsc. Sci.* **103**:111–133.

Delachambre, J., 1971, La formation des canaux cuticulaires chez l'adulte de *Tenebrio molitor,* étude ultrastructurale et remarques histochimiques, *Tissue Cell* **3**:499–520.

Filshie, B. K., and Smith, D. S., 1980, A proposed solution to a fine structural puzzle: The organization of gill cuticle in a crayfish (*Panulirus*), *Tissue Cell* **12**:209–226.

Frank, F. C., 1958, On the theory of liquid crystals, *Discuss. Faraday Soc.* **25**:19–28.

Friedel, M. G., 1922, Les états mésomorphes de la matière, *Ann. Phys.* (Paris) **18**:273–474.

Furneaux, P. J. S., and Mackay, A. L., 1972, Crystalline protein in the chorion of insect eggshells, *J. Ultrastruct. Res.* **38**:343–359.

Gharagozlou-van Ginneken, I. D., and Bouligand, Y., 1973, Ultrastructures tégumentaires chez un crustace copepode *Cletocamptus retrogressus, Tissue Cell* **5**:413–439.

Giraud, M. M., 1977, Rôle de complexe chitino-protéique et de l'anhydrase carbonique dans la calcification tégumentaire de *carcinus maenas L.* Thèse de Doctorat de 3e cycle, l'université Pierre et Marie Curie, Paris 6, Biologie Animale.

Giraud, M. M., Castanet, J., Meunier, F. J., and Bouligand, Y., 1978, The fibrous structure of coelancanth scales: A twisted plywood, *Tissue Cell* **10**:671–686.

Gordon, J. E., 1968, *The New Science of Strong Materials, or Why You Don't Fall Through the Floor,* Pelican Books, London.

Gray, G. W., 1962, *Molecular Structure and the Properties of Liquid Crystals,* Academic Press, London.

Hackman, R. H., 1974, Chemistry of the insect cuticle, In: *The Physiology of Insecta,* 2nd ed., edited by M. Rockstein, vol. 6, pp. 215–270, Academic Press, New York.

Hackman, R. H., and Goldberg, M., 1976, Comparative chemistry of arthropod cuticular proteins, *Comp. Biochem. Physiol.* **55B**:201–206.

Hamodrakas, S., Jones, C. W., and Kafatos, F. C., 1982, Secondary structure predictions for silkmoth chorion proteins, *Biochim. Biophys. Acta* **700**:42–51.

Harris, W. F., 1977, Disclinations, *Sci. Amer.* **237**:130–146.

Hepburn, H. R. (ed.), 1976, *The Insect Integument,* Elsevier Scientific Publishing Co., Amsterdam.

Hinton, H. E., 1969, Respiratory systems of insect eggshells, *Annu. Rev. Entomol.* **14**:343–368.

Hinton, H. E., 1981, *Biology of Insect Eggs,* Pergamon Press, Oxford.

Jones, C. W., Rosenthal, N., Rodakis, G. C., and Kafatos, F. C., 1979, The chorion multigene families. II. The evolution of the two major families as inferred from protein and cDNA clone sequences, *Cell* **18**:1317–1332.

Kafatos, F. C., 1980, Recombinant DNA studies of the structure, evolution and developmental expression of structural gene families, *Amer. J. Trop. Med. Hyg.* **29**(5) Suppl.: 1111–1116.

Kafatos, F. C., Regier, J. C., Mazur, G. D., Nadel, M. R., Blau, H. M., Petri, W. H., Wyman, A. R., Gelinas, R. E., Moore, P. B., Paul, M., Efstratiadis, A., Vournakis, J. N., Goldsmith, M. R., Hunsley, J. R., Baker, B., Nardi, J., and Koehler, M., 1977, The eggshell of insects: Differentiation-specific proteins and the control of their synthesis and accumulation during development. In *Results and Problems in Cell Differentiation,* vol. 8, edited by W. Beermann, pp. 45–145, Springer-Verlag, Berlin.

King, R. C., and Aggarwal, S. K., 1965, Oogenesis in *Hyalophora cecropia, Growth* **29**:17–83.

Livolant, F., 1977, L'organisation spatiale du DNA dans les chromosomes de *Prorocentrum micans* (dinoflagellé). Étude Cytologique et Cytophysique, Thése de doctorat de 3e Cycle, l'Université Pierre et Marie Curie, Paris 6; Biologie Animale.

Livolant, F., Giraud, M. M., and Bouligand, Y., 1978, A goniometric effect observed in sections of twisted fibrous materials, *Biol. Cellulaire* **31**:159–168.

Mazur, G. D., Regier, J. C., and Kafatos, F. C., 1980, The silkmoth chorion: Morphogenesis of surface structures and its relation to synthesis of specific proteins, *Develop. Biol.* **76**:305–321.

Mazur, G. D., Regier, J. C., and Kafatos, F. C., 1982, Morphogenesis of the silkmoth chorion: Modification of the early framework (manuscript in preparation).

Neville, A. C., 1975, *Biology of the Arthropod Cuticle,* Springer-Verlag, New York.

Neville, A. C., and Caveney, S., 1969, Scarabaeid beetle exocuticle as an optical analogue of cholesteric crystals, *Biol. Rev.* **44**:531–562.
Neville, A. C., and Luke, B. M., 1971, A biological system producing a self-assembling cholesteric protein liquid crystal, *J. Cell Sci.* **8**:93–109.
Paul, M., and Kafatos, F. C., 1975, Specific protein synthesis in cellular differentiation. II. The program of protein synthetic changes during chorion formation by silkmoth follicles, and its implementation in organ culture, *Develop. Biol.* **42**:141–159.
Paul, M., Goldsmith, M. R., Hunsley, J. R., and Kafatos, F. C., 1972, Cellular differentiation and specific protein synthesis: Production of eggshell proteins by silkmoth follicular cells, *J. Cell Biol.* **55**:653–680.
Regier, J. C., Kafatos, F. C., Goodfliesh, R., and Hood, L., 1978a, Silkmoth chorion proteins: Sequence analysis of the products of a multigene family, *Proc. Nat. Acad. Sci. USA* **75**:390–394.
Regier, J. C., Kafatos, F. C., Kramer, K. J., Heinrikson, R. L., and Keim, P. S., 1978b, Silkmoth chorion proteins: Their diversity, amino acid composition and the amino terminal sequence of one component, *J. Biol. Chem.* **253**:1305–1314.
Regier, J. C., Mazur, G. D., and Kafatos, F. C., 1980, The silkmoth chorion: Morphological and biochemical characterization of four surface regions, *Devel. Biol.* **76**:286–304.
Regier, J. C., Mazur, G. D., Kafatos, F. C., and Paul, M., 1982, Morphogenesis of silkmoth chorion: Initial framework formation and its relation to synthesis of specific proteins, *Develop. Biol.* (in press).
Richards, A. G., 1951, *The Integument of Arthropods*, University of Minnesota Press, Minneapolis.
Robinson, C., 1961, Liquid crystalline structures in polypeptide solutions, *Tetrahedron* **13**:219–234.
Rodakis, G. C., 1978, The chorion of the lepidoptera *Antheraea polyphemus:* A model system for the study of molecular evolution. Ph.D. Thesis, Dept. of Biology, University of Athens, Athens, Greece.
Sakaguchi, B., Chikushi, H., and Doira, H., 1973, Observations of the eggshell structures controlled by gene action in *Bombyx mori, J. Fac. Agric. Kyushu Univ.* **18**:53–62.
Smith, D. S., Telfer, W. H., and Neville, A. C., 1971, Fine structure of the chorion of a moth, *Hyalophora cecropia, Tissue Cell* **3**:477–498.
Telfer, W. H., and Smith, D. S., 1970, Aspects of egg formation, *Symp. R. Ent. Soc. Lond.* **5**:117–134.
Tsitilou, S. G., Regier, J. C., and Kafatos, F. C., 1980, Selection and sequence analysis of a cDNA clone encoding a known chorion protein of the A family, *Nuc. Acids Res.* **8**:1987–1997.
Vogel, S., and Bretz, W. L., 1972, Interfacial organisms: passive ventilation in the velocity gradients near surfaces, *Science* **175**:210–211.
Weis-Fogh, T., 1970, Structure and formation of insect cuticle, *Symp. R. Ent. Soc. Lond.* **5**:165–185.
Wilkins, M. H. F., 1963, X-ray diffraction studies of the molecular configurations of nucleic acids. In: *Aspects of Protein Structure*, edited by G. N. Ramachandran, Academic Press, New York.
Willis, J. H., Regier, J. C., and Debrunner, B. A., 1981, The metamorphosis of arthropods, In: *Insect Endocrinology and Nutrition, a Tribute to G. S. Fraenkel*, edited by G. Bhaskaran, S. Friedman, and J. G. Rodriguez, Plenum Press, New York.

II

The Ultrastructure of Developing Cells

7

The Cytoplasmic Architecture of the Insect Egg Cell

DIETER ZISSLER AND KLAUS SANDER

"The secret of development lies in the composition of the fertilized egg; from it, all the rest follows of necessity."
C. P. Raven (1959)

1. Introduction

Investigations on pterygote insect oogenesis have yielded a multitude of data on the architecture and development of early and late oocytes (for reviews see King, R. C., 1970; Telfer and Smith, 1970; Mahowald, 1972; Anderson, 1974; Telfer, 1975; see also Miya *et al.*, 1969; Truckenbrodt, 1970; Büning, 1972). Oogenesis has also been studied in some Apterygota (Cone and Scalzi, 1967; Cantacuzène and Martoja, 1972; Palévody, 1972, 1973; Matsuzaki, 1973; Bilinsky, 1976, 1977, 1979; Klag, 1977, 1978). Yet the end product of oogenesis was usually exempted from investigation; the newly laid insect egg thus remains largely unknown with respect to ultrastructure. This may mainly be due to technical reasons: the prevalence of yolk and the extremely resistant egg covers of insect eggs provide considerable handicaps for fixation and sectioning. Descriptions of the egg shell are therefore much more frequent than publications dealing with the egg cell.

DIETER ZISSLER AND KLAUS SANDER • Institut für Biologie I (Zoologie), der Albert-Ludwigs-Universität; Albertstrasse 21a, D 7800 Freiburg i.Br., Federal Republic of Germany.

The external morphology of the egg has been studied in some insect species with the scanning electron microscope (Hinton, 1970, 1981; Sakaguchi *et al.*, 1973; Klug *et al.*, 1974; Turner and Mahowald, 1976). The ultrastructure of the egg covers (egg shell), which may vary greatly among species, has been investigated with the transmission electron microscope by R. C. King and Koch (1963), R. C. King and Aggarwal (1965), Richards and P. E. King (1967), P. E. King *et al.* (1969), Richards (1969), Quattropani and Anderson (1969), Furneaux *et al.* (1969), Furneaux and MacKay (1972), Truckenbrodt (1971), Smith *et al.* (1971), Chauvin and Barbier (1972), Barbier and Chauvin (1972a,b), Cruickshank (1972a), Chauvin *et al.* (1973), R. C. King and Cassidy (1973), Chauvin *et al.* (1974), Barbier and Chauvin (1974a,b), Dorn (1976), Kafatos *et al.* (1977), Blau and Kafatos (1979), and Margaritis *et al.* (1980). As for the egg cell itself, most ultrastructural descriptions were restricted to certain selected regions (R. C. King *et al.*, 1966; Overton and Raab, 1967; Wolf, 1969; Haget, 1970, 1972; Schwalm *et al.*, 1971). Attempts at analyzing the entire cytoarchitecture and the whole inventory of structures in the newly laid insect egg were made so far only in *Drosophila melanogaster* (Okada and Waddington, 1959), *Smittia* species (Zissler and Sander, 1973; 1977), *Bombyx mori* (Miya, 1978).

Egg cell structure must be intimately related to embryonic pattern formation, as expressed succinctly by Raven (see head quotation). In insects, contrary to earlier beliefs, the pattern of the future body is not predetermined in the ooplasm by a detailed mosaic of localized determinants: the embryonic pattern can be altered in a global way, even in the most advanced insect groups, during the period between egg deposition and the blastoderm or germ band stages (for reviews see Krause and Sander, 1962; Sander, 1976, 1981). This was strikingly demonstrated, for instance, in chironomid species, where early embryos may be caused to produce "double abdomens" or "double heads;" in the former case, the anterior part of the egg cell becomes reprogrammed to form a posterior partial pattern of reversed axial polarity, while in the latter the posterior egg regions form a reversed anterior partial pattern (for a review see Kalthoff, 1979). If both events occur together, a completely reversed embryo results (Rau and Kalthoff, 1980).

These and other experimental results (see Sander, 1976) led us to investigate the cytoplasmic architecture of egg cells from species that are amenable to experimental and/or genetic manipulation of embryonic pattern formation. It is our intention to study the spatial distribution of egg components that might serve as ooplasmic determinants of pattern formation, and to follow their rearrangements during early development, if such occur. The egg of the chironomid midge *Smittia* spec., besides offering the adventages of experimental reprogramming just described, is attractive because of its very small size, which might ultimately permit an exhaustive description of the ultrastructure of the entire egg cell. For comparison, we studied some sections of the much larger egg of *Chironomus anthracinus*. The pteromalid wasp *Nasonia* (*Mormoniella*) *vitripennis* was considered interesting because of its genetics and embryonic lethal mutations (Saul *et al.*, 1965). The egg of the cucujid beetle *Oryzaephilus surinamensis* provides mainly technical advantages. Compared to other beetle eggs, it is small, and fairly transparent. Moreover, the egg cell shows a longitudinal gradient of basophilia that might be significant for embryonic pattern formation (Haget, 1953; for reviews see Sander, 1976, 1981).

Unless indicated otherwise, the data presented in this review are from eggs that had not yet formed pole cells [P_0 stage; (Kalthoff and Sander, 1968)]; the *Nasonia* eggs were obtained from mated females, but it is not known whether they were fertilized or not. The egg cover was punctured in a single spot to provide access for the fixatives. The sections shown in the figures were taken far away from the site of puncture, so they should be free from artifacts caused by major structural derangement.

2. The General Elements of Egg Architecture

The general architecture of an insect egg can be described using semithin sections of *Smittia* (Figures 1-4) as an example. The outer cover is an egg shell consisting of several layers. The cell surface, the oolemma, has the characteristics of a unit membrane and may sometimes be folded. Inside the oolemma lies a largely yolk-fee periplasm, and this in turn envelops a central system consisting of yolk particles and endoplasm. In the posterior egg pole, the periplasm contains specific structures, the polar granules, collectively known as the oosome (Figures 1,4); these mark the germ line in a variety of insect eggs (Mahowald, 1971a; Beams and Kessel, 1974; Eddy, 1975). We shall describe the three major components, periplasm, endoplasm, and yolk, and discuss specific structures and organelles residing in them.

2.1. Periplasm and Endoplasm

Periplasm and endoplasm consist of a cytoplasmic matrix that is rich in free ribosomes and encloses a multitude of organelles and other components. The periplasm, which lacks yolk particles, is somewhat thicker at the egg poles, particularly at the anterior pole in *Smittia*, where it may protrude inward as a cone-shaped plug (Figure 3). The endoplasm consists of a network of strands varying in thickness that here and there form "cytoplasmic islands" of different sizes (Figures 1, 2). Some of these islands contain cleavage nuclei. Peripherally, the endoplasm merges into the periplasm without any change in the ultrastructural aspect.

Among the inclusions in the cytoplasmic matrix, polysomes were seen in P_0 stage eggs of *Nasonia* but could not be detected in *Smittia* eggs, where at this stage polysomes appear to be rare on biochemical evidence, too (Jäckle, 1979). The most numerous organelles are the mitochondria, which appear rather evenly distributed. In sectioned eggs of *Smittia*, a few constricted mitochondria are seen, and these may be in stages of division. Better evidence for mitochondrial division is available from *Nasonia* and *Oryzaephilus*, where some mitochondria show internal partitions while the outer membrane is continuous (Figures 5, 6). This configuration is known to represent early division stages (Baxter, 1971).

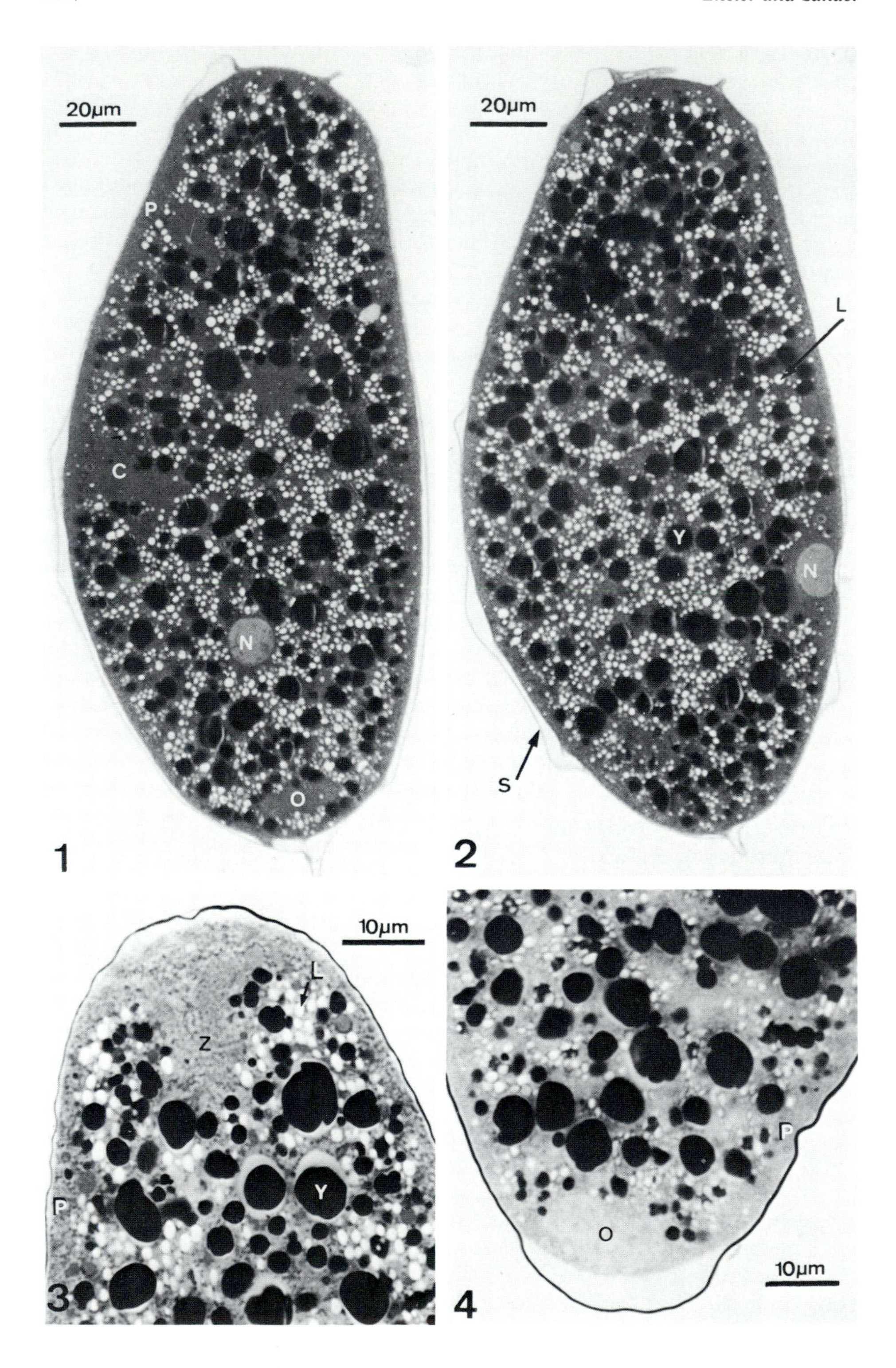
20µm
P
C
N
O
1
20µm
L
Y
N
S
2
10µm
L
Z
P
Y
3
P
O
10µm
4

Other structures found are cisternae and vesicles of smooth and rough ER, sometimes continuous with annulate lamellae. Isolated microtubules were observed only very rarely, while microtubules associated with other cell organelles seem to be more frequent (sections 2.2 and 3.2.2). Dictyosomes were found occasionally in *Smittia* and more frequently in *Oryzaephilus* and *Nasonia*.

2.2. Nuclei

Smittia eggs 80–100 min old are undergoing the first cleavage mitosis or contain two nuclei (Figures 1,2). In the three binucleate eggs that were analyzed in detail, both nuclei were located in the posterior half of the egg, one centrally in the yolk system about 50 μm from the posterior egg pole (Figure 1) and the other near the dorsal surface of the egg in a cytoplasmic island connected with the periplasm (Figure 2). Sectioned nuclei are oval to nearly circular, with diameters up to about 10 μm (Figures 7,8). The nuclear envelopes show many pores. The perinuclear cistern is of narrow but uniform width throughout. The nuclei are surrounded by a halo of 0.2 μm thickness and reduced electron density. Over wide stretches, cisternae separate this halo from the surrounding cytoplasm, which contains numerous mitochondria. The cisternae are equidistant from the nuclear envelope, and are usually arranged in a single layer. Within the nucleus, lamellar fragments consisting of two closely applied unit membranes can frequently be seen. Both pecularities may be related to the mode of formation of the nuclear envelope. We assume that after anaphase, the star- or rosette-shaped configurations described in section 3.2.1 organize the construction of the nuclear envelope by producing cisternae of endoplasmic reticulum (ER) which wrap in several layers around the chromatin. While the most interior cisternae fuse to form the typical nuclear envelope, some remnants of the other cisternae may persist and thus cause the "empty" halo. However, we cannot exclude the possibility that the halo is due to microtubules, as seen around the cleavage nuclei of the pole cell stage and the nuclei of the pole cells in *Smittia* (section 3.7), or to microfilaments that are known to envelop the oocyte nuclei in *Drosophila virilis* (Kinderman and R. C. King, 1973) and *Heteropeza pygmeae* (Went *et al.*, 1978); these structures might have been lost

---

Figures 1 and 2. Parallel longitudinal sections (thickness 1 μm) through the same *Smittia* egg fixed about 100 min after oviposition. These show the gross distribution of the main constituents and the positions of the nuclei (N). The anterior pole is oriented toward the top; the dorsal side toward the right. Sequential glutaraldehyde-OsO_4 fixation, methylene blue-azure II staining. C, cytoplasmic island; L, lipid; O, oosome; P, periplasm; S, egg shell; Y, proteid yolk. Reproduced from Zissler and Sander (1977).

Figures 3 and 4. Semithin sections (1 μm) of *Smittia* eggs photographed with the light microscope, as in Figures 1 and 2.

Figure 3. A longitudinal section through the anteriormost 25 μm of the egg. A somewhat paraboloid cone of periplasm (Z) intrudes into the anterior face of the yolk-endoplasmic system near the anterior egg pole. Fixation and staining as in Figures 1 and 2. L, lipid; P, periplasm; Y, proteid yolk.

Figure 4. A longitudinal section through the posterior pole region, showing the lens-shaped oosome (O). Permanganate fixation, staining as above, P, periplasm. Reproduced from Zissler and Sander (1973).

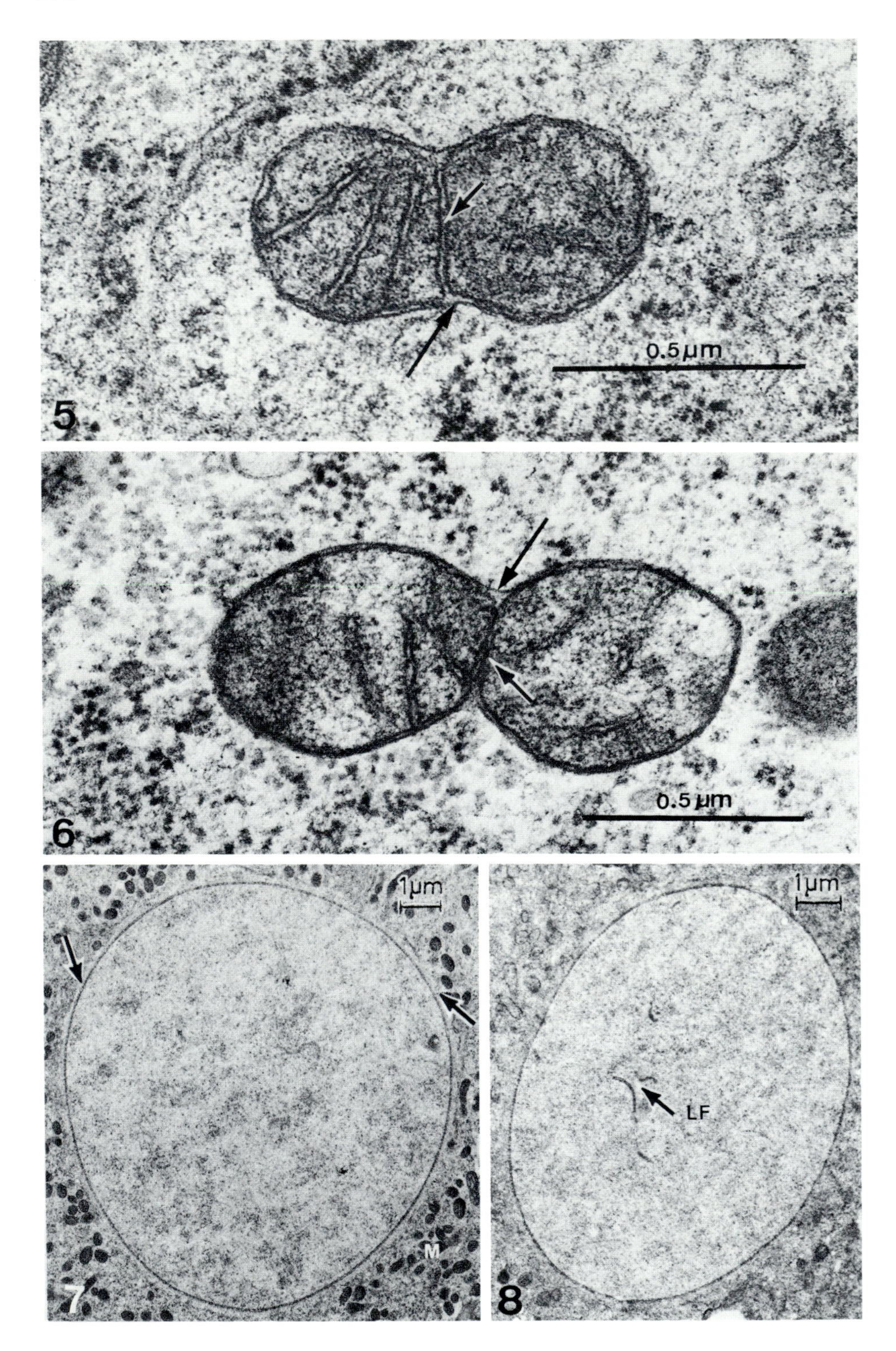
0.5 µm
5
0.5 µm
6
1µm
M
7
1µm
LF
8

during fixation. Associations and interactions of the nuclear envelope with microtubules and microfilaments occur widely (for reviews see Franke and Scheer, 1974; Wunderlich *et al.*, 1976).

The *Nasonia* eggs so far studied with the electron microscope contained two nuclei, in agreement with observations using the light microscope (Bull, 1981). As in *Smittia* (section 3.2.1), the nuclear envelope appears to be reconstructed after mitosis from star- to rosette-shaped configurations. From these configurations numerous ER cisternae extend, forming a densely packed multilayered envelope around the chromatin (Figure 22). This envelope resembles the multilayered nuclear envelope that Wolf (1969; 1980) observed in the egg of the gall midge *Wachtliella persicariae*, particularly in the prospective somatic region. The cleavage nuclei of *Nasonia* show some pecularities (Figures 5–8). During telophase, the future daughter nuclei move apart, and dense material is deposited in the equatorial region of the mitotic spindle (Figures 9, 10). The nuclei look very compact at this stage and may represent a specific state connected with nuclear movement during cleavage. This interpretation is supported by the finding that the nuclear pores are lacking the central granule, which indicates that nucleocytoplasmic exchange does not occur during this stage (Wunderlich *et al.*, 1976). Such compact nuclei are frequently seen during pole cell formation, and the nuclei approaching the peripheral region of the egg cell are also in the compact state. Shortly before the blastoderm stage, when they have entered the periplasm, the typical perinuclear cistern is fully developed. The nuclei are then pear-shaped, and in the constricted region the nucleolar material is seen as dense lumps. This particular part of the nucleus is connected with a cytaster (Figure 10).

2.3. Storage Substances

The meshes of the endoplasm are filled by storage substances that can be classified as proteid yolk bodies, lipid droplets, and fields of glycogen (R. C. King, 1960). The contributions that these substances make to the total volume of the yolk-endoplasm system is much lower in *Smittia* and *Nasonia* than in the "yolk-rich" (Krause, 1939) egg of *Oryzaephilus*. The bulk of the yolk belongs to different categories in different species. Proteid yolk bodies dominate in *Smittia* and *Oryzaephilus*, while the yolk system of *Nasonia* is characterized by extensive amounts of glycogen; thus the "endoplasm-rich" eggs of *Smittia* and *Nasonia* differ with respect to the prevalent yolk substances. Glycogen occurs in fields that are not separated from the surrounding cytoplasm

←

Figures 5 and 6. "Partitioned" mitochondria. The contents of the mitochondria appear divided into two cristae-containing compartments separated by a partition (short arrows), while the whole is surrounded by a continuous outer membrane (long arrows). Figure 5 was taken from the egg of *Nasonia (Mormoniella) vitripennis*, while Figure 6 is from the egg of *Oryzaephilus surinamensis*.
Figures 7 and 8. Sections through nuclei from the central (Figure 7) and the peripheral (Figure 8) regions of the egg at the two-cell stage of *Smittia*. Arrows in Figure 7 point to the perpendicular halo. LF, lamellar fragments; M, mitochondria. Reproduced from Zissler and Sander (1977).

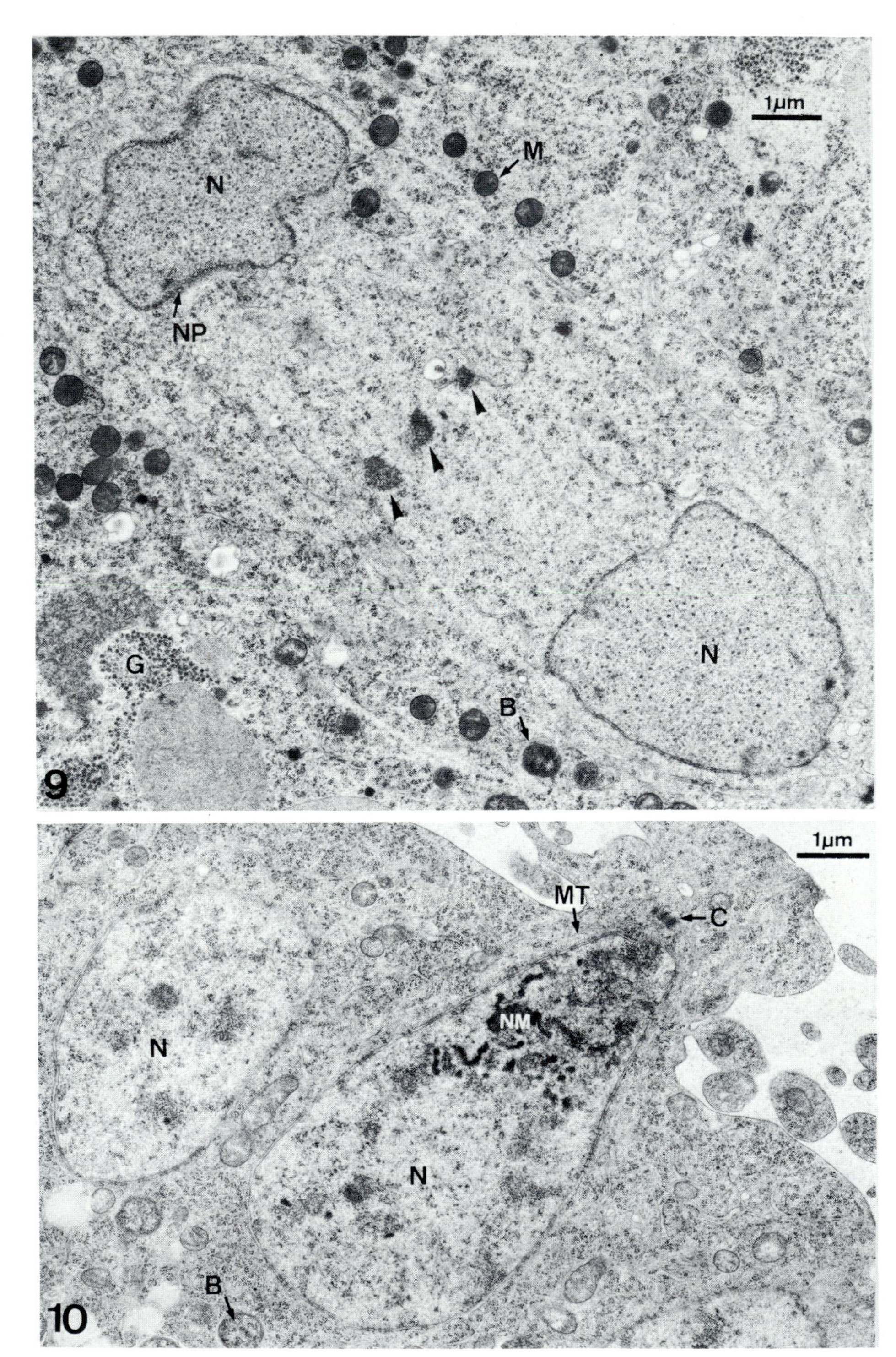
1µm
M
N
NP
G
B
N
9
1µm
MT
C
NM
N
N
B
10

by membranes or any other limiting structures. These fields vary in size and may measure up to 14 × 25 μm in *Nasonia*. The globular lipid droplets are surrounded by a monolayer, which, following Yatsu and Jacks (1972) and Kwiatkowska (1973), may be considered as a "half unit membrane."

The proteid yolk bodies differ in shape and especially in structure among the species described. The diameter of the spherical or ovoid proteid yolk particles in the egg of *Smittia* ranges from 1.5 to 8 μm. The contents of these particles are uniform or consist of several fragments of electron dense material frequently surrounded by a less dense granular matrix. The structure of the dense material may be amorphous and/or crystalline, both states occuring side by side (Figure 11). Sections in appropriate planes demonstrate that the crystals consist of hexagonal subunits that aggregate to form honeycomb arrangements (Figure 12). Crystalline proteid yolk is also known in *Aedes aegypti* (Roth and Porter, 1964), *Drosophila melanogaster* (R. C. King *et al.*, 1966), and many non-insect species (Nørrevang, 1968). All *Smittia* yolk bodies are delimited on their outer faces by a trilaminar membrane. This is in agreement with the conclusion of Giorgi (1974) that in *Drosophila* the membranes of the yolk platelets come from two sources, the oolemma as an "external" source and the Golgi apparatus and/or the endoplasmic reticulum as an "internal" source. In the *Nasonia* egg some proteid yolk bodies resemble those found in *Smittia*; as shown by P. E. King *et al.* (1972), they carry crystalline inclusions which were also noted by Cassidy and R. C. King (1972) and R. C. King and Cassidy (1973) in *Habrobracon*. Besides these structures, we found some spherical or ovoid structures not covered by a membrane but consisting of a dense, granular or fibrillar substance enclosing less dense areas (Figure 13). The majority of the proteid yolk bodies are membrane bound and filled by a granular mass of rather low electron density that surrounds one or several small and dense core structures (Figure 14). The yolk bodies of *Oryzaephilus* differ from those described so far. They measure from 5 to 8 μm and sometimes even 12 μm in diameter, are spherical or lobated, and appear to be surrounded by a membrane. They contain dense material shaped as large lumps, or a more or less coarse network embedded in a matrix of particles that resemble glycogen in size and structure (Figure 15).

The distribution of storage substances seems completely random in the egg cells of *Smittia* and *Nasonia*. We did not recognize a specific distribution pattern as observed in the ooplasm of *Musca domestica* (Engels, 1971), where large assemblies of glycogen occur near the surface and smaller assemblies near the center. In *Oryzaephilus* we gained the impression that the yolk globules containing large lumps of dense material are more frequent toward the

←

Figure 9. Two cleavage nuclei (N) of *Nasonia* immediately after division. Dense bodies are formed at the equatorial region of the spindle. These resemble chromatin rather than midbodies, but their identity remains uncertain. G, glycogen; M, mitochondrion; B, bacteroid; NP, nuclear pore.

Figure 10. Nuclei (N) in the periplasm of *Nasonia* shortly before the blastoderm stage. The "neck" of the pear-shaped nucleus is connected to a cytaster. From the centriole (C) some microtubules (MT) extend toward the nucleus and along its flanks. NM, nucleolar material; B, bacteroid.

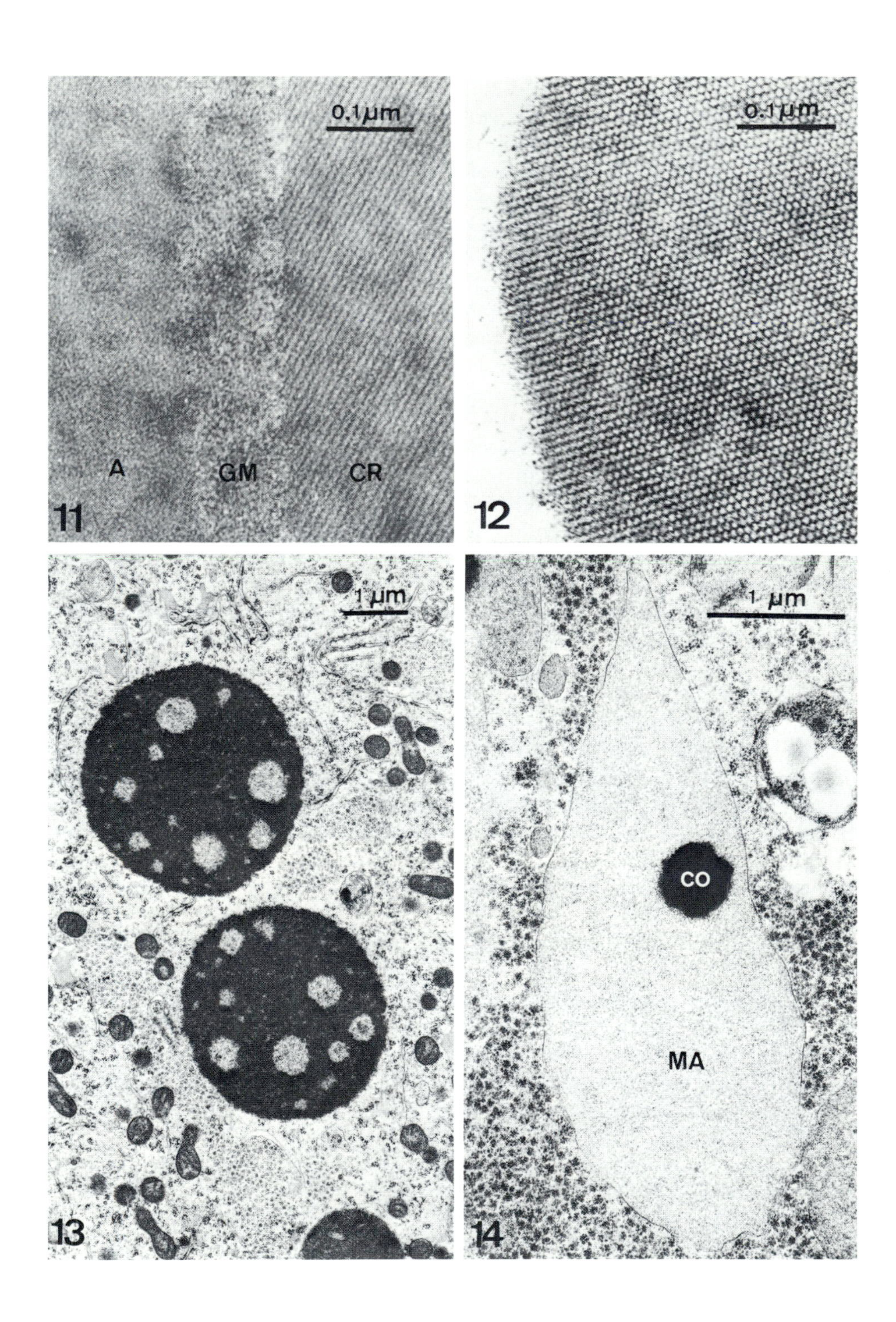
0.1 μm
A
GM
CR
11
0.1 μm
12
1 μm
13
1 μm
CO
MA
14

poles, while in the middle of the egg those globules dominate that seem to contain a network of dense material. This impression, however, requires further confirmation.

3. Special Structures

Besides the ubiquitous structures just mentioned, the insect egg cells described here contain a number of special structures that can be classified under the following subheadings.

3.1. Vesicular Structures

This category, which is especially well represented in the *Smittia* egg, includes a number of structures not easily characterized. A common feature is an "envelope" consisting of a single membrane (Figures 16–18). The interior is of multivesicular appearance (Figure 16) or may contain electron-dense material that, at least in some instances, shows the crystalline structure known from proteid yolk bodies (Figure 17). Other specimens in this category might be related to lysosomes (Figure 18). If these different structures should be linked by common origin and/or function, their occurrence during early cleavage could mean that some yolk bodies are not fully matured at oviposition or that yolk degradation is already under way soon afterward. Giorgi and Jacob (1977) also suggest a role for lysosomal enzymes in the maturation of yolk material.

3.2. Perinuclear Structures

3.2.1. Star- or Rosette-Shaped Configurations

Endoplasm and periplasm of *Smittia* contain star- or rosette-shaped configurations consisting of a rather dense central body measuring about 0.5–1 μm in diameter, from which ER cisternae radiate; the dense body has a finely granulated matrix filled with vesicular or tubular elements (Figures 19,20). These ER configurations are also found in pole cells and blastoderm cells. Their

←

Figures 11–14. Proteid yolk bodies.

Figures 11 and 12. Parts of yolk bodies from *Smittia* eggs, the former showing amorphous (A) and crystalline (CR) regions embedded side by side in a granular matrix (GM), the latter displaying the hexagonal arrangement of subunits in the crystalline regions. Reproduced from Zissler and Sander (1977).

Figures 13 and 14. Sections from *Nasonia* eggs. Figure 13 represents the less frequent type of proteid yolk particle. It consists of dense globules not covered by a membrane but speckled with less dense areas. Figure 14 shows an ovoid yolk body of the more frequent type. It consists of a membrane-bound granular mass of low electron density (MA) with a dense central core (CO). Similar bodies in *Habrobracon* were interpreted to be accessory nuclei by Cassidy and R. C. King (1972). However, in *Nasonia* they are lacking the double membrane and contain a crystalline core.

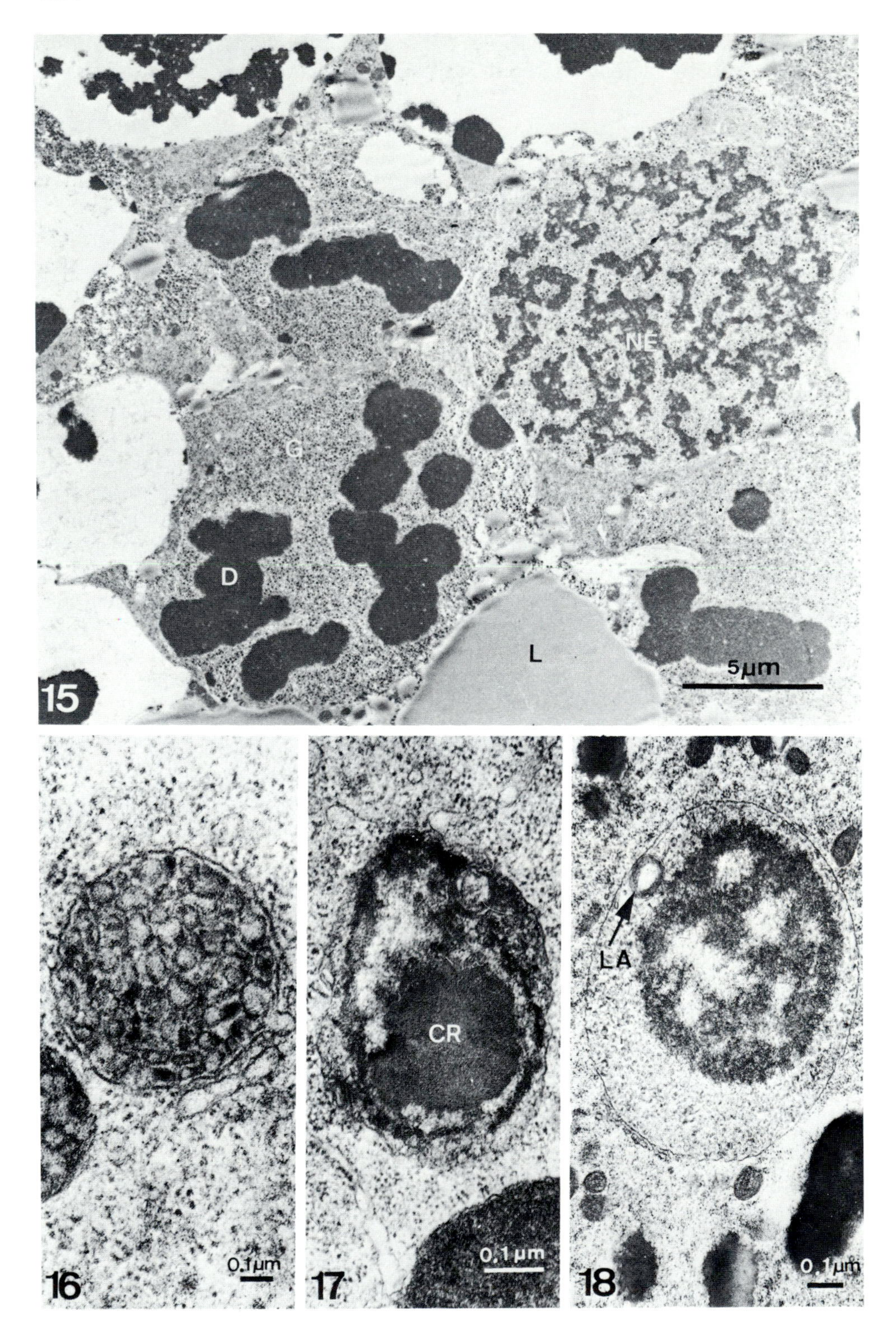
NE
G
D
L
5μm
15
0.1μm
16
CR
0.1 μm
17
LA
0.1μm
18

position and shape may indicate that they produce membrane cisternae serving to construct the nuclear envelopes after mitotic divisions (Figure 21). We earlier proposed the name "nuclear membrane organizing centers" for these configurations (Zissler and Sander, 1977), but they might serve other functions, too; this is indicated by their proximity to the "accessory nuclei" (section 3.2.3). Similar configurations occur in the *Nasonia* egg (Figure 22). Here they usually cluster in groups incorporated into membrane whorls that may represent concentrically stacked cisternae enveloping chromatin particles from dividing cleavage nuclei. In *Nasonia,* too, these structures seem to supply membrane material for the future nuclear envelopes. This view is supported by observations from spermatogenesis. In the secondary spermatocytes of *Nasonia,* Hogge and P. E. King (1975) have described similar structures as "rosettes." These authors assume that the rosettes produce ER, and that this contributes to the membrane system that aggregates around the chromosomes during the second meiotic division. The membrane system, in turn, gives rise to the typical envelopes of the daughter nuclei.

3.2.2. Centrioles and Cytasters

The centrioles in *Smittia* and *Nasonia* contain the usual circular set of nine triplet tubules. They carry a "hub and spokes" cartwheel at one end, as is typical for centrioles (Petersen and Berns, 1980), and are surrounded by clouds of osmiophilic material (Figures 23–25).

As already mentioned (section 3.2.1), in *Nasonia* the nuclear space is surrounded by many cisternae and star-shaped configurations at metaphase. Subsequently, the stacks of cisternae assume spindle shape and the number of membrane layers decreases. The cisternae are clearly connected with the ER, especially at the two poles of the spindle-shaped configuration where some stacks spread out into a network. Centrioles can be seen in the center of this network, and an astral ray arrangement of cisternae indicates the presence of a cytaster (Figure 23). Cytasters surrounding centrioles were also found shortly before the blastoderm stage. They occur close to the nuclei on the side facing the egg surface. At this stage, the centriole is possibly associated with a satellite, a procentriole, or even a second centriole oriented at right angles (Figure 10), as described for the blastoderm cells of *Drosophila* (Fullilove and Jacobson, 1971).

Wolf (1980) has shown in *Wachtliella persicariae* that the first cytaster arises a few minutes after oviposition in the cone of periplasm at the anterior

←

Figure 15. The proteid yolk bodies of *Oryzaephilus* contain dense material (D) embedded as large lumps or in the shape of a network (NE) in a matrix of particles that resemble glycogen in size and structure (G). L, lipid.

Figures 16–18. Vesicular bodies enclosed by a single membrane, seen in newly laid eggs of *Smittia.* The interior contains multiple vesicles (Figure 16) or electron-dense material (Figure 17) showing the crystalline structure (CR) characteristic of proteid yolk bodies. Alternatively, the bodies may contain lamellar structures (LA) in addition to granular and vesicular materials (Figure 18). Reproduced from Zissler and Sander (1977).

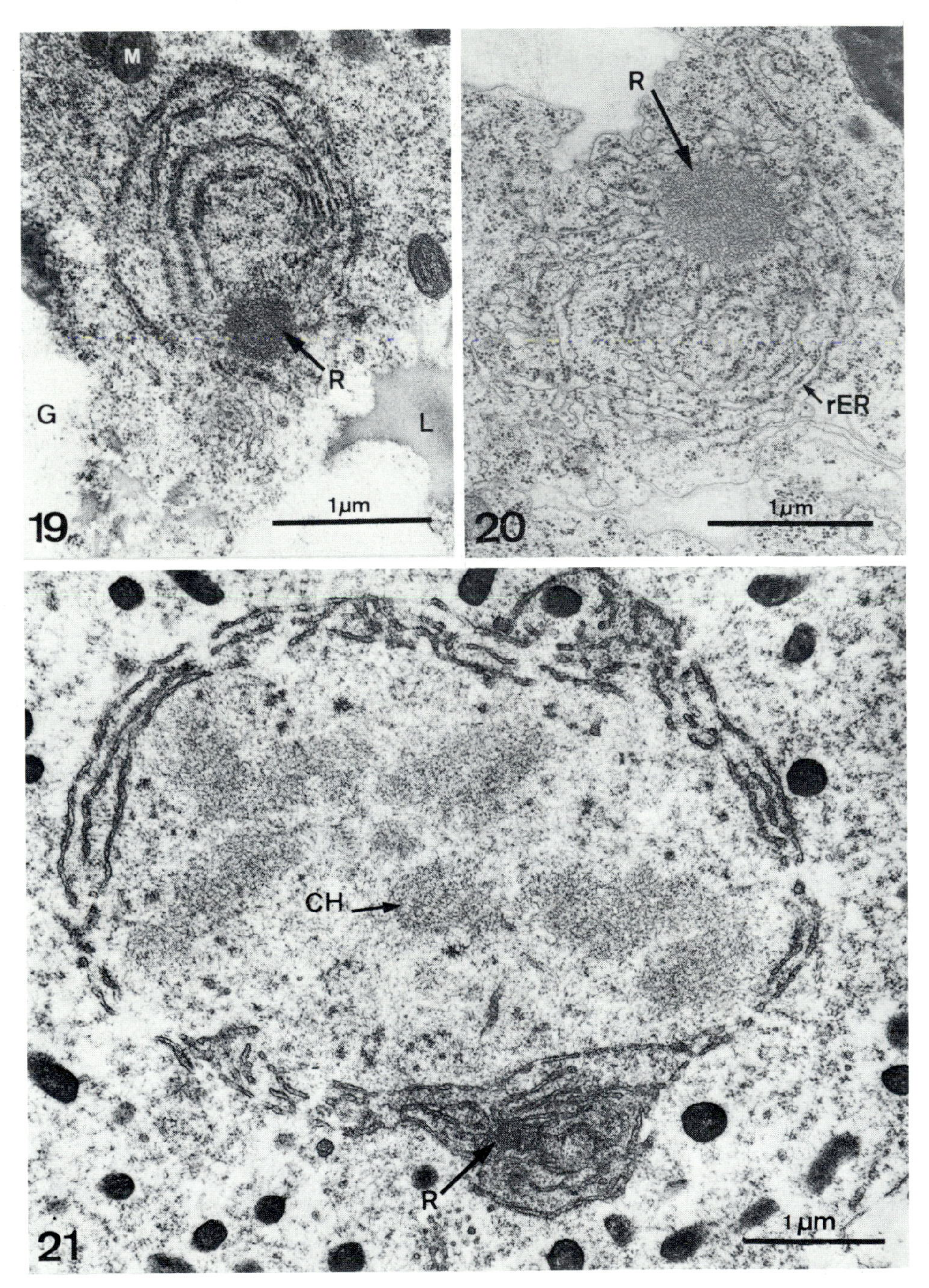

Figures 19–21. Star- or rosette-shaped configurations (R). These are shown in the endoplasm of the newly laid egg (Figure 19) and a pole cell stage (Figure 20), and contacting the incipient envelope of a metaphase nucleus in the blastoderm of *Smittia* (Figure 21). CH, chromatin; G, glycogen (extracted during preparation); L, lipid; M, mitochondrion; ER, rough endoplasmic reticulum. Figure 19 reproduced from Zissler and Sander (1977).

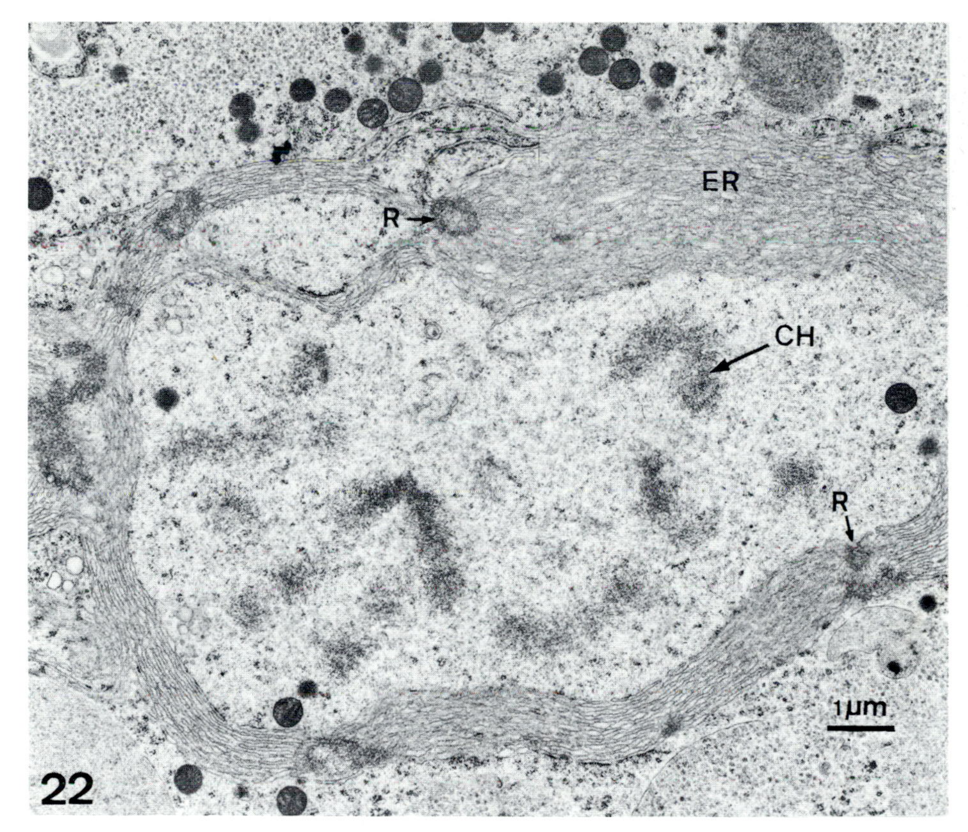

Figure 22. Star- or rosette-shaped configurations (R) among stacks of endoplasmic reticulum (ER) that surround a metaphase chromosome configuration of a pole cell stage *Nasonia* egg. CH, chromatin; M, mitochondrion.

egg pole; this occurs even in unfertilized eggs where no cleavage mitoses ensue. This observation parallels our finding of an isolated cytaster in the anterior pole plasm of the newly deposited *Smittia* egg, quite remote from the nuclei which are restricted to the posterior third of the egg cell at that stage (Zissler and Sander, 1973). In *Wachtliella,* when the cytaster fails to make contact with a nucleus, it nevertheless starts dividing and gives rise to an increasing number of anucleate cytasters (Wolf, 1980).

3.2.3. Circular or Spherical Configurations of ER Cisternae

Periplasm and endoplasm of *Smittia* occasionally harbor circular or spherical configurations of cisternae that are usually connected to rough or smooth ER. Three distinct shapes can be recognized (Figures 26–28): (1) The configurations of the first type contain granular to fibrous substances, and their coverings consist of two membranes separated by a cistern, the outer membrane being studded with ribosomes. Representations like those in Figure 26 suggest that such structures are derived from ER whorls. (2) Bodies of similar

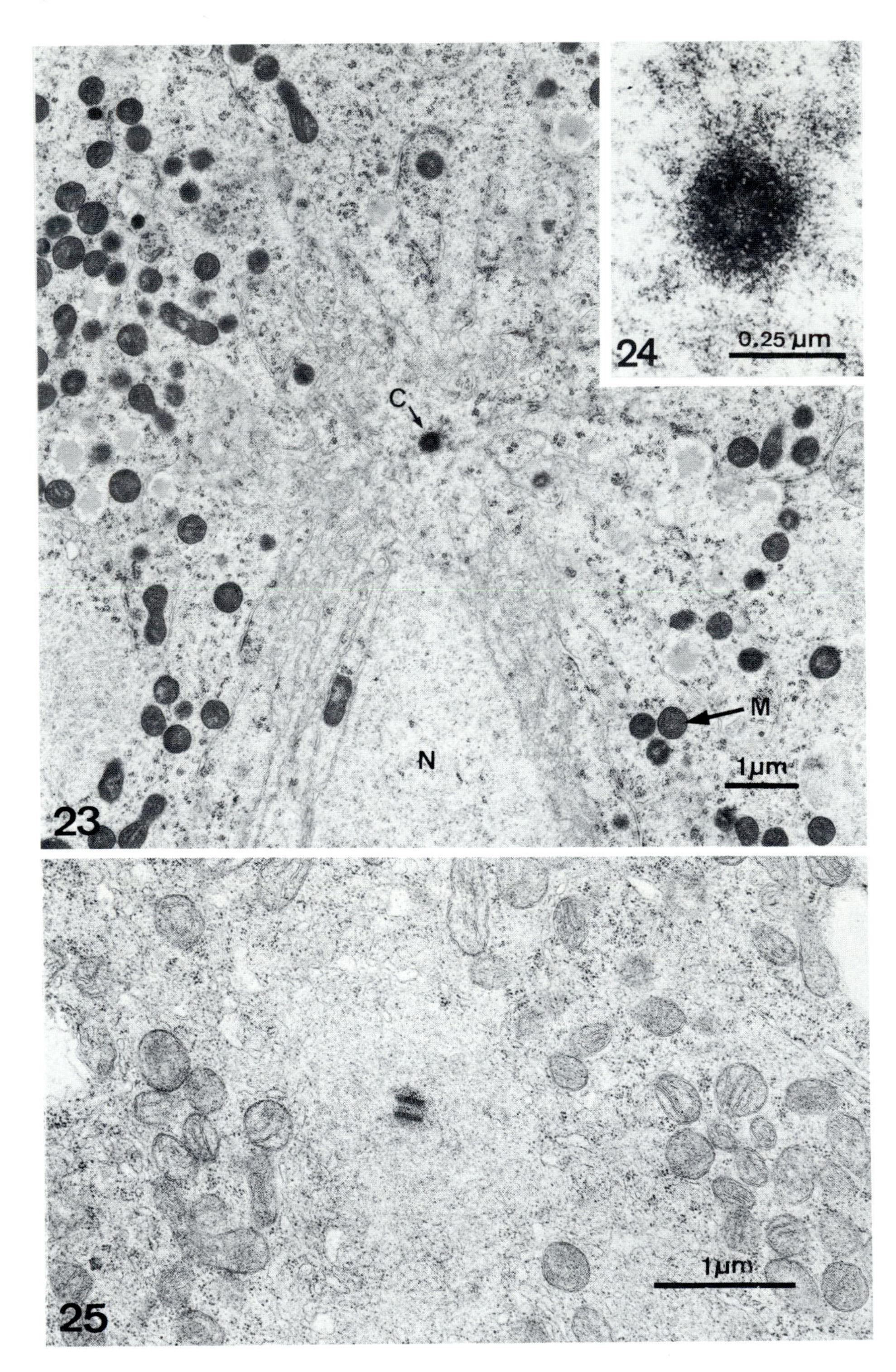
24
0.25 µm
C
M
N
1µm
23
25
1µm

appearance but with patchy contents (Figure 27) were found in the periplasm and in the marginal region of the anterior pole plasm. These bodies are delimited by a number of ER vesicles that later on might fuse to form a double envelope, or might represent the remnants of a disintegrated double membrane. (3) In pole cell stages of *Smittia*, the egg cell may harbor bodies similar to the first type described above (Figure 28). Here, an electron translucent matrix with electron dense inclusions is surrounded by a double membrane showing nuclear pores. These bodies are mostly located close to cleavage nuclei. They are surrounded by stacked ER, usually the rough type. These ER stacks are frequently linked to annulate lamellae and to the star- or rosette-shaped configurations assumed to function as nuclear membrane organizing centers (see section 3.2.1).

Whether these types of bodies represent structures of different origin or rather different stages in a single chain of development remains to be established. Types 1 and 2 are possibly involved in the formation or degradation of proteid yolk, as proposed earlier (Zissler and Sander, 1973), but type 2 might also be related in some way to the accumulations of condensed material to be described below (section 3.5).

The structures of the third type are located close to typical nuclei, and this suggests that they represent accessory nuclei. They may be derived from early cleavage nuclei or, at least in some instances, from the oocyte nucleus. Accessory nuclei are surrounded by a double membrane with pores, and contain electron-dense inclusions in a translucent matrix. Descriptions of the ultrastructures of such structures from oocytes of hymenopterans have been provided by Hopkins (1964), P. E. King and Richards (1968), P. E. King and Fordy (1970), P. E. King *et al.* (1971), and Cassidy and R. C. King (1972). Palévody (1972) described those found in the oocytes of the collembolan *Folsoma candida*. Wülker and Winter (1970) found that in the midges *Chironomus thummi* and *Ch. melanotus* these structures arise as protrusions that become severed from the oocyte nucleus, and this origin seems probable also in the light of recent investigations by Meyer *et al.* (1979) on oocytes of the wood gnats *Phryne cincta* and *P. fenestralis*. Hopkins (1964), P. E. King and Fordy (1970), and Cassidy and R. C. King (1972) have shown that the accessory nuclei in general are Feulgen-negative, while the electron-dense inclusions contain RNA and proteins. Meyer *et al.* (1979) observed that the accessory nuclei have a much greater pore density than the nuclei from which they probably originate and conclude that they are metabolically more active. The peripheral position of the accessory nuclei in the oocyte of the bumble bee, *Bombus terrestris*, about the time of yolk incorporation caused Hopkins (1964) to assume that they might control this process in some way without becoming converted into proteid yolk bodies themselves. Cruickshank (1964; 1972a,b) saw some indications that the accessory nuclei participate in forming the vitelline membrane. But in *Smittia*,

---

Figures 23–25. Centrioles.

Figure 23. A cleavage nucleus of *Nasonia* (N) with a cytaster and centriole (C).

Figure 24. The cross-sectioned centriole at higher magnification.

Figure 25. A longitudinal section through a centriole from the pole cell stage of *Smittia*.

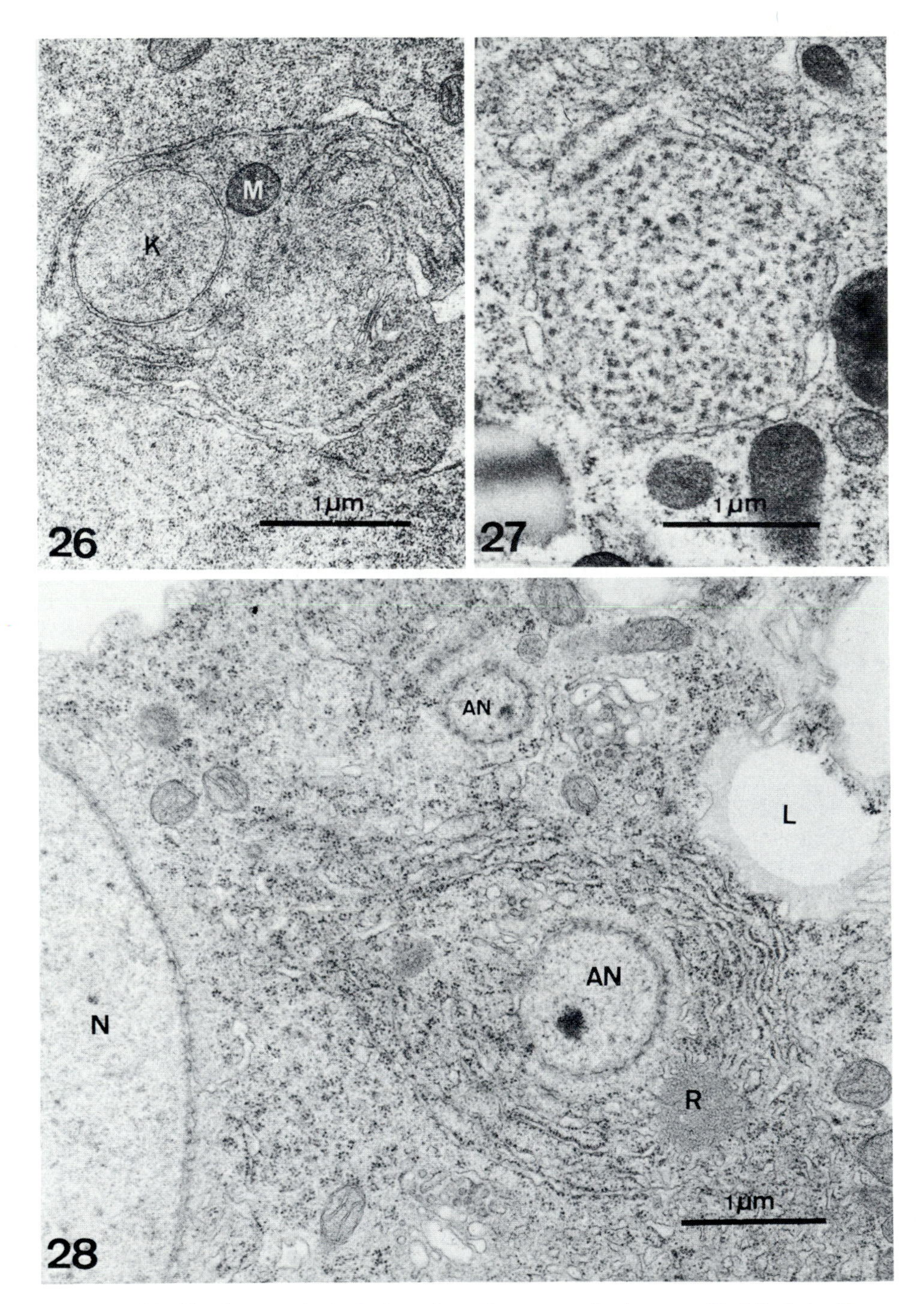

Figures 26–28. Circular or spherical cistern-bound structures in the egg cell of *Smittia*. Figures 26 and 27 (reproduced from Zissler and Sander 1973; 1977) are from newly laid eggs, while Figure 28 is from the pole cell stage. Figure 26 resembles a karyomerelike body (K) (see Zissler and Sander, 1977). Figure 27 shows a body delimited by ER vesicles, and Figure 28 displays "accessory nuclei" (AN). L, lipid; M, mitochondrion; N, cleavage nucleus; R, star- or rosette-shaped configuration.

where similar bodies are found after egg deposition and even in pole cell stages, they must serve some later functions.

3.3. Helical RNA Structures

In *Nasonia vitripennis* and *Chironomus anthracinus* we observed helical structures measuring about 200–400 nm in length, with a coil diameter of about 35 nm (Figures 29–32). They frequently were situated close to nuclei. Such helices have been observed so far in three eggs from *Nasonia*; one was newly laid (Figure 31), another was at pole cell formation, and the third was approaching the blastoderm stage (Figures 29,30). Helices were also found in one newly laid egg of *Ch. anthracinus* (Figure 32). Some helices resemble the configurations of dense particles or dense helical filaments that Mahowald (1962) found at the periphery of polar granules in the pole cells of *Drosophila*. Such structures also have been seen in mycoplasms (Maniloff *et al.*, 1965), in rapidly differentiating plant cells (Echlin, 1965), and in other animal cells (Waddington and Perry, 1963; Pfuderer and Swartzendruber, 1966; Mahowald, 1971b; 1975). Behnke (1963) believed them to be helically arranged polysomes, while Weiss and Grover (1968) described these structures as ribosomes arranged at equidistant intervals along the course of a sinistral helix, winding around a flexible central column roughly 22 nm in diameter. Other helical structures observed by us resemble a coiled filament (Figure 32) like the helices described by Pappas (1956) and Stevens (1967) in *Amoeba*. Wise and coworkers (1972) showed that the helices of *Amoeba* contain RNA and may represent some form of messenger RNA–protein complex transported from the nucleus to the cytoplasm through the nuclear pores. Their low frequency in *Amoeba* was explained by assuming a rapid breakup into subunits immediately after emergence from the nucleus. This might explain why we, too, failed to see these structures more frequently.

3.4. Oosome

An oosome is found in the cytoplasm near the posterior egg pole in all three species. In the light microscope it appears as a lens-shaped body in *Smittia* and *Nasonia*, while in *Oryzaephilus* it is more cup-shaped. The oosome in all three species consists of an electron-dense substance that is not organized in polar granules as in higher dipterans, such as *Drosophila* (Mahowald, 1971a) or the kelp fly, *Coelopa frigida* (Schwalm *et al.*, 1971; Schwalm, 1974); rather, this substance forms a network within a well circumscribed cytoplasmic region, as in some cecidomyids (Wolf, 1967, 1969; Mahowald, 1975). In all three species the oosome region is rich in ribosomes. In *Smittia* other organelles are largely absent, but in *Nasonia* some cisternae studded with ribosomes extend through the entire oosome region and create the impression that the granular material is subdivided into separate fields (Figure 33). In *Oryzaephilus* the oosome appears less well demarcated, since cisternae and vesicles of ER, lipid droplets, and small patches of glycogen are interspaced among the granular material in this region.

It has been known for a long time that the oosome material becomes incorporated into the prospective germ cells. However, in *Smittia* part of the oosome

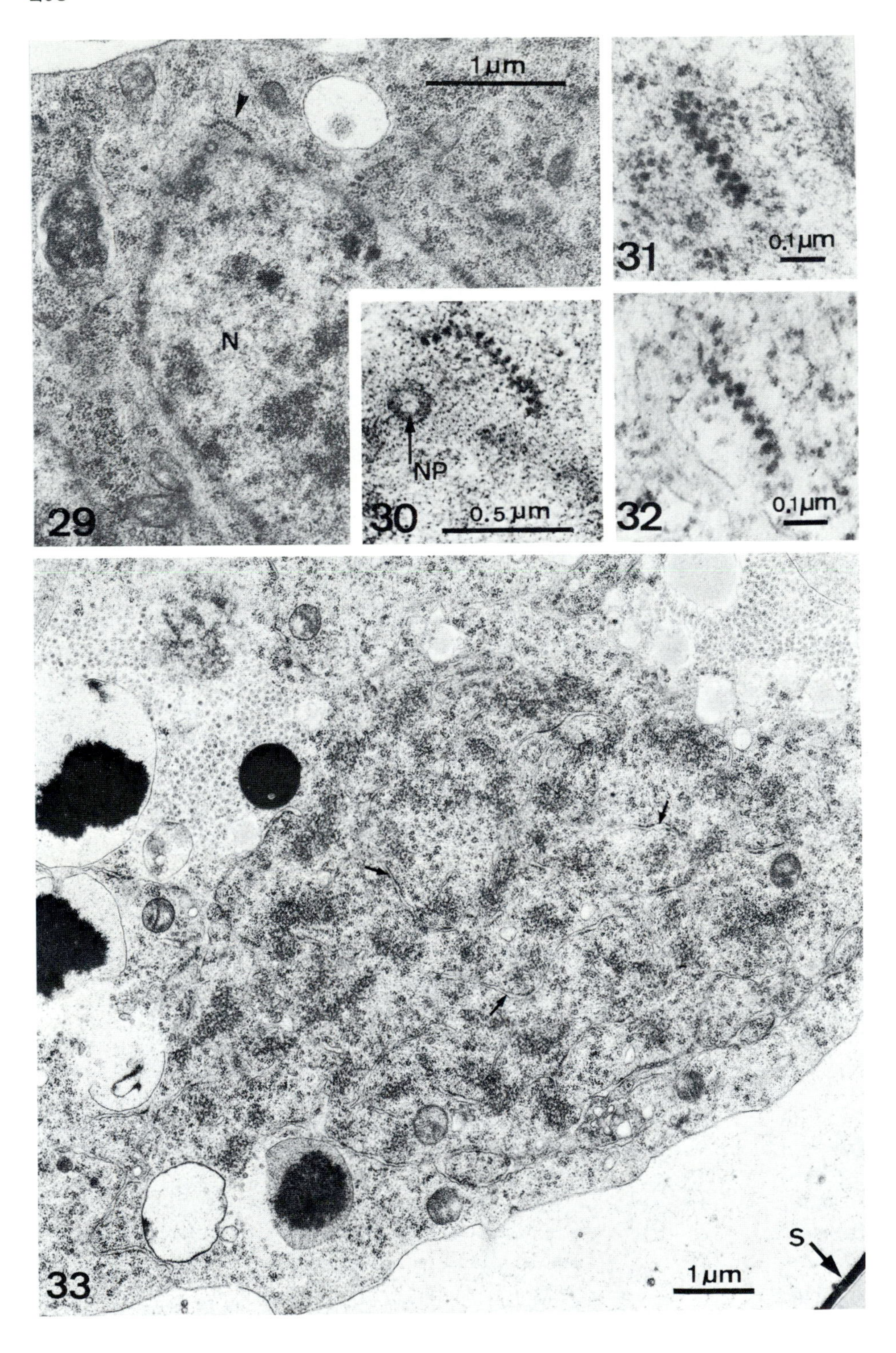
1 µm
N
29
0.1 µm
31
NP
30
0.5 µm
0.1 µm
32
33
1 µm
S

material is left behind in the egg cytoplasm when the first pole cell segregates (Figure 34). The final number of four pole cells is reached by two divisions of the primary pole cell (Kalthoff and Sander, 1968).

3.5. Accumulations of Condensed Material

The cytoplasm of the eggs of *Smittia* and *Oryzaephilus* (Figure 35) contains granular areas that apparently represent a network free of organelles but containing patches of condensed material. By the time pole cells form in *Smittia,* the number of such accumulations seems to increase. These structures resemble the oosome material localized in the posterior egg pole. In some instances we found in *Smittia* structurally similar accumulations containing a dense core (Figure 36). Whether accumulations with and without dense cores represent sections at different levels of the same type of body, or whether they are different structures remains to be determined. This question applies also to the bodies described in section 3.2.3, which show patchy contents enveloped by ER cisternae. Comparable cytoplasmic aggregates of an electron-dense substance are known from previtellogenic oocytes in several dragonflies. According to Kessel and Beams (1969), the dense basophilic masses of cytoplasm contain RNA. They first appeared as numerous fine granules that are more electron opaque and smaller than ribosomes. These then aggregated, forming large compact masses. These masses were comparable to the "yolk nuclei" or "Balbiani bodies" described by earlier investigators of insect oocytes (see Raven, 1961). Later on, annulate lamellae appeared in the dense masses. In the dragonfly *Sympetrum frequens,* the granules were large in size (20 nm), and dispersed later without forming annulate lamellae (Matsuzaki, 1971). In active previtellogenic oocytes of the dragonfly *Cordulia aenea,* a substance is extruded from the chromosomes into the cytoplasm, where it subsequently forms masses of fibers and granules (Halkka and Halkka, 1975). These nuclear extrusions develop ultrastructurally into two components, the so-called nematosomes (Grillo, 1970) and the dense masses. This observation suggests that the nematosomes are storage structures for long-lived mRNA.

3.6. Microfilaments

In the newly laid egg of *Chironomus anthracinus* we found tapering bundles of microfilaments (Figure 37). These bundles (2–3 μm long and 0.3–0.4 μm wide) are embedded in the cytoplasmic matrix without any apparent contact to other cell organelles. They resemble the stacks of protein fibrils

←

Figures 29–32. Helical structures assumed to contain RNA. In Figure 29 (arrowhead) and Figure 30 these are shown close to the pores (NP) of a nucleus (N). Both figures are taken from *Nasonia* eggs shortly before blastoderm formation. Figure 31 is from a P_0 stage of *Nasonia,* and Figure 32 is from the same stage in *Chironomus anthracinus.*

Figure 33. Oosome region in a newly laid egg of *Nasonia.* The oosome material seems to be subdivided in separate fields by some cisternae (arrows). S, egg shell.

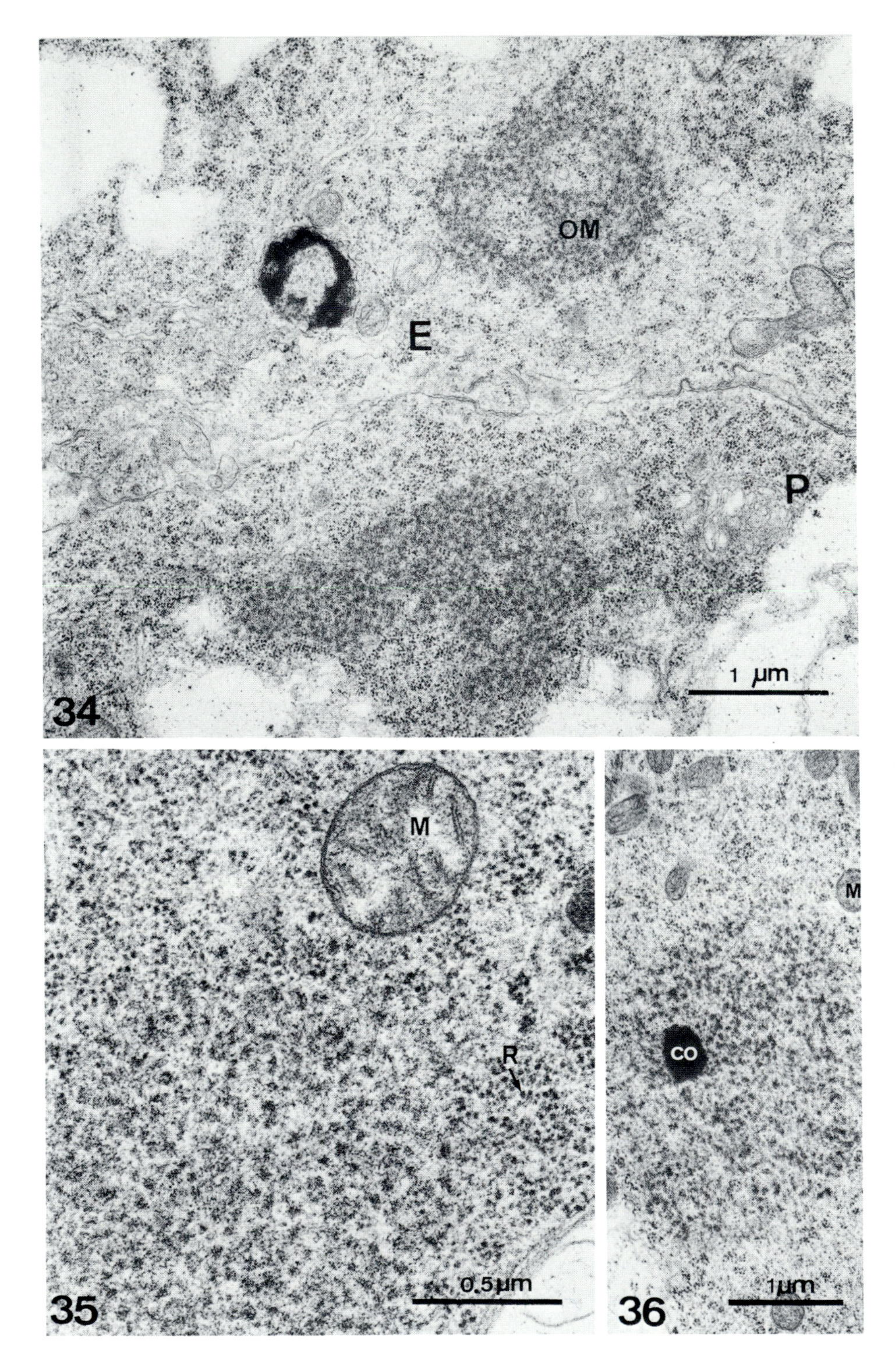
OM
E
P
1 μm
34
M
R
0.5 μm
35
M
CO
1 μm
36

observed by Kinderman and R. C. King (1973) in the cytoplasm and lining the rims of the ring canals connecting the oocyte and nurse cells in *Drosophila virilis*. These authors favor the idea that the microfilaments detach from the rims and subsequently attach to the oocyte nucleus. Here they form an amorphous mantle about 1 μm thick. Kinderman and King suggest that this mantle may insulate the nucleoplasm of the oocyte from substances that trigger transcription on the nurse cell chromosomes. Went *et al.* (1978) have demonstrated microfilaments in the oocytes and nurse chambers of the pedogenetic gall midge *Heteropeza pygmaea*. Here the microfilaments form parallel strands extending between the cell membrane and the nuclear envelope, and also a juxtanuclear sheet consisting of a network of nonparallel filaments. Went *et al.* suggest that these microfilaments are involved in pulsating movements of oocyte nuclei, which can be observed concomitantly with constrictions of these nuclei *in vitro*. In the gall midge *Wachtliella persicariae*, Wolf (1978; 1980) has presented evidence that the cleavage nuclei are connected by 10-nm filaments to cytasters, which cause movement of the cleavage nuclei relative to the surrounding ooplasm.

3.7. Filament Complexes

The nuclei of *Smittia* eggs at the pole cell stage are surrounded by microtubules, and similar areas in the pole cell of *Chironomus anthracinus* are occupied by filaments that consist of solid 14-nm diameter rods covered by a diffuse layer that increases the overall diameter to about 25 nm. Such filaments are found very frequently in the periplasm and endoplasm of the posterior third of the egg cell of *Chironomus* (the only egg region checked so far), and they always are aggregated in complexes (Figure 38). These are probably produced from the ER, since they are always found in between ER cisternae (Figure 39). So far we have not seen them associated with cleavage nuclei in the way they associate with pole cell nuclei (Figure 40). Possibly the association of these filaments with nuclei is germline specific, since it was also found in spermatocytes by Elsenhans (1978). Similar filaments 30–50 nm in diameter and surrounded by diffuse material occur in the central syncytium of the germarium in the megalopteran *Sialis flavilatera* (Büning, 1979).

3.8. Bacteroid Microorganisms

The presence of microorganisms in insect eggs has long been noted (see Buchner, 1965). The "A-bodies" observed by R. C. King about 1957 in the egg chambers of *Drosophila* (see R. C. King, 1970, Fig. VI-4) represent, to our

←

Figure 34. Pole cell stage of *Smittia* demonstrating part of the oosome material (OM) left behind in the egg cytoplasm (E) after budding off the primary pole cell (P).

Figures 35 and 36. Accumulations of condensed material in the egg cytoplasm of *Oryzaephilus* (Figure 35), and similar accumulations with a core structure (CO) in the periplasm of a pole cell stage egg of *Smittia* (Figure 36). M, mitochondrion; R, ribosome.

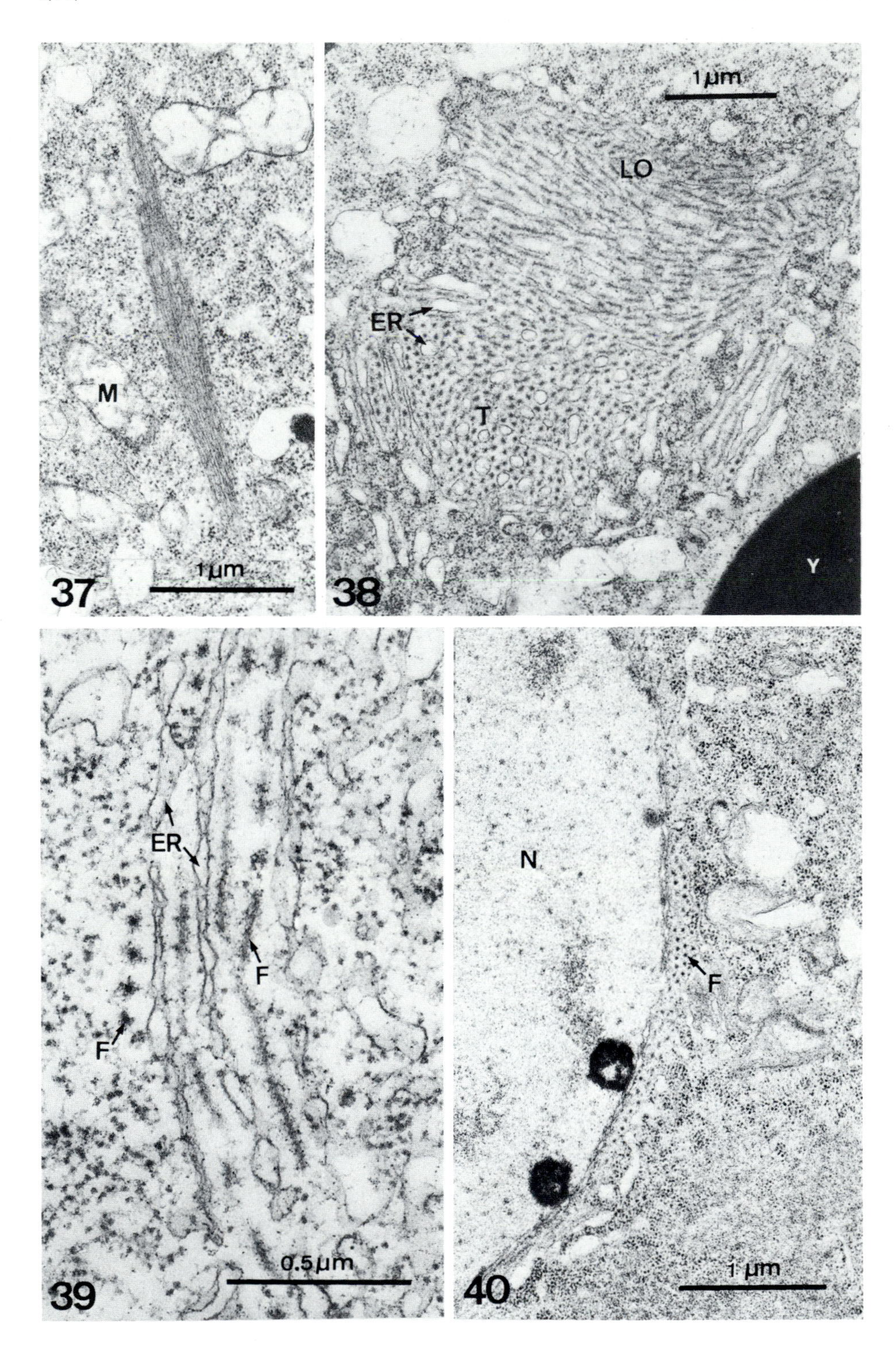
M
1 µm
37
1 µm
LO
ER
T
Y
38
ER
F
F
0.5 µm
39
N
F
1 µm
40

knowledge, the first observation at the EM level of microorganisms transmitted through the ovary. Because of difficulties encountered in classification (see, e.g., Brooks, 1970; Körner, 1979; Houk and Griffiths, 1980), we subsume these microorganisms under the term bacteroids. One type of bacteroid (Figure 41) consists of spherical or oval bodies enclosed by three membranes. Of these the most external, thought to be derived from the plasmalemma of the host cell, is separated from the other two membranes by a space. The structure of the cell wall indicates that these bacteroids should be classified as Gram-negative (Freer and Salton, 1971). Within their membranes the bacteroids contain a network of fibrils, probably DNA, and clustered particles, presumably ribosomes. Some bacteroids seem to be dividing, while others appear in groups of two or three enclosed in a common envelope, a configuration likely to result from recent divisions (Figures 42,43). These bacteroids are distributed more or less evenly over the whole egg cell, but there may be some localized accumulations. They are also found within the pole cells and the incipient blastoderm cells. We found this type of bacteroid in the eggs of *Nasonia* and *Oryzaephilus;* the bacteroids in *Oryzaephilus* are probably identical with the microorganisms observed by Louis (1970) in the cytoplasm of oocytes and trophocytes, while in hymenopterans, bacteria were first described by Cassidy and R. C. King (1972). The egg of *Oryzaephilus* contains yet another type of bacteroid, or perhaps a pleomorphic variant of the bacteria found in the ovary by Louis (1970). This type is located in the extracellular space between the oolemma and the egg shell, as noted already by Koch (1931; 1937). In *Chironomus anthracinus* we found bacteroids lacking the outer membrane presumably derived from the host cell (Figure 44). In *Smittia* no bacteroids have been found so far, although these eggs have been investigated in the most detail.

3.9. Miscellaneous Structures

We have made detailed studies of more than fifty sectioned *Smittia* eggs, and therefore we hope to have recognized most major components of this cell visible with present electron microscope techniques. In the other species mentioned in section 1, we have gained an overview of the entire egg architecture with the aid of composite EM photographs prepared from a median section through the egg cell, but as yet we are just beginning to establish an inventory of specific structures. Besides those listed above, we occasionally found some other structures on whose origins or functions we cannot even speculate. From these, we show a highly organized lamellar body of *Nasonia vitripennis* (Figure 45) and a tubular body from *Chironomus anthracinus* (Figure 46).

←

Figure 37. A bundle of microfilaments in the egg of *Chironomus.* M, mitochondrion.
Figures 38–40. Filament complexes in the egg of *Chironomus.* Figure 38 is from the periplasm of a pole cell stage egg and shows filaments cut transversely (T) and longitudinally (LO); Figure 39 is the same, at higher magnification. Figure 40 is of transversely cut filaments near the envelope of a pole cell nucleus. ER, ER-cistern; Y, yolk; F, filaments.

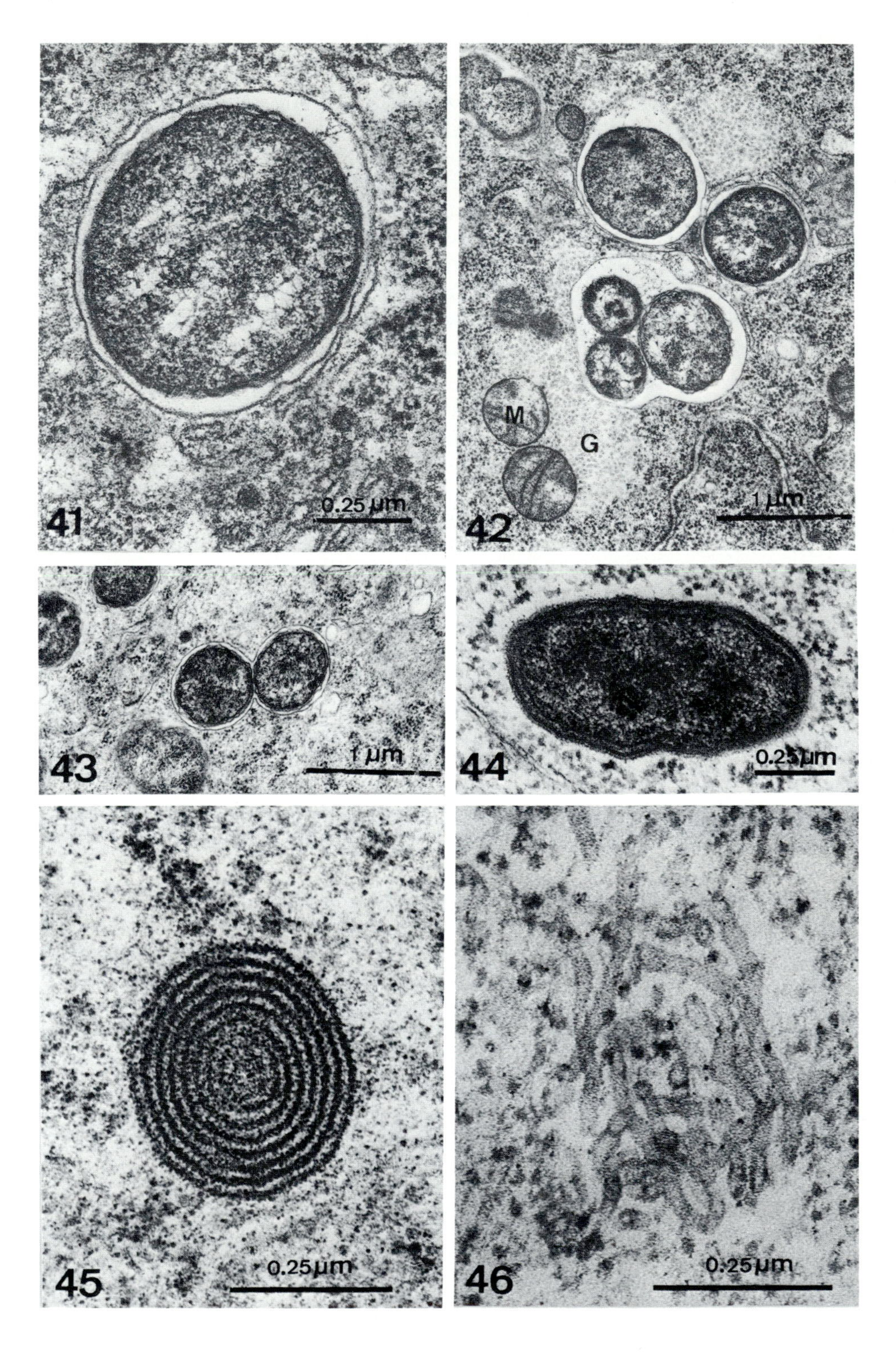
41
0.25 µm
42
M
G
1 µm
43
1 µm
44
0.25 µm
45
0.25 µm
46
0.25 µm

4. Ultrastructure and Pattern Formation

As for our original aim, to identify and characterize structures essential for embryonic pattern formation, we have met with little success so far. Kalthoff's recent experimental work (see Kalthoff, 1979) indicates that ribonucleoproteid particles act as anterior determinants. The oosome-like substances found in the egg cytoplasm outside the posterior pole region (see section 3.5) might represent the structural counterpart of some such RNP particles. The fact that these particles are not restricted to the anterior pole is not at variance with a possible function as anterior determinants, since these determinants (or at least the effects of their inactivation) appear distributed in graded fashion over the entire anterior half of the egg (Kalthoff, 1971). Moreover, the presence of an assumed determinant need not *per se* evoke the respective developmental response, as is shown by the fact that in *Smittia* a considerable fraction of oosomal granules remain behind when the primary pole cell buds off, yet these assumed pole cell determinants fail to initiate further budding of pole cells. To elucidate the problem of ooplasmic determinants will require further investigations, particularly on the quantitative differences in the organelle inventory among different regions, and on eggs experimentally treated so as to predictably alter the course of embryonic patterning.

To sum up, the data presented in this review demonstrate that considerable differences in ultrastructure exist among the egg cells of different insect species. However, the differences observed are not as drastic as those among sperm cells (Phillips, 1970). Being motile elements, the sperm cells may be subject to more strongly divergent requirements than the egg cells, which in all species are immobile and filled by storage substances. Types, amounts, and distribution of these substances and of egg architecture in general reflect the conditions of oogenesis as well as the future course of embryogenesis (Bier, 1970).

ACKNOWLEDGMENTS

We are indebted to Mrs. Sabine Collatz for excellent technical help and to Mrs. Margrit Scherer for typing and retyping the manuscript. Eggs of *Nasonia* and *Chironomus* were kindly provided by Alice Bull, Ulla Elsenhans, Renate Rössler, and Wolfgang Wülker. Our investigations were supported by the Deutsche Forschungsgemeinschaft, Sonderforschungsbereich 46.

Figures 41–44. Bacteroid microorganisms from *Oryzaephilus* (Figures 41, 42), *Nasonia* (Figure 43), and *Chironomus* (Figure 44). G. glycogen; M, mitochondrion.

Figures 45 and 46. Miscellaneous structures in the cytoplasm of newly laid eggs. Figure 45 shows a lamellar body from *Nasonia vitripennis,* and Figure 46 a tubular body from *Chironomus anthracinus.*

References

Anderson, E., 1974, Comparative aspects of the ultrastructure of the female gamete, *Int. Rev. Cytol.* Suppl. **4**:1–70.

Barbier, R., and Chauvin, G., 1972a, Etude expérimentale de la perméabilité des enveloppes de l'oeuf et de la cuticle sérosale chez *Tinea pellionella* L. (Lépidoptère Tineidae), *C. R. Acad. Sci. Paris* **275D**:1004–1006.

Barbier, R., and Chauvin, G., 1972b, Origine et structure des enveloppes de l'oeuf et mise en place de la cuticle sérosale chez *Monopis rusticella* Clerck (Lépidoptère Tineidae), *C. R. Acad. Sci. Paris* **274D**:1079–1082.

Barbier, R., and Chauvin, G., 1974a, The aquatic egg of *Nymphula nymphaeata* (Lepidoptera: Pyralidae). On the fine structure of the egg shell, *Cell Tissue Res.* **149**:473–479.

Barbier, R., and Chauvin, G., 1974b, Ultrastructure et rôle des aéropyles et des enveloppes de l'oeuf de *Galleria mellonella, J. Insect Physiol.* **20**:809–820.

Baxter, R., 1971, Origin and continuity of mitochondria. In *Origin and Continuity of Cell Organelles*, edited by J. Reinert and H. Ursprung, pp. 46–64, Springer-Verlag, Berlin.

Beams, H. W., and Kessel, R. G., 1974, The problem of germ cell determinants, *Int. Rev. Cytol.* **39**:413–479.

Behnke, O., 1963, Helical arrangement of ribosomes in the cytoplasm of differentiating cells of the small intestine of rat foetuses, *Exp. Cell Res.* **30**:597–598.

Bier, K. H., 1970, Oogenesetypen bei Insekten und Vertebraten, ihre Bedeutung für die Embryogenese und Phylogenese, *Zool. Anz.* Suppl. **33**:7–29.

Bilinski, S., 1976, Ultrastructure studies on the vitellogenesis of *Tetrodontophora bielanensis* (Waga) (Collembola), *Cell Tissue Res.* **168**:399–410.

Bilinski, S., 1977, Oogenesis in *Acerentomon gallicum* Jonescu (Protura). Previtellogenic and vitellogenic stages, *Cell Tissue Res.* **179**:401–412.

Bilinski, S., 1979, Oogenesis in *Campodea sp.* (Diplura): The ultrastructure of the egg chamber during vitellogenesis, *Cell Tissue Res.* **202**:133–143.

Blau, H. M., and Kafatos, F. C., 1979, Morphogenesis of the silkmoth chorion: Patterns of distribution and insolubilization of the structural proteins, *Develop. Biol.* **72**:211–225.

Brooks, M. A., 1970, Comments on the classification of intracellular symbionts of cockroaches and a description of the species. *J. Invert. Pathol.* **16**:249–258.

Buchner, P., 1965, *Endosymbiosis of Animals with Plant Microorganisms*, Wiley, New York.

Bull, A. L., 1981, Stages of living embryos in the jewel wasp *Mormoniella* (*Nasonia*) *vitripennis* (Walker) *Int. J. Insect Morphol. Embryol.* (in press).

Büning, J., 1972, Untersuchungen am Ovar von *Bruchidius obtectus* Say. (Coleoptera-Polyphaga) zur Klärung des Oocytenwachstums in der Prävitellogenese. *Z. Zellforsch.* **128**:241–282.

Büning, J., 1979, The telotrophic-meroistic ovary of Megaloptera. I. The ontogenetic development. *J. Morphol.* **162**:37–66.

Cantacuzène, A., and Martoja, R., 1972, Origine des enclaves vitellines de l'oocyte d'un insecte Thysanoure, *Petrobius maritimus, C. R. Acad. Sci. Paris* **274D**:1723–1726.

Cassidy, J. D., and King, R. C., 1972, Ovarian development in *Habrobracon juglandis* (Ashmead) (Hymenoptera:Braconidae). I. The origin and differentiation of the oocyte-nurse cell complex, *Biol. Bull.* **143**:483–505.

Chauvin, G., and Barbier, R., 1972, Perméabilité et ultrastructure des oeufs de deux Lépidoptères Tineidae: *Monopis rusticella* et *Trichophaga tapetzella, J. Insect Physiol.* **18**:1447–1462.

Chauvin, G., Barbier, R., and Bernard, J., 1973, Ultrastructure de l'oeuf de *Triatoma infestans* Klug (Heteroptera, Reduviidae), formation des cuticules embryonnaires, rôle des enveloppes dans le transit de l'eau, *Z. Zellforsch.* **138**:113–132.

Chauvin, G., Rahn, R., and Barbier, R., 1974, Comparaison des oeufs des lépidoptères *Phalera bucephala* L. (Ceruridae), *Acrolepia assectella* Z. et *Plutella maculipennis* Curt. (Plutellidae): Morphologie et ultrastructures particulières du chorion au contact du support vegetal, *Int. J. Insect Morphol. Embryol.* **3**:247–256.

Cone, M. V., and Scalzi, H. A., 1967, An ultrastructural study of oogenesis in the silverfish *Lepisma* sp. (Thysanura), *J. Cell Biol.* **35**:163A.

Cruickshank, W. J., 1964, Formation and possible function of the "accessory yolk nuclei" in *Anagasta (=Ephestia) kühniella, Nature* (London) **201**:734–735.

Cruickshank, W. J., 1972a, Ultrastructural modifications in the follicle cells and egg membranes during development of flour moth oocytes, *J. Insect Physiol.* **18**:485–498.

Cruickshank, W. J., 1972b, The formation of "accessory nuclei" and annulate lamellae in the oocytes of the flour moth *Anagasta kühniella, Z. Zellforsch.* **130**:181–192.

Dorn, A., 1976, Ultrastructure of embryonic envelopes and integument of *Oncopeltus fasciatus* Dallas (Insecta, Heteroptera), *Zoomorphologie* **85**:111–131.

Echlin, P., 1965, Helical polysomes in pollen mother cells; *J. Cell Biol.* **24**:150–153.

Eddy, E. M., 1975, Germ plasm and the differentiation of the germ cell line, *Int. Rev. Cytol.* **43**:229–280.

Elsenhans, U. 1978, Untersuchungen zur Ultrastruktur der Gonaden von *Chironomus anthracinus* (Diptera): Normalentwicklung der Hoden im 4. Larvenstadium der Puppe und der Imago. Dissertation, Fakultät Biologie, Universität Freiburg, pp. 1–34.

Engels, W., 1971, Verteilungsmuster und Ultrastruktur des Glykogens in der Oocyte von *Musca domestica, Wilhelm Roux's Arch.* **167**:294–298.

Franke, W. W., and Scheer, U., 1974, Structures and functions of the nuclear envelope, In *The Cell Nucleus,* edited by H. Busch, vol. II, pp. 219–341, Academic Press, New York.

Freer, J. H., and Salton, M. R. J., 1971, The anatomy and chemistry of gram-negative cell envelopes. In *Microbial Toxins,* edited by G. Weinbaum, S. Kadis, and S. J. Ajl, vol. IV, *Bacterial Endotoxins,* pp. 67–126, Academic Press, New York.

Fullilove, S., and Jacobson, A. G., 1971, Nuclear elongation and cytokinesis in *Drosophila montana, Develop. Biol.* **26**:560–577.

Furneaux, P. J. S., and MacKay, A. L., 1972, Crystalline protein in the chorion of insect egg shells, *J. Ultrastruct. Res.* **38**:343–359.

Furneaux, P. J. S., James, C. R., and Potter, S. A., 1969, The egg shell of the house cricket (*Acheta domesticus*): An electron-microscope study, *J. Cell Sci.* **5**:227–249.

Giorgi, F., 1974, Multiple origin of the membrane of yolk platelets in oocytes of *Drosophila melanogaster, J. Submicrosc. Cytol.* **6**:120.

Giorgi, F., and Jacob, J., 1977, Recent findings in oogenesis of *Drosophila melanogaster.* III. Lysosomes and yolk platelets, *J. Embryol. Exp. Morphol.* **39**:45–57.

Grillo, M. A., 1970, Cytoplasmic inclusion bodies resembling nucleoli in sympathetic neurons of adult rats, *J. Cell Biol.* **45**:100–117.

Haget, A., 1953, Analyse experimentale des facteurs de la morphogenèse embryonnaire chez le Coléoptère *Leptinotarsa, Bull. Biol. France Belg.* **87**:125–217.

Haget, A., 1970, Premières données infrastructurales sur la surface et le périplasme de l'oeuf du Doryphore, *Leptinotarsa decemlineata* Say., *C. R. Acad. Sci. Paris* **271D**:1303–1306.

Haget, A., 1972, Characteristiques ultrastructurales du pôle postérieur et de l'oosome, dans l'oeuf jeune du Coléoptère *Leptinotarsa decemlineata* Say., *C. R. Acad. Sci. Paris* **275D**:2737–2740.

Halkka, L., and Halkka, O., 1975, Accumulation of gene products in the oocytes of the dragonfly *Cordulia aenea* L., *J. Cell Sci.* **19**:103–115.

Hinton, H. E., 1970, Insect eggshells, *Sci. Am.* **223(8)**:84–91.

Hinton, H. E., 1981, *Biology of Insect Eggs,* Pergamon Press, Oxford.

Hogge, M. A. F., and King, P. E., 1975, The ultrastructure of spermatogenesis in *Nasonia vitripennis* (Walker) (Hymenoptera: Pteromalidae), *J. Submicrosc. Cytol.* **7**:81–96.

Hopkins, C. R., 1964, The histochemistry and fine structure of the accessory nuclei in the oocyte of *Bombus terrestris, Quart. J. Microsc. Sci.* **105**:457–480.

Houk, E. J., and Griffiths, G. W., 1980, Intracellular symbiotes of the Homoptera, *Annu. Rev. Entomol.* **25**:161–187.

Jäckle, H., 1979, Degradation of maternal poly(A)-containing RNA during early embryogenesis of an insect (*Smittia* spec., Chironomidae, Diptera), *Wilhelm Roux's Arch. Dev. Biol.* **187**: 179–193.

Kafatos, F. C., Regier, J. C., Mazur, D. G., Nadel, M. R., Blau, H. M., Petri, W. H., Wyman, A. R., Gelinas, R. E., Moore, P. B., Paul, M., Efstratiadis, A., Vournakis, J. N., Goldsmith, M. R., Hunsley, J. R., Baker, B. K., Nardi, J., and Koehler, M., 1977, The eggshell of insects: Differentiation-specific proteins and the control of their synthesis and accumulation during development. In *Biochemical Differentiation in Insect Glands,* edited by W. Beerman, pp. 45–145, Springer-Verlag, Berlin.

Kalthoff, K., 1971, Position of targets and period of competence for UV-induction of the malformation "double abdomen" in the egg of *Smittia* spec. (Diptera, Chironomidae), *Wilhelm Roux's Arch.* **168:**63–84.

Kalthoff, K., 1979, Analysis of a morphogenetic determinant in an insect embryo (*Smittia* spec., Chironomidae, Diptera). In *Determinants of Spatial Organization, 37th Symp. Soc. Develop. Biol.,* edited by S. Subtelny and I. R. Konigsberg, pp. 97–126, Academic Press, New York.

Kalthoff, K., and Sander, K., 1968, Der Entwicklungsgang der Missbildung "Doppelabdomen" im partiell UV-bestrahlten Ei von *Smittia parthenogenetica* (Dipt., Chironomidae), *Wilhelm Roux's Arch.* **161:**129–146.

Kessel, R. G., and Beams, H. W., 1969, Annulate lamellae and "yolk nuclei" in oocytes of the dragonfly, *Libellula pulchella, J. Cell Biol.* **42:**185–201.

Kinderman, N. B., and King, R. C., 1973, Oogenesis in *Drosophila virilis.* I. Interactions between the ring canal rims and the nucleus of the oocyte, *Biol. Bull.* **144:**331–354.

King, P. E., and Fordy, M. R., 1970, The formation of "accessory nuclei" in the developing oocytes of the parasitoid hymenopterans *Ophion luteus* (L.) and *Apanteles glomeratus* (L.), *Z. Zellforsch.* **109:**158–170.

King, P. E., and Richards, J. G., 1968, Accessory nuclei and annulate lamellae in hymenopteran oocytes, *Nature (London)* **218:**488.

King, P. E., Ratcliffe, N. A., and Copland, M. J. W., 1969, The structure of the egg membranes in *Apanteles glomeratus (L.) (Hymenoptera: Braconidae), Proc. R. Ent. Soc. London (A)* **44:** 137–142.

King, P. E. Ratcliffe, N. A., and Fordy, M. R., 1971, Oogenesis in a braconid, *Apanteles glomeratus* (L.), possessing an hydropic type of egg, *Z. Zellforsch.* **119:**43–57.

King, P. E., Rafai, J., and Richards, J. G., 1972, Formation of protein yolk in the eggs of a parasitoid hymenopteran, *Nasonia vitripennis* (Walker) (Pteromalidae:Hym.), *Z. Zellforsch.* **123:**330–336.

King, R. C., 1960, Oogenesis in adult *Drosophila melanogaster.* IX. Studies on cytochemistry and ultrastructure of developing oocytes, *Growth,* **24:**265–323.

King, R. C., 1970, *Ovarian Development in Drosophila melanogaster,* Academic Press, New York.

King, R. C., and Aggarwal, S. K., 1965, Oogenesis in *Hyalophora cecropia, Growth* **29:**17–84.

King, R. C., and Cassidy, J. D., 1973, Ovarian development in *Habrobracon juglandis* (Ashmead) (Hymenoptera: Braconidae). 2. Observations on the growth and differentiation of the component cells of the egg chamber and their bearing upon the interpretation of radiosensitivity data from *Habrobracon* and *Drosophila, Int. J. Insect Morphol. Embryol.* **2:**117–136.

King, R. C., and Koch, E. A., 1963, Studies on the ovarian follicle cells of *Drosophila, Quart. J. Microsc. Sci.* **104:**297–320.

King, R. C., Bentley, R. M., and Aggarwal, S. K., 1966, Some of the properties of the components of *Drosophila* ooplasm, *Amer. Natur.* **100:**365–367.

Klag, J., 1977, Differentiation of primordial germ cells in the embryonic development of *Thermobia domestica* Pack. (Thysanura): An ultrastructural study, *J. Embryol. Exp. Morphol.* **38:**93–114.

Klag, J., 1978, Oogenesis in *Acerentomon gallicum* Jonescu (Protura): An ultrastructural analysis of the early previtellogenesis stages, *Cell Tissue Res.* **189:**365–374.

Klug, W. S., Campbell, D., and Cummings, M. R., 1974, External morphology of the egg of *Drosophila melanogaster* Meigen (Diptera: Drosophilidae), *Int. J. Insect Morphol. Embryol.* **3:** 33–40.

Koch, A., 1931, Die Symbiose von *Oryzaephilus surinamensis* L. (Cucujidae, Coleoptera), *Z. Morphol. Ökol. Tiere* **23:**389–424.

Koch, A., 1937, Symbiosestudien, II. Experimentelle Untersuchungen an *Oryzaephilus surinamensis* L. (Cucujidae, Coleopt.), *Z. Morphol. Ökol. Tiere* **32:**137–180.

Körner, H. K., 1979, Probleme der Klassifizierung intrazellulärer Symbionten von wirbellosen Tieren, *Naturwiss. Rundsch.* **32:**363–365.

Krause, G., 1939, Die Eitypen der Insekten, *Biol. Zentralbl.* **59:**495–536.

Krause, G., and Sander, K., 1962, Ooplasmic reaction systems in insect embryogenesis, *Adv. Morphogen.* **2:**259–303.

Kwiatkowska, M., 1973, Half unit membranes surrounding osmiophilic granules (lipid droplets) of the so-called lipotubuloid in *Ornithogalum, Protoplasma* **77:**473–476.

Louis, C., 1970, Presence de micro-organismes intracytoplasmiques dans les ovarioles d'*Oryzaephilus* surinamensis (L.) et leur transport par les cordons trophocytaires, *7th Int. Conf. Electron Microscopy* (Grenoble), pp. 657–658, Societé Française de Microscopie Electronique, Paris.

Mahowald, A. P., 1962, Fine structure of pole cells and polar granules in *Drosophila melanogaster, J. Exp. Zool.* **151:**201–215.

Mahowald, A. P., 1971a, Origin and continuity of polar granules. In *Origin and Continuity of Cell Organelles,* edited by J. Reinert and H. Ursprung, pp. 158–169, Springer-Verlag, Berlin.

Mahowald, A. P., 1971b, Polar granules of *Drosophila.* IV. Cytochemical studies showing loss of RNA from polar granules during early stages of embryogenesis, *J. Exp. Zool.* **175:**345–352.

Mahowald, A. P., 1975, Ultrastructural changes in the germ plasm during the life cycle of *Miastor* (Cecidomyidae, Diptera), *Wilhelm Roux's Arch.* **175:**223–240.

Mahowald, A. P., 1975, Ultrastructural changes in the germ plasm during the life cycle of *Miastor* (Cecidomyidae, Diptera), *Wilhelm Roux's Arch. Dev. Biol.* **175:**223–240.

Maniloff, J., Morowitz, H. J., and Barrnett, R. J., 1965, Studies of the ultrastructure and ribosomal arrangements of the pleuropneumonialike organism A5969, *J. Cell Biol.* **25:**139–150.

Margaritis, L. H., Kafatos, F. C., and Petri, W. H., 1980, The eggshell of *Drosophila melanogaster.* I. Fine structure of the layers and regions of the wild type eggshell, *J. Cell Sci.* **43:**1–35.

Matsuzaki, M., 1971, Electron microscopic studies on the oogenesis of dragonfly and cricket, with special reference to the panoistic ovaries, *Develop. Growth Differ.* **13:**379–398.

Matsuzaki, M., 1973, Oogenesis in the springtail, *Tomocerus minutus* Tullberg (Collembola: Tomoceridae), *Int. J. Insect Morphol. Embryol.* **2:**335–349.

Meyer, G. F., Sokoloff, S., Wolf, B. E., and Brand, B., 1979, Accessory nuclei (nuclear membrane balloons) in the oocytes of the dipteran *Phryne, Chromosoma* **75:**89–99.

Miya, K., 1978, Electron microscope studies on the early embryonic development of the silkworm, *Bombyx mori.* I. Architecture of the newly laid egg and the changes by sperm entry, *J. Fac. Agr., Iwate Univ.* **14:**11–34.

Miya, K., Kurihara, M., and Tanimura, J., 1969, Electron microscope studies on the oogenesis of the silkworm, *Bombyx mori.* I (Fine structure of oocyte and nurse cells in the early developmental stages), *J. Fac. Agr., Iwate Univ.* **9:**221–237.

Nørrevang, A., 1968, Electron microscopic morphology of oogenesis, *Int. Rev. Cytol.* **23:**113–186.

Okada, E., and Waddington, C. H., 1959, The submicroscopic structure of the *Drosophila* egg, *J. Embryol. Exp. Morphol.* **7:**583–597.

Overton, J., and Raab, M., 1967, The development and fine structure of centrifuged eggs of *Chironomus thummi, Develop. Biol.* **15:**271–287.

Pálevody, C., 1972, Presence de noyaux accessoires dans l'ovocyte du collembole *Folsomia candida* Willem (Insecte Apterygote), *C. R. Acad. Sci. Paris* **274D:**3258–3261.

Pálevody, C., 1973, Differentiation du noyau de l'ovocyte au cours de la prophase meiotique chez les Collemboles (Insectes, Apterygotes). Etude ultrastructurale, *C. R. Acad. Sci. Paris* **277D:** 2201–2203.

Pappas, G. D., 1956, Helical structures in the nucleus of *Amoeba proteus, J. Biophys. Biochem. Cytol.* **2:**221–223.

Peterson, S. P., and Berns, M. W., 1980, The centriolar complex, *Int. Rev. Cytol.* **64:**81–106.

Pfuderer, P., and Swartzendruber, D. C., 1966, The configuration of isolated polysomes, *J. Cell Biol.* **30:**193–197.

Phillips, D. M., 1970, Insect sperm: Their structure and morphogenesis, *J. Cell Biol.* **44:**243–277.

Quattropani, S. L., and Anderson, E., 1969, The origin and structure of the secondary coat of the egg of *Drosophila melanogaster, Z. Zellforsch.* **95:**495–510.

Rau, K. G., and Kalthoff, K., 1980, Complete reversal of anteroposterior polarity in a centrifuged insect embryo, *Nature* (London) **287:**635–637.

Raven, C. P., 1959, *An Outline of Developmental Physiology,* Pergamon Press, New York.

Raven, C. P., 1961, *Oogenesis: The Storage of Developmental Information,* Pergamon Press, New York.

Richards, J. G., 1969, The structure and formation of the egg membranes in *Nasonia vitripennis* (Walker) (Hymenoptera, Pteromalidae), *J. Microsc.* (Oxford) **89:**43–53.

Richards, J. G., and King, P. E., 1967, Chorion and vitelline membranes and their role in resorbing eggs of the hymenoptera, *Nature* (London) **214:**601–602.

Roth, T. F., and Porter, K. R., 1964, Yolk protein uptake in the oocyte of the mosquito *Aedes aegypti* L., *J. Cell Biol.* **20:**313–332.

Sakaguchi, B., Chikushi, H., and Doira, H., 1973, Observations of the eggshell structures controlled by gene action in *Bombyx mori, J. Fac. Agric. Kyushu Univ.* **18:**53–62.

Sander, K., 1976, Specification of the basic body pattern in insect embryogenesis, *Adv. Insect Physiol.* **12:**125–238.

Sander, K., 1981, Pattern generation and pattern conservation in insect ontogenesis—problems, data and models. In *Progress in Developmental Biology*, edited by H. W. Sauer, pp. 101–119, Gustav Fischer Verlag, Stuttgart.

Saul, G. B., Whiting, P. W., Saul, S. W., and Heidner, C. A., 1965, Wild-type and mutant stocks of *Mormoniella, Genetics* **52:**1317–1327.

Schwalm, F. E., 1974, Autonomous structural changes in polar granules of unfertilized eggs of *Coelopa frigida* (Diptera), *Wilhelm Roux's Arch.* **175:**129–133.

Schwalm, F. E., Simpson, R., and Bender, H. A., 1971, Early development of the kelp fly, *Coelopa frigida* (Diptera): Ultrastructural changes within the polar granules during pole cell formation, *Wilhelm Roux's Arch.* **166:**205–218.

Smith, D. S., Telfer, W. H., and Neville, A. C., 1971, Fine structure of the chorion of a moth, *Hyalophora cecropia, Tissue Cell* **3:**477–498.

Stevens, A. R., 1967, Machinery for exchange across the nuclear envelope. In *The Control of Nuclear Activity*, edited by L. Goldstein, pp. 837–871, North Holland Publishing Company, Amsterdam.

Telfer, W. H., 1975, Development and physiology of the oocyte nurse cell syncytium, *Adv. Insect Physiol.* **11:**223–319.

Telfer, W. H., and Smith, D. S., 1970, Aspects of egg formation. In *Insect Ultrastructure*, edited by A. C. Neville, Symposia of the Royal Entomological Society of London, No. 5, Blackwell Scientific Publications, Oxford.

Truckenbrodt, W., 1970, Über die Oogenese der Termite *Kalotermes flavicollis* Fabr. I. Zur Feinstruktur des Terminalfilaments, Germariums und der Prophaseregion, *Z. Zellforsch.* **108:** 339–356.

Truckenbrodt, W., 1971, Untersuchungen am Ei-Chorion der Termite *Kalotermes flavicollis* Fabr. unter normalen Bedingungen und nach Behandlung des Weibchens mit Colcemid (Insecta, Isoptera), *Z. Morphol. Ökol. Tiere* **69:**48–81.

Turner, F. R., and Mahowald, A. P., 1976, Scanning electron microscopy of *Drosophila* embryogenesis: I. The structure of the egg envelopes and the formation of the cellular blastoderm, *Develop. Biol.* **50:**95–108.

Waddington, C. H., and Perry, M. M., 1963, Helical arrangement of ribosomes in differentiating muscle cells, *Exp. Cell Res.* **30:**599–600.

Weiss, P., and Grover, N. B., 1968, Helical array of polyribosomes, *Proc. Natl. Acad. Sci. USA* **59:**763–768.

Went, D. F., Fux, T., and Camenzind, R., 1978, Movement pattern and ultrastructure of pulsating oocyte nuclei of the paedogenetic gall midge, *Heteropeza pygmaea* Winnertz (Diptera: Cecidomyidae), *Int. J. Morphol. Embryol.* **7:**301–314.

Wise, G. E., Stevens, A. R., and Prescott, D. M., 1972, Evidence of RNA in the helices of *Amoeba proteus, Exp. Cell Res.* **75:**347–352.

Wolf, R., 1967, Der Feinbau des Oosoms normaler und zentrifugierter Eier der Gallmücke *Wachtliella persicariae* L. (Diptera), *Wilhelm Roux's Arch.* **158:**459–462.

Wolf, R., 1969, Kinematik und Feinstruktur plasmatischer Faktorenbereiche des Eies von *Wachtliella persicariae* L. (Diptera). I. Das Verhalten ooplasmatischer Teilsysteme im normalen Ei, *Wilhelm Roux's Arch.* **162:**121–160.

Wolf, R., 1978, The cytaster, a colchicine-sensitive migration organelle of cleavage nuclei in an insect egg, *Develop. Biol.* **62:**464–472.

Wolf, R., 1980, Migration and division of cleavage nuclei in the gall midge *Wachtliella persicariae*. II. Origin and ultrastructure of the migration cytaster, *Wilhelm Roux's Arch.* **188:**65–73.

Wülker, W., and Winter, G., 1970, Untersuchungen über die Ultrastruktur der Gonaden von *Chironomus* (Dipt.). 1. Normalentwicklung der Ovarien im 4. Larvenstadium, *Z. Zellforsch.* **106:**348-370.

Wunderlich, F., Berezney, R., and Kleinig, H., 1976, The nuclear envelope: An interdisciplinary analysis of its morphology, composition and functions. In *Biological Membranes*, edited by D. Chapman and D. F. H. Wallach, pp. 241-333, Academic Press, New York.

Yatsu, L. Y., and Jacks, T. J., 1972, Spherosome membranes: Half unit-membranes, *Plant Physiol.* **49:**937-943.

Zissler, D., and Sander, K., 1973, The cytoplasmic architecture of the egg cell of *Smittia* spec. (Diptera, Chironomidae). I. Anterior and posterior pole regions, *Wilhelm Roux's Arch.* **172:**175-186.

Zissler, D., and Sander, K., 1977, The cytoplasmic architecture of the egg cell of *Smittia* spec. (Diptera, Chironomidae). II. Periplasm and yolk-endoplasm, *Wilhelm Roux's Arch.* **183:** 233-248.

8

Morphological Analysis of Transcription in Insect Embryos

VICTORIA FOE, HUGH FORREST, LINDA WILKINSON, AND CHARLES LAIRD

1. *Introduction*

One of the motivations for analyzing transcription during embryogenesis is the expectation that gene expression is important in programming and effecting development. In particular, it would be of interest to learn whether or not transcriptional events are involved in cell determination. Another motivation is to understand mechanisms of transcriptional control. These mechanisms are especially amenable to analysis in the early embryonic stages of insects because it is at these stages that transcriptional induction generally occurs. This review focuses on an examination of the above issues based on morphological analyses of transcription in the fruit fly, *Drosophila melanogaster,* and in the milkweed bug, *Oncopeltus fasciatus.*

2. *Timing of Transcriptional Activation with Respect to Cell Determination*

The major morphological events of early development are very similar for these two insects. After fertilization, nuclei and their surrounding cytoplasmic islands multiply and divide within the yolky egg interior. Most nuclei with their surrounding cytoplasm then migrate to the peripheral cytoplasm to form

VICTORIA FOE • Department of Biochemistry and Biophysics, University of California Medical School, San Francisco, California 94143, USA.
HUGH FORREST • Department of Zoology, University of Texas, Austin, Texas 78712, USA.
LINDA WILKINSON AND CHARLES LAIRD • Department of Zoology, University of Washington, Seattle, Washington 98195, USA.

a syncytial blastoderm. After additional divisions these blastoderm nuclei become partitioned by cell membranes to form a cellular blastoderm (see Foe, 1975; Zalokar and Erk, 1976). Developmental studies with *Drosophila melanogaster* indicate that the developmental fates of premigration nuclei are not yet determined, as shown by transplantation of nuclei with adhering cytoplasm (Okada *et al.*, 1974). In contrast, some syncytial blastoderm nuclei transplanted with their surrounding cytoplasm are determined for anterior and posterior fates (Kauffman, 1980). This extent of determination is maintained and possibly increased after the completion of cell membrane formation (Chan and Gehring, 1971; Wieschaus and Gehring, 1976).

What transcriptional events occur at the critical blastoderm stage in insect embryogenesis? This question was first approached using the incorporation of radioactive nucleosides to measure newly synthesized RNA. For the several insects examined in this way, it appeared that synthesis of total RNA increased at the syncytial blastoderm stage except for the pole (presumptive germ) cells [Lockshin, 1966 (three beetle species); Pietruschka and Bier, 1972 (*Musca domesticus*); Zalokar, 1976 (*Drosophila*)]. For two insects, synthesis of ribosomal RNA was found to increase subsequent to cellularization [Harris and Forrest, 1967 (*Oncopeltus*); Hansen-Delkeskamp, 1968 (*Achaeta domestica*)]. Two difficulties with this approach are that the results do not distinguish clearly between nuclear synthesis and mitochondrial synthesis, and they do not permit detailed analysis of nonribosomal RNA synthesis, which is expected to include informational RNAs. For example, a study by Fausto-Sterling *et al.* (1974) indicated a high level of transcriptional activity in *Drosophila* embryos prior to nuclear migration. It was later concluded that this early RNA synthesis was mitochondrial in origin rather than nuclear, based on analysis of stratified egg components after centrifugation (Zalokar, 1976). However, even this elegant study by Zalokar was not able to distinguish between ribosomal and nonribosomal transcription.

Such a distinction between different classes of RNA has been made by Anderson and Lengyel (1979), who measured rates of synthesis of poly-$(A)^+$ and poly-$(A)^-$ RNAs, with the former class representing nonribosomal RNAs. However, the large numbers of embryos needed for each developmental stage, as well as the brevity of each stage in *Drosophila*, make it difficult to determine how much of the newly synthesized RNA originated from syncytial blastoderm nuclei. Thus, it is not clear from biochemical studies whether an increase in informational RNA synthesis occurs before the completion of cell membrane formation.

A molecular approach was designed specifically to detect nonribosomal RNA in individual embryos (Lamb, 1975). This approach used polyuridylic acid, which had been shown to hybridize to polyadenylic acid tracts in nonribosomal RNA (Jones *et al.*, 1973). Radiolabeled polyuridylic acid was hybridized *in situ* to nuclei and cells from individual *Drosophila* embryos (Lamb, 1975; Lamb and Laird, 1976). Autoradiographic analysis revealed that in general the concentration of nuclear poly-$(A)^+$ RNA began to increase after nuclear migration and before cellularization. No increase was observed in pole cells at the blastoderm stage, in agreement with the data of Zalokar (1976). However, these results are subject to the criticism that some nonribosomal RNA molecules are not polyadenylated and would therefore not be detected by this assay. In addition, some poly-(A)$^+$ tracts appear not to be attached to longer RNA molecules (Slater *et al.*, 1973), thus raising the possibility that synthesis of poly-(A)$^+$ rather than informational RNA was being measured.

An alternative approach to examining the timing and properties of early transcription is a direct electron microscopic examination of transcription using the chromatin spreading procedures developed by Miller and Beatty (1969). This approach is not subject to many of the above problems and thus can complement and extend biochemical and molecular results.

For two insects, the Hemipteran, *Oncopeltus fasciatus*, and the Dipteran, *Drosophila melanogaster*, electron microscopic analyses have been reported in which embryos were examined and compared to detect transcriptional regulation (Foe, 1975; McKnight and Miller, 1976). These studies were carried out on individual staged embryos, thus permitting a direct examination of transcriptional activity at specific embryonic stages.

2.1. Morphological Properties of Ribosomal and Nonribosomal Transcription Units

The morphological approach required the identification of, and distinction between, chromatin regions of ribosomal and nonribosomal RNA synthesis. Analysis of transcriptional activity on spread chromatin led to the detection of transcription units, which were defined as chromatin regions bounded by sites of transcription initiation and termination (see section 3) (Foe *et al.*, 1976; Laird *et al.*, 1976). Active ribosomal transcription units (rTUs) have distinguishing characteristics of defined length, tandem repeat occurrence, nonbeaded appearance of their chromatin, and usually a high number of nascent ribonuclear protein fibers arranged in a very apparent gradient of lengths (Figure 1A). Active nonribosomal transcription units are less easily identified than are rTUs. Their fibers are often widely spaced such that a gradient of lengths is not always readily apparent (Figure 1B). A quantitative analysis of fiber lengths and spacing (Laird *et al.*, 1976) can be used to reveal whether or not an array of fibers has specific initiation and termination sites (Figure 2). Arrays with such sites are inferred to be functional transcription units. The properties that usually distinguish nonribosomal from ribosomal fiber arrays are their solitary occurrence, the beaded appearance of their chromatin, their length, and their relatively low density of nascent ribonuclear protein fibers.

2.2. Timing of Transcriptional Activation in Oncopeltus

Although the identification of embryonic stages has been well worked out for *Drosophila* (see Zalokar and Erk, 1976), less information is available for *Oncopeltus* (see Butt, 1949; Foe, 1975). Micrographs of some of the various stages of early *Oncopeltus* development are therefore included here. For *Oncopeltus*, the designation of embryonic stages involved light and electron micro-

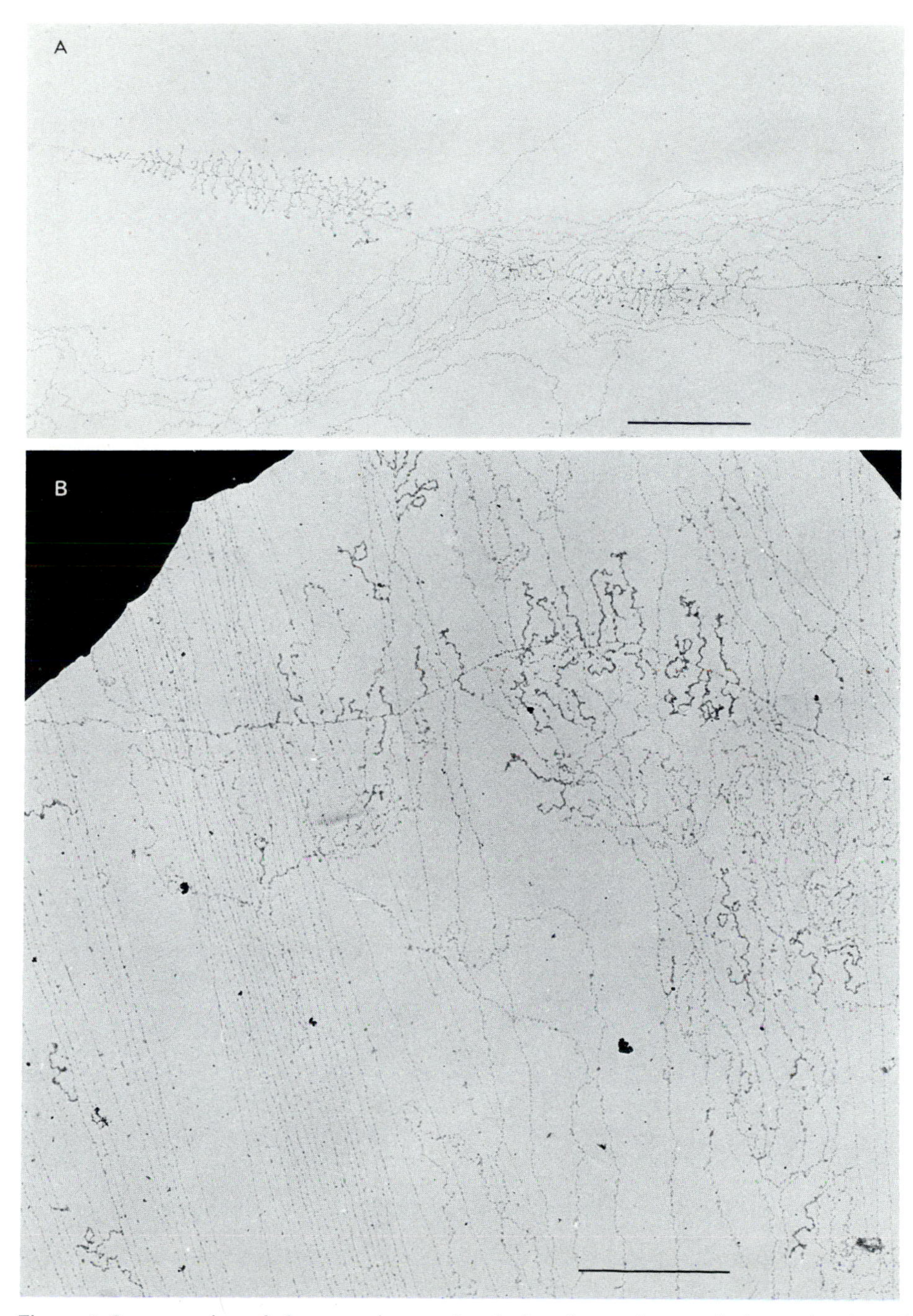

Figure 1. Interpretation of electron micrographs: Active ribosomal transcription units (A) and nonribosomal transcription units (B) from *Oncopeltus* embryos. The morphological properties that distinguish these two classes of transcription are given in section 2.1. Interpretive drawing of the longest fiber array in (B) is given in Figure 2. (Scale bars = 1 μm.)

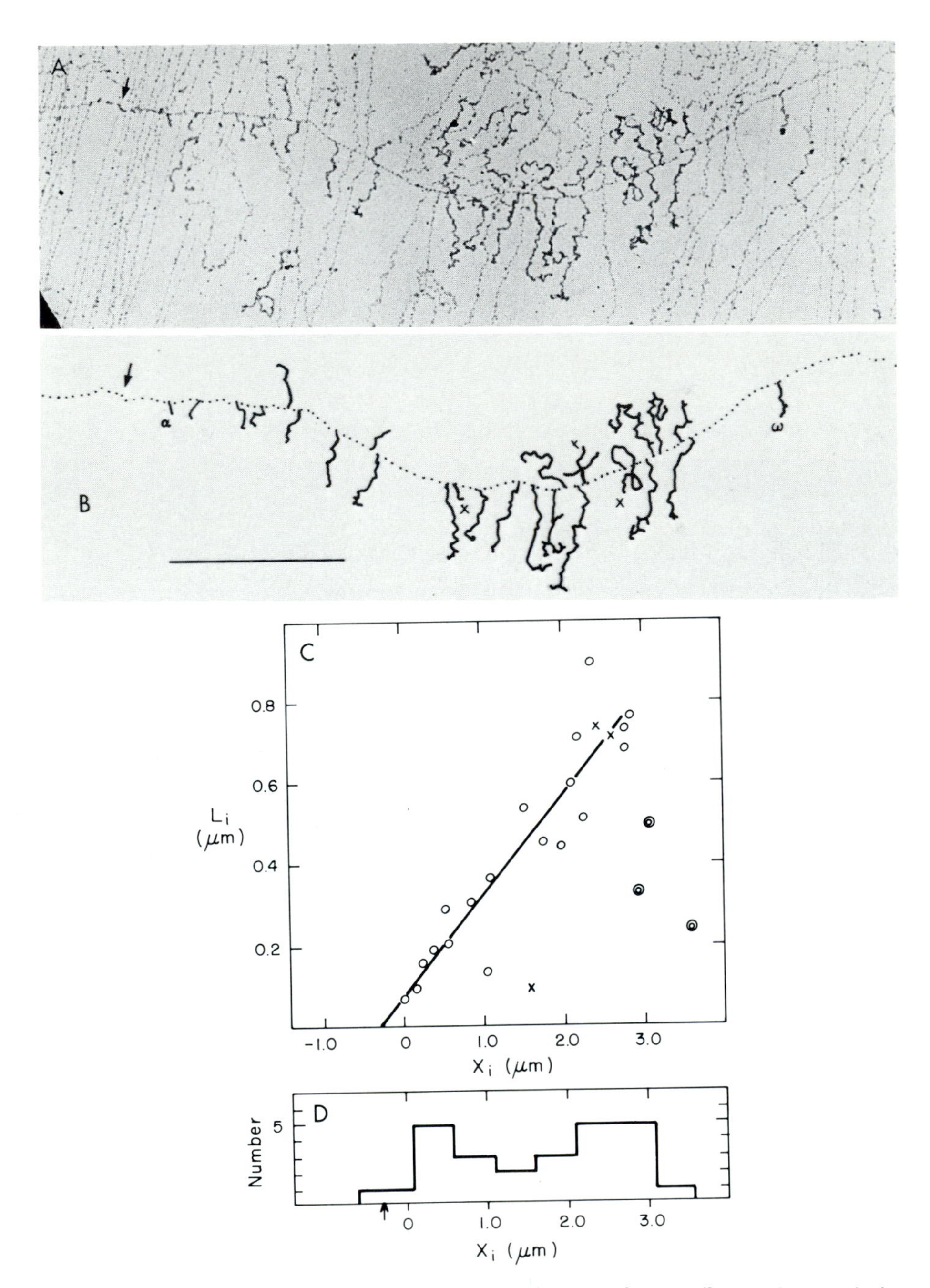

Figure 2. (A) Interpretation of an electron micrograph: An active nonribosomal transcription unit from an *Oncopeltus* embryo.
(B) A drawing of the fiber array shown in (A). Dotted and solid lines represent chromatin and nascent RNP fibers, respectively. Arrow indicates the inferred size of transcriptional initiation estimated from (C). The three fibers marked "X" had unclear attachment sites. These were measured, but the values were not included in the least-square analysis shown in (C). Fibers marked "α" and "ω"

scopic examination of sections prepared from embryos (Figures 3, 4). Between 18 and 23 hr after fertilization (at 21°C), nuclei migrate from the yolky egg interior to the peripheral cytoplasm. The syncytial blastoderm stage lasts until at least 28 hr, by which time incomplete membranes are observed between adjacent nuclei. Membrane formation between adjacent blastoderm nuclei appears to be completed between 28 and 34 hr.

Nuclei from these various stages were spread for electron microscopy and examined in order to compare their transcriptional activity (Foe, 1975). Well-spread chromatin at each nuclear periphery was scanned for the presence of ribosomal and nonribosomal TUs (Figure 5). Active ribosomal TUs were observed at 38 hr but not 32 hr after fertilization (Foe, 1977). A dramatic increase in ribosomal transcriptional activity was detected between 38 and 68 hr, times that corresponded approximately to early gastrula and neurula stages, respectively. By 116 hr, transcriptional activity had dropped to the low level observed in gastrula stage embryos. For most of the embryonic stages examined, a remarkable consistency was apparent between morphological and biochemical data on rates of synthesis of ribosomal RNA (Figure 6). It thus appears that during early development of *Oncopeltus*, the major induction of rRNA synthesis occurs shortly after the completion of the cell membranes.

When does synthesis of nonribosomal RNA occur during the early embryonic stages of *Oncopeltus*? Foe (1975) reported that no arrays of ribonucleoprotein (RNP) fibers were observed on nonribosomal chromatin from premigration nuclei. In contrast, such arrays were detected in nuclei from syncytial blastoderm and from cellular blastoderm stages (Figure 7). (It must be noted that transcription at low rates of initiation was not measured in the morphological studies on *Oncopeltus*. The criterion of scoring transcriptional arrays with three or more contiguous RNP fibers on chromatin was used in order to distinguish transcriptional activity from twists in chromatin strands—Foe *et al.*, 1976). The activity of Form II (nonribosomal) RNA polymerase in postmigration nuclei was at least four times higher than in premigration nuclei (Figure 7). Together, these data indicate that nonribosomal RNA synthesis is induced after the migration of the nuclei to the peripheral cytoplasm and before the completion of the cell membranes.

2.3. *Timing of Transcriptional Activation in Drosophila*

The electron microscopic analysis of transcriptional activity in *Drosophila* embryos (McKnight and Miller, 1976) reveals similarities and differences be-

---

are the most apex-proximal and apex-distal fibers that were measured for this array. (Scale bar = 1 μm.)

(C) Graph of fiber length (L_i) and position (X_i) on chromatin; $X = 0$ is the position of fiber α; the least-squares line is described by $L = 0.08 + 0.25X$ (correlation coefficient = 0.93). Fibers marked "X" and the three most distal fibers (double circles) were not included in the least-squares analysis. The abscissal intercept at −0.3 μm indicates the inferred site of transcription initiation. The slope of 0.25 indicates that the length of nascent RNP fibers is about one quarter that of the transcribed chromatin.

(D) Frequency distribution of fibers shown in (A).

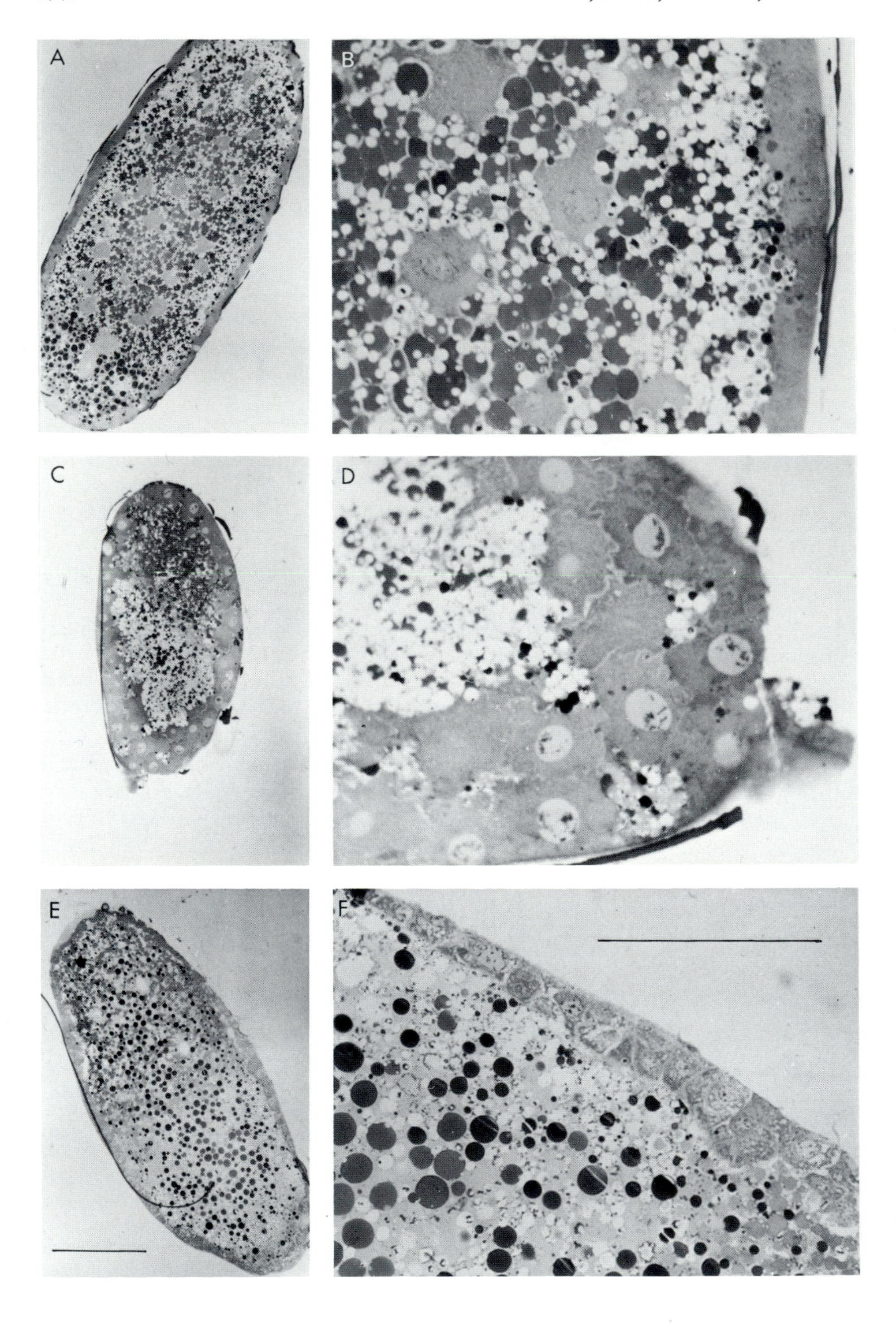
A
B
C
D
E
F

tween *Oncopeltus* and *Drosophila*. Nonribosomal transcription in *Drosophila* embryos was reported to occur at low levels in premigration nuclei and at five-fold increased levels after migration. (In the premigration nuclei, 1.1% of the measured chromatin was reported to be transcriptionally active. However, these measurements were made on chromatin from nuclei that had at least one well-defined array of nascent fibers, and thus this value may be an overestimate.) These observations are similar to those for *Oncopeltus* in that they suggest that the major increase in nonribosomal transcriptional activity occurs after the nuclei have migrated to the peripheral cytoplasm.

How does the timing of this major increase relate to the time of cellularization in *Drosophila?* We conclude that the increase in nonribosomal transcription in *Drosophila* occurs *before* the completion of cellularization. In embryos of the stage used by McKnight and Miller (1976) and referred to as "cellular blastoderm," the partitioning of blastoderm nuclei by membranes was not complete. The molecular analysis of poly-(A) RNA by Lamb and Laird (1976) lends support to this conclusion. Prior to *in situ* hybridization, *Drosophila* embryos were squashed, and the contents were examined under a light microscope. Embryos with "membranes extending at least halfway into the cytoplasm" had nuclei that were not enclosed by cell membranes (Lamb and Laird, 1976). Since this stage of embryogenesis is similar to that defined as cellular blastoderm in the McKnight and Miller study, it appears that both morphological and molecular data support the conclusion that the increase in nonribosomal RNA synthesis occurs prior to the completion of the cell membranes.

2.4. *Implications and Prospects*

These studies, then, indicate that for two insects the major induction of nonribosomal RNA synthesis during early embryogenesis occurs close to the stage in which determination is initiated in *Drosophila*. Furthermore, comparison of *Drosophila* and *Oncopeltus* shows that the pattern of transcriptional induction is markedly stage- rather than time-specific, since the time scale of their developmental stages differs by a factor of ten. It will be of interest to learn whether this transcriptional induction is a cause or a consequence of the restriction of developmental fates that begins at this syncytial blastoderm stage.

←

Figure 3. Low- and high-power light micrographs of *Oncopeltus* embryos. Epon sections (2 μm thick) were stained by Richardson's method (Foe, 1975).

(A,B) An 18-hr (pre-nuclear migration) embryo showing nuclei with cytoplasmic islands. The peripheral cytoplasm contains no nuclei.

(C,D) A 23-hr embryo (post-nuclear migration, syncytial blastoderm). Nuclei have migrated to the peripheral cytoplasm. The irregularly shaped partitions between nuclei at this stage appear to be discontinuous membranes (see Figure 4C).

(E,F) A 34-hr embryo (cellular blastoderm). The blastoderm cells are separated by membranes. Differences in the sizes of the embryos and in the thicknesses of the layers of peripheral cytoplasm are due to the different planes of sections. A,C, and E, and B,D, and F are at the same magnifications. (Scale bars, E = 0.2 mm, F = 0.1 mm.)

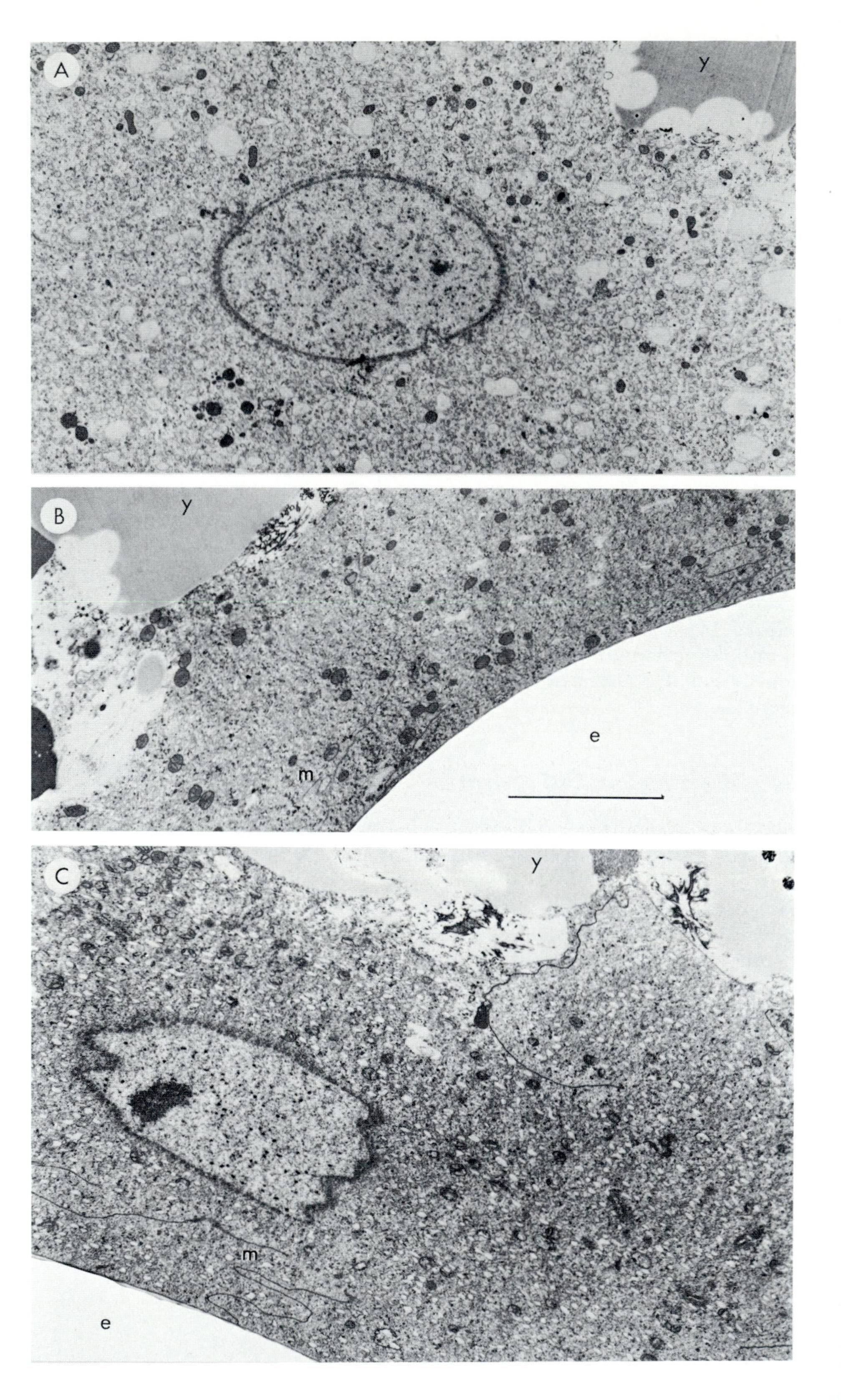
A
y
B
y
e
m
C
y
m
e

What may be inferred about the mechanism of transcriptional induction? The arrival of nuclei in peripheral cytoplasm is correlated with the major activation of transcription (with the exception of nuclei that form pole cells, to be discussed below). Some possible causes of this activation are the presence of transcriptional inducers in peripheral cytoplasm, the loss of transcriptional repressors in the egg interior as a consequence of nuclear migration to the periphery, or a programed number of nuclear divisions. Since the migrating nuclei are accompanied by cytoplasm from the egg interior as they migrate and reach the periphery (Scriba, 1964), the presence of positive inducers in the nonpolar peripheral cytoplasm seems more likely than the loss of transcriptional inhibition upon nuclear migration. An assay for RNA synthesis in premigration nuclei prematurely put in contact with peripheral cytoplasm would be useful in distinguishing among these possibilities.

The presence of regional differences in peripheral cytoplasm is indicated by a delay in transcriptional induction in pole cells. The suggestion that pole plasm inhibits transcriptional activation, i.e., maintains transcriptional quiescence, is supported by both biochemical and molecular data (Lamb and Laird, 1976; Zalokar, 1976). Pole cells appear to be rigidly determined by pole plasm, including polar granulae (Illmensee and Mahowald, 1974, 1976), before the major transcriptional induction occurs (Zalokar, 1976). It would thus seem that major transcriptional induction is not a specific cause or an immediate consequence of germ cell determination.

This consideration of pole cells points out one major limitation of the morphological analysis of transcription as it has been used thus far with whole embryos. No distinction has yet been made between the two cell types (pole and nonpole blastoderm cells) or between blastoderm cells from different regions in the embryo. Examination of transcriptional patterns is possible using only a few microdissected nuclei. Coupling this approach with probes for specific chromatin regions may be useful in analyzing further the relationship between developmental determination and transcriptional programs.

3. *Inferred Mechanisms of Transcriptional Control*

The electron microscope allows a direct examination of individual nascent RNP fibers. From such examination, inferences can be made about the regulatory processes that control RNA synthesis. Visualized transcription arrays provide insights into questions concerning initiation, termination, chromatin structure, replication, and RNP processing.

←

Figure 4. Electron micrographs showing pre- (A,B) and post- (C) nuclear migration stages of *Oncopeltus* embryos. Epon-embedded sections (approximately 60 nm thick) were stained with uranyl acetate and lead citrate (Foe, 1975). All micrographs are at the same magnification.

(A) A cytoplasmic island containing a cleavage nucleus from an 18-hr embryo. There are no cell membranes apparent within or at the edge of the cytoplasm of the island.

(B) The peripheral cytoplasm of an 18-hr embryo lacks nuclei but contains long, discontinuous, membranelike structures (m). (Scale bar = 5 μm.)

(C) A nucleus in the peripheral cytoplasm of a 23-hr embryo. Irregular membranes (m) occur near the nucleus but do not appear to enclose it (see also Figure 3B). "y" and "e" indicate yolk spheres and exterior, respectively.

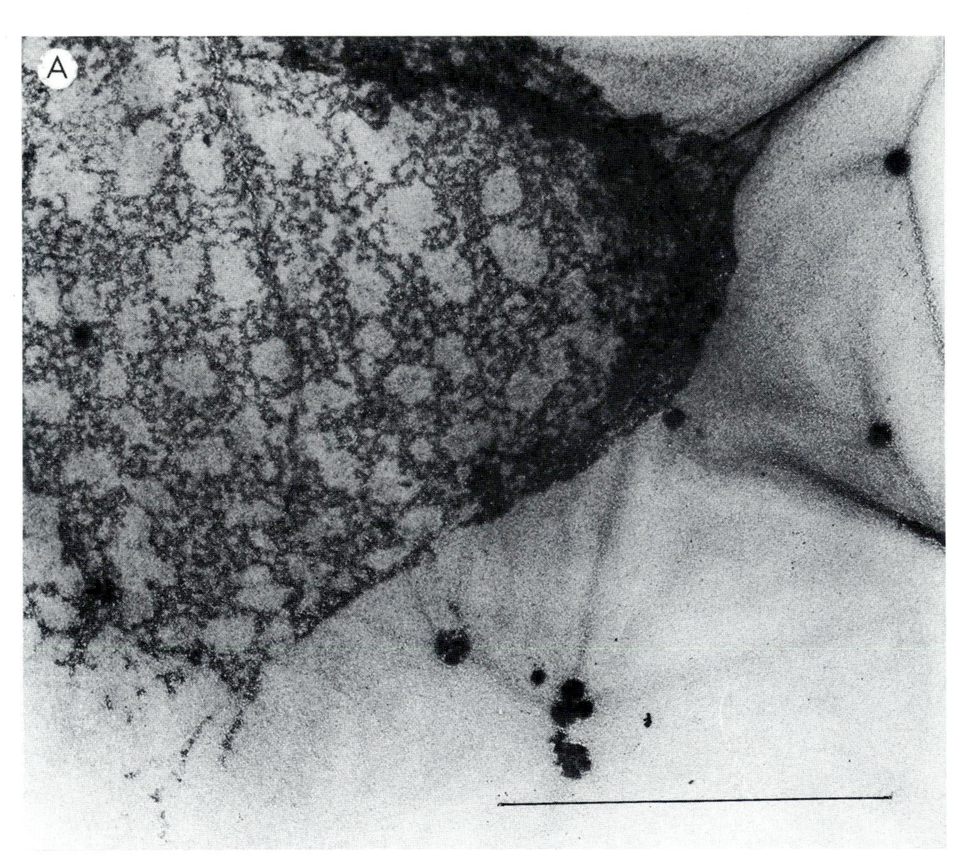
A

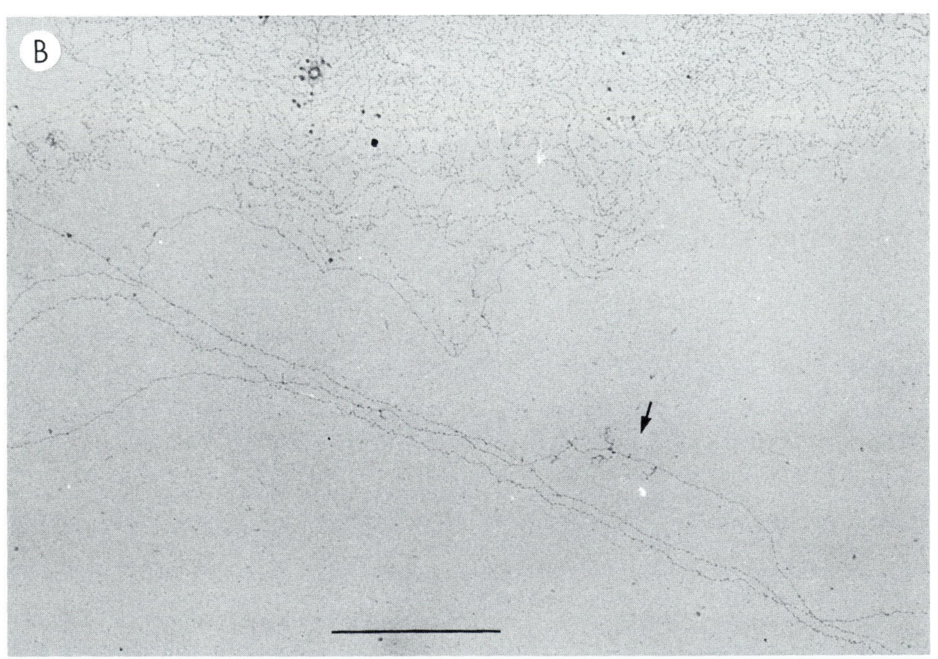
B

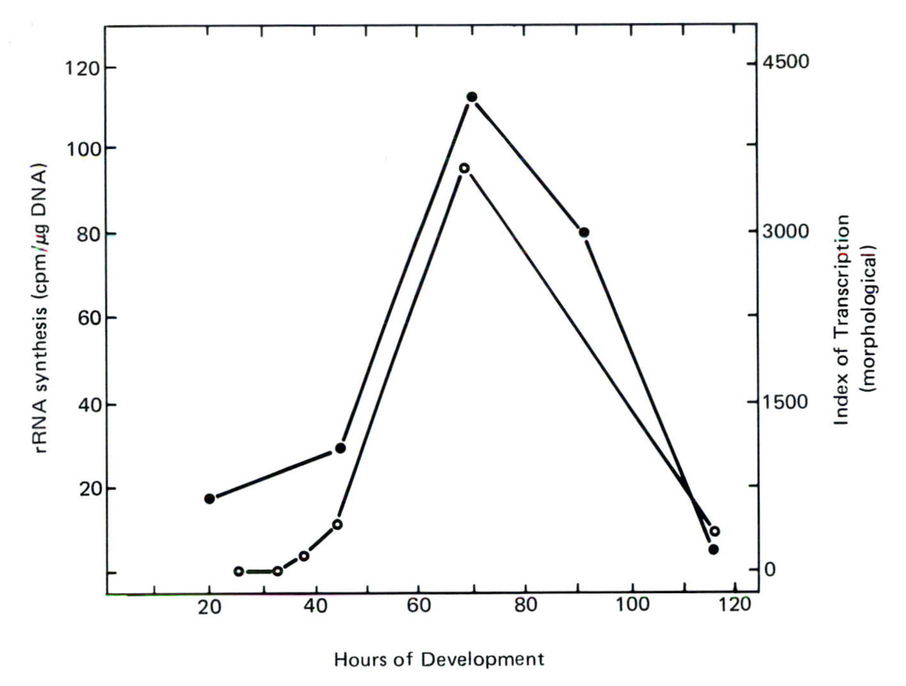

Figure 6. Comparison of morphological and biochemical assays of rRNA synthesis in *Oncopeltus* embryos. Closed circles represent the rate of incorporation of ^{3}H-uridine into rRNA per unit of DNA (data of Harris and Forrest, 1967). Open circles represent data for ribosomal transcriptional activity derived by multiplying the mean number of transcripts per ribosomal transcription unit by the percent of nuclei in which at least one ribosomal transcription unit was observed (see Foe, 1977).

3.1. Specific Initiation and Termination Sites

The biochemical evidence that nonribosomal RNA in eukaryotes is heterogeneous in size (see Darnell *et al.*, 1973) is consistent with the presence of many homogeneous populations of RNA, the sum of which give rise to "heterogeneous RNA." Alternatively, it could indicate the existence of multiple or random initiation and termination sites for transcription of specific regions (Robertson and Dickson, 1975). Morphological analysis of transcription indicates that the former explanation is the more likely.

By measuring the lengths and spacings of fibers from an array, one can determine whether nascent fibers originated from the same chromatin region. If all of the transcripts in a TU were to initiate at a single specific site, and if they were assembled uniformly in RNP fibers, then they should form a single gradient of fiber lengths. The intersection of this gradient with the chromatin axis would represent the initiation site.

←

Figure 5. A nucleus and its extruded chromatin (A) from an *Oncopeltus* embryo. The Miller procedure (Miller and Beatty, 1969) visualizes arrays of nascent RNP fibers (arrow) on well-spread chromatin (B). (Scale bars, A = 10 μm, B = 1 μm.)

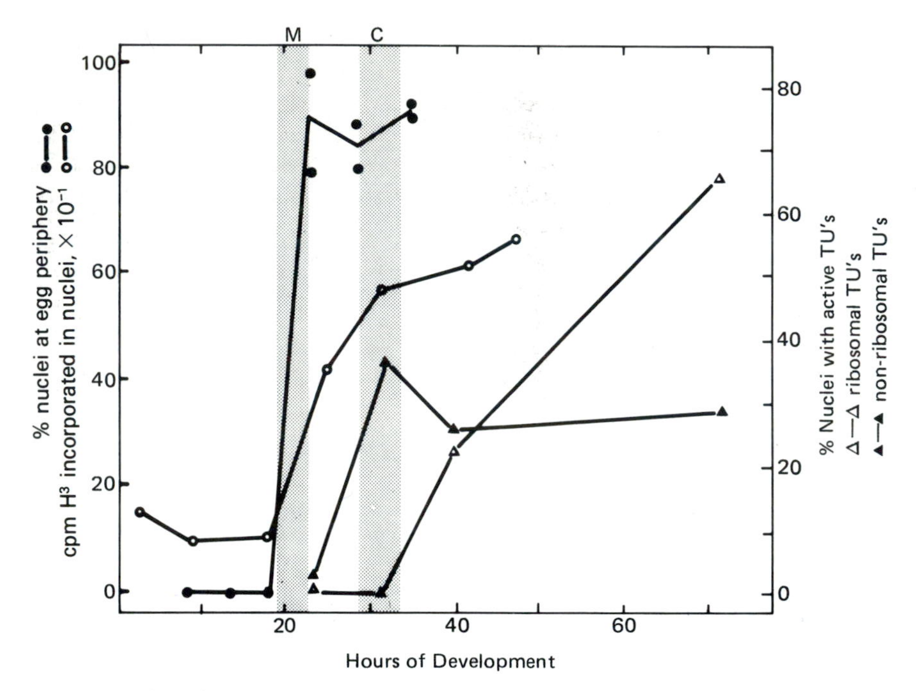

Figure 7. Relationship between transcriptional activity and developmental stage in *Oncopeltus*. The increase in percentage of nuclei found at the egg periphery, determined from examination of sectioned embryos, indicates nuclear migration leading to a syncytial blastoderm. The shaded strip (M) indicates experimental uncertainty in the timing of nuclear migration. ^{3}H-UTP incorporation owing to the action of endogenous Form II RNA polymerase increases prior to and parallel with the increase in active nonribosomal TUs observed in nuclear chromatin spreads. Shaded area (C) indicates experimental uncertainty as to the time of completion of cell membranes. The biochemical and morphological data together indicate that synthesis of nonribosomal RNA increases during the syncytial blastoderm stage prior to induction of ribosomal RNA synthesis.

All of the rTUs and most of the analyzed nonribosomal TUs have been interpreted as having specific initiation sites (Figure 2) (Foe *et al.*, 1976; Laird and Chooi, 1976; Laird *et al.*, 1976). A few nonribosomal arrays had complex fiber patterns that did not exhibit a single continuous length gradient. These arrays probably also had specific initiation sites, although additional interpretations, discussed below, were required to explain the discontinuities in the gradient.

Although the lengths and spacings of fibers in most transcribed arrays are consistent with specific and unique initiation sites, closely spaced multiple sites or a small initiation region could not be ruled out. For example, the early promotor region of T7 DNA has been shown to have three polymerase binding sites spaced within a 300-base region (Bordier and Dubochet, 1974; Darlix and Dausse, 1975). This amount of DNA is equivalent to only two nucleosomes. Because of fiber length variability, such close-spaced, multiple initiation sites would not be resolved in chromatin spreads. It is also possible that different cells use alternate initiation sites for a specific chromatin region. These considerations point out the desirability of identifying site-specific transcription in individual cells.

A specific termination site, as opposed to random termination of RNA synthesis, leads to the prediction that fiber frequency (fibers per μm chromatin) would decrease abruptly toward the distal end of an array. The fiber frequency distributions for TUs from *Drosophila* and *Oncopeltus* were found to be consistent with specific termination sites (Foe *et al.*, 1976; Laird and Chooi, 1976; Laird *et al.*, 1976). However, the presence of fibers distal to the main array indicated that termination efficiencies were variable. Formally, the possibility also exists that fiber release and termination of transcription do not occur at the same site. If fibers moved substantially past the site where polymerization ceases, the gradient of fiber lengths should plateau at the distal end. This was not observed for nonribosomal RNA. Thus, most analyzed fiber arrays have been interpreted as being delineated by specific sites of transcription initiation and termination.

3.2. Processing of Nascent Transcripts

It is known that RNP molecules are sometimes cleaved subsequent to transcription. For example, ribosomal RNA is transcribed as one molecule and is later cut into smaller fragments (Scherrer and Darnell, 1962). In the bacterium, *Escherichia coli,* this processing of rRNA appears to occur during transcription. Does such processing occur in eukaryotic cells before RNP is released from the chromatin? Some fiber arrays on chromatin of *D. melanogaster* and *O. fasciatus* showed discontinuities in their length gradients (Foe *et al.*, 1976; Laird and Chooi, 1976). The simplest explanation is that each discontinuity in fiber lengths represented specific cleavage of RNP fibers. After cleavage, the shortened RNP fiber remained attached to the RNA polymerase as transcription continued along the chromatin. Recent biochemical evidence indicates that processing of nascent transcripts can occur (Nevins and Darnell, 1978), lending credence to this interpretation of the morphological analysis. DNA transcribed in this way would be expected to generate complex genetic complementation maps (see Laird and Chooi, 1976).

3.3. Morphology of Transcriptionally Active Chromatin

With the preparative conditions used for visualizing nuclear contents of *Oncopeltus* and *Drosophila,* the chromatin of moderately active nonribosomal TUs is beaded (nu chromatin). Chromatin of active ribosomal TUs in *Oncopeltus* is unbeaded (rho chromatin) (Foe *et al.*, 1976). (For *Drosophila* rTUs the chromatin configuration is usually obscured by the high frequency of nascent fibers.) What do these morphologies represent in terms of chromatin structure? An answer to this question provides insight into the size and regulation of transcription units.

The beads observed on fiberfree chromatin of *Drosophila* and *Oncopeltus* are interpreted to be analogous to chromatin beads or nucleosomes (Oudet *et al.*, 1975), reported for other organisms (Olins and Olins, 1974). This interpretation is based on the approximate sizes, periodicities, staining properties, and general occurrence of the beads throughout the chromatin (Foe *et al.*, 1976; Laird *et al.*, 1976). Nucleosomes usually appear connected by thin (3.0 nm) filaments of presumably B-structure DNA. Although it is not clear whether nucleosomes are separated *in vivo* by a filament (see Daneholt, this volume), nucleosome frequencies can be used to calculate the packaging of DNA in chromatin observed after spreading. Assuming that the chromatin subunit of approximately 200 base pairs that is detected biochemically (Kornberg, 1974) corresponds to the subunit and filament that are detected morphologically, fiberfree nu chromatins from *D. melanogaster* and *O. fasciatus* were estimated to have DNA packaging ratios of 1.9 and 2.3 μm DNA per μm chromatin, respectively (Laird *et al.*, 1976).

The beads observed between fibers on the chromatin of moderately active nonribosomal transcriptional units are similar to nucleosomes in their staining properties and their mean size and are interpreted as nucleosomes. For both *Drosophila* and *Oncopeltus,* there are 0.6–0.7 times as many nucleosomes on active TUs as there are on fiberfree chromatin. This reduced frequency for active versus inactive chromatin corresponds to a deficiency of about two nucleosomes per nascent fiber. Assuming that the DNA of some nucleosomes has been completely extended, and that the remaining nucleosomes contain the same amount of DNA as those in nontranscribed chromatin, we calculated that the DNA packing ratio in transcribed nu chromatin is 1.6 for both *Drosophila* and *Oncopeltus* (Laird *et al.*, 1976). It is interesting that the same DNA packing ratio was calculated by Lamb and Daneholt (1979; see Daneholt, this volume) for a specific region of active chromatin in *Chironomus*. In this case, however, nucleosomes were absent or obscured by the high frequency of nascent RNP fibers.

The packing ratio of DNA in active chromatin, together with the measured distances between transcriptional initiation and termination sites, led to estimates of the size of nonribosomal transcription units. Average values for *Drosophila* and *Oncopeltus* were about 20,000 base pairs of DNA (Foe *et al.*, 1976; Laird and Chooi, 1976). These values thus provided relatively direct confirmation of the large estimates derived from biochemical studies (see Darnell, *et al.*, 1973).

The presence of nucleosomes on moderately active nonribosomal TUs implies that histones can be present on transcriptionally active chromatin. A similar inference has been made from biochemical data (Axel *et al.*, 1975; Gottesfeld *et al.*, 1975; Lacy and Axel, 1975; Weintraub and Groudine, 1976). To test whether the beads of transcriptionally active regions observed in the EM represent nucleosomes, McKnight *et al.* (1977) used antihistone antibodies on chromatin spreads. Their preliminary data indicated a similar labeling by antihistone H2b of beads in active regions compared with inactive regions, suggesting that at least this histone is present on active nucleosomes. Transcriptional activation, therefore, can be more subtle than complete removal of repressing histones from a chromatin region (Bonner and Huang, 1974).

The chromatin of active ribosomal transcription units has a lower DNA packing ratio than does other chromatin. Based on the observed average length of rTUs, and the known length of B-structure DNA necessary to encode for ribosomal precursor RNA, the chromatin of rTUs is estimated to have a DNA packing ratio of 1.1 and 1.2 for *Drosophila* and *Oncopeltus*, respectively (Laird *et al.*, 1976).

What is the significance of the difference between the rho chromatin of active ribosomal TUs, and the nu chromatin of active nonribosomal TUs? Rho chromatin is not nu chromatin that has been extended by the presence and activity of many RNA polymerase molecules. This is inferred from the observation that in *Oncopeltus*, rTUs with few or no fibers have chromatin with a smooth morphology and low DNA packing ratio. Rho chromatin is probably associated with proteins, as inferred from its stainability with phosphotungstic acid and its thickness relative to that expected for naked DNA (Foe *et al.*, 1976). Proteins appear to be associated with rDNA *in vivo*, rather than being bound during the preparation for electron microscopy. This conclusion was based on observations of T4 DNA added to embryonic lysates that were prepared for electron microscopy (Foe, 1977). T4 DNA was conspicuously thinner than the rho chromatin in the same preparation. Thus, it seems likely that rho chromatin is coated with proteins *in vivo*.

In *Oncopeltus*, rho chromatin is first detected at or just prior to the time that ribosomal RNA synthesis begins (Foe, 1977). Segments of fiberfree rho chromatin were observed that were of the length and tandem repeat occurrence expected for rTUs. Such fiberless rTUs were found occasionally in nuclei from all developmental stages, but were most frequent in the gastrula stage (38 and 44 hr) embryos. Prior to this time, no such segments of rho chromatin were observed. Inactive rTUs were presumably in nu chromatin conformation. It was postulated that the segments of fiberfree rho chromatin represent ribosomal genes before or shortly after transcription had been initiated on them and that the transition from nu to rho chromatin is part of the induction process (Foe *et al.*, 1976; Foe, 1977). Thus, the difference in morphology of nu and rho chromatin appears to relate to the transcription of ribosomal and nonribosomal DNA, rather than solely to the polynucleotide sequence in underlying DNA.

One important distinction between ribosomal and nonribosomal RNA synthesis is the involvement of different polymerases. Some feature of chromatin appears to be responsible for this specificity. For some organisms, RNA polymerase I transcribes purified DNA indiscriminately (Roeder *et al.*, 1970); however, from a chromatin substrate it preferentially transcribes ribosomal DNA (Matsui *et al.*, 1975). Thus, it is tenable that RNA polymerase I and II recognize differences *in vivo* that are reflected under electron microscopic preparative conditions as distinctive morphologies and DNA packing ratios of nu and rho chromatin.

3.4. Replication of Transcriptionally Active Regions

There are both developmental and structural reasons for considering the relationship between replication and transcription. The partitioning of nascent transcripts into progeny cells, the effect of replication on transcriptional control signals, and the physical interaction between DNA and RNA polymerases

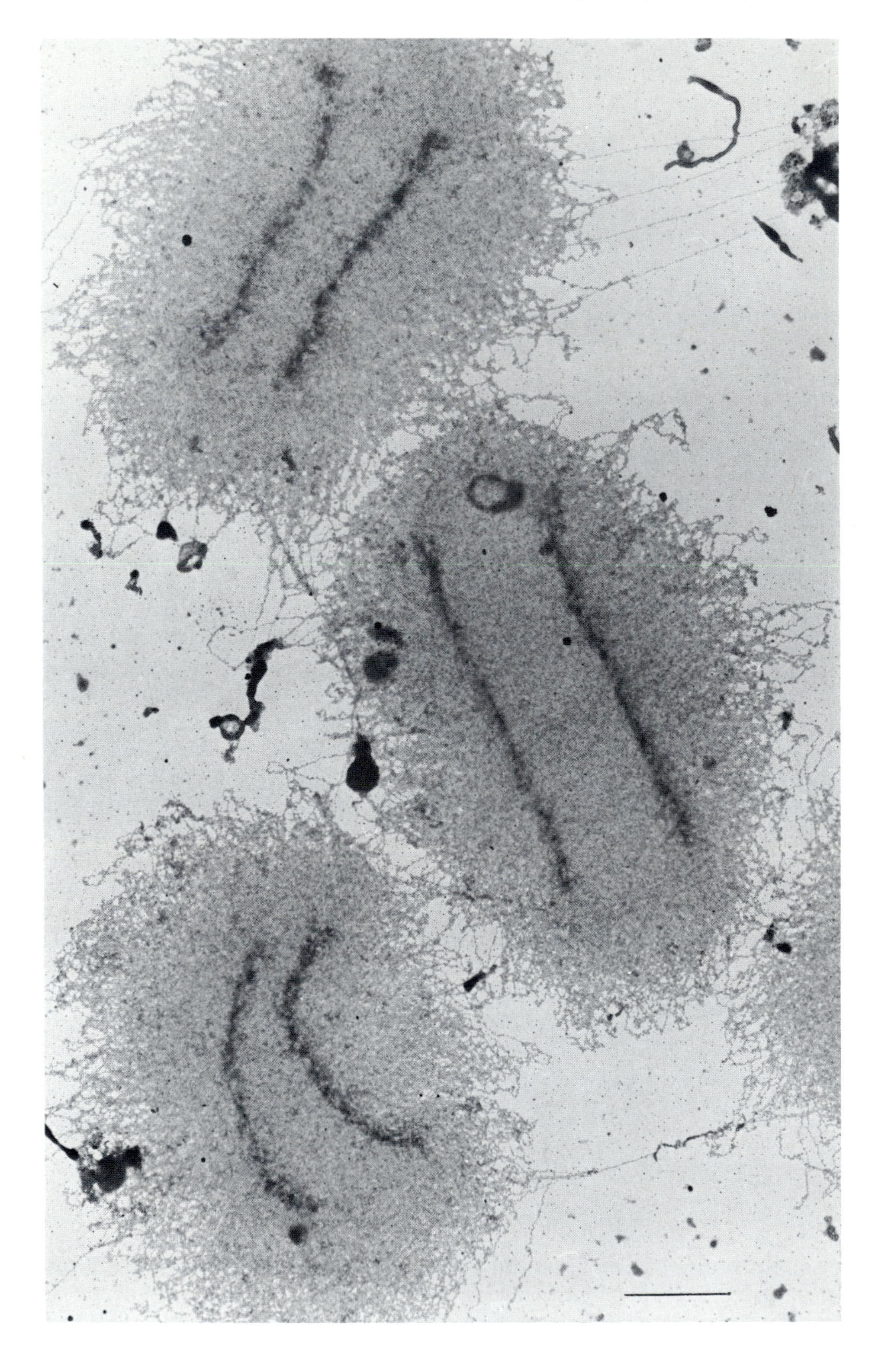

are examples of the kinds of phenomena that are potentially approachable with a morphological analysis of transcription.

McKnight and Miller (1979) have observed a number of transcription units on what appears to be newly replicated chromatin. Of special interest for the issues posed above are cases in which a replication fork occurs within or at the edge of an active transcription unit. This situation was found for a number of rTUs. Such configurations suggest that DNA and RNA polymerases sometimes encounter one another, as opposed to there being a prereplication clearing of the chromatin. Replication was inferred to begin usually within the spacer region and proceed bidirectionally toward the adjacent transcribed rDNA. Four examples were seen in which the proximal but not the distal region of an rTU was replicated. The unreplicated region of the rTU contained nascent RNP fibers, while the replicated region was fiberfree.

Six examples were found in which a replication fork apparently had traveled through the spacer region toward the distal end of an rTU and had stopped at the transcription termination site. No examples were seen in which a fork had progressed past the inferred termination site into the active rTU. Assuming that replication is bidirectional, McKnight *et al.* (1977) concluded that replication forks can progress through an active TU only by moving in the same direction as the RNA polymerases. After replication, initiation of transcription appears to be temporarily halted. In one case where reinitiation was observed, it appeared to be symmetrical on the two duplicated strands of chromatin (Foe, unpublished observation).

From a developmental standpoint, it presumably is replication of nonribosomal chromatin that is of interest. It is important to learn whether replication can asymmetrically alter the control of transcription and thus lead to daughter nuclei of differing information content. A number of paired nonribosomal TUs lying side by side were observed with strikingly similar patterns of nascent RNP fibers (McKnight and Miller, 1979). Because of the inferred postreplicative state of the nuclei from which the chromatin was spread, these pairs were interpreted as representing transcription on newly replicated chromatin. It was further concluded that the nascent RNP fibers had been initiated after replication because of the relatively high frequency of fibers. (If the paired fiber arrays had resulted from partitioning of one single array by replication of an active TU, then one would expect the fiber distribution to be either asymmetric, if nascent fibers went preferentially with the parental sense strand, or symmetrical but at one half the average fiber frequency if random partitioning occurred.) If the fibers on the paired arrays arose independently after replication, one can determine the precision of transcriptional control at a specific chromatin region. The similar lengths of members of a pair of fiber arrays indicated that the same initiation and termination sites were used on each array. The similar fiber frequencies within a pair indicated that regulation of initia-

←

Figure 8. Interpretation of electron micrograph: Condensed mitotic chromosomes from *Oncopeltus* embryo (pre-nuclear migration stage). Chromatin spreads, prepared and stained with phosphotungstic acid as described by Miller and Beatty (1969) reveal condensed chromosomes from nuclei in mitotic division. The double axis of each condensed chromosome is interpreted as indicating that chromosomes have duplicated but have not separated. (Scale bar = 2 μm.)

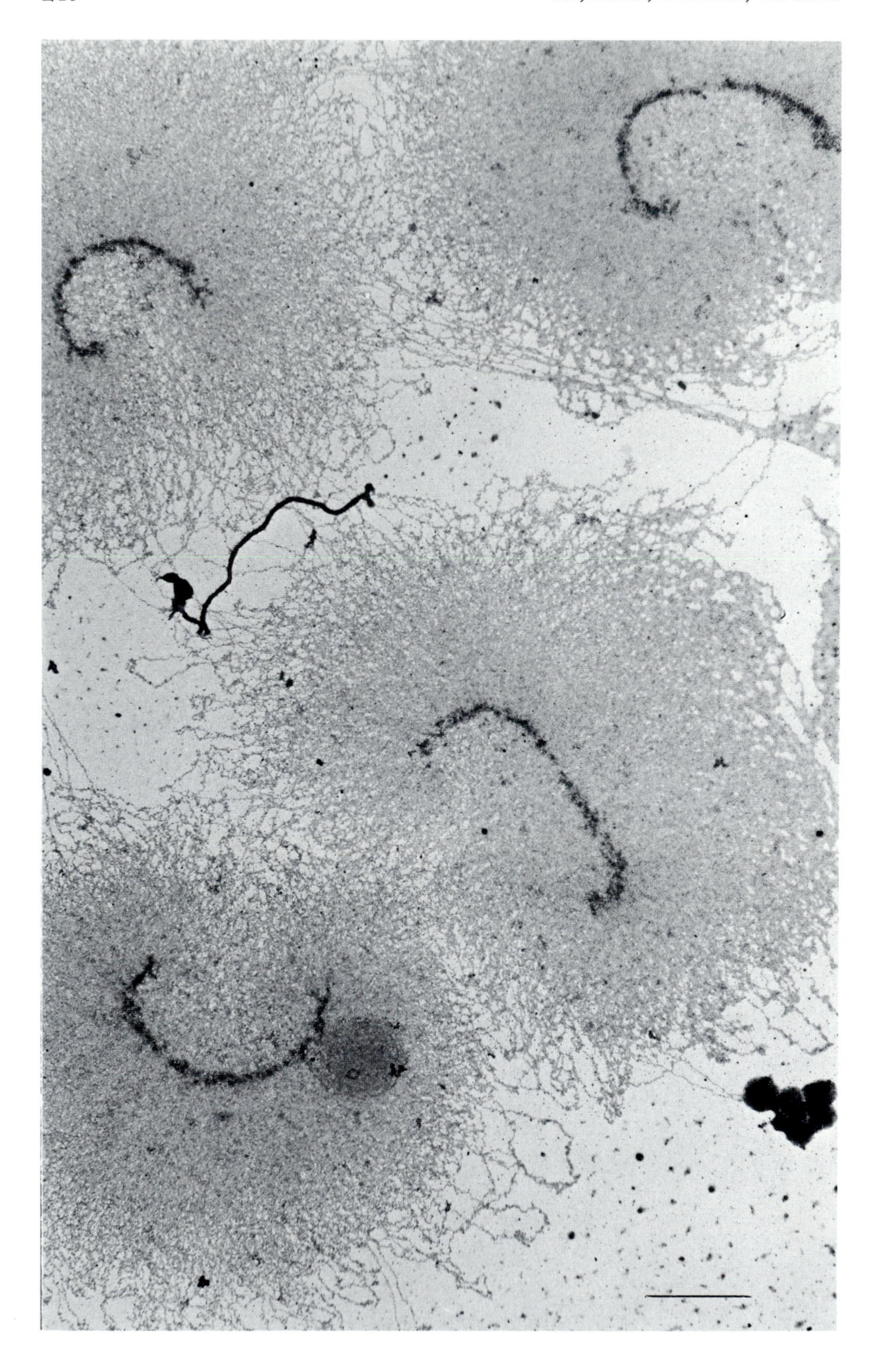

tion and/or elongation rates was highly localized. On some pairs of TUs, the fiber distribution was nonuniform. Internal, fiberfree regions were found at the same location within both members of a pair. These gaps could represent fluctuations in the initiation of transcription in which the two units responded similarly to the control mechanisms.

The question of whether replication can asymmetrically alter the control of nonribosomal transcription is not entirely answered by the study of McKnight and Miller (1979). The designation of pairs of fiber arrays as being homologous and on sister chromatids depended on their similarity in fiber patterns. Disruptions in initiation and/or termination could distort fiber patterns and thereby remove the basis for inferring homology of underlying chromatin segments. Conclusions at this level must await further examples in which partially replicated chromatin strands maintain their association at replication forks.

3.5. *Assembly of RNA into Ribonuclear Protein*

The nascent fibers on chromatin represent RNA and associated proteins. What can be learned about this association? The ratio of the length of RNP fiber to the length of chromatin transcribed is similar for all fibers of a given array, but different among arrays (Laird *et al.*, 1976). RNP fibers also exhibit substructure, such as beads (Martin *et al.*, 1973) and folds. How specific is this substructure?

Beyer *et al.* (1980) have examined subunit structures of nascent RNP fibers of *Drosophila* for sequence specificity. The 24-nm RNP subunits that occur along the length of RNP fibers varied in number and arrangement on RNP fibers of different TUs. Within a TU, however, subunits were found at approximately the same inferred sequence on most fibers of an array. Particularly striking was a pair of nonribosomal TUs assumed to be on newly replicated chromatids. The pattern of subunit location was very similar for these two TUs with presumably identical sequences. Thus, it seems that the formation or stability of at least some subunit structure in RNP is sequence specific.

4. *Higher-Order Chromosome Structure*

The chromatin spreading technique is also useful in examining the structure of condensed mitotic chromosomes (Figures 8, 9). Chromatin loops, as predicted by the work of Cook and Brazell (1975), Worcel and Benyajati (1977), and Laemmli *et al.* (1977), are visible in mitotic chromosomes of *Oncopeltus* (Figure 10). Especially interesting is the axis that extends the length of the chromosomes. This axis may represent either proteins of a central scaffold (Laemmli *et al.*, 1977; but see also Okada and Comings, 1980) or a specialized structure that relates to the holocentric nature of *Oncopeltus* chromosomes (Wilson,

←

Figure 9. Interpretation of electron micrograph: Mitotic chromosomes from *Oncopeltus* embryo. The single axis of each condensed chromosome is interpreted as indicating that the chromosomes have separated at anaphase (see legend to Figure 8). (Scale bar = 2 μm.)

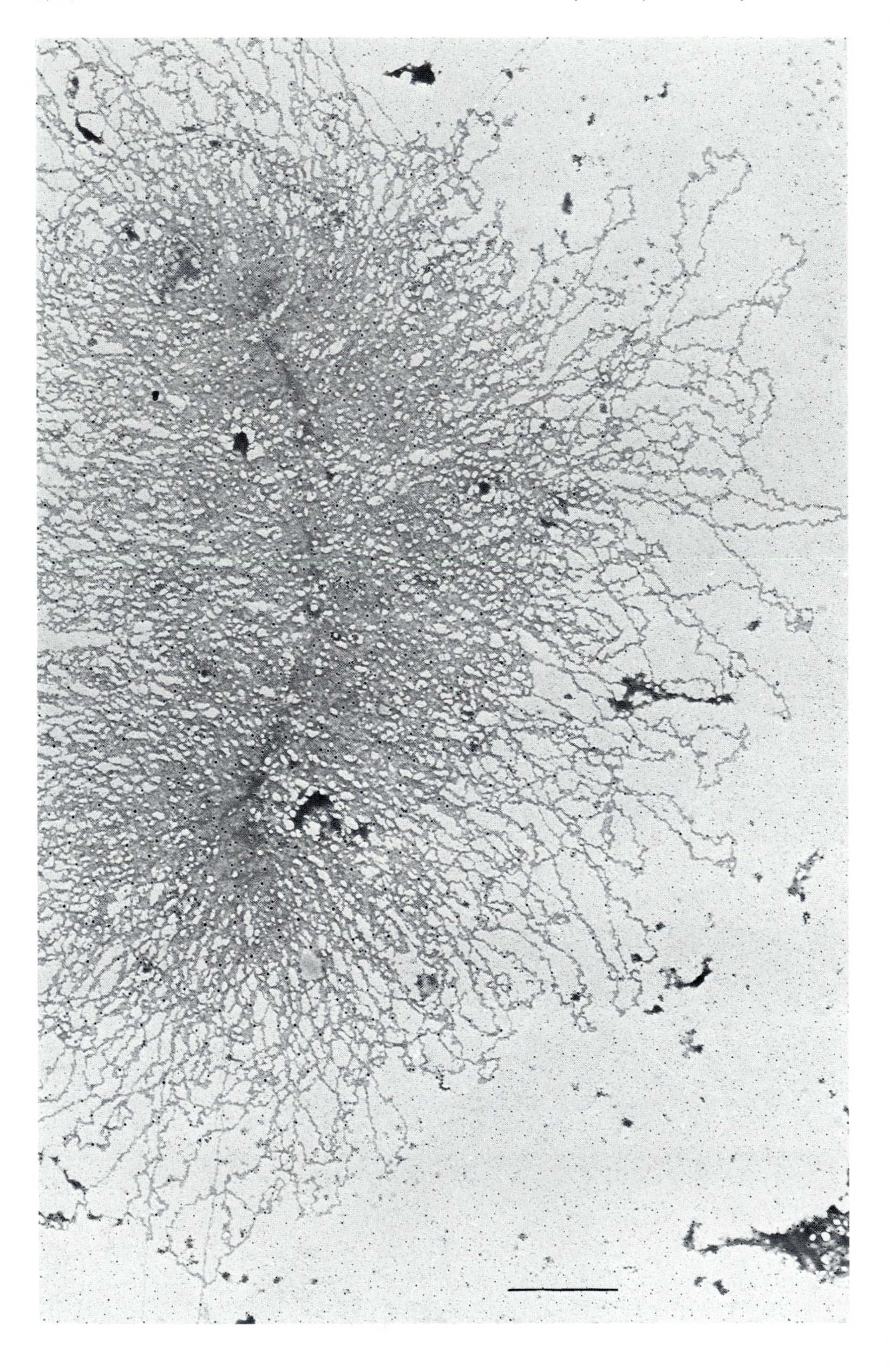

1912; Comings and Okada, 1972). In either case, the embryonic mitotic chromosomes of *Oncopeltus* appear to be well suited for further morphological and biochemical analysis. Particularly important to understanding RNA synthesis is whether or not chromosome condensation alters transcriptional control. In *Bombyx*, it appears that some nascent RNP fibers do remain attached to condensed meiotic chromosomes (Rattner *et al.*, 1980). The embryonic mitotic chromosomes examined for *Oncopeltus* are from stages prior to the major transcriptional induction, and thus this question cannot yet be answered for embryonic development.

5. Summary and Conclusions

Germ cell determination preceeds the major activation of RNA synthesis in pole cells of *Drosophila*. In other blastoderm nuclei, major transcriptional activation and developmental determination begin at the syncytial blastoderm stage. It appears that either the transcriptional causes or consequences of determination are different in the nuclei of pole cells and nonpole cells. In *Oncopeltus*, induction of nonribosomal RNA synthesis clearly precedes that for ribosomal RNA. This implies a difference in the mechanism of transcriptional induction for these two classes of RNA. It may be fruitful to assay for the transcriptional activation of premigration nuclei by the peripheral cytoplasm. The close relationship between transcriptional activation and restriction of developmental fates of embryonic nuclei is intriguing, but the cause–effect relationship remains to be determined.

Transcriptional control acts during embryogenesis at specific sites of initiation, termination, and possibly processing. In addition, changes in the structure of ribosomal chromatin apparently provide a gross level of control that facilitates transcriptional initiation.

Transcription units have large sizes and specific control sites that provide support for conclusions derived from biochemical data. Properties of transcription units also provide possible explanations for complex genetic complementation maps.

Assembly of RNA and protein into RNP fibers appears to be partially sequence specific. Morphological analysis can complement biochemical studies of RNP structure.

DNA replication and transcription occur concurrently, but with restrictions in replication polarity and disruptions in transcriptional initiation. A replication fork cannot enter an active ribosomal gene in a direction opposite to the polarity of RNA polymerase elongation. DNA polymerase can, however, enter a ribosomal transcription array from the proximal end, following or

---

Figure 10. Interpretation of electron micrograph: Anaphase chromosome from *Oncopeltus* embryo. Loops of thick (20–40 nm) chromatin appear at the periphery. These chromosomes differ from those discussed previously (Laemmli *et al.*, 1978; Okada and Comings, 1980) in that *Oncopeltus* chromosomes are holocentric. It is possible that this property accounts for the apparent chromosome axis. (Scale bar = 1 μm.)

dislodging ribonuclear protein fibers. Apparently, the two types of polymerases do not pass one another on ribosomal chromatin.

Whole mount preparations of condensed mitotic chromosomes of insects are suitable for structural and functional analysis. The early mitotic divisions during insect embryogenesis provide especially useful chromosome material.

ACKNOWLEDGMENTS

Richard Fehon, Martin Hammond, Gerold Schubiger, and Laurie Smith made especially important contributions to this review. The authors' research has been supported by grants from the N.S.F., the N.I.H. and the Robert A. Welch Foundation, Houston, Texas.

References

Anderson, K. V., and Lengyel, J. A., 1979, Rates of synthesis of major classes of RNA in *Drosophila* embryos, *Develop. Biol.* **70**:217–231.

Axel, R., Cedar, H., and Felsenfeld, G., 1975, The structure of the globin genes in chromatin, *Biochemistry* **14**:2489–2495.

Beyer, A. L., Miller, O. L., McKnight, S. L., 1980, Ribonucleoprotein structure in nascent hnRNA is non random and sequence-dependent, *Cell* **20**:75–84.

Bonner, J., Huang, R. C., 1964, Role of histones in chromosomal RNA synthesis. In *The Nucleohistones,* edited by J. Bonner and P. Ts'o, pp. 251–261, Holden-Day, San Francisco.

Bordier, C., and Dubochet, J., 1974, Electron microscopic localization of the binding sites of *Escherichia coli* RNA polymerase in the early promotor region of T7 DNA, *Eur. J. Biochem.* **44**:617–624.

Butt, F. H., 1949, Embryology of the milkweed bug *Oncopeltus fasciatus* (Hemiptera), Memoir 283, Cornell University, Agricultural Experimental Station.

Chan, L.-N., and Gehring, W., 1971, Determination of blastoderm cells in *Drosophila melanogaster, Proc. Nat. Acad. Sci. USA* **68**:2217–2221.

Comings, D. E., and Okada, T. A., 1972, Holocentric chromosomes in *Oncopeltus:* kinetochore plates are present in mitosis but absent in meiosis, *Chromosoma* **37**:177–192.

Cook, P. R., and Brazell, I. A., 1975, Supercoils in human DNA *J. Cell Sci.* **19**:261–279.

Darlix, J.-L., and Dausse, J.-P., 1975, Localization of *Escherichia coli* RNA polymerase initiation sites in T7 DNA early promoter region, *FEBS Lett.* **50**:214–218.

Darnell, J. E., Jelinek, W. R., and Molloy, G. R., 1973, Biogenesis of mRNA: Implications for genetic regulation in mammalian cells, *Science* **181**:1215–1221.

Fausto-Sterling, A., Zheutlin, L. M., and Brown, P. R., 1974, Rates of RNA synthesis during early embryogenesis in *Drosophila melanogaster, Develop. Biol.* **40**:78–83.

Foe, V. E., 1975, Activation of transcriptional units during the embryogenesis of *Oncopeltus fasciatus,* Ph.D. Thesis, Dept. of Zoology, University of Texas, Austin.

Foe, V. E., 1977, Modulation of ribosomal RNA synthesis in *Oncopeltus fasciatus:* An electron microscopic study of the relationship between changes in chromatin structure and transcriptional activity, *Cold Spring Harbor Symp.* **42**:723–740.

Foe, V. E., Wilkinson, L. E., and Laird, C. D., 1976, Comparative organization of active transcription units in *Oncopeltus fasciatus, Cell* **9**:131–146.

Gottesfeld, J. M., Murphy, R. F., and Bonner, J., 1975, Structure of transcriptionally active chromatin, *Proc. Nat. Acad. Sci. USA* **72**:4404–4408.

Hansen-Delkeskamp, E., 1968, Synthese ribosomaler RNA in normalen und embryolosen Keimen von *Acheta domestica* L., *Wilhelm Roux's Arch. Dev. Biol.* **161**:23–29.

Harris, S. E., and Forrest, H. S., 1967, RNA and DNA synthesis in developing eggs of the milkweed bug, *Oncopeltus fasciatus* (Dallas), *Science* **156:**1613–1615.

Illmensee, L., and Mahowald, A. P., 1974, Transplantation of posterior polar plasm in *Drosophila:* Induction of germ cells at the anterior pole of the egg, *Proc. Nat. Acad. Sci. USA* **71:**1016–1020.

Illmensee, L., and Mahowald, A. P., 1976, The autonomous function of germ plasm in a somatic region of the *Drosophila* egg, *Exp. Cell Res.* **97:**127–140.

Jones, K. W., Bishop, J. O., Brito-da-Cunha, A., 1973, Complex formation between poly-r (U) and various chromosomal loci in *Rhynchosciara, Chromosoma* **43:**375–390.

Kauffman, S. A., 1980, Heterotropic transplantation in the syncytial blastoderm of *Drosophila:* Evidence for anterior and posterior nuclear commitments, *Wilhelm Roux's Arch. Dev. Biol.* **189:**135–145.

Kornberg, R. D., 1974, Chromatin structure: A repeating unit of histones and DNA, *Science* **184:** 868–871.

Lacy, E., and Axel, R., 1975, Analysis of DNA of isolated chromatin subunits, *Proc. Nat. Acad. Sci. USA* **72:**3978–3982.

Laemmli, U. K., Cheng, S. M., Adolf, K. W., Paulson, J. R., Brown, J. A., and Baumbach, W. R., 1977, Metaphase chromosome structure: The role of nonhistone proteins, *Cold Spring Harbor Symp.* **42:**351–360.

Laird, C. D., and Chooi, W. Y., 1976, Morphology of transcription units in *Drosophila melanogaster, Chromosoma* **58:**193–218.

Laird, C. D., Wilkinson, L. E., Foe, V. E., and Chooi, W. Y., 1976, Analysis of chromatin-associated fiber arrays, *Chromosoma* **58:**169–192.

Lamb, M. M., 1975, The appearance of poly-A containing RNA during embryogenesis in *Drosophila melanogaster*, Ph.D. Thesis, Dept. of Zoology, University of Washington.

Lamb, M. M., and Daneholt, B., 1979, Characterization of active transcription units in Balbiani rings of *Chironomus tentans, Cell* **17:**835–848.

Lamb, M. M., and Laird, C. D., 1976, Increase in nuclear poly(A)-containing RNA at syncytial blastoderm in *Drosophila melanogaster* embryos, *Develop. Biol.* **52:**31–42.

Lockshin, R. A., 1966, Insect embryogenesis: Macromolecular synthesis during early development, *Science* **154:**775–776.

Martin, T., Billings, P., Levey, A., Ozarslan, S., Quinlan, T., Swift, H., and Urbas, L., 1973, Some properties of RNA: Protein complexes from the nucleus of eukaryotic cells, *Cold Spring Harbor Symp.* **38:**921–932.

Matsui, S., Fuke, M., Ballai, N. R., and Busch, H., 1975, Reconstitution of nucleolar chromatin-fidelity of rRNA readouts, *J. Cell Biol.* **67:**266a.

McKnight, S. L., and Miller, O. L., 1976, Ultrastructural patterns of RNA synthesis during early embryogenesis of *Drosophila melanogaster, Cell* **8:**305–319.

McKnight, S. L., and Miller, O. L., 1979, Post-replicative non ribosomal transcription units in *D. melanogaster* embryos, *Cell* **17:**551–563.

McKnight, S. L., Bustin, M., and Miller, O. L., 1977, Electron microscopic analysis of chromosome metabolism in the *Drosophila melanogaster* embryo, *Cold Spring Harbor Symp.* **42:**741–754.

Miller, O. L., Jr., and Beatty, B. R., 1969, Visualization of nucleolar genes, *Science* **164:**955–957.

Nevins, J. R., and Darnell, J. E., 1978, Steps in the processing of Ad2 mRNA: Poly $(A)^+$ nuclear sequences are conserved and poly (A) addition proceeds splicing, *Cell* **15:**1477–1493.

Okada, M., Kleinman, I. A., and Schneiderman, H. A., 1974, Chimeric *Drosophila* adults produced by transplantation of nuclei into specific regions of fertilized eggs, *Develop. Biol.* **39:**286–294.

Okada, T. A., and Comings, D. E., 1980, A search for protein cores in chromosomes: Is the scaffold an artifact? *Amer. J. Hum. Genet.* **32:**814–832.

Olins, A. L., and Olins, D. E., 1974, Spheroid chromatin units (Nu bodies), *Science* **183:**330–332.

Oudet, P., Gross-Bellard, M., and Chambon, P., 1975, Electron microscopic and biochemical evidence that chromatin structure is a repeating unit, *Cell* **4:**281–300.

Pietruschka, F., and Bier, K., 1972, Autoradiographische Untersuchungen zur RNS- und Protein-Synthese in der frühen Embryogenese von *Musca domestica, Wilhelm Roux's Arch. Dev. Biol.* **169:**56–69.

Rattner, J. B., Goldsmith, M., and Hamkalo, B. A., 1980, Chromatin organization during meiotic prophase of *Bombyx mori, Chromosoma* **79:**215–224.

Robertson, H., and Dickson, E., 1975, RNA processing and the control of gene expression, *Brookhaven Symp. Biol.* **26**:240–266.

Roeder, R. G., Reeder, R. H., and Brown, D. D., 1970, Multiple forms of RNA polymerase in *Xenopus laevis:* Their relationship to RNA synthesis *in vivo* and their fidelity of transcription *in vitro, Cold Spring Harbor Symp.* **35**:727–735.

Scherrer, K., and Darnell, J. E., 1962, Sedimentation characteristics of rapidly labelled RNA from HeLa cells, *Biochem. Biophys. Res. Commun.* **7**:486–490.

Scriba, M. E. L., 1964, Beeinflussung der frühen Embryonalentwicklung von *Drosophila melanogaster* durch Chromosomenaberrationen, *Zool. Jb. Abt. Anat. Ontog.* **81**:435–490.

Slater, I., Gillespie, D., and Slater, D. W., 1973, Cytoplasmic adenylylation and processing of maternal RNA, *Proc. Nat. Acad. Sci. USA* **70**:406–411.

Weintraub, H., and Groudine, M., 1976, Chromosomal subunits in active genes have an altered conformation, *Science* **193**:848–856.

Weischaus, E., and Gehring, W., 1976, Clonal analysis of primordial disc cells in the early embryo of *Drosophila melanogaster, Develop. Biol.* **50**:249–263.

Wilson, E. B., 1912, Studies on chromosomes. VIII. Observations on the maturation-phenomena in certain hemiptera and other forms, with considerations on synapsis and reduction, *J. Exp. Zool.* **13**:345–449.

Worcel, A., and Benyajati, C., 1977, Higher order coiling of DNA in chromatin *Cell* **12**:83–100.

Zalokar, M., 1976, Autoradiographic study of protein and RNA formation during early development of *Drosophila* eggs, *Develop. Biol.* **49**:425–437.

Zalokar, M., and Erk, I., 1976, Division and migration of nuclei during early embryogenesis of *Drosophila melanogaster, J. Microsc. Biol. Cell.* **25**:97–106.

9

The Morphogenesis of Imaginal Discs of *Drosophila*

DIANE K. FRISTROM AND WAYNE L. RICKOLL

1. *Introduction*

Imaginal discs, unique to holometabolous insects, are developmental precursors to many external adult structures. Discs arise early in embryonic development as small invaginations of the epidermis. From then on, the discs follow developmental sequences distinct from those of larval tissues. As the larval epidermis passes through successive molts, with the secretion and apolysis of cuticle, the imaginal discs remain undifferentiated but grow rapidly by cell division. At the end of larval development, each disc contains thousands of cells in an epithelial monolayer that has a characteristic shape and pattern of folds (Figure 1). A layer of mesenchymelike adepithelial cells is loosely associated with the basal surface of the epithelium. Each disc is set aside early in development as a group of around fifty cells to form a specific region of the adult insect. Each leg disc, for example, gives rise to one leg and the surrounding thorax (derived from the disc epithelium) and part of the leg musculature (derived from the adepithelial cells). The fates of particular regions within each disc become specified later in larval development and detailed fate maps of the discs have been established (Bryant, 1978); for example, the central "knob" of the leg disc gives rise to most of the tarsus (Figures 2, 3).

The metamorphosis of discs into adult structures can be divided into two phases. First, morphogenesis, the process by which a disc takes on the shape of the adult structure, and second, the differentiation of cells with specialized functions, such as bristle and sensory cells. The study of the differentiation of the vast array of insect cells is an enormous field, much of which is reviewed

DIANNE K. FRISTROM AND WAYNE L. RICKOLL • Department of Genetics, University of California, Berkeley, California 94720, USA.

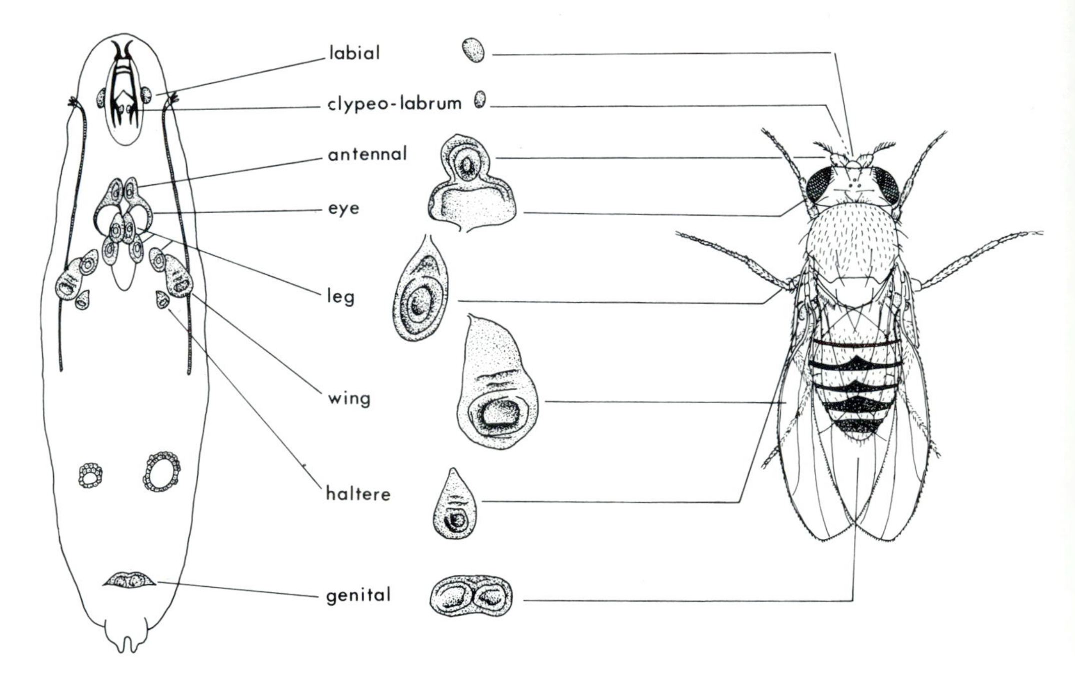

Figure 1. A diagram showing the distribution and morphology of imaginal discs in the third instar larva and the external adult structures that they form.

elsewhere in this volume. This chapter will be confined to the first phase of imaginal disc metamorphosis, morphogenesis.

The basic cellular processes leading to changes in tissue shape are shared by all multicellular organisms and can be divided into the following three categories:

(1) Addition or deletion of cells by:
   (a) localized or oriented cell division, e.g., in plant development (Green, 1980) and in the formation of folds in larval imaginal discs (Vijverberg, 1974).
   (b) localized cell death, e.g., in sculpturing of the vertebrate limb (Saunders, 1966) and in some aspects of imaginal disc development (Fristrom, 1969; Spreij, 1971).
(2) Changes in cell shape, e.g., in sea urchin development (Gustafson and Wolpert, 1963) and many examples of epithelial morphogenesis in vertebrates (Wessells *et al.*, 1971; Spooner *et al.*, 1973).
(3) Cell movements, e.g., the migration of neural crest cells and movements of epithelial sheets (Trinkaus, 1976).

Rapid advances are being made in understanding the cellular mechanisms of these processes, as in the role of cytoskeletal elements such as microfilaments and microtubules in cell shape changes and morphogenetic movements. Ultrastructural observations are playing a central role in this progress. However, little is known of how such processes are regulated within a single cell or within the organism as a whole. How a relatively simple group of cells, such as a blastula or an imaginal disc, acquires the complex and specific shape of the adult is a central problem of developmental biology.

Imaginal discs offer many advantages for the study of morphogenesis. First, each disc is a discrete entity that gives rise to a small fraction of the adult and is, therefore, a simpler system than an embryo destined to form an entire organism. Second, the three basic developmental processes (cell division, morphogenesis, and differentiation) tend to overlap in most organisms, especially the processes of cell division and morphogenesis. In imaginal discs these processes are by and large separated in time. Cell division is virtually completed during larval development. Then, with the rise in ecdysteroid titer triggering metamorphosis, there is a period (the prepupal or pharate pupal period) largely devoted to morphogenetic processes, a period of remodeling of the concentrically folded discs into the basic shape of the adult appendage. Finally, most cellular differentiation takes place during the pupal period. Thus, an imaginal disc at the onset of metamorphosis is comparable in many ways to the blastula stage of a vertebrate embryo; but it is a blastula that already contains almost all the cells of the adult organism, and that undergoes complete morphogenesis in response to a specific hormonal stimulus before commencing cellular differentiation.

Imaginal discs are also amenable to a variety of experimental manipulations. They develop *in vitro* in a chemically defined medium in the presence of 20-hydroxyecdysone (Martin and Schneider, 1978). They can also be cultured *in vivo* by transplantation to a host larva (Hadorn, 1978). In spite of their small size, a wide variety of biochemical investigations can be carried out using mass-isolation procedures combined with *in vitro* culture (Silvert and Fristrom, 1980). Finally, the highly sophisticated genetic techniques available in the study of *Drosophila* open a variety of avenues for the study of the genetic regulation of developmental processes.

In this chapter, we describe the morphogenesis of two types of imaginal discs, leg discs and eye discs. Both types of discs start out as simple monolayers of undifferentiated cells that undergo strikingly different morphogenetic processes. First, we discuss leg discs as an example of discs that give rise to the appendages of the adult. Here the change in shape is dramatic, as the concentrically folded disc is transformed into an elongated segmented leg. Second, we describe the formation of the highly organized compound eye from an eye disc. In this case, the overall change in shape of the tissue is relatively minor (Figure 1), but morphogenesis involves the organization of cells within the disc epithelium to form the precise and complex pattern of ommatidia. We then describe the ultrastructural characteristics of disc cells with particular emphasis on cytoskeletal structures and intercellular junctions and their possible roles in morphogenetic processes.

2. *Morphogenetic Processes in Imaginal Discs*

2.1. Leg Discs

The morphogenetic transformation of the leg disc that occurs at the beginning of metamorphosis is shown in the scanning electron micrographs in Figure 2 and is depicted schematically in Figure 3. The larval leg disc is a sac consisting

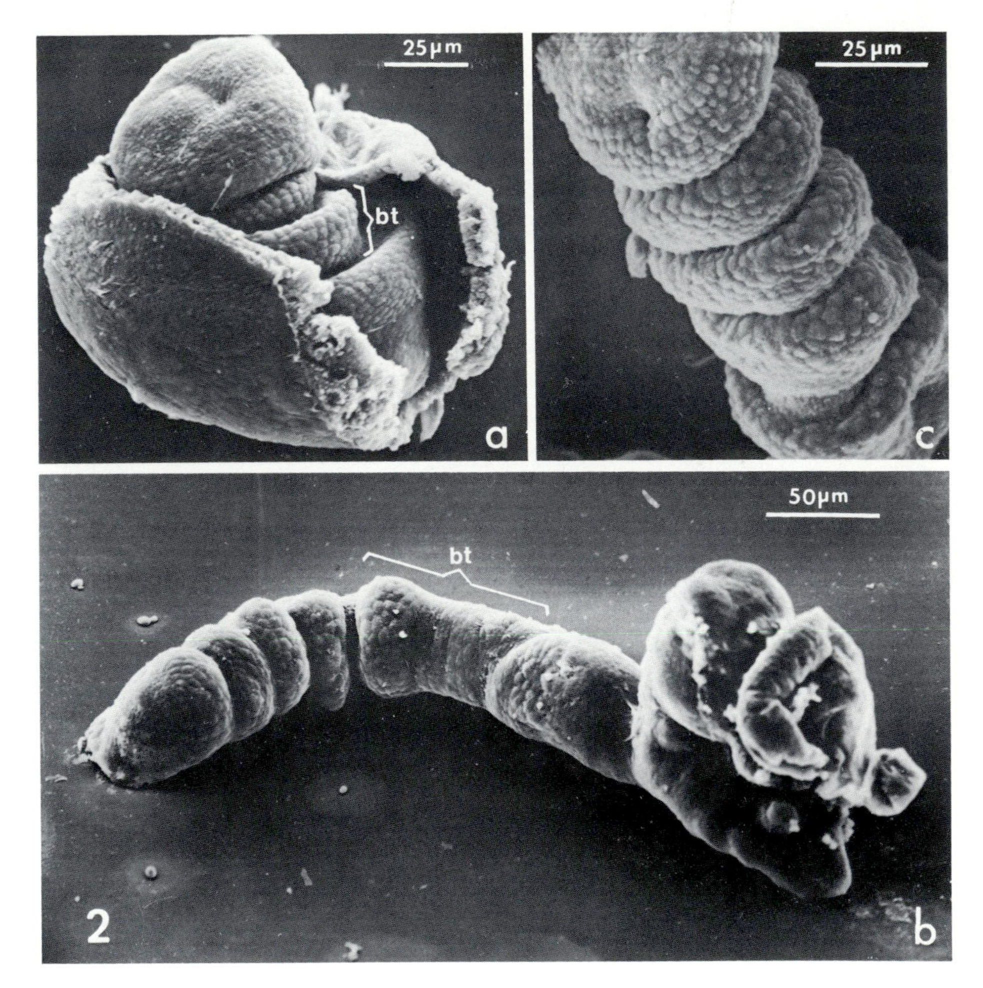

Figure 2. Scanning electron micrographs of evaginating leg discs. (a) A leg disc that has just begun to evaginate (the peripodial membrane has been partly removed), (b) a fully evaginated leg disc, and (c) the distal tarsal segments (2 to 5) of a fully evaginated leg disc. Note the similarity of cell shape in (a) and (c). bt, basitarsus (first tarsal segment).

on one side of a folded columnar epithelium, the disc proper, and on the other side a squamous epithelium, the peripodial membrane. The apical surface of the disc epithelium faces a lumen (Figure 3b) that is continuous with the subcuticular space of the larval epidermis. The basal surface forms the outside of the disc and is bathed by larval hemolymph. The disc proper is composed of a series of concentrically arranged folds of increasing diameter. The unfolding and extension of the disc, referred to as "evagination," produces an elongate tubular structure such that the center of the larval disc forms the distal tip of the leg and the periphery of the disc forms the proximal end of the leg (Figure 3). As the disc elongates it also narrows, particularly at the proximal end (compare the diameters indicated by the line a–a' in Figures 3a,c). As evagination occurs, the elongating appendage escapes from the peripodial membrane so

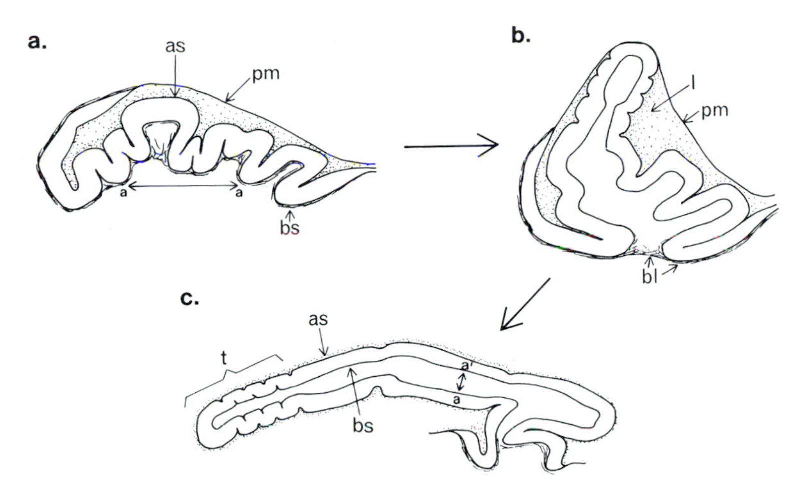

Figure 3. Diagrams of sagittal sections of evaginating leg discs. (a) Unevaginated, (b) partly evaginated, and (c) fully evaginated. Note the decrease in diameter (a–a′) of the disc in (c) compared to that in (a). as, apical surface; bs, basal surface; bl, basal lamina; l, lumen filled with extracellular matrix; pm, peripodial membrane; t, tarsal segments 2–5. Adepithelial cells (not shown) lie between the basal surface and the basal lamina.

that the apical surface of the epithelium is on the outside (Figure 3c). At the same time there is a general flattening of the epithelium and some localized morphological changes, such as the segmentation of the tarsus. We will first discuss the cellular events that result in the elongation and narrowing of the tissue during evagination.

Although cell division and cell death may occur to a limited extent during metamorphosis, they are not causes of evagination. Evagination can take place *in vitro* in the complete absence of cell division [(i.e., in the presence of inhibitors of DNA synthesis and in the presence of colcemid (Fristrom *et al.*, 1977)]. Cell death is also extremely limited during evagination (Fristrom and Fristrom, 1975). Thus, the number of cells in the disc remains effectively constant during evagination. Given a constant number of cells maintained in a single cohesive layer, there are only two ways to effect the observed changes in tissue shape: either by pronounced changes in cell shape or movement of cells into new positions. These alternatives are illustrated schematically in Figure 4a. Scanning electron micrographs in which cell boundaries could be clearly discerned show that significant changes in cell shape do not occur (Figures 2a,c). Furthermore, cell counts made in the region of the basitarsus (a cylinder of tissue comparable to that diagramed in Figure 4a) showed that the number of cells in the long axis of the segment more than doubles, while the number of cells encircling the segment decreases (Fristrom, 1976) with the total number of cells remaining constant.

These results led to the inescapable, but somewhat disturbing, conclusion that evagination results from cell movement. Disc cells, like those of other epithelia, do not look as though they can move. Each cell is closely connected on all lateral sides to neighboring cells. These connections take the form of highly

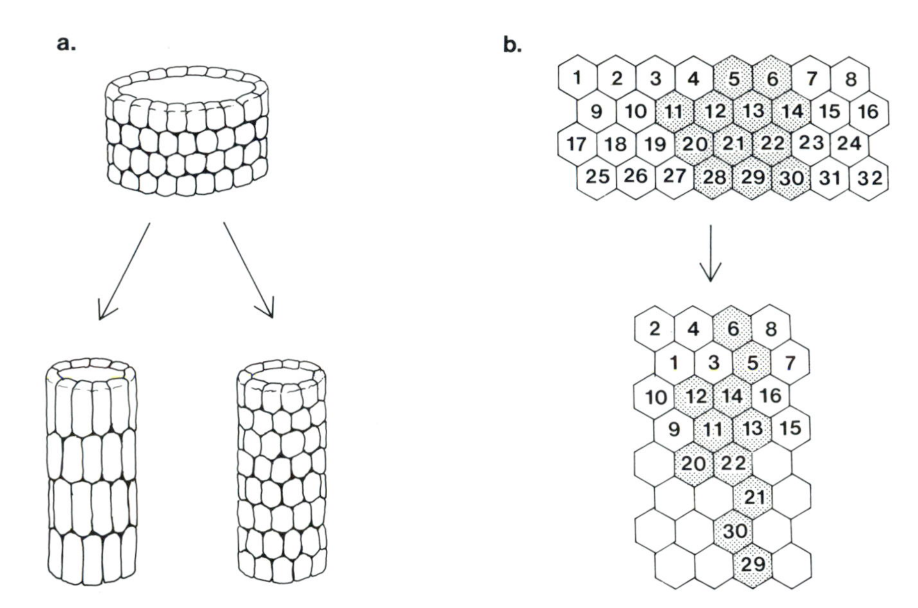

Figure 4. Schematic representations of changes in tissue shape. (a) Alternative ways of changing the shape of a tube of tissue composed of a constant number of cells in a monolayer; left, cell shape change; right, cell rearrangement. (b) One scheme for the rearrangement of cells that could produce a change in tissue shape without disrupting the continuity of a mosaic patch (shaded).

specialized junctions (section 3.4) that appear to provide a physical impediment to cell movements. Furthermore, structures one normally associates with cell motility, such as lamellopodia or filopodia, are absent. Finally, the classical observations on genetic mosaics in *Drosophila* showed that clonally related cells remain associated in patches (Stern, 1936) and formed the basis of the early belief that no cell movements occur during disc morphogenesis. However, it is not necessary for individual cells to move very far in order to accomplish the kind of shape change associated with evagination. Small movements of many cells can result in a change in tissue shape without separating clonally related cells, as shown in Figure 4b. The mechanism of such movements could well be fundamentally different from the mechanisms that propel migratory cells over long distances.

In light of the observations on leg evagination, the term "cell rearrangement" was proposed to refer to small movements of many cells within epithelial sheets where close associations are maintained between cells (Fristrom and Fristrom, 1975; Fristrom, 1976). Until then, this type of cell movement had not been described in developing epithelia although it had been predicted by Waddington (1962) and a similar phenomenon had been described in epithelial cells in tissue culture (Steinberg, 1973). Since then, cell rearrangements have been described in neurulation in newts (Jacobson and Gordon, 1976), nephric duct elongation in *Ambystoma* (Poole and Steinberg, 1977), and gastrulation in *Xenopus* (Keller, 1978). Thus, cell rearrangement may be a widespread phenomenon in epithelial morphogenesis and probably occurs to some extent in all situations involving movements of epithelial sheets (Falk and King, 1964; Trinkaus, 1976).

Although cell rearrangement is largely responsible for evagination, changes in cell shape also occur during leg disc morphogenesis. The surface area clearly increases as the compact imaginal discs are transformed into adult structures. This is most pronounced in wing discs (Figure 1), but also occurs to some extent in leg discs. The increase in surface area is largely accomplished by cell flattening. In both wing and leg discs, cell flattening begins during evagination as the epithelium changes from columnar to cuboidal (Auerbach, 1936; Fristrom, 1969; Poodry and Schneiderman, 1970; Fristrom and Fristrom, 1975) and continues during later stages of development. Cell flattening does not by itself result in evagination and can occur in a number of situations where evagination is inhibited (Fekete *et al.*, 1975; Fristrom and Fristrom, 1975; Fristrom, *et al.*, 1981). However, evagination without concommitant cell flattening has not yet been observed, an so it is conceivable that cell flattening may be a necessary aspect of cell rearrangement.

Superimposed on the general processes of cell flattening and cell rearrangement are some localized changes in cell shape that give the various appendages their characteristic morphology. These localized processes can be regarded as the "detailed sculpturing" of morphogenesis. An example is seen in the pronounced grooves that arise in the central knob of the leg disc during evagination and form the boundaries between the distal tarsal segments (Figure 2c, 3c). The cells in the groove become short and wedge-shaped, i.e., the apical ends of the cells are constricted in the proximodistal direction (see also, Poodry, 1980a). Other examples of the detailed sculpturing of discs are described by Fristrom *et al.* (1981).

2.2. Eye Discs

The morphogenetic processes resulting in the complex repetitive pattern of ommatidia that make up the compound eye appear to differ strikingly from those that produce leg evagination. The *Drosophila* eye consists of a lattice of approximately 800 units or ommatidia, each composed of twenty-two cells comprising four different cell types: four cone cells that secrete the facet, a light focusing lens; eight photoreceptor cells that transmit light-generated impulses to the brain; a sheath of six pigment cells that optically isolates each ommatidium; and, a sensory hair group. The pattern of the ommatidial field is extremely precise both with respect to the regularity of the lattice and in the organization of cells within each ommatidium. For example, the photoreceptor cells that form the center of each ommatidium are always arranged as shown in Figure 5, with seven peripheral cells making contact with a central eighth cell.

The ommatidial field is laid down sequentially from posterior to anterior, beginning sometime in the first half of the third instar with the clustering of a few cells at the posterior tip of the eye disc (Steinberg, 1943). The patterned area extends slowly anteriorly until shortly after puparium formation when ommatidial organization is complete. At any stage in this process the anterior

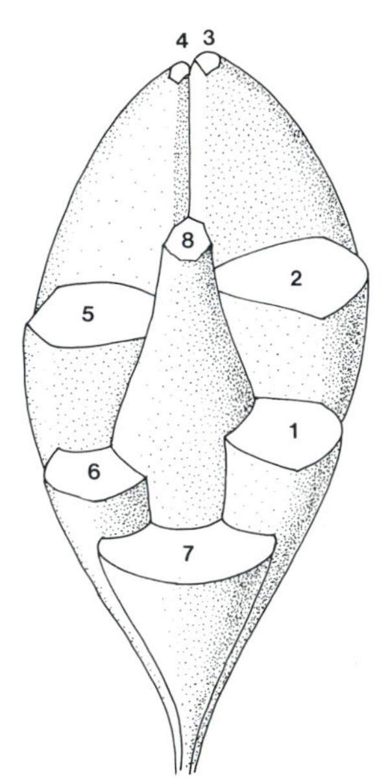

Figure 5. Cut-away diagram showing the arrangement of photoreceptor cells in an ommatidium. After Ready *et al.*, 1976.

boundary of the ommatidial field is marked by a dorsoventral furrow (Melamed and Trujillo-Cenoz, 1975; Ready *et al.*, 1976) (Figure 6). Immediately ahead of the furrow is a region of intense mitotic activity. Immediately behind the furrow, preclusters of photoreceptor cells can be recognized in sections. These preclusters are incomplete, containing only 5–6 photoreceptor cells (Ready *et al.*, 1976). Further posteriorly, there is a second band of mitotic activity (Ready *et al.*, 1976; Campos-Ortega and Hofbauer, 1977) posterior to which the ommatidial precursors contain their full complement of cells and extend axons into the optic nerve (Figure 6). At this point, ommatidial organization is complete, and a square array of cell clusters is visible by phase contrast or differential interference microscopy (Figure 7a).

We have recently discovered a new and relatively simple technique for visualizing the ommatidial pattern as soon as it is established (Fristrom and Fristrom, 1982). The plant lectin peanut agglutinin (PNA) binds selectively to the apical ends of the photoreceptor cells as shown by labeling discs with ferritin-conjugated PNA (Figure 7b). Discs labeled with fluorescein-conjugated PNA show a characteristic square array of fluorescent spots, with each spot marking the center of one ommatidium (Figure 7c). The anterior boundary of the pattern of spots corresponds precisely with the position of the dorsoventral furrow. Thus, the dorsoventral furrow, aptly termed the morphogenetic furrow by Ready *et al.* (1976), sweeps across the eye disc, leaving in its wake the pattern of the ommatidial field.

The processes that take place in the morphogenetic furrow are of particular interest because this is where the pattern of ommatidia is laid down. A number of clonal analyses, culminating with the work of Lawrence and Green (1979), have shown beyond reasonable doubt that the fate of eye cells is determined

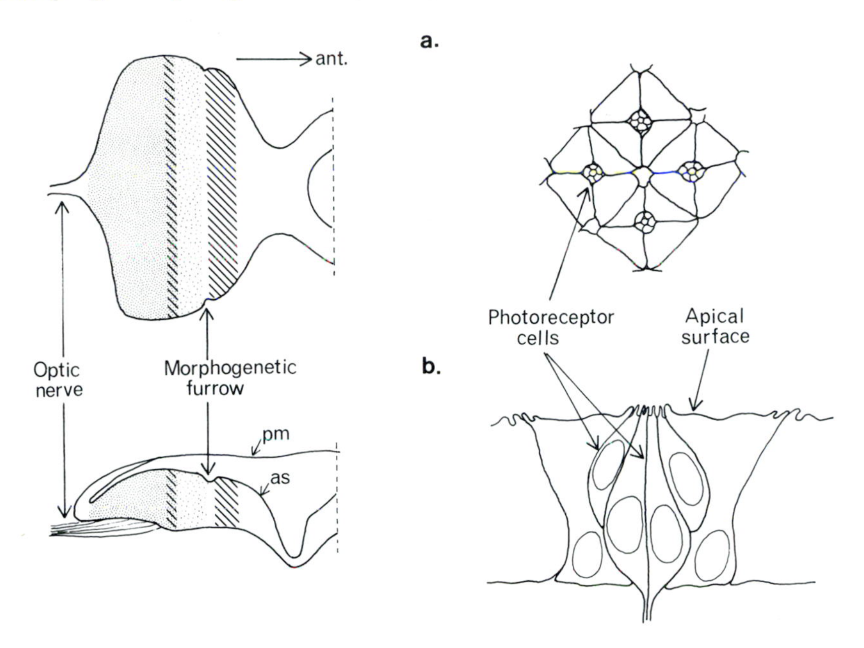

Figure 6. Schematic representations of the cellular organization in a late third instar eye disc. (a) Dorsal surface, and (b) sagittal views. Lightly stippled regions, area of the disc where preclusters of photoreceptor cells have formed. Heavily stippled regions, area where ommatidial organization is complete. Striated regions, bands of mitotic activity. Right, details of photoreceptor clusters. As, apical surface; pm, peripodial membrane.

by a process of "recruitment" rather than by clonal relationships. In simple terms, one can think of the cells ahead of the morphogenetic furrow as being determined only with respect to "eyeness." As these cells enter the furrow many of them become specified to form a particular type of eye cell, e.g., photoreceptor or cone cell. PNA labeling shows that each new row of preclusters is laid down with a precise center-to-center spacing that is staggered with respect to the preceding row (Figure 7c). This suggests that specification as a particular type of eye cell may be a function of position with respect to the preceding row of ommatidial clusters. Recruitment as a photoreceptor cell is immediately manifested as a change in cell shape. The apical ends of the cells become constricted so that photoreceptors occupy a relatively small per cent of the apical surface of the disc (Figure 6). This constriction accounts for the pointlike pattern of spots obtained with PNA labeling (Figure 7c). As the photoreceptor cells constrict, the adjacent cells (presumptive cone cells) expand in surface area so that there is no net change in shape of the epithelium. The photoreceptors are also characterized by large numbers of longitudinally oriented microtubules, apical microvilli, and vesicles (Figure 7b). The basal ends of the photoreceptors are long and tapered and eventually extend into axons entering the optic nerve. The overall effect of these changes in cell shape is the formation of characteristic clusters of photoreceptor cells (Figures 6, 7) that have been likened to bundles of spring onions (Waddington and Perry, 1960).

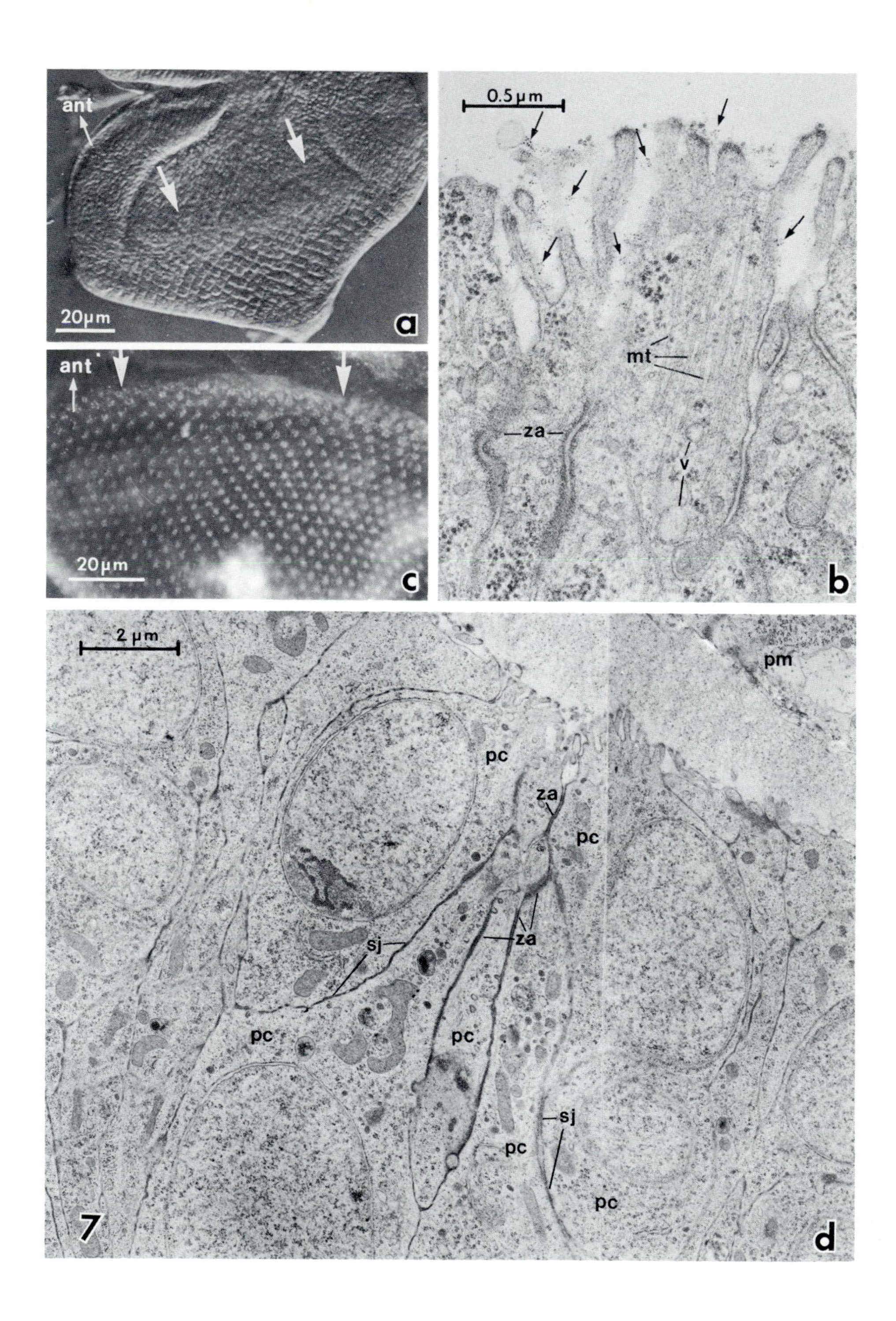
ant
20μm
a
0.5μm
mt
za
v
b
ant
20μm
c
2 μm
pm
pc
za
pc
sj
za
pc
pc
sj
pc
pc
7
d

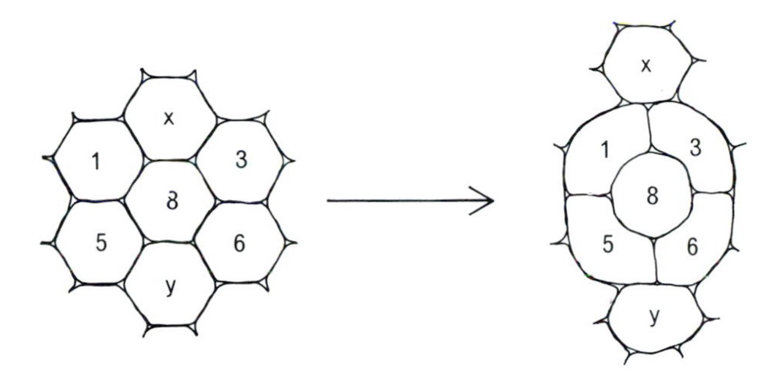

Figure 8. Hypothetical scheme for the rearrangement of eye disc cells into a precluster of photoreceptor cells.

It is apparent that subtle cell rearrangements must also be involved in the formation of the ommatidial bundles. Let us suppose, for example, that the initial five-cell precluster of photoreceptors is formed from a hexagonally arranged group of cells as shown in Figure 8. Two of the cells (x and y) lose their contact with the future central photoreceptor (8) as four presumptive photoreceptor cells (1, 3, 5, and 6) come together around the central cell. Furthermore, the entire complex adopts a smooth outline, suggesting that adhesive interactions between photoreceptor cells are stronger than those between photoreceptors and surrounding cells. This is consistent with the observed increase in junctional contacts between photoreceptor cells (Figure 7d). This example shows the minimum rearrangement necessary to produce the observed distribution of cells in a precluster from an undifferentiated hexagonal array. Clonal data suggest that actual rearrangements may be more extensive, since an individual cell from a small mosaic patch can be separated by one or two ommatidia from the rest of the patch (Lawrence and Green, 1979). Other cell rearrangements occur in the later stages of ommatidial organization (after the second wave of mitoses), as the remaining photoreceptor cells insert into the complex to give the final arrangement shown in Figure 5.

In summary, morphogenesis of both leg and eye discs involves cell rearrangements and cell shape changes. Cell rearrangements play the predominant role in leg disc evagination, whereas cell shape changes, as well as rearrangements are important in the ommatidial organization in eye discs. We now turn to examine the ultrastructural characteristics of discs in an attempt to discover the cellular mechanisms of these processes.

←

Figure 7. Ommatidial organization in the late third instar eye disc. (a) A differential interference micrograph. Each "square" in the posterior part of the disc contains over twenty cells and will give rise to one ommatidium. (b) An electron micrograph showing the binding of ferritin-conjugated PNA (arrows) to the apical surface of photoreceptor cells. (c) A light micrograph showing the distribution of fluorescein-conjugated PNA. (d) A montage of two low power electron micrographs showing a cluster of photoreceptor cells (pc). Note the extensive junctional complexes between photoreceptor cells. mt, microtubules; pm, peripodial membrane; sj, septate junction; v, vesicles; za; zonula adherens; white arrows, position of morphogenetic furrow.

3. *The Ultrastructure of Imaginal Discs during Morphogenesis*

Having established that cell movements and shape changes take place during disc morphogenesis, we can now ask "what kinds of subcellular machinery do the cells possess to carry out such processes?" Various aspects of the ultrastructure of *Drosophila* imaginal discs have been described by a number of authors, including Perry (1968), Wehman (1969), Poodry and Schneiderman (1970; 1971), Mandaron and Sengel (1973), Fristrom and Fristrom (1975), Fristrom (1976), Reed *et al.* (1975), Rheinhardt *et al.* (1977), and Eichenberger-Glinz (1979). Much of this material has already been reviewed by Ursprung (1972) and Poodry (1980a,b). After a brief review of the general ultrastructure of the disc epithelium we will restrict our discussion to structures that may play a role in morphogenesis.

3.1. General Ultrastructure of the Disc Epithelium

Before morphogenesis begins, different discs and different regions within discs are very similar in ultrastructure. The cytoplasm of larval discs is typical of undifferentiated cells, containing little endoplasmic reticulum, numerous free ribosomes, and small mitochondria with few cristae. The nuclei are smooth surfaced, oval, and occupy most of the width of the cell. Each cell has a small supranuclear Golgi apparatus and an apically located pair of centrioles. Electron lucent lipid granules occur in most cells and are particularly concentrated at the basal ends of the cells of the wing pouch (Fristrom, 1969), making the wing pouch one of the few larval disc tissues readily identifiable at the ultrastructural level.

As in other epithelia, there is a marked apicobasal polarity to disc cells. As we have seen (section 2.1, Figure 3), the apical surface of the epithelium faces the lumen of the disc and becomes the outer surface of the adult fly. A cortical network of microfilaments (section 3.2.1) underlies the apical cell surface and extends into the microvilli (Figures 9a, 10a). The apical surface is actively involved in endocytosis and secretion. Small, coated vesicles (Figure 9a) and larger, smooth-surfaced vesicles are involved in adsorptive endocytosis, as demonstrated by the uptake of ferritin-conjugated concannavalin A (ConA)) (Fristrom, unpublished). Lysosomes and multivesicular vacuoles, presumably involved with the digestion of vesicle contents, are also present (Figure 9a). The apical surface of mature larval discs secretes a fibrous extracellular matrix (ecm) that frequently fills the lumen of the disc. This material is not secreted in vesicles but appears to be spun out from the apical surface in long filamentous strands (Figure 9a, 10a), particularly from the dense plaques at the tips of microvilli. We have concluded, based on staining by ruthenium red, binding of ConA, and sensitivity to trypsin, that the ecm is composed mainly of glycoproteins. During the prepupal period, secretion of the apical ecm is replaced by secretion of a chitin-containing cuticle. (Poodry and Schneiderman, 1970; Reed *et al.*, 1975).

The basal surface of the disc epithelium is relatively quiescent (Figure 9b).

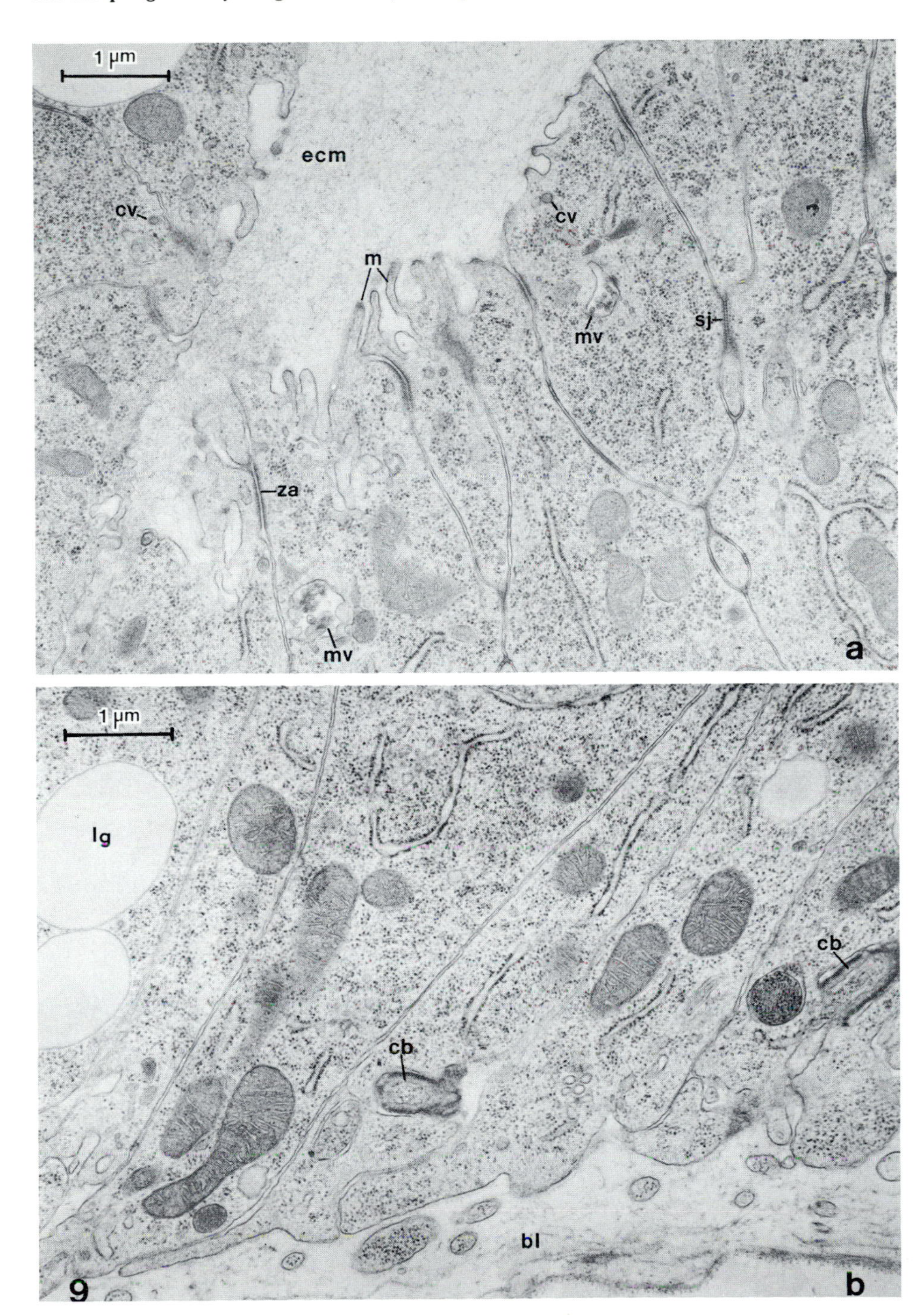

Figure 9. Low-power micrographs showing many of the structures typical of the imaginal disc epithelium. (a) Apical area. (b) basal area. bl, basal lamina; cb, cytoplasmic bridge; cv, coated vesicle; ecm, extracellular matrix; lg, lipid granule; m, microvilli; mv, multivesicular vacuole; sj, septate junction; za, zonula adherens.

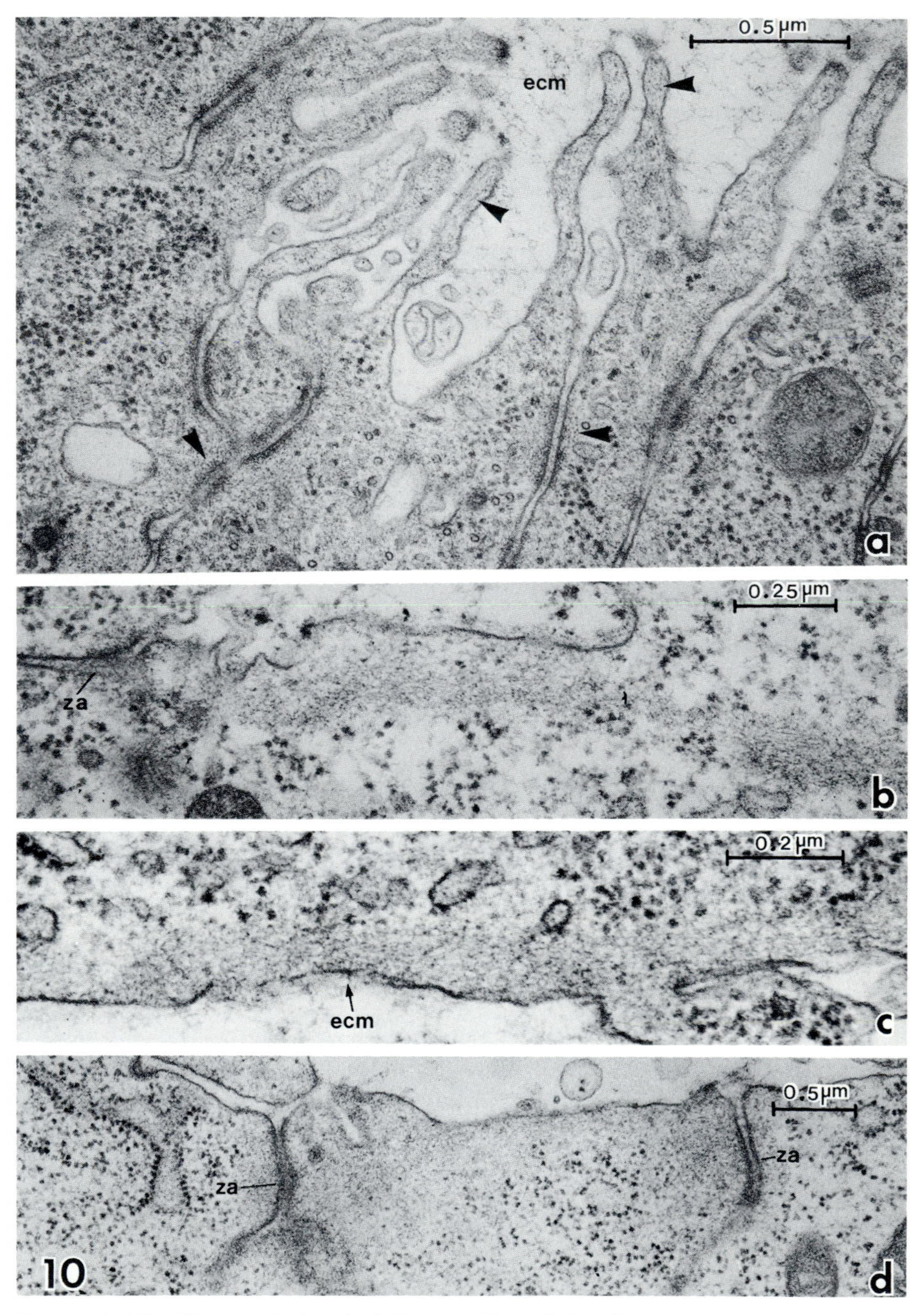

Figure 10. Microfilaments in imaginal discs. (a) The apical surface. Note the microfilaments in the microvilli and adjacent to the zonulae adhaerentes (arrows). (b) A section grazing a zonula adherens (za) shows the band of junction-associated filaments. (c) Microfilaments at the basal surface of an evaginated disc. (d) The apical surface of a disc treated with 10 μg/ml cytochalasin B. Note the absence of microvilli and the accumulation of a granular material (presumably derived from the disassembly of microfilaments) near the zonulae adhaerentes. ecm, extracellular matrix.

The surface is smooth and lined by basal lamina that, like the apical ecm, binds ConA, stains with ruthenium red, and is digested by trypsin. During evagination of wing and leg discs, the cells in the center of the disc pull away from the basal lamina leaving a loose network of basal lamina material at the open end of the appendage (Figure 3). These cells then secrete an extremely tenuous layer of extracellular material (Figure 10c) that does not bind ConA and is, therefore, pesumably of a composition different from the larval basal lamina (Fristrom, unpublished).

Another general characteristic of the disc epithelium is the distribution of a well-developed set of specialized intercellular junctions. Apically, a zonula adherens (belt desmosome or intermediate junction) encircles every cell. Below this there is a broad region of septate junctions interconnecting all lateral cell surfaces. Plaquelike gap junctions occur mainly interspersed among the septate junctions, and at a lower frequency throughout the lateral cell surface below the level of the septates. Intercellular bridges representing the remnants of the preceding cytokinesis, are found close to the basal surface (Figure 9b; see also, Poodry, 1980b).

In the remainder of the chapter we will consider the possible roles of the cytoskeleton (microfilaments and microtubules) in cell shape changes and cell movements and the modulation of intercellular junctions during these processes. Many of the functional interpretations presented here are highly speculative but will, we hope, stimulate further investigations.

3.2. Microfilaments

In a wide variety of nonmuscle cells, cell movements and cell shape changes are mediated, at least in part, by filamentous actin, identified ultrastructurally as 5-nm microfilaments (Pollard and Weihing, 1974). Actin filaments are disrupted by the drug cytochalasin B (CB), a phenomenon that has helped to identify actin-mediated processes in morphogenesis (Wessells *et al.*, 1971; Lin *et al.*, 1978). In a number of instances, contraction of microfilaments leads directly to a change in cell shape (Cloney, 1966; Wessells *et al.*, 1971). A classical example is the apical constriction of cells that occurs in a wide variety of morphogenetic processes (Schroeder, 1973), including neurulation (Baker and Schroeder, 1967; Burnside, 1971) salivary gland morphogenesis (Spooner and Wessells, 1972), and tubular gland formation in the oviduct (Wrenn, 1971). Such apical constrictions appear to be caused by contraction of circumferentially distributed microfilaments in what has become widely known as the "purse string effect". Actin performs a variety of cellular functions in addition to producing cytoplasmic contractions. Actin filaments occur as a cortical network in most cells, where they stabilize the cell surface, mediate the translocation of membrane components (Edelman, 1976), and modulate changes in the conformation of the surface (Mooseker and Tilney, 1975; Taylor and Condeelis, 1979; Stendahl *et al.*, 1980). Thus, nonmuscle actin is functionally diverse and may play many different roles in morphogenesis.

In imaginal discs, actin represents 5%–10% of the cytoplasmic proteins (Kuniyuki, 1976; Fristrom *et al.*, 1977). Microfilaments seen in thin sections represent a much smaller fraction of the cell's cytoplasm, probably because filamentous actin is poorly preserved by standard electron microscopic procedures (Heuser and Salpeter, 1979). As in many other epithelia, the filaments that we do see are predominantly located at the apical and basal cell surfaces.

3.2.1. The Distribution of Microfilaments

Microfilaments at the apical surface form a tenuous cortical network that extends into the microvilli (Figure 10a). A belt of circumferentially oriented microfilaments associated with the zonula adherens also encircles the apical end of each cell (Figures 10a,b) as is typical of this type of junction (Schroeder, 1973; Staehlin, 1974; Hull and Staehelin, 1979). At the basal surface of a disc that is beginning to evaginate there are occasional patches of filamentous material associated with protrusions or blebs. As evagination proceeds, the basal surface becomes relatively flat and smooth again. The basal filaments become more extensive (Figure 10c) and can occasionally be seen to stretch across the entire basal surface of a cell and enter plaques on the lateral cell surfaces (Fristrom and Fristrom, 1975). Both the apical and basal filaments are disrupted by CB, resulting in a loss of microvilli and accumulation of granular material at the apical periphery (Figure 10d) and in basal protrusions.

3.2.2. The Function of Microfilaments

The presence of a circumferential band of CB-sensitive microfilaments associated with the zonula adherens suggests that a "purse string" contraction might well account, in part, for the apical constriction of the photoreceptors similar to that seen in the various examples of vertebrate morphogenesis cited above. As the apical perimeter of the cell decreases, the length of the zonula adherens and its associated filaments increases (compare Figure 7b,d with 11b) as one would expect if the total amount of junctional material is conserved.

The precise role played by actin filaments in cell rearrangement is still obscure. However, because cell rearrangement is a CB-sensitive and energy requiring process (Fekete *et al.*, 1975), we believe that microfilaments are involved in the translocation of cells. We originally proposed (Fristrom and Fristrom, 1975) that circumferential contraction of the basal filaments squeezed cells out of their original positions into new ones. We have since rejected this hypothesis for a number of reasons (Fristrom and Chihara, 1978). It is apparent from time lapse movies that it is the apical (not the basal) surface that is actively contractile. The entire surface appears to pulsate with oscillating contractions which become more pronounced as evagination proceeds. In order for evagination to occur, the net effect of the cell movements must be directional, but we have not yet been able to follow individual cells for a long enough time to establish patterns of movement associated with changes in tissue shape. It is not clear how apical contractions result in relative movements of cells. One hypothesis holds that oscillating contractions combined with changes in intercellular adhesivity tend to separate cells with reduced adhesiveness allowing other cells to intervene. The role of the basal filaments in evagination is even more obscure.

These filaments do not appear to contract during evagination, and they play a passive cytoskeletal role in smoothing and stiffening the basal cell surface as part of the cell flattening process that accompanies evagination.

3.3. Microtubules

The distribution of microtubules in imaginal discs was described in some detail by Fristrom and Fristom (1975). Longitudinally oriented microtubules are particularly prominent in mid third instar leg and wing discs. These tubules appear to originate at or near small dense plaques at the apical surface (Figure 11a) and extend toward but do not reach the basal cell surface. Longitudinally oriented tubules become shorter and less numerous toward the end of the third instar, apparently depolymerizing from their basal ends. We also noted that evagination can occur in concentrations of colcemid that disrupt microtubules (Fristrom and Fristrom, 1975). Thus, microtubules are apparently not required for evagination (cell rearrangement) although several lines of evidence suggest that some microtubule depolymerization is necessary to allow evagination to occur (Fristrom and Fristrom, 1975; Fristrom *et al.*, 1981).

In contrast to the appendage-forming discs, there is a dramatic increase in the number of longitudinally oriented microtubules in the photoreceptor cells of the eye disc at the onset of ommatidial organization. These tubules also appear to originate from the apical cell surface but are not always associated with dense plaques (Figure 7b). Longitudinal microtubules persist through subsequent differentiation of the photoreceptor complex (Perry, 1968), and their formation may represent the first signs of photoreceptor differentiation, rather than being necessary for eye disc morphogenesis.

A set of microtubules that may have a morphogenetic function occurs in the vicinity of the zonula adherens (Wehman, 1969; Fristrom and Fristrom, 1975). These tubules appear to encircle the cell just inside the adherens- associated filaments. Thus, in perpendicular sections we see circular profiles of microtubules adjacent to the junction (Figures 11b,c), whereas tangential sections show longitudinal profiles (Figure 11d). One possible arrangement of these tubules is shown schematically in Figure 12. The association of microtubules with the adherens junction is unusual. It does not appear to occur in vertebrate tissues (Burnside, personal communication), and it may be unique to insect epithelia (see also, micrographs in Poodry and Schneiderman, 1970; Sedlack and Gilbert, 1976).

We have noted before that disc cells are subject to pronounced changes in apical circumference, the photoreceptor and cone cells being exteme examples. The photoreceptor cells become highly constricted at their apical ends and have long adherens junctions with associated microfilaments. These highly constricted cells have, however, no circumferentially oriented microtubules (Figure 7b). Cone cells, in contrast, become expanded at their apical ends and almost invariably have microtubules associated with the adherens junction, as do most other disc cells. We can envision two possible roles for the circumferential microtubules. They may serve as antagonists to the contraction of circumferential microfilaments and thereby prevent the apical constriction of some cells,

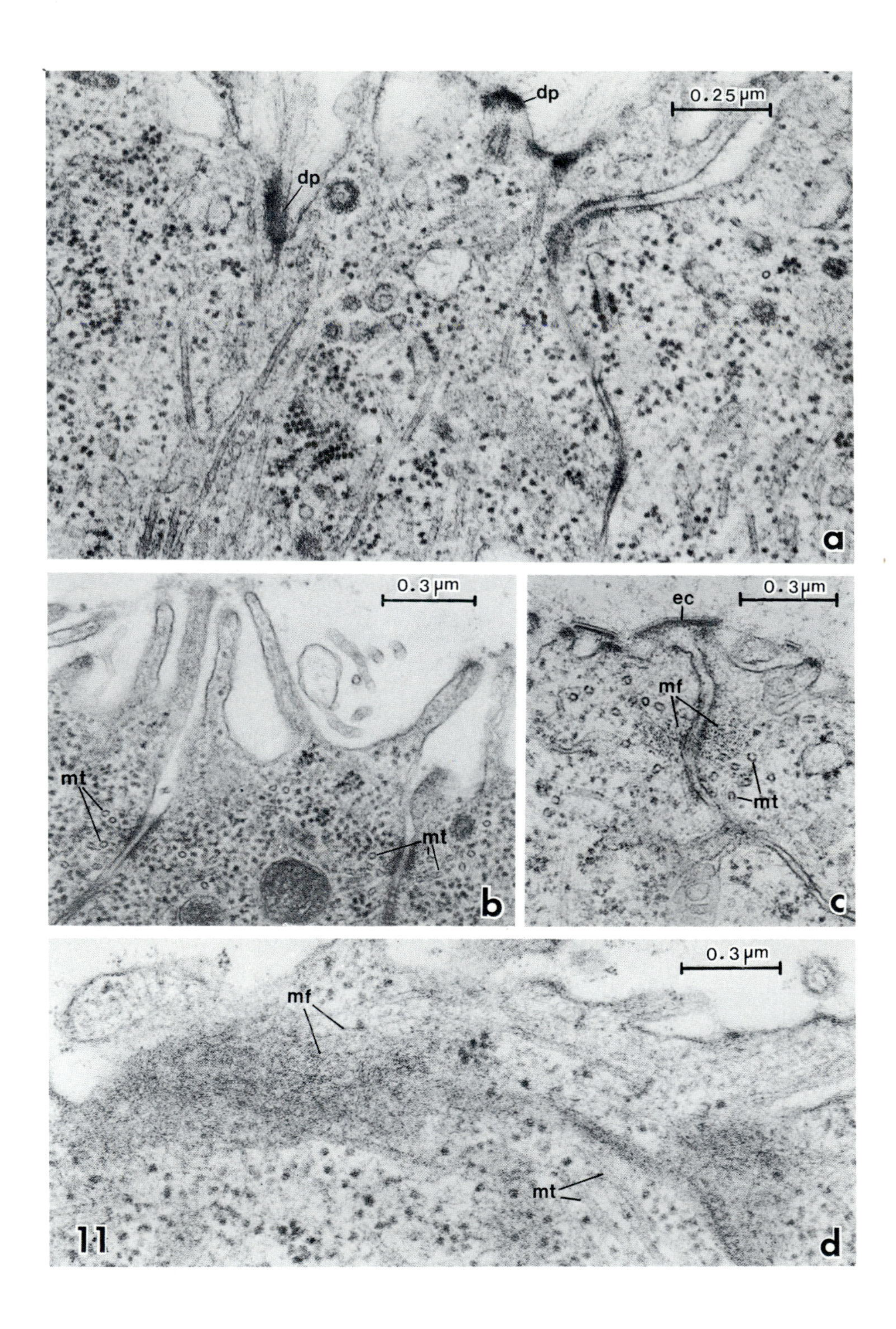

dp
0.25 μm
dp
a
0.3 μm
mt
mt
b
0.3 μm
ec
mf
mt
c
0.3 μm
mf
mt
11
d

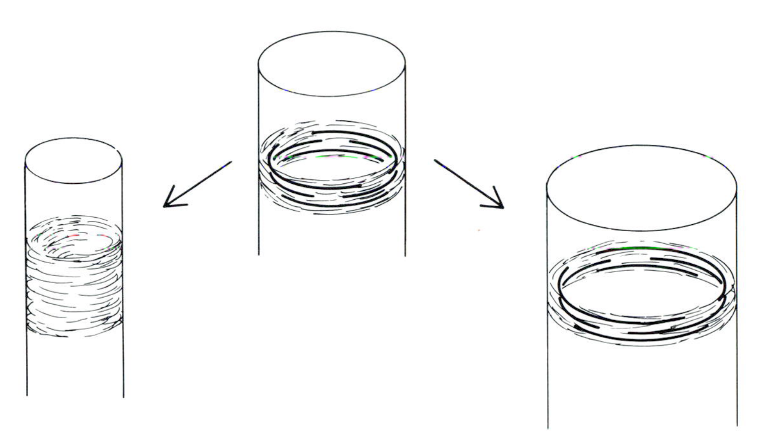

Figure 12. A schematic representation of the circumferentially oriented microfilaments (fine lines) and microtubules (thick lines) associated with the zonula adherens. A decrease in the apical perimeter of a cell (left) appears to involve a "purse-string" contraction of the microfilaments and loss of the microtubules. An increase in apical perimeter (right) might involve the elongation and/or sliding of the microtubules to produce an outwardly directed force on the cell's perimeter.

or they may, by elongating and/or sliding past each other, actively expand the cell's perimeter (Figure 12).

3.4. Intercellular Junctions

The three major types of intercellular junctions in discs (zonulae adhaerentes, septate junctions, and gap junctions) persist throughout morphogenesis and maintain close associations between adjacent cells. This raises some intriguing problems concerning the behavior of junctions as cells undergo rearrangements and changes in shape. Such problems have in the past been explained by the notion that junctions must be continuously assembled and disassembled as cells slide past each other (Fristrom, 1976; Keller, 1978). The rapid assembly and disassembly of the small plaquelike gap junctions seems feasible, but the continuous breakdown and reformation of junctions that completely circumscribe the cell, such as the zonula adherens and the septate junction, is difficult to visualize. In particular, the septate junction, with its complex arrays of septa bridging the intercellular space, appears to provide a formidable impediment to cell rearrangement. We will consider the structure and distribution of each of the major types of junction with special reference to possible

←

Figure 11. Microtubules and microfilaments in imaginal discs. (a) Longitudinally oriented microtubules approach dense plaques (dp) at the apical surface of a mid third instar disc. (b) Circumferentially oriented microtubules (mt) in an unevaginated disc. (c) Circumferentially oriented microtubules (mt) in an evaginated disc. (Note also the transversely sectioned microfilaments (mf) near the zonula adherens). (d) An oblique section near the zonula adherens shows long strands of microfilaments (mf) and microtubules (mt). (ec) epicuticle.

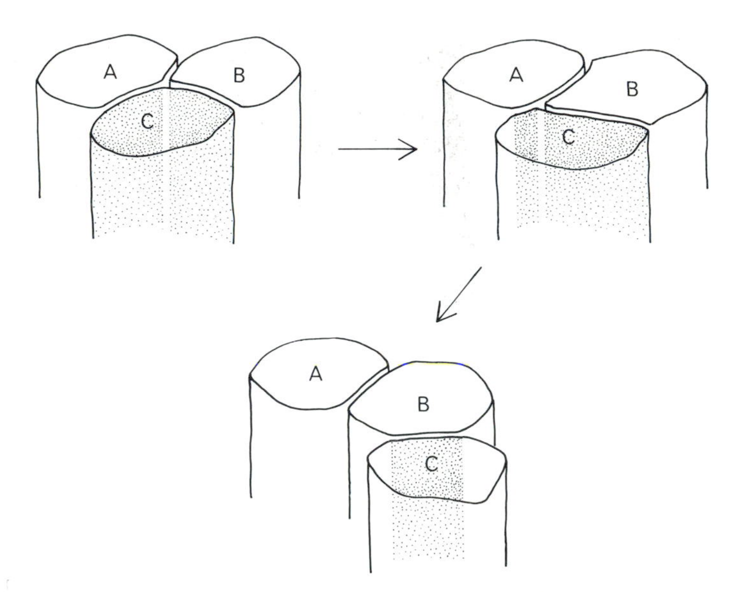

Figure 13. The rearrangement of three cells, A, B, and C. As cell B inserts itself between A and C, the A–C domain (stippled area of contact between cells A and C) decreases while the B–C domain increases. When the insertion of cell B is complete (bottom) the A–C domain disappears.

adjustments of the junctions as cells change neighbors. In analyzing this problem it is useful to refer to the simple situation diagrammed in Figure 13 showing the changing areas of contact between three cells as one cell (B) inserts itself between two others (A and C). The intersection between the three cells can be regarded as the region where neighbor exchanges occur, and the behavior of junctions at such intersections will be considered in detail.

3.4.1. Septate Junctions.

The septate junction is characteristic of invertebrate epithelia (Satir and Gilula, 1973; Noirot-Timothée and Noirot, 1980) and appears to be the functional counterpart of the vertebrate tight junction, providing a barrier to the extracellular passage of molecules across the epithelium (Staehelin, 1974; Filshie and Flower, 1977; Noirot-Timothée *et al.*, 1978; Wood and Kuda, 1980). The junction bridges adjacent cells with parallel rows of septa alternating with dense interseptal material. The septa follow tortuous paths that run predominantly parallel to the surface of the epithelium, isolating the apical from the basal intercellular space.

The septate junction of imaginal discs is of the pleated sheet type (Figure 14a). In freeze-fracture replicas, long, multistranded rows of intramembranous particles mark the course of individual septa. The particles usually appear on the protoplasmic (P) face (Figures 14b,c) and in some regions faint indications of complementary pits can be discerned on the external (E) face (Figure 14c). The relationship between the images obtained by freeze fracture and in thin sections are reviewed in Figure 15. In freeze-fracture replicas of imaginal discs, the fracture plane frequently passes along interfaces where the lateral surfaces of three cells intersect. As the rows of P face particles approach such an intersection they invariably turn downward and run parallel to the intersection (perpendicular to the surface of the epithelium) for some distance (Figure 14b). From two to ten rows of septate particles may lie along a three-cell intersection. E faces are devoid of septate associated particles except for single irregular rows that lie along the angle of flexion at all three-cell intersections (Figure 14b). This distribution of P and E face particles is essentially identical to that described by Noirot-Timothée and Noirot (1980) for the septate junctions of a variety of insect tissues.

Because the rows of intramembranous particles correspond to the position of the intercellular septa (Gilula *et al.*, 1970; Noirot-Timothée *et al.*, 1978) (Figure 15), we can conclude that as individual septa approach an intersection they turn abruptly toward the basal surface and run close to and parallel to the intersection, as shown schematically in Figure 16. Thus, individual septa connect pairs of adjacent cells but do not cross an intersection to connect with a third cell. This arrangement results in a small triangular channel free of septa at the corners of cells. When sectioned longitudinally, this channel can be seen to be occupied by a regularly spaced series of lens-shaped plates (Figure 14a), as described by Noirot-Timothée and Noirot (1980). These structures may contribute to the permeability barrier at three-cell intersections. In summary, the lateral surface of each cell can be divided into a number of domains equal to the number of adjacent cells it contacts. Each domain is crossed by long continuous stretches of septa dividing the apical from the basal intercellular space and is closed laterally by the downturning of the septa at the lateral boundary of the domain (Figure 17). Thus, the septate junction as a whole can function as a transepithelial barrier even though a single septum does not encircle the entire cell (Fristrom, 1982).

This model of the organization of the septate junction has important implications for morphogenesis. With reference to Figure 13, one can see that there is a progressive increase in the width of the domain between cells A and B and a corresponding decrease in the width of the A–C domain. A change in width of domains will also occur during cell shape changes, e.g., all domains will decrease in width as a cell decreases its apical perimeter. We can imagine that as changes in the width of a domain occur, the septa, with their associated particles, might redistribute themselves in the place of the membrane so that septa will always stretch across the entire domain. Thus, we envision a "pushing together" or "stretching out" of septa as the junction accommodates to changing areas of cell contact (Figure 17). In the case of domains that continue to narrow until the cells lose contact (e.g., cells A and C in Figure 13), we have to assume that the septa bridging these cells eventually breakdown, but we have not been able to follow this process. The extracellular lens-shaped structures described above are found in evaginating discs, but their behavior during cell rearrangement is unknown. They may be "pushed along" as the relative position of a three-cell intersection changes or they may break down and reform.

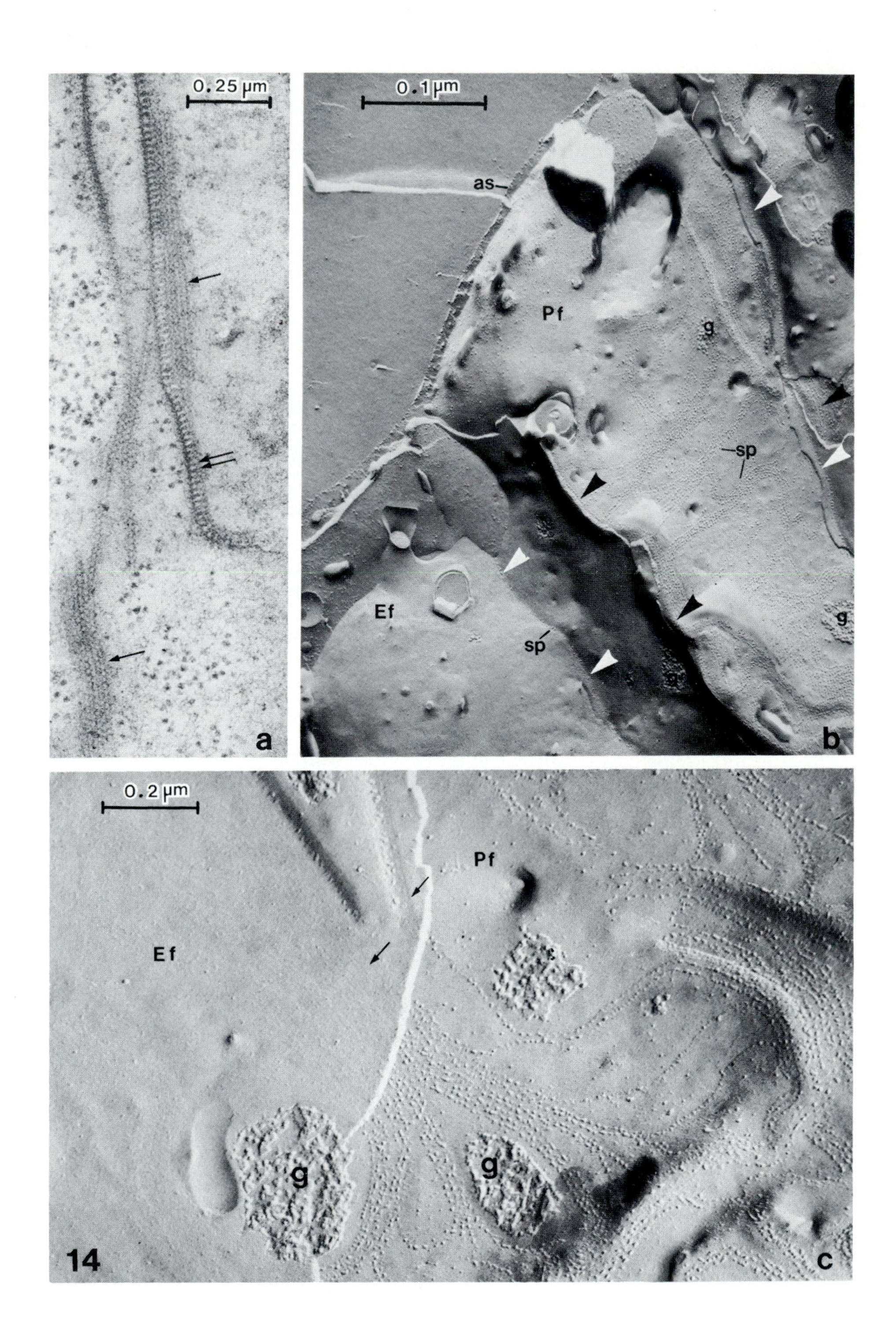
0.25 µm
a
0.1 µm
as
Pf
g
sp
Ef
sp
g
b
0.2 µm
Pf
Ef
g
g
14
c

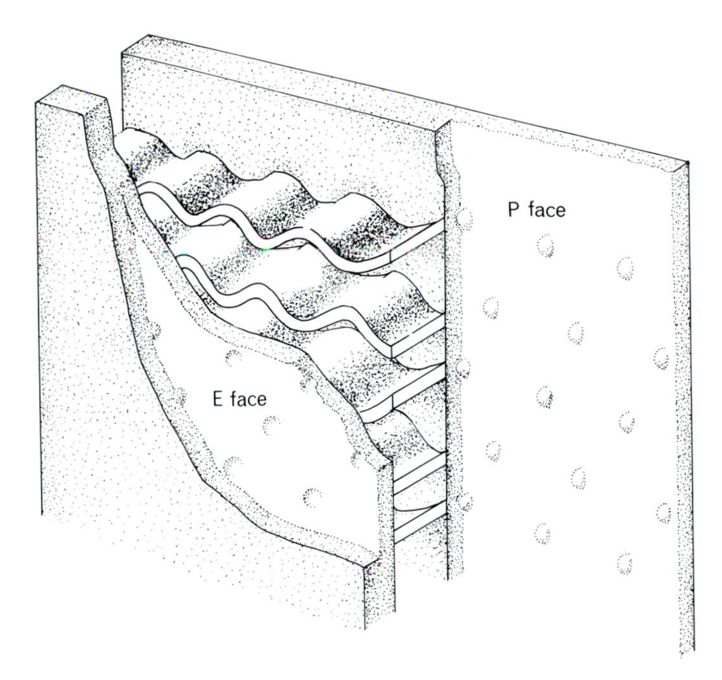

Figure 15. A diagram of the pleated sheet septate junction showing the relationship between the intercellular septa seen in thin sections and the intramembranous particles seen in freeze fracture. The particles are shown on the P (protoplasmic) face and complementary pits on the E (external) face. After Gilula *et al.*, 1970.

The most important implication of this model is that cell rearrangements and cell shape changes do not require the *continuous* breakdown and reformation of septate junctions but the redistribution of existing septa in the plane of the membrane so as to maintain a transepithelial barrier. Presumably, some turnover of junctional elements occurs as well, but this can take place in a relatively leisurely fashion without endangering the integrity of the transepithelial barrier. There are a number of parallels that can be drawn between the distribution of septate juctions as described here and that of tight junctions. From the observations of Friend and Gilula (1972) and Staehelin (1973), the tight junction also appears to cross intercellular domains and turn downward at three-cell intersections. Furthermore, Hull and Staehelin (1976) have shown that tight

←

Figure 14. Intercellular junctions in imaginal discs. (a) A section showing both the intercellular "pleated sheet" arrangement of septa between two cells (single arrows) and a series of lens-shaped structures at a three-cell intersection (double arrows). (b) A freeze-fracture micrograph showing the distribution of septate particles. Rows of protoplasmic (P) face septate particles (sp) run more or less parallel to the apical surface (as), and then turn downward and run parallel to three-cell intersections (black arrows). A three-cell intersection on an external (E) face (white arrows) is marked by a single row of septate particles. (c) A freeze-fracture micrograph showing rows of P face particles and complementary E face pits or grooves (arrows). Gap junction particles are equally distributed between P and E faces. Ef, external face; Pf, protoplasmic face (see Figure 15); g, gap junction.

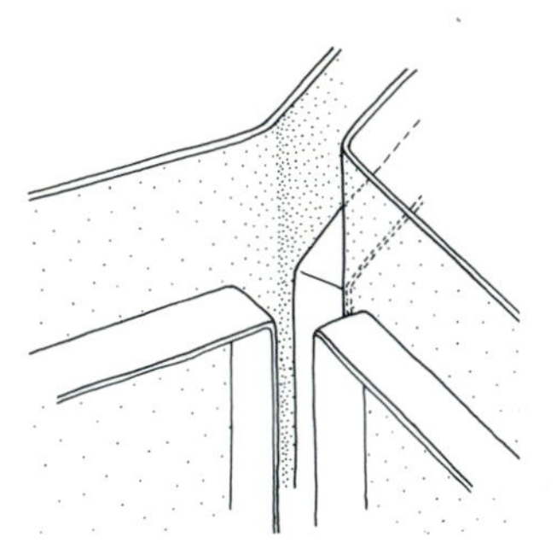

Figure 16. Diagram showing the proposed arrangement of septa at a three-cell intersection. For simplicity, the septa are drawn straight rather than pleated.

junctions "stretch out" in cells of increasing circumference. Thus, it is possible that our model for the redistribution of septate junctions may be generally applicable to occluding junctions in tissues undergoing morphogenesis.

3.4.2. Zonulae Adhaerentes

The zonula adherens is a continuous beltlike junction that circumscribes the apical ends of many types of epithelial cells (Staehelin, 1974). This junction is characterized by a dense fibrous mat on the cytoplasmic surface of opposing cell membranes (Figure 10a), but there is no highly organized intercellular structure as in septate junctions. As we have seen, a circumferential band of 5-nm microfilaments is closely associated with the cytoplasmic surface of the junction (Figures 10a,b, 11c). The zonula adherens serves principally as a site of strong intercellular adhesion (Staehelin, 1974). Because of the adhesive nature of the junction, its beltlike arrangement and its close association with microfilaments, this junction is ideally suited to act as a transmitter of contractile forces across the epithelium.

Because of its adhesive character and beltlike distribution, the zonula adherens appears to present a severe impediment to cell rearrangements. However, if one postulates that the adhesion between cells at the zonula adherens

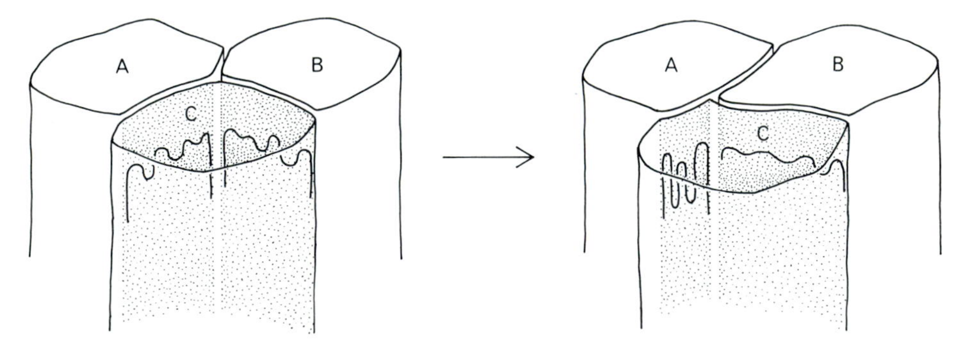

Figure 17. Model for the redistribution of septa during cell rearrangement. Wavy lines represent the distribution of two septa across the A–C and B–C domains (see Figure 13). We propose that septa adjust to the changing width of the domain by becoming increasingly convoluted (as in the A–C domain) or stretched out (as in the B–C domain).

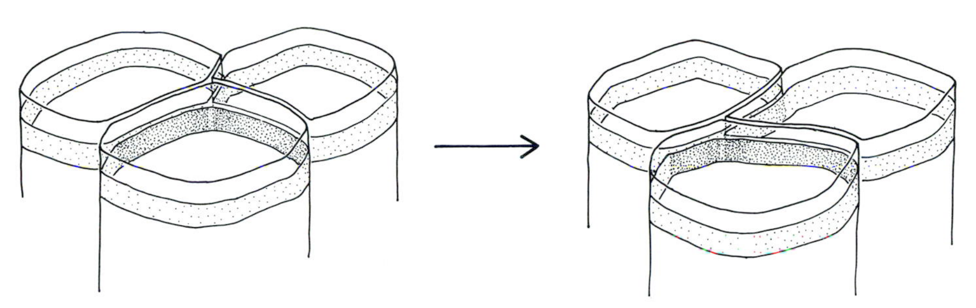

Figure 18. Schematic diagram showing the distribution of the zonulae adhaerentes during the rearrangement of three adjacent cells.

is such as to resist a tensile force (i.e., a force tending to pull cells apart by acting at right angles to the membrane of the junction) but that the junction provides little resistance to a shearing force (i.e., a force acting parallel to the plane of the junction membrane), the opposing junctional membranes would be able to slide laterally with respect to each other. Since the adherens junctions of adjacent cells are in perfect register (Figure 18) and the intercellular material is relatively structureless, one can imagine that the junctional membrane of one cell might slide past a three-cell intersection to oppose the junctional membrane of an adjacent cell (Figure 18).

Another aspect of junction redistribution is seen as cells change shape. As we have seen (section 3.2), the depth of the zonula adherens of a photoreceptor cell increases as the circumference of the cell (and the junction) decreases. This implies that the junctional membrane and its associated components are very fluid, becoming redistributed as cells change shape, and is yet another example of the conservation of junctional elements.

3.4.3. *Gap Junctions*

Gap junctions are specialized regions of the cell surface where opposing regions of the plasma membranes of adjacent cells come into close contact with each other and establish well-insulated cytoplasmic channels between cells that serve as sites of intercellular communication (Peracchia, 1980). Intercellular communication *via* gap junctions is undoubtedly vital for the coordination of developmental processes. It is, therefore, no surprise to find that this most common of all intercellular junctions appears very early in the development of multicellular animals (Ducibella and Anderson, 1975; de Laat *et al.*, 1980; Rickoll and Counce, 1980).

Gap junctions become progressively more numerous during the larval development of imaginal discs (Eichenberger-Glinz, 1979) and are present throughout disc morphogenesis (Figures 14b,c). In freeze-fracture replicas, the gap junctions of discs show the structural characteristics typical of arthropod gap junctions (Gilula, 1980). The intramembranous particles associated with regions of close membrane apposition are irregular in size and frequently present as fused aggregates. In any given junction some of the particles adhere to the P face and some to the E face (Figure 14c).

Let us now consider the possible fates of gap junctions during cell rearrangement. One method of gap junction removal often observed in differentiating tissues is the endocytosis of regions of gap junction-containing membranes (Ginzberg and Gilula, 1979; Larsen *et al.*, 1979). This method of gap junction removal is not seen in discs, and indeed seems more appropriate for tissues in which gap junctions are being permanently removed. The appearance and disappearance of gap junctions has also been related to the aggregation and dispersal of gap junction particles within the plane of the membrane (Lane and Swales, 1978). The identification of gap junctions in thin sections is correlated with the presence of condensed particle arrays. The dispersal and reaggregation of gap junction particles provides an attractive mechanism for rapid gap junction turnover in cells undergoing rearrangement, but we have no evidence that it occurs in discs. It is possible that gap junctions disperse and reassemble so rapidly that intermediate stages are not seen. Or perhaps the membranes of the junction separate while the characteristic particle lattices are retained, as is the case in some experimentally separated cells. (Goodenough and Gilula, 1974). This would require that the resulting "half junctions" move as units through the membrane until they come into register with other "half junctions" on the opposing cell surface to reestablish a functional gap junction.

We should point out here that gap junction morphology appears to be particularly sensitive to variations in physiological and fixation conditions (Raviola *et al.*, 1980). Thus, the presence in freeze fracture micrographs of the typical arrays of "gap" particles does not always indicate the presence of functional gap junctions or even necessarily regions of close membrane apposition (Goodenough and Gilula, 1974). Thus, new approaches to the analysis of gap junction modulation during cell rearrangement are indicated. We would predict, however, gap junctions will follow the biologically conservative pattern shown by the septate; namely, the redistribution of intramembranous components within the plane of the membrane.

4. Future Directions

The observations described in this chapter summarize many years of work on imaginal disc morphogenesis using a variety of microscopic techniques. Much information is available now that was not available a decade ago when *in vitro* evagination was first observed. However, many new questions have been raised, and research goals that originally centered on describing the process of evagination have now turned to focus on the subcellular and molecular mechanisms of morphogenetic processes in discs.

Evidence accumulated in our laboratory indicates that ecdysone acts to induce evagination by modulating transcription and hence protein synthesis. This raises the question of how newly arisen or altered polypeptides result in directed cell movements and changes in cell shape. Much recent research indicates that cell surface components are vital in cell-cell interactions and morphogenetic processes (Bertolotti *et al.*, 1980; Brabec *et al.*, 1980; Shinnick and Lerner, 1980; Subtelney and Wessells, 1980), and we have preliminary evidence that this is also true in discs. For example, we have shown, using two-dimensional gel electrophoresis, that ecdysone has substantial effects on the synthesis of membrane proteins that are probably located at the cell surface (Rickoll and Fristrom, in preparation). Thus, much of the future effort of this laboratory will be focused on identifying and localizing cell surface proteins that have a potential morphogenetic function. To that end, we are engaged in the production of monoclonal antibodies against membrane proteins synthesized both in the presence and absence of ecdysone. Such antibodies, when conjugated with tracers such as fluorescein or ferritin will allow precise geographic localization of specific cell surface components. Localization will be extremely important not only to confirm the surface location of a particular antigen, but the actual distribution might give clues as to the function of that antigen. How provocative it would be, for example, to find an ecdysone-stimulated antigen localized at the zonula adherens or the septate junction! Further clues as to the function of a given antigen might be obtained by the effect of its antibody on *in vitro* evagination and the distribution of the antibody in evagination-defective mutants already available. A long-term goal, now possible because of recombinant DNA technology, will be to induce mutations in the genes encoding surface proteins. One would expect at least some of these mutants to be defective in evagination. An ultrastructural analysis of the phenotypes of such mutants might again provide clues as to the function of the gene product. Thus, we see the future trend of imaginal disc ultrastructure as a powerful accessory to modern techniques of experimental biology rather than as an end in itself. We hope this kind of approach will eventually lead to an understanding of some of the cellular and molecular processes that shape discs into their adult morphologies. In view of the basic structural similarities in epithelial structure in vertebrates and invertebrates, it is likely that the broad strategies of epithelial morphogenesis will also be similar, and that results obtained with discs will be applicable to other developing epithelia.

Acknowledgments

We would like to thank Bernadette Jaroch-Hagerman for patiently and expertly typing the many drafts of this manuscript and Dr. J. W. Fristrom for his helpful comments and additions. Previously unpublished work was supported by UHPHS Grant GM19937 to Dr. J. W. Fristrom.

References

Auerbach, C., 1936, The development of the legs, wings and halteres in wild type and some mutant strains of *Drosophila melanogaster*, *Trans. R. Soc. Edinburgh* **58**:787–815.

Baker, P. C., and Schroeder, T. E., 1967, Cytoplasmic filaments and morphogenetic movement in the amphibian neural tube, *Develop. Biol.* **15**:432–450.

Bertolotti, R., Rutishauser, U., and Edelman, G. M., 1980, A cell surface molecule involved in aggregation of embryonic liver cells, *Proc. Nat. Acad. Sci. USA* **77**:4831–4835.

Brabec, R. K., Peters, B. P., Bernstein, I. A., Gray, R. H., and Goldstein, I. J., 1980 Differential lectin binding to cellular membranes in the epidermis of the newborn rat, *Proc. Nat. Acad. Sci. USA* **77**:477–479.

Bryant, P. J., 1978, Pattern formation in imaginal discs. In *The Genetics and Biology of Drosophila*, edited by M. Ashburner and T. R. F. Wright, vol. 2C, pp. 230–335, Academic Press, London.

Burnside, B., 1971, Microtubules and microfilaments in newt neurulation, *Develop. Biol.* **26**: 416–441.

Campos-Ortega, J. A., and Hofbauer, A., 1977, Cell clones and pattern formation: On the lineage of photoreceptor cells in the compound eye of *Drosophila*, *Wilhelm Roux's Arch. Dev. Biol.* **181**:227–245.

Cloney, R. A., 1966, Cytoplasmic filaments and cell movements: Epidermal cells during ascidian metamorphosis, *J. Ultrastr. Res.* **14**:300–328.

de Laat, S. W., Tertoolen, L. G. J., Dorresteijn, A. W. C., and van der Biggelaar, J. A. M., 1980, Intercellular communication patterns are involved in cell determination in early molluscan development, *Nature* (London) **287**:546–548.

Ducibella, T., and Anderson, E., 1975, Cell shape and membrane changes in the eight-cell mouse embryo: Prerequistes for morphogenesis of blastocytes, *Develop. Biol.* **47**:45–58.

Edelman, G. M., 1976, Surface modulation in cell recognition and cell growth, *Science* **192**:218–226.

Eichenberger-Glinz, S., 1979, Intercellular junctions during development and in tissue cultures of *Drosophila melanogaster:* An electron-microscopic study, *Wilhelm Roux's Arch. Dev. Biol.* **186**:333–349.

Falk, G. J., and King, R. C., 1964, Studies on the developmental genetics of the mutant *tiny* of *Drosophila melanogaster* for egg production, *Growth* **28**:291–324.

Fekete, E., Fristrom, D., Kiss, I., and Fristrom, J. W., 1975, The mechanism of evagination of imaginal discs of *Drosophila melanogaster*. II. Studies on trypsin-accelerated evagination, *Wilhelm Roux's Arch. Dev. Biol.* **178**:123–138.

Filshie, B. K., and Flower, N. E., 1977, Junctional structures in *Hydra*, *J. Cell Sci.* **23**:151–172.

Friend, D. S., and Gilula, N. B., 1972, Variations in tight and gap junctions in mammalian tissues, *J. Cell Biol.* **53**:758.

Fristrom, D., 1969, Cellular degeneration in the production of some mutant phenotypes in *Drosophila melanogaster*, *Mol. Gen. Genet.* **103**:363–379.

Fristrom, D., 1976, The mechanics of evagination in imaginal discs of *Drosophila melanogaster*. III. Evidence for cell rearrangement. *Develop. Biol.* **54**:163–171.

Fristrom, D., 1982, Septate junctions in imaginal discs of *Drosophila:* A model for the redistribution of septa during cell rearrangement, *J. Cell Biol.* in press.

Fristrom, D., and Chihara, C., 1978, The mechanisms of evagination of imaginal discs of *Drosophila melanogaster*. V. Evagination of disc fragments, *Develop. Biol.* **66**:564–570.

Fristrom, D., and Fristrom, J. W., 1975, The mechanism of evagination of imaginal discs of *Drosophila melanogaster*. I. General considerations, *Develop. Biol.* **43**:1–23.

Fristrom, D., and Fristrom, J. W., 1982, Cell surface binding sites for peanut agglutinin in the differentiating eye disc of *Drosophila*, *Develop. Biol.* in press.

Fristrom, J. W., Fristrom, D. K., Fekete, E., and Kuniyuki, A. H., 1977, The mechanism of evagination of imaginal discs of *Drosophila melanogaster*, *Amer. Zool.* **17**:671–684.

Fristrom, D., Fekete, E., and Fristrom, J. W., 1981, Imaginal disc development in a non-pupariating lethal mutant in *Drosophila melanogaster*, *Wilhelm Roux's Arch. Dev. Biol.* **190**:11–21.

Gilula, N. B., 1980, Cell-to-cell communication and development. In *The Cell Surface: Mediator of Developmental Processes*, edited by S. Subtelny and N. K. Wessells, pp. 23–41, Academic Press, New York.

Gilula, N. B., Branton, D., and Satir, P., 1970, The septate junction: A structural basis for intercellular coupling, *Proc. Nat. Acad. Sci. USA* **67**:213–220.

Ginzberg, R. D., and Gilula, N. B., 1979, Modulation of cell junctions during differentiation of the chicken otocyst sensory epithelium, *Develop. Biol.* **68**:110–129.

Goodenough, D. A., and Gilula, N. B., 1974, The splitting of hepatocyte gap junctions and zonulae occludentes with hypertonic disaccharides, *J. Cell Biol.* **61**:575–590.

Green, P. B., 1980, Organogenesis—A biophysical view, *Annu. Rev. Plant Physiol.* **31**:51–82.

Gustafson, T., and Wolpert, L., 1963, The cellular basis of morphogenesis and sea urchin development, *Int. Rev. Cytol.* **15**:139–214.

Hadorn, E., 1978, Transdetermination. In *The Genetics and Biology of Drosophila*, edited by M. Ashburner and T. R. F. Wright, vol. 2C, pp. 555–617, Academic Press, London.

Heuser, J. E., and Salpeter, S. R., 1979, Organization of acetylcholine receptors in quick-frozen, deep-etched, and rotary replicated *torpedo* postsynaptic membrane, *J. Cell Biol.* **82**:150–173.

Hull, B. E., and Staehelin, L. A., 1976, Functional significance of the variations in the geometrical organization of tight junction networks, *J. Cell Biol.* **68**:688–704.

Hull, B. E., and Staehelin, L. A., 1979, The terminal web: A reevaluation of its structure and function, *J. Cell Biol.* **81**:67–82.

Jacobson, A. G., and Gordon, R., 1976, Changes in the shape of the developing vertebrate nervous system analyzed experimentally, mathematically and by computer simulation, *J. Exp. Zool.* **197**:191–246.

Keller, R. E., 1978, Time-lapse cinemicrographic analysis of superficial cell behavior during and prior to gastrulation, *J. Morphol.* **157**:223–247.

Kuniyuki, A. H., 1976, Ecdysone-induced protein synthesis in the imaginal discs of *Drosophila melanogaster*, Ph.D. Thesis, Department of Genetics, University of California, Berkeley.

Lane, N. J., and Swales, L. S., 1978, Changes in the blood-brain barrier of the central nervous system in the blowfly during development, with special reference to the formation and disaggregation of gap and tight junctions. II. Pupal development and adult flies, *Develop. Biol.* **62**: 415–431.

Larsen, W. J., Tung, H., Murray, S. A., and Swenson, C. A., 1979, Evidence for the participation of actin microfilaments and bristle coats in the internalization of gap junction membrane, *J. Cell Biol.* **83**:576–587.

Lawrence, P. A., and Green, S. M., 1979, Cell lineage in the developing retina of *Drosophila*, *Develop. Biol.* **71**:142–152.

Lin, S., Lin, D. C., and Flanagan, M. D., 1978, Specificity of the effects of cytochalasin B on transport and motile processes, *Proc. Nat. Acad. Sci. USA* **75**:329–333.

Mandaron, P., 1976, Ultrastructure des disques de patte de *drosophile* cultivés *in vitro*. Evagination, sécrétion de la cuticule nymphale et apolysis, *Wilhelm Roux's Arch. Dev. Biol.* **179**:185–196.

Mandaron, P., and Sengel, P., 1973, Effect of cytochalasin B on the evagination *in vitro* of leg imaginal discs, *Develop. Biol.* **32**:201–207.

Martin, P., and Schneider, I., 1978, *Drosophila* organ culture. In *The Genetics and Biology of Drosophila*, edited by M. Ashburner and T. R. F. Wright, vol. 2a, pp. 219–264, Academic Press, London.

Melamed, J., and Trujillo-Cenoz, O., 1975, The fine structure of the eye imaginal discs in muscoid flies, *J. Ultrastruct. Res.* **51**:79–93.

Mooseker, M. S., and Tilney, L. G., 1975, Organization of an actin filament-membrane complex: Filament polarity and membrane attachment in the microvilli of intestinal epithelial cells, *J. Cell Biol.* **67**:725–743.

Noirot-Timothée, C., and Noirot, C., 1980, Septate and scalariform junctions in arthropods, *Int. Rev. Cytol.* **63**:97–140.

Noirot-Timothée, C., Smith, D. S., Cayer, M. L., and Noirot, C., 1978, Septate junctions in insects: Comparison between intercellular and intramembranous structures, *Tissue Cell* **10**:125–136.

Peracchia, C., 1980, Structural correlates of gap junction permeation, *Int. Rev. Cytol.* **66**:81–146.

Perry, M. M., 1968, Further studies on the development of the eye of *Drosophila melanogaster*. I. The ommatidia, *J. Morph.* **124**:227–248.

Pollard, T. D., and Weihing, R. R., 1974, Actin and myosin and cell movement, *CRC Crit. Rev. Biochem.* **37**:537–583.

Poodry, C. A., 1980a, Imaginal discs: Morphology and development. In *The Genetics and Biology of Drosophila*, edited by M. Ashburner and T. R. F. Wright, vol. 2D, pp. 407–441, Academic Press, London.

Poodry, C. A., 1980b, Epidermis: Morphology and development. In *The Genetics and Biology of Drosophila*, edited by M. Ashburner and T. R. F. Wright, vol. 2D, pp. 443–497, Academic Press, London.

Poodry, C. A., and Schneiderman, H. A., 1970, The ultrastructure of the developing leg of *Drosophila melanogaster*, *Wilhelm Roux's Arch. Dev. Biol.* **166**:1–44.

Poodry, C. A., and Schneiderman, H. A., 1971, Intercellular adhesivity and pupal morphogenesis in *Drosophila melanogaster*, *Wilhelm Roux' Arch. Dev. Biol.* **168**:1–9.

Poole, T. J., and Steinberg, M. S., 1977, SEM-aided analysis of morphogenetic movements: Development of the amphibian pronephric duct. *Scanning Electron Microscopy* **2**:43–52.

Raviola, E., Goodenough, D. A., Raviola, G., 1980, Structure of rapidly frozen gap junctions, *J. Cell Biol*, **87**:273–279.

Ready, D. F., Hanson, T. E., and Benzer, S., 1976, Development of *Drosophila* retina, a neurocrystalline lattice, *Develop. Biol.* **53**:217–240.

Reed, C. T., Murphy, C., and Fristrom, D., 1975, The ultrastructure of the differentiating pupal leg of *Drosophila melanogaster, Wilhelm Roux' Arch. Dev. Biol.* **178**:285–302.

Rheinhardt, C. A., Hodgkin, N. M., and Bryant, P. J., 1977, Wound healing in the imaginal discs of *Drosophila*. I. Scanning electron microscopy of normal and healing wing discs, *Develop. Biol.* **60**:238–257.

Rickoll, W. L., and Counce, S. J., 1980, Morphogenesis in the embryo of *Drosophila melanogaster*—germ band extension, *Wilhelm Roux' Arch. Dev. Biol.* **188**:163–177.

Satir, P., and Gilula, N. B., 1973, The fine structure of membranes and intercellular communication in insects, *Annu. Rev. Entomol.* **18**:143–166.

Saunders, J. W., 1966, Death in embryonic systems, *Science* **154**:604–612.

Schroeder, T. E., 1973, Cell constriction: Contractile role of microfilaments in division and development, *Amer. Zool.* **13**:949–960.

Sedlak, B., and Gilbert, L. I., 1976, Epidermal cell development during the pupal-adult metamorphosis of *Hyalophora cecropia, Tissue Cell* **8**:637–648.

Shinnick, T. M., and Lerner, R. A., 1980, The cbpA gene: Role of the 26,000-dalton carbohydrate-binding protein in intercellular cohesion of developing *Dictyostelium discoideum* cells, *Proc. Nat. Acad. Sci.* USA **77**:4788–4792.

Silvert, D. J., and Fristrom, J. W., 1980, Biochemistry of imaginal discs: Retrospect and prospect, *Insect Biochem.* **10**:341–355.

Spooner, B. S., and Wessells, N. K., 1972, An analysis of salivary gland morphogenesis: Role of cytoplasmic microfilaments and microtubules. *Develop. Biol.* **27**:38–57.

Spooner, B. S., Hilfer, S. R., and Hay, E. D. (Organizers), 1973, Symposium on: Factors controlling cell shape during development *Amer. Zool* **13**:941–1135.

Spreij, T. E., 1971, Cell death during the development of the imaginal discs of *Calliphora erythrocephala, Neth. J. Zool.* **21**:221–264.

Staehelin, L. A., 1973, Further observations on the fine structure of freeze-cleaved tight junctions. *J. Cell Sci.* **13**:763.

Staehelin, L. A., 1974, Structure and function of intercellular junctions, *Int. Rev. Cytol.* **39**:191.

Steinberg, A. G., 1943, The development of the wild type and bar eyes of *Drosophila melanogaster, Can. J. Res.* **21,D**:277–283.

Steinberg, M. S., 1973, Cell movement in confluent monolayers: A re-evaluation of the causes of "contact inhibition", In *Locomotion of Tissue Cells*, pp. 333–341, *Ciba Foundation Symp. 14*, Elsevier-North Holland, Amsterdam.

Stendahl, O. I., Hartwig, J. H., Brotschi, E. A., and Stossel, T. P., 1980, Distribution of actin-binding protein and myosin in macrophages during spreading and phagocytosis, *J. Cell Biol.* **84**: 215–224.

Stern, C., 1936, Somatic crossing-over and segregation in *Drosophila melanogaster, Genetics* **21**: 625–730.

Subtelney, S., and Wessells, N. K. (eds.), 1980, *The Cell Surface: Mediator of Developmental Processes*. In *The 38th Symposium of the Society for Developmental Biology*, Academic Press, New York.

Taylor, D. L., and Condeelis, J. S., 1979, Cytoplasmic structure and contractility in amoeboid cells, *Int. Rev. Cytol.* **56**:57–144.

Trinkaus, J. P., 1976, On the mechanism of metazoan cell movements. In *The Cell Surface in Animal Embryogenesis and Development*, edited by G. Poste and G. L. Nicholson, pp. 225–329, North Holland Publ. Co., Amsterdam.

Ursprung, H., 1972, The fine structure of imaginal discs. In *The Biology of Imaginal Discs*, edited by H. Ursprung and R. Nöthiger, pp. 93–107, Springer-Verlag, Berlin.

Vijverberg, A. J., 1974, A cytological study of the proliferation patterns in imaginal discs of *Calliphora erythrocephala* Meigen during larval and pupal development, *Neth. J. Zool.* **24(2)**: 171–217.

Waddington, C. H., 1962, *New Patterns in Genetics and Development*, p. 176, Columbia University Press, New York.

Waddington, C. H., and Perry, M. M., 1960, The ultrastructure of the developing eye of *Drosophila*, *Proc. R. Soc. London Set. B.* **153**:155–178.

Wehman, H. J., 1969, Fine structure of *Drosophila* wing imaginal discs during early stages of metamorphosis, *Wilhelm Roux' Arch. Dev. Biol.* **163**:375–390.

Wessells, N. K., Spooner, B. S., Ash, J. F., Bradley, M. O., Luduena, M. A., Taylor, E. L., Wrenn, J. T., and Yamada, K. M., 1971, Microfilaments in cellular and developmental processes, *Science* **171**:135–143.

Wood, R. L., and Kuda, A. M., 1980, Formation of junctions in regenerating hydra: Septate junctions, *J. Ultrastruct. Res.* **70**:104–117.

Wrenn, J. T., 1971, An analysis of tubular gland morphogenesis in chick oviduct, *Develop. Biol.* **26**:400–415.

III

The Ultrastructure of the Development, Differentiation, and Functioning of Specialized Tissues and Organs

10

Fine Structure of the Cuticle of Insects and Other Arthropods

BARRY K. FILSHIE

1. Introduction

The integument, or outer covering of insects is composed of the cuticle, which is a multilayered extracellular composite material, and an underlying epidermis, from which the several layers of the cuticle are secreted. During the life cycle of an insect the process of formation and shedding of the cuticle occurs several times, which permits increases in body size and changes in form.

In addition to forming the outer covering, the cuticle is invaginated to form the linings of the esophagus, the foregut, the hindgut and rectum, the tracheae and tracheoles, and various glands of epidermal origin that open onto the surface. In other regions of the body, solid invaginations called apodemes serve for the attachment of some muscles and for the support of certain internal organs. All the external appendages and outgrowths such as wings, legs, mouthparts, scales, and hairs are also covered with cuticle.

The basis of our knowledge of the structure of the cuticle has come from studies by light microscopy combined with staining reactions and histochemical experiments. These findings have been summarized in numerous reviews, for example Wigglesworth (1948; 1957), Richards (1951), and Hackman (1971). Accordingly, it has been shown that the cuticle is extremely variable in its gross structure, grading from very thin flexible membranes, as in the intersegmental regions, to thick horny plates, as in the elytra of beetles. The structure and physical properties of a particular cuticle can be correlated with its specific function. There are two main horizontal subdivisions of the cuticle (Figures 1, 2), an

BARRY K. FILSHIE • Commonwealth Scientific and Industrial Research Organization, Division of Entomology, Canberra, Australia.

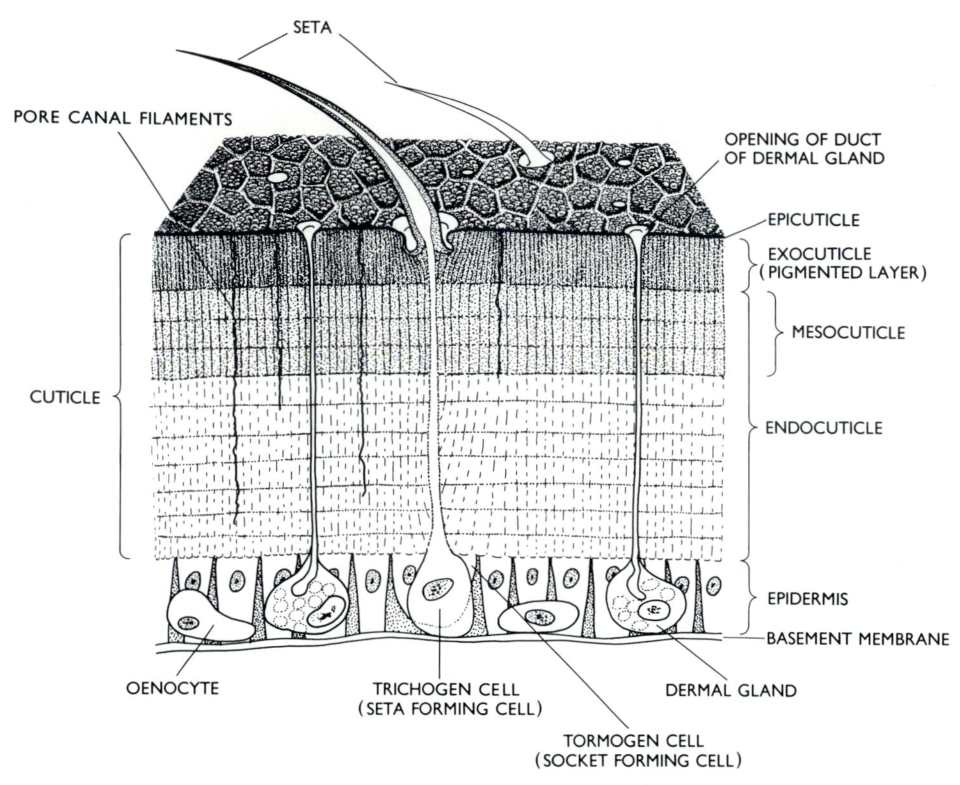

Figure 1. Generalized diagram of the structure of the insect integument (from Hackman, 1971).

outer, thin, nonchitinous layer known as the *epicuticle,* and an inner layer, composed mainly of protein and the polysaccharide α-chitin, known as the *procuticle* (using the convention established by Richards, 1951). In hard cuticles, the preecdysial portion (that laid down before shedding of the old cuticle) is sclerotized immediately after ecdysis and is then known as the *exocuticle.* The soft cuticle that is subsequently laid down beneath the exocuticle in the period between moults is known as the *endocuticle.* There are exceptions to the above rule for delineating exocuticle and endocuticle. In some cases, only part of the preexuvial cuticle becomes sclerotized, while in other, such as the labium of adult crickets, the outer part is sclerotized on emergence, and with age the entire cuticle becomes progressively tanned (see Richards, 1967).

A third subdivision of the procuticle, the *mesocuticle,* interposed between exocuticle and endocuticle, may be demonstrated by certain staining reactions. It is not known whether the mesocuticle is different chemically from, or merely a physical modification of, endocuticle, which reacts differently to the staining solutions.

It has long been known from polarized light microscopy (Biedermann, 1903) and X-ray diffraction (Meyer and Pankow, 1935), that the procuticle and its derivatives are fibrillar, but in all but a few cases, such as the "Balken" of beetles (Biedermann, 1903), this fibrous organization is beyond the resolving power of the light microscope. Traversing most cuticles are numerous channels, called *pore canals*, which may contain solid material of filaments. Light microscopy has shown that sometimes these canals are extensions of the underlying epidermal cells.

Ever since the resolution of the light microscope was surpassed by that of the early experimental electron optical instruments, electron microscopy has been used as an important tool for study of the structure of the insect cuticle. It is of incidental interest, historically, that insect cuticle (the wing of *Musca domestica*) was the very first unprepared biological material from which electron micrographs were recorded (Driest and Müller, 1935, and see Ruska, 1980). Several papers were subsequently published by Richards and his co-workers (see references quoted in Richards, 1951) that yielded useful information on the structure of tracheae, tracheoles, and various setae, and also on the structural basis for the iridescent colors exhibited by certain butterfly scales. However, it was not until techniques of plastic embedding and ultrathin sectioning were established and improved in the 1950s, that ideas of the basic structure of the cuticle gained from light microscopy began to be modified by the results of electron microscopy.

The aim of this chapter is to demonstrate the indispensible role that the electron microscope has played in revealing the fine structure of the cuticle. Neville (1975) has reviewed this subject extensively up to 1972, so the subsequent contributions will be discussed more fully.

2. The Chitin-Containing Cuticle

2.1. Microfibrils and Matrix

Fibrils such as those seen by light microscopy in "Balken" of beetles are larger by two or three orders of magnitude than the smallest filaments, known as *microfibrils*, which can be resolved by electron microscopy. Evidence for microfibrils was first presented by Richards and Korda (1948), who examined whole mounts of disrupted tracheae. The "microfibers" they revealed were 10–30 nm in diameter and extremely long, like those of cellulose. We now know that these "microfibers" are composed of aggregates of microfibrils.

Locke (1960; 1961) demonstrated a fine fibrillar structure in sections of intact cuticle, but the individual microfibrils could not be resolved satisfactorily in the material he examined. The first clear micrographs of microfibrils came from cross sections of a wasp ovipositor (Rudall, 1965), where there is a high degree of fibrillar orientation along the axis of the ovipositor. Here it was demonstrated that microfibrils stained poorly, are of uniform diameter (approximately 5 nm), and are embedded in an electron-dense matrix (Figure 3).

Rudall (1965) considered the microfibrils as being composed of chitin and the matrix of protein. He provided support for this assumption (1967) by showing that the inner layers of the puparium of *Calliphora* exhibited reduced overall staining intensity during their digestion by the molting fluid. It had previously been demonstrated by X-ray diffraction (Fraenkel and Rudall, 1947) that pro-

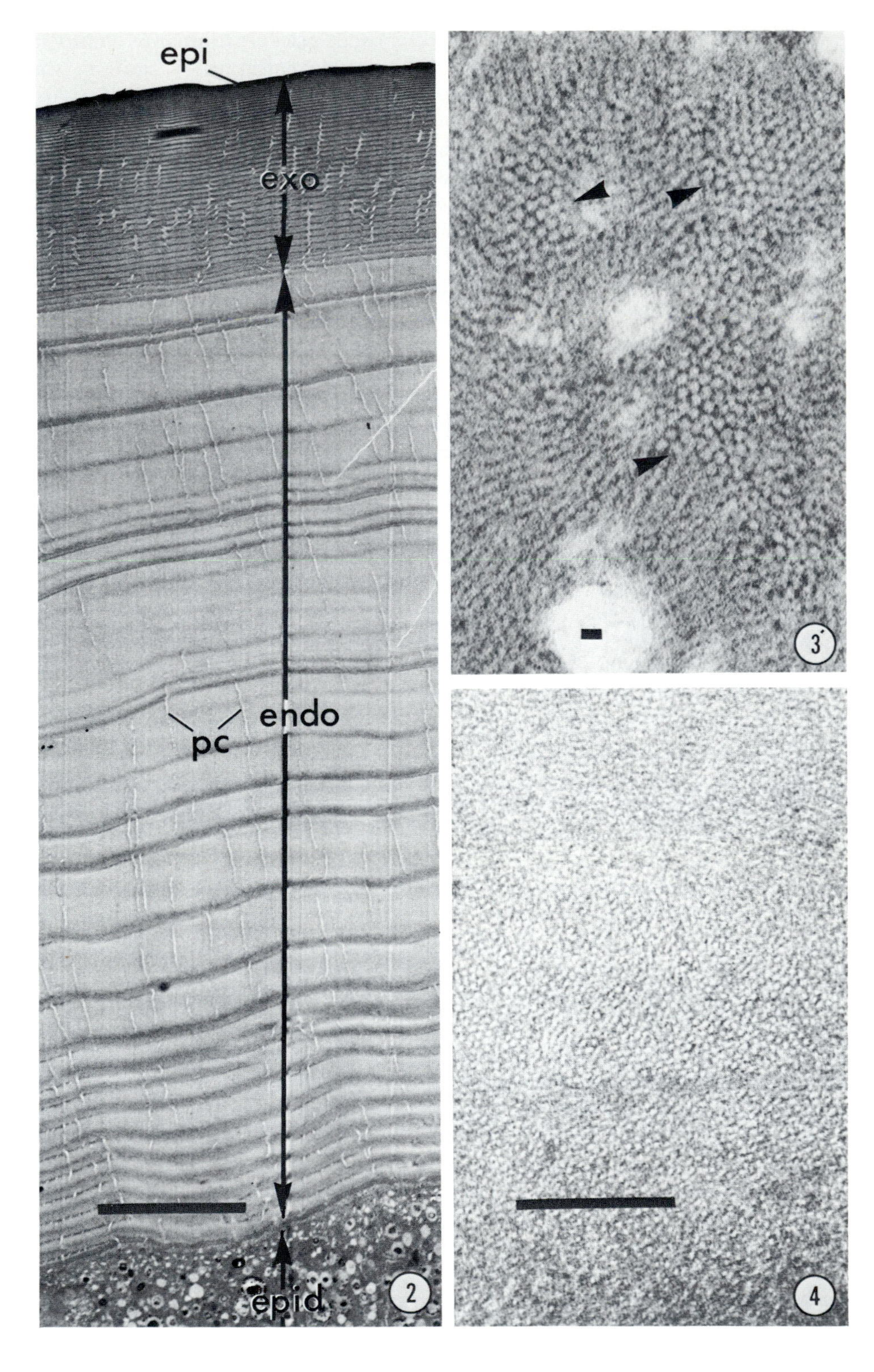
epi
exo
endo
pc
epid
2
3
4

tein was digested more rapidly than was chitin from these layers. So, by inference, the electron-dense material was protein and the electron-lucent microfibrils were chitin. Neville (1970) reached the same conclusion after examining the relative staining properties of pure protein (resilin) and resilin–chitin mixtures in insect rubberlike cuticle. Microfibrils were found only in the chitin-containing regions. Neville *et al.* (1976) calculated the volume fractions of microfibrils and matrix from electron micrographs of rubberlike cuticle as being 20.6% (but see later discussion). This value agreed reasonably well with measurements before and after acid hydrolysis, which showed that the proportion of chitin increased from 13% to 24% of the total dry weights of samples measured between 2 days prior to adult ecdysis and 20 days after it (Neville, 1963).

Despite the above evidence, there is at least one exception to the assignment of a chitinous composition to microfibrils. The procuticle of the alloscutum of the adult female cattle tick *Boophilus microplus* contains only 3.8% chitin on a water-free basis (Hackman, 1974), and yet electron-lucent microfibrils about 4 nm in diameter can be demonstrated in thin sections (Figure 4). It is estimated from electron micrographs that these microfibrils occupy 30%–50% of the volume of the procuticle (Filshie, unpublished data), which is a much higher figure than would be expected if the microfibrils were composed only of chitin. It seems that in this instance a high proportion of the microfibrillar component may be protein. There are no other published data on the volume fraction of microfibrils from cuticles of very low chitin content. In tracheae, there is some evidence that the microfibrils comprising the taenidia may also contain a high proportion of protein relative to chitin (see discussion, Chapter 12, section 2.3.5).

To calculate volume fractions of microfibrils from electron micrographs, it must be possible to measure their diameters accurately. The diameters of microfibrils have been estimated from transverse sections of cuticles of thirteen arthropod species (eleven insects, one arachnid, and one crustacean) by Neville *et al.* (1976). They determined that the mean diameter was relatively constant among species, the spread of these means being between 2.4 and 3.1 nm, with an average of 2.8 nm. They used electron micrographs of sections that had been treated with potassium permanganate, which, it was claimed, stained the matrix, but left the microfibril unstained (see also section 2.3). Most of their micrographs were taken at electron-optical magnifications around 50,000. The micrographs were magnified onto the screen of a Nikon profile projector and they measured the diameter of individual microfibrils and the intermicrofibrillar spacings in areas of regular hexagonal packing of microfibrils in the electron-dense matrix. The latter measurements were checked against lattice parameters

←

Figure 2. Transverse section of the femur cuticle of the Australian plague locust (*Chortoicetes terminifera*). The exocuticle (exo) is composed of helicoidally packed sheets of microfibrils, which appear as close-spaced regular light and dark bands. In the endocuticle (endo), the broad light bands are layers in which microfibrils are oriented in preferred directions. The intervening narrow dense bands are regions of helicoidal cuticle. epi, epicuticle; epid, epidermis; pc, pore canals. (Bar = 10 μm.)
Figure 3. Transverse section of the ovipositor of the wasp, *Sirex noctilio*, showing cross sections of microfibrils hexagonally packed in groups (arrows). (Bar = 10 nm.)
Figure 4. Microfibrils in the procuticle of the alloscutum of the adult female cattle tick *Boophilus microplus* (Bar = 10 nm.)

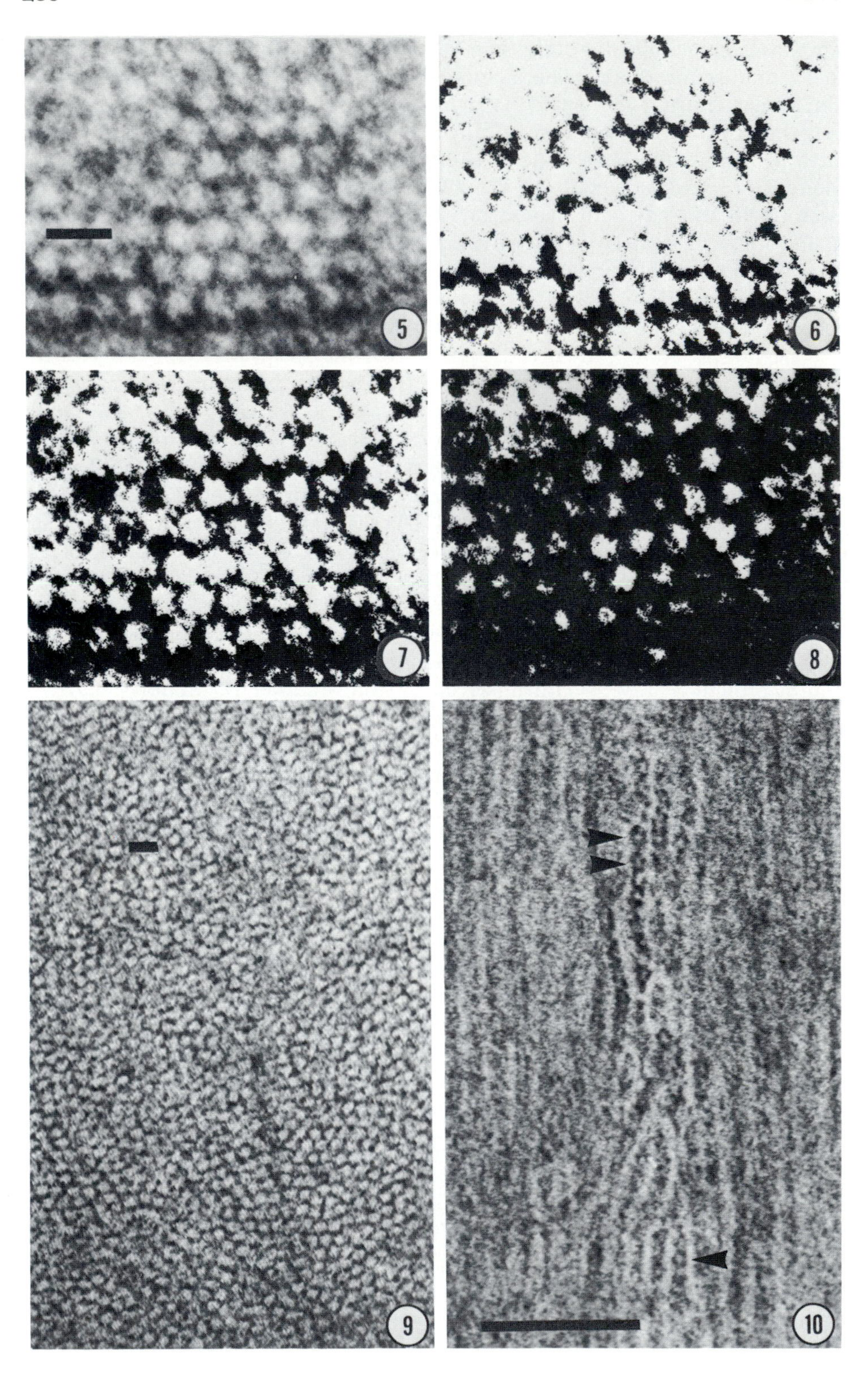
5
6
7
8
9
10

measured from optical diffractograms taken from the micrographs. The lattice spacings measured by the two methods were in agreement.

The measurements of the diameters were also remarkably constant (the extremes being 2.3 nm minimum and 3.5 maximum), but these results should be assessed critically in the light of possible systematic and subjective errors of measurement not discussed by the authors. For instance, in high resolution micrographs the microfibrils are seen to be poorly stained, not unstained as stated by the authors. Nor are the edges well defined, as the accompanying micrograph (Figure 5) demonstrates. This micrograph shows cross sections of microfibrils magnified 1,000,000 times (electron-optical magnification: 250,000) and close to true focus. The microfibril boundaries are diffuse partly because of a lack of contrast in the image and partly because of an overall graininess, the latter being the result of the precipitation of relatively large particles of staining compounds (in this case probably manganese dioxide and lead). To measure the outer boundary of the microfibrils, an observer must make an arbitrary assessment of the grey level within the boundary that represents the edge. This assessment can have a profound effect on the value obtained for the diameter, as shown in Figures 6–8, where the micrograph (Figure 5) has been converted into two-tone images by printing onto high contrast graphic arts film. Using progressively longer printing exposures, the boundary is represented by progressively less dense grey levels, and so the apparent diameter of microfibrils decreases. The result of 2.8 nm obtained for the diameter of the microfibril by Neville *et al.* (1976) should (I believe) be taken as the lower limit.

All other published values for microfibril diameter are greater than 2.8 nm. In the sternite exocuticle of the adult mealworm *Tenebrio molitor*, the figure given by Neville (1970) was 5.0 nm. Microfibrils of the exocuticle of the giant water bug *Hydrocyrius columbiae* measured approximately 4.5 nm in diameter, with a center-to-center spacing of about 6.5 nm (Neville and Luke, 1969b). In the spider *Cupiennius salei* the smallest value found for microfibril diameter was 3.5 nm (Barth, 1973). Axial beading, with a period between 6.0 and 12.0 nm was also observed in longitudinal profiles of these microfibrils. Filshie and Hadley (1979) demonstrated microfibrils 4.0–4.5 nm in diameter in sections of the endocuticle of the dorsal plate of a scorpion (*Hadrurus arizonensis*) (Figure 9). In the outer or hyaline exocuticle (Figure 10) complex microfibrils were described with small electron-lucent cores of about 2-nm diameter surrounded by electron-dense coats, giving overall diameters slightly larger than are found in the endocuticle. Again, for the reasons outlined above, these values are of uncertain accuracy.

A promising method for the study of the fine structure of "native" micro-

←

Figure 5. High resolution micrograph of a group of transversely sectioned microfibrils from the sternite exocuticle of *Tenebrio molitor*. (Bar = 10 nm.)

Figures 6–8. Two-tone prints of Figure 5 at increasing exposures to demonstrate the effect on the apparent diameter (For explanation, see text.)

Figure 9. Transverse section of microfibrils in preferred orientation in the dorsal plate endocuticle of the scorpion (*Hadrurus arizonensis*). (Bar = 10 nm.)

Figure 10. Microfibrils of an unusual structure in the hyaline exocuticle of the dorsal plate of *Hadrurus*. Cross sections are indicated by arrows. (Bar = 100 nm.)

fibrils was presented by Rudall (1969), but has not been pursued in more recent times. Thin layers of endocuticle were stripped from the cuticle of locust tibia and negatively stained with uranyl acetate. Rudall measured the diameter of the microfibrils as being around 5.0 nm. A regular axial beading of stain was seen also, with a repeat close to the 3.1-nm axial period found in X-ray patterns of intact cuticle and attributed to a regular distribution of protein along the chitin chains.

Measurements have also been made of chitin microfibrils isolated from cuticle by alkaline hydrolysis, acid treatment, and ultrasonic disruption and negatively stained with uranyl acetate (Rudall, 1976). Uniform fibrils about 3.0 nm in diameter were described, but my measurements of the published micrographs yield values closer to 4 nm. In spider cuticle, similarly isolated microfibrils had a minimum diameter of 3.5 nm (Barth, 1973). Chitin microfibrils dispersed in lithium thiocyanate, sonicated, and negatively stained (Neville, 1975), were swollen and about 7.5 nm in diameter. Considering the harsh chemical and physical treatments required to isolate chitin microfibrils, they should not be likened too closely to microfibrils of intact cuticles.

The conclusion to be drawn from the above discussion is that, although the microfibril in intact cuticle contains a high proportion of chitin, electron microscopy has not provided convincing evidence that any part or all of it is pure chitin. Therefore, volume fractions of microfibrils and matrix derived from electron micrographs of intact cuticles cannot be used to calculate the relative amounts of chitin and protein present. Similarly, X-ray data obtained from *pure* chitin samples cannot be used to calculate the dimensions of "chitin crystallites" within cuticular microfibrils, as did Neville *et al.* (1976).

The most recent model for the structure of the microfibril deduced from X-ray diffraction studies of intact and deproteinized cuticles, is that of Blackwell and Weih (1980). On the assumption that the microfibril consists of a core of crystalline chitin surrounded by a coat of protein, the model that best fitted their X-ray diagrams was a core 3.8 nm in diameter surrounded by protein subunits, giving a total diameter for the microfibril of 7.25 nm, the latter being also the spacing of microfibrils in the hexagonal lattice.

2.2. Packing Arrangements of Microfibrils and Matrix in Procuticle

Sections of the chitin-containing cuticles of insects and other arthropods usually contain lamellae, which appear as alternating light and dark bands running parallel or nearly parallel to the cuticular surface. These were originally seen by light microscopy, but are very obvious in electron micrographs also (Figure 2). When sections are cut at angles other than perpendicular to the cuticular surface, arcuate patterns are seen to be superimposed on these lamellae (Figure 11).

In order to explain the appearance of lamellae and arcuate patterns, two sharply contrasting models have been proposed, based on different fibrous organizations of the cuticle. According to the first model proposed by Drach

(1953), the dark bands contain filaments oriented horizontally, and the light, intermediate zones contain arced filaments lying in parallel oblique planes and meeting neighboring dark bands. Locke's model (Locke, 1961) is essentially the same as Drach's, with the plane of the arced filaments perpendicular rather than oblique to the bands (Figure 12a).

According to the second model, proposed by Bouligand (1965), arced, interband fibers do not exist. The appearance of arcs in sections is produced by the projections of short lengths of fibers lying in horizontal planes. Fibers are parallel to one another in a single plane, and the fiber orientation within successive planes rotates anticlockwise by a small constant angle downward through the cuticle (Figure 12b). A 180° rotation of the fiber direction produces a single lamella. This helicoidal model, as it is commonly called, has been applied to many other biological materials and is also the molecular configuration found in cholesteric liquid crystals (see reviews by Bouligand, 1972; 1978).

There have been numerous discussions in the literature over the past fifteen years on the validity of one or the other model. The following references support the helicoidal model for arthropod cuticle: Neville and Caveney, 1969; Neville and Luke, 1969a,b; 1971; Neville *et al.*, 1969; Noirot and Noirot-Timothée, 1969; Neville, 1970; Delachambre, 1971; 1975; Neville and Berg, 1971; Bruck and Stockem, 1972a,b; Green and Neff, 1972; Pace, 1972; Zelazny and Neville, 1972; Zacharuk, 1972; Barth, 1973; Gharagozolou-van Ginneken and Bouligand, 1973; 1975; Gharagozolou-van Ginneken, 1974; 1976; Altner, 1975; Gubb, 1975; Wigglesworth, 1975b; Heghadl *et al.*, 1977a,b; Credland, 1978; Livolant, *et al.*, 1978; Filshie and Hadley, 1979; Filshie and Smith, 1980. The following authors have raised objections to the helicoidal model and have favored the Drach hypothesis; Rudall, 1969; Dennell, 1973, 1974, 1975, 1976a,b, 1978; Mutvei, 1974, 1977; and Dalingwater, 1975a,b; 1977.

At present, it seems the evidence favors the helicoidal model, that is, if any model can be applied universally to the arthropod cuticle. The objections raised to the helicoidal model, mainly from observations of apparently curved fibers oriented other than parallel to the lamellae, have been explained in terms of artifacts produced during sample preparation or misinterpretation of the observations (see Bouligand, 1978; Livolant, *et al.*, 1978).

Bouligand (1965) first applied the helicoidal model to the crab cuticle, where the individual fibers, generally known as macrofibrils, comprising the helicoidally stacked, horizontal layers are large (about 50 nm in diameter according to Neville, 1975). When extended to the cuticle of insects, where there are no obvious macrofibrils (but see following discussion), the corresponding layers were shown to be composed of microfibrils that stack with a spacing of approximately 6.5 nm (Neville and Luke, 1969b). They found, in the exocuticle of the giant water bug, that each lamella (180° twist of the helicoid) consisted of twenty-two to twenty-five layers of microfibrils, representing between 7° and 8° average rotation of microfibrillar direction per layer.

Neville and Luke extended Bouligand's hypothesis to take into account the thick layers seen in endocuticles, in which all the microfibrils are oriented in a preferred direction (Figure 2). These layers are clearly visible in the phase contrast and polarizing microscopes and, together with thin layers of helicoidal

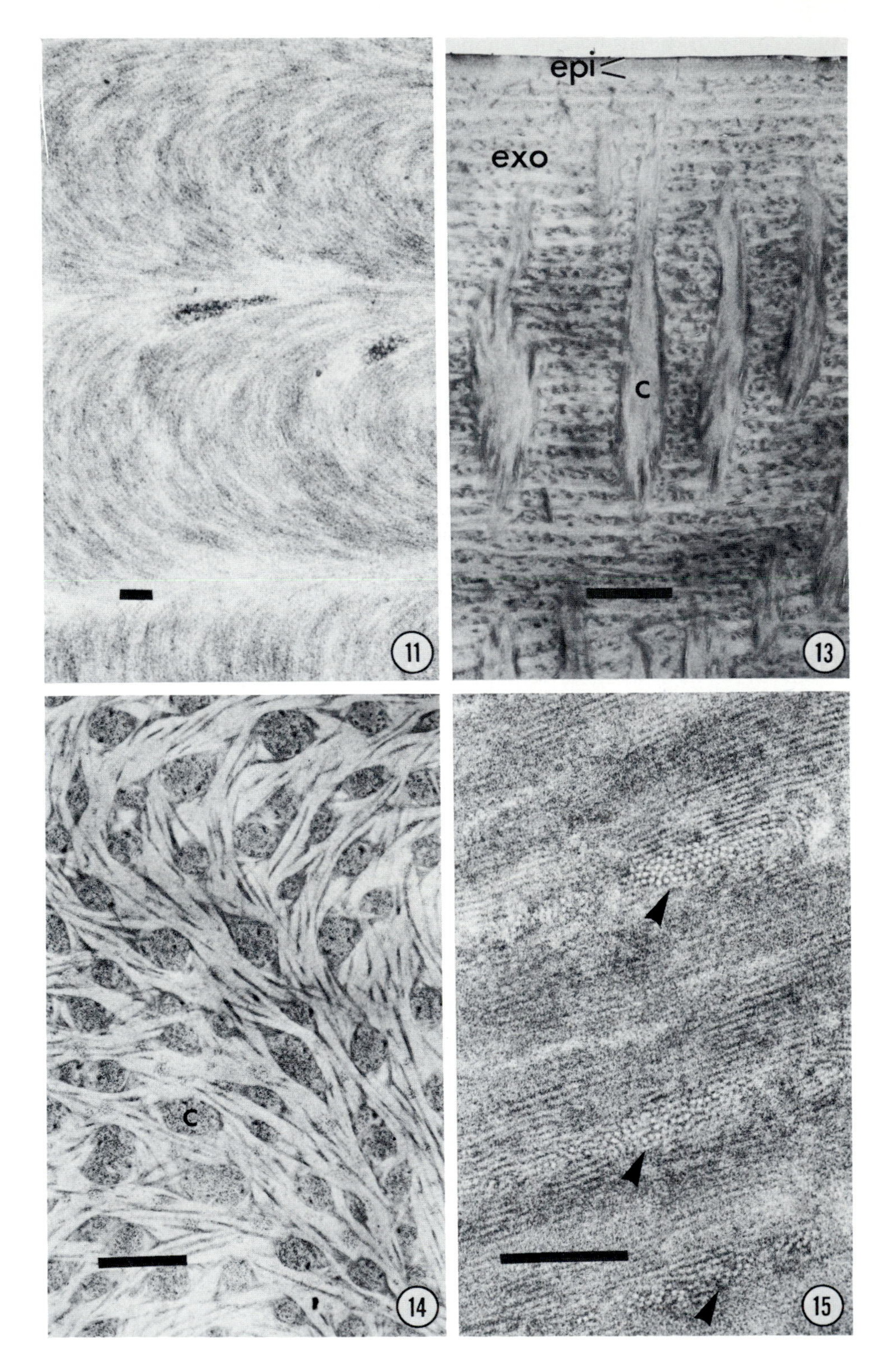
epi
exo
c
11
13
c
14
15

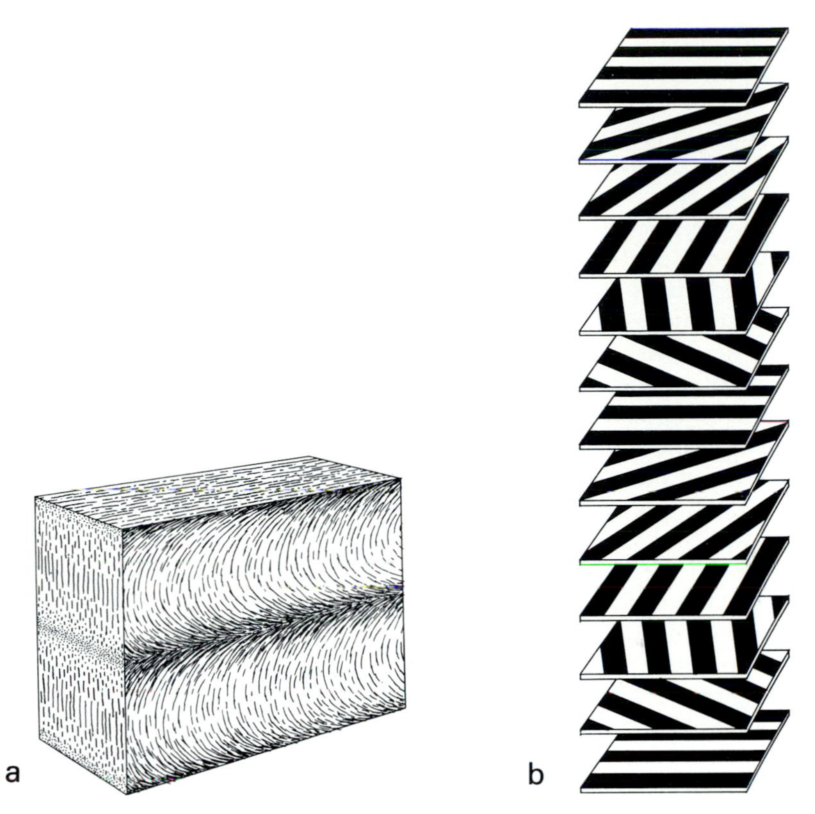

Figure 12. Diagrams of two interpretations of the arcuate patterns in cuticular lamellae. (a) Continuous curved filaments pass from one lamella to the next (redrawn after Locke, 1961). (b) Filaments are packed parallel to one another in sheets, the latter being stacked helicoidally. In angled sections, projections of small lengths of straight microfibrils lead to arcuate patterns (after Bouligand, 1965).

cuticle, form the daily growth layers of insect cuticle described by Neville (1963, and see discussion and later references in Neville, 1975). Neville and Luke (1969b) showed that, interposed between the thick layers of microfibrils with preferred orientation were lamellae or part-lamellae of helicoidal cuticle, the amount of twist of the latter determining the difference in direction of microfibril orientation between the unidirectional layers. They further predicted that their "two-system" model (helicoidal and/or unidirectional layers) could

←

Figure 11. Arcuate patterns in lamellae of the *Lucilia cuprina* larval procuticle, sectioned at 45° to vertical. (Bar = 0.1 μm.)

Figure 13. Nearly transverse section of the outer part of the sternite cuticle of *Tenebrio molitor*. The exocuticle (exo) contains horizontal lamellae and vertical columns (c) of oriented microfibrils. epi, epicuticle. (Bar = 1 μm.)

Figure 14. Nearly horizontal section of the sternite exocuticle of *Tenebrio* showing cross sections of vertical columns of microfibrils (c) and apparent macrofibrils curving between the columns. (Bar = 1 μm.)

Figure 15. High magnification of a transverse section of *Tenebrio* sternite exocuticle showing three groups of hexagonally packed microfibrils cut in cross section (arrows). (Bar = 100 nm.)

be used to describe the packing arrangements in all insect cuticles. The two-system model appears to hold for all arthropod cuticles examined to date except the orthogonal "plywood" cuticles such as that found in the walking leg endocuticle and articular membrane of a spider (Barth, 1973). Here the change in orientation between the orthogonal layers is abrupt, without any interposed helicoidal layers.

It has been pointed out (Neville and Luke, 1969b, Bouligand, 1972) that the proposed models will rarely, if ever, be manifested, in perfect fashion in fully formed arthropod cuticles, because of stresses and deformations placed upon (1) exocuticles during their postecdysial expansion prior to sclerotization and (2) endocuticles and procuticles during intermolt growth. Observed deviations from the models have usually been interpreted in this way. Nevertheless, there are some observations for which this explanation remains unconvincing.

One example is the abdominal sternite exocuticle of *Tenebrio molitor*. The superficial appearance of this cuticle is that it is lamellate and fits the helicoidal model. In addition to the horizontal lamellae are numerous columns of vertically oriented microfibrils that can be seen in both vertical and horizontal sections of the cuticle (Figures 13, 14). At high magnification the microfibrils of the horizontal lamellae can be clearly seen in accurate transverse sections (Figure 15). The microfibrils nearly always appear in flattened bundles of up to a hundred or more in hexagonal or near-hexagonal array. The bundles may contain seven or eight layers of microfibrils clearly resolved (see also Caveney, 1970, Figure 3c). In horizontal sections, the bundles are seen in longitudinal profile, gently curving between the vertical columns of cuticle (Figure 16). It can be calculated for a perfect heliocoid with a half period of the dimensions of the lamellar spacing in *Tenebrio* exocuticle (ca. 0.2 μm) and a thickness of 6.5 nm for each layer of the helicoid (the center-to-center spacing of microfibrils), that the rotation of microfibrillar direction between successive layers is approximately 6°. In sections about 50 nm thick of such a helicoidal structure, microfibrils in layers tilted more than 7.5° from perpendicular would not be seen as individual profiles but as electron-lucent bands. This is brought about because the center of any microfibril within the layer at the top of the section overlaps the center of its neighbor at the bottom (taking the intermicrofibrillar spacing as 6.5 nm) (Figure 17). Therefore, if *Tenebrio* exocuticle were a perfect helicoid, only in one or two adjacent layers of microfibrils per lamella would microfibrils be seen as discrete cross-sectioned profiles. In reality, as seen from Figure 15, up to seven or eight layers of discrete microfibrils are visible. Further relevant information may be gained from sections cut approximately 45° to the cuticular surface (Figure 18). Here, instead of the smooth parabolic arcs predicted from the helicoidal model, one sees only three or four discrete orientations of microfibrils between lamellae. A simplified sketch is provided in Figure 19. A similar structure can be seen in the micrographs of the femur exocuticle of *Hydrocyrius columbiae* published by Neville and Luke (1969b, Figures 19, 20). These authors interpret the deviation from parabolic patterns as being reorientation caused by adult expansion forces.

An alternative explanation, which seems to fit the available data, is that the unit fiber of the helicoid is not a single microfibril but a bundle of microfibrils (macrofibril) and further, that the macrofibrils are formed *de novo*, rather than by reorientation. Formation of the observed macrofibrils during adult expansion, from a strictly helicoidal configuration of microfibrils, would require not a mere reorientation of the structure but more a recrystallization. This problem will be resolved only by a careful examination of the pharate adult cuticle, before expansion and hardening. The accompanying micrographs (Figures 13–16, 18) are partly tanned exocuticles, from newly emerged adults in which cuticle expansion has been completed.

Another arthropod cuticle that appears to contain macrofibrils is that of the alloscutum of the adult female cattle tick (*Boophilus microplus*) (Figures 20–22). As mentioned earlier, this cuticle contains less than 4% chitin, so the microfibrillar component, which occupies 30%–50% of the volume of the cuticle, probably contains far less chitin than its counterpart in insects. This cuticle is also capable of stretching (approximately fifteen times in projected surface area), so its structure is very likely adapted in some (as yet unknown) way for this purpose. The macrofibrils can be shown from sections (Figure 20, insert) and from freeze fracture replicas (Figure 22) to be composed of microfibrils aligned axially.

An apparent variant of the helicoidal model has been reported for the cuticles of two copepods, *Cletocamptus retrogressus* (Gharagozolou-van Ginnekan and Bouligand, 1973) and *Triops cancriformis* (Rieder, 1972a,b), and a damsel fly gill (Neville, 1975). Recently, for the gill of a crayfish, *Panulirus argus*, Filshie and Smith (1980) showed, by combined studies of sectioning and freeze fracturing (Figures 23–25), that rows of apparently vertically oriented channels seen in transverse sections of these cuticles are not real structures, as interpreted by earlier authors, but banding along microfibrils resulting from the regular disposition of particles attached to microfibrils. A model of our interpretation is shown in Figure 26.

2.3. Pore Canals

The cuticles of most insects and other arthropods contain transverse canal networks that serve to retain direct connections between the epidermis and various levels within the cuticle and to the surface. The pore canals, as they are known in the chitin-containing part of the cuticle, are connected to epicuticular channels in the epicuticle. Where both are present, they always have a varied structure and are always connected to one another. The structure of the epicuticular channels and their connections with pore canals are discussed in section 3.4.

The structure, shape and possible functions of pore canals have been reviewed by Neville (1975). Pore canals are, initially, direct cytoplasmic extensions of the epidermal cells, but often these extensions are withdrawn, leaving an apparently clear channel containing one or more electron-dense pore canal filaments. In some instances, microfibrils and matrix material fill or partially fill the channel, so that these microfibrils become oriented perpendicular to the rest of those in the cuticle. An extreme example of the latter is seen in the sternite exocuticle of *Tenebrio* (Figure 13). Relatively large, straight cylinders of ori-

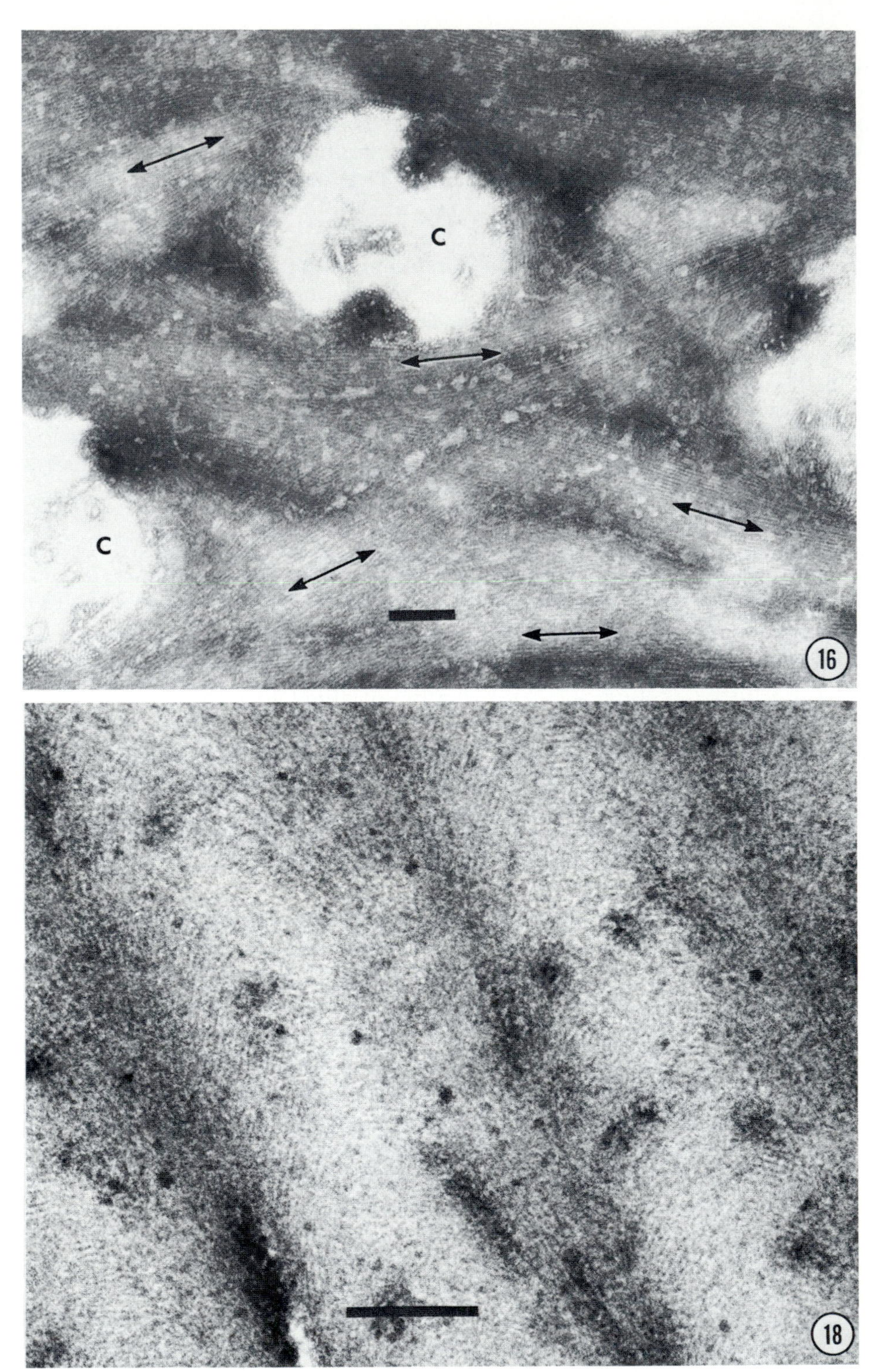
C
C
16
18

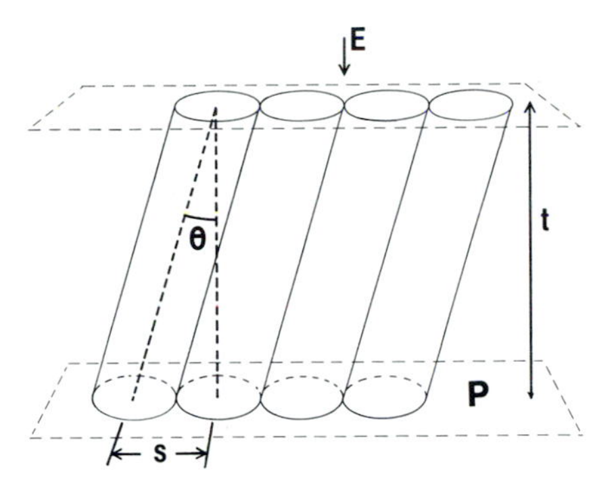

Figure 17. Diagram to demonstrate the approximate angle of tilt of planes of microfibrils in sections, such that they fail to be visualized as single microfibrils, but as electron-lucent bands. This angle (Θ) is given by; tan $\Theta \simeq \frac{s}{t}$, where t = section thickness and s = microfibrillar spacing. The electron optical axis, E, is normal to the section plane, P.

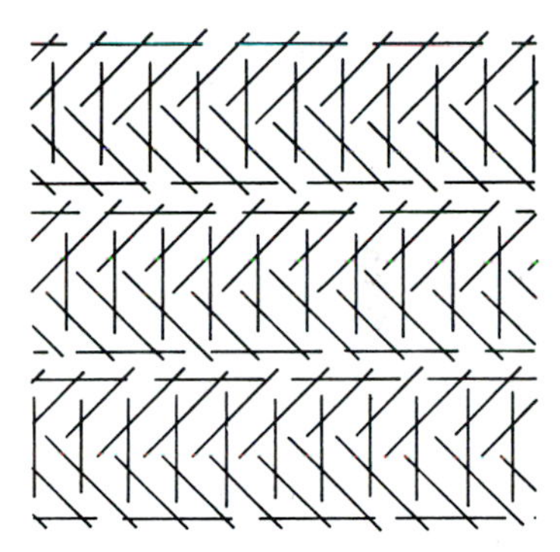

Figure 19. Simplified interpretation of the microfibrillar pattern seen in Figure 18. For explanation, see text.

ented fibrous cuticle replace the cytoplasmic extensions that are present during formation of the pharate adult cuticle, but that progressively retract as the cuticle is laid down (Delachambre, 1971).

In most cases, the shape of the pore canals is determined by the fibrous architecture of the cuticle. In fact, the shape of the pore canals was used to predict the microfibrillar packing within some arthropod cuticles (Neville and Luke, 1969a; Neville *et al.*, 1969). Subsequently, Neville and Luke (1969b) developed a staining method utilizing potassium permanganate and lead citrate, which clearly revealed microfibrils in sections of the same cuticles and confirmed the earlier predictions. The above authors showed that, in cross section, pore canals are generally flattened in the same direction as the orientation of microfibrils within horizontal planes so that, in a helicoidal cuticle, the pore canals take the form of twisted ribbons. Through layers of the endocuticle with preferred orientation of microfibrils, pore canals form straight ribbons that twist

←

Figure 16. High magnification of a section similar to Figure 14, stained with potassium permanganate and lead citrate to reveal bundles of microfibrils (double ended arrows) curving between the vertical columns (c). (Bar = 100 nm.)

Figure 18. Section cut at 45° to the surface of *Tenebrio* sternite exocuticle showing parts of three lamellae and the microfibrillar pattern within them. A simplified interpretation of this pattern is depicted in Figure 19. (Bar = 100 nm.)

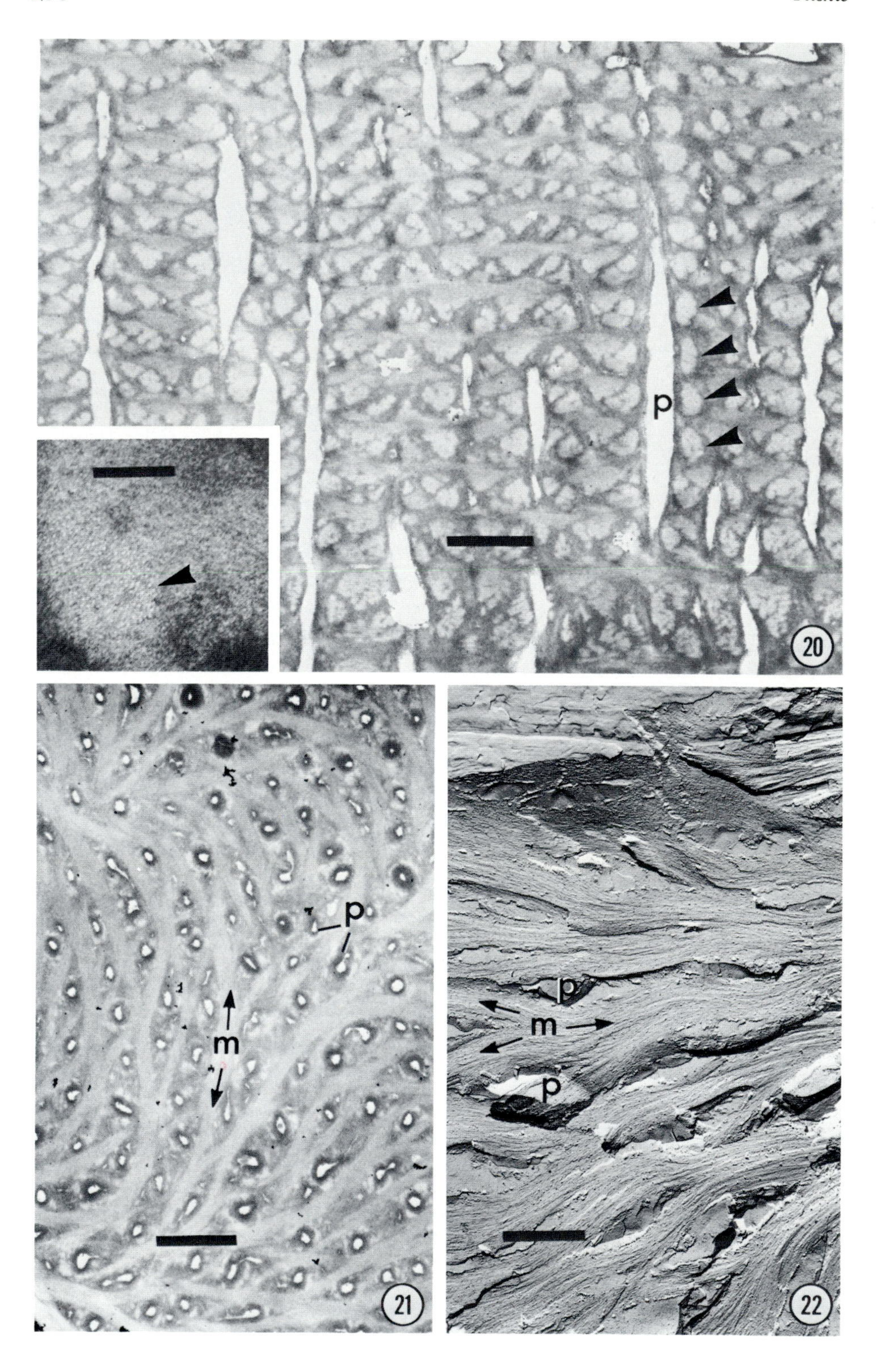
p
20
p
m
21
p
m
p
22

only where they pass through helicoidal lamellae situated between the preferred layers. In some instances, pore canals were found to be helical (Kennaugh, 1965; Neville *et al.*, 1969). As an explanation for this phenomenon, Neville (1975) suggested a secondary coiling of the pore canals resulting from shrinkage of the cuticle in a direction normal to the surface.

Very little additional information on the structure of pore canals has been published since Neville's (1975) review. He lists several cuticles as lacking pore canals, namely rubberlike cuticle in locust wing hinges, the pleural arch of fleas, free-living copepods, the wax-secreting cuticle of *Calpodes ethlius,* and gut cuticle and corneal cuticle of *Lampyris* (see Neville's Chapter 2 for references). Pore canals are also absent from the larval cuticle of *Lucilia cuprina* (Filshie, 1970b) and probably from other dipteran larval cuticles. In tracheal cuticle, pore canals have never been identified with certainty (see Chapter 12, section 2.3.6). Kayser-Wegmann (1976) has examined differences between the larval and the pupal cuticles of *Pieris brassicae*. The pore canals of the larva contain only a single filament, but in the pupa each canal contains a bundle of 10–20 fibers.

In the cuticle of the scorpion *Hadrurus arizonensis* (Filshie and Hadley, 1979), the pore canals are more complex than those described for other arthropods. The pore canal in the endocuticle of this species has fibrous cuticle associated with it (Figure 27). Between the endocuticle and the inner exocuticle (Figure 28) is a membrane that delimits these two layers and evaginates to form the lining of the pore canal in the inner exocuticle. A constriction in the pore canal occurs at the junction between the inner and the outer exocuticle (Figure 29), and a much narrower pore canal continues through the outer exocuticle to the epicuticle. In the cuticles of the spiders *Cupiennius salei* (Barth, 1970) and *Latrodectus hesperus* (Hadley, personal communication), the pore canals possess the same general structure as those of insects. In the cuticle of the mite *Caloglyphus mycophagus* (Kuo and McCully, 1978), the pore canals of the adult are unbranched and apparently are not connected to epicuticular channels. Those of the hypopus are branched and connected to epicuticular filaments.

2.4. Lipid Distribution in the Procuticle

Wigglesworth (1975a) developed a method that purported to demonstrate saturated lipids in sections prepared for the electron microscope. Saturated

Figure 20. Transverse section of the cuticle of the alloscutum of the adult female cattle tick, 72 hr after adult emergence and before the final rapid engorgement. Horizontal lamellae are apparently composed of macrofibrils (arrows) of varied thicknesses, oriented horizontally. Pore canals (p) are oriented vertically. (Bar = 1 μm.) The insert is a cross section of a single macrofibril containing cross sections of microfibrils (arrows). (Bar = 100 nm.)

Figure 21. Horizontal section of a tick cuticle similar to that of Figure 20, showing pore canals (p) and macrofibrils (m). (Bar = 1 μm.)

Figure 22. Horizontal freeze fracture replica of a cuticle similar to that of Figure 21. Macrofibrils (m) can be seen to be composed of microfibrils. p, pore canals. (Bar = 1 μm.)

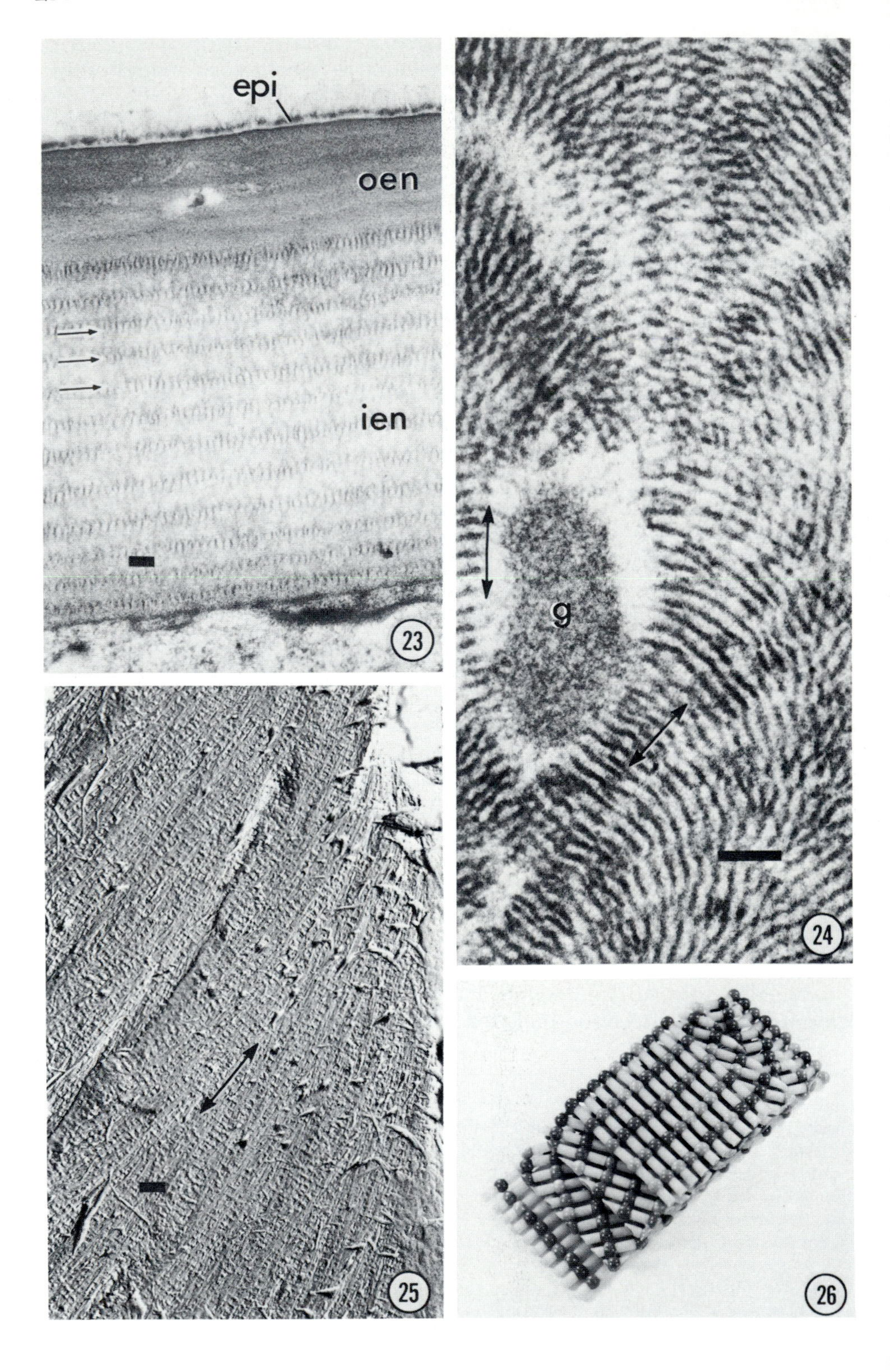
epi
oen
ien
g
23
24
25
26

lipids are extracted by the normal preparative methods, which consist of fixation in aldehydes and osmium tetroxide followed by dehydration in organic solvents and embedding in epoxy resins. Unsaturated lipids, on the other hand, bind to osmium and are revealed as electron-dense deposits. Wigglesworth incorporated unsaturated lipid (monoterpene hydrocarbons) into the existing lipid in the tissue, increasing the amount of osmium bound in the subsequent fixation. He applied the method to a study of the distribution of lipid in the epicuticle and the lamellate endocuticle of *Rhodnius* (Wigglesworth, 1975b,c). In the treated endocuticle, it was claimed that observed electron-dense banding, superimposed on the arcuate fibrous pattern, represented a lamellar distribution of lipid. A similar effect (but without the addition of hydrocarbons) had been noted previously by Bouligand (1972), who claimed that the bands resulted from periodic variations in section thickness caused by microtomy artifacts. He claimed that where microfibrils are pointing out of the block face being cut, they are turned up by the edge of the knife, but that where microfibrils are lying in the plane of the block face they are merely separated from one another, leading to corrugation of the surface of the section. If the corrugations on the upper and lower surfaces are out of phase, the effect observed in the microscope is periodic light and dark banding, which may vary in position with respect to the arcuate banding. Subsequently, Gordon and Winfree (1978) proposed that the banding may arise from lateral displacement of either microfibrils or matrix through the section by interaction of the knife with the material.

It can be demonstrated by cutting sections of soft cuticles, in which the lamellae undulate with respect to the cutting direction, that Wigglesworth's interpretation was incorrect. In the accompanying micrograph (Figure 30) of blowfly (*Lucilia cuprina*) larval cuticle, electron-dense bands are discontinuous with respect to the arcuate lamellae and appear to be related to the cutting direction, which is indicated by the knife mark. This evidence supports the proposals of Bouligand (1972) and Gordon and Winfree (1978) that the dense bands are microtomy artifacts.

Previously, Wigglesworth (1970; 1971) described methods for detecting lipid "laminae" at the light microscope level. At present, these "laminae" cannot be compared with the lamellae observed in electron microscope sections, because the respective bandings appear to arise in two quite distinct ways.

Figure 23. Transverse section of lobster (*Panulirus argus*) gill cuticle. The inner endocuticle (ien) is subdivided into lamellae by layers of alternating electron-transparent and electron-dense bands oriented perpendicular to the surface (arrows). Microfibrils are not visible in sections (epi, epicuticle; oen, outer endocuticle). (Bar = 0.1 μm.)

Figure 24. Horizontal section of lobster gill cuticle in the region of a granular inclusion (g). Microfibrils are not visible, but the dense bands indicate their organization into ribbons. The bands are perpendicular to the fiber direction, the latter being indicated by double ended arrows. (Bar = 0.1 μm.)

Figure 25. Near-horizontal freeze fracture replica of the inner endocuticle of the lobster gill cuticle. Microfibrils are clearly visible aligned in the direction of the arrow, with regularly arranged rows of particles oriented perpendicular to the fiber direction. The particles are interpreted as being equivalent to the dense bands seen in sections (Figures 23, 24). (Bar = 0.1 μm.)

Figure 26. Model of an interpretation of the structure of the inner endocuticle of the lobster gill (after Filshie and Smith, 1980).

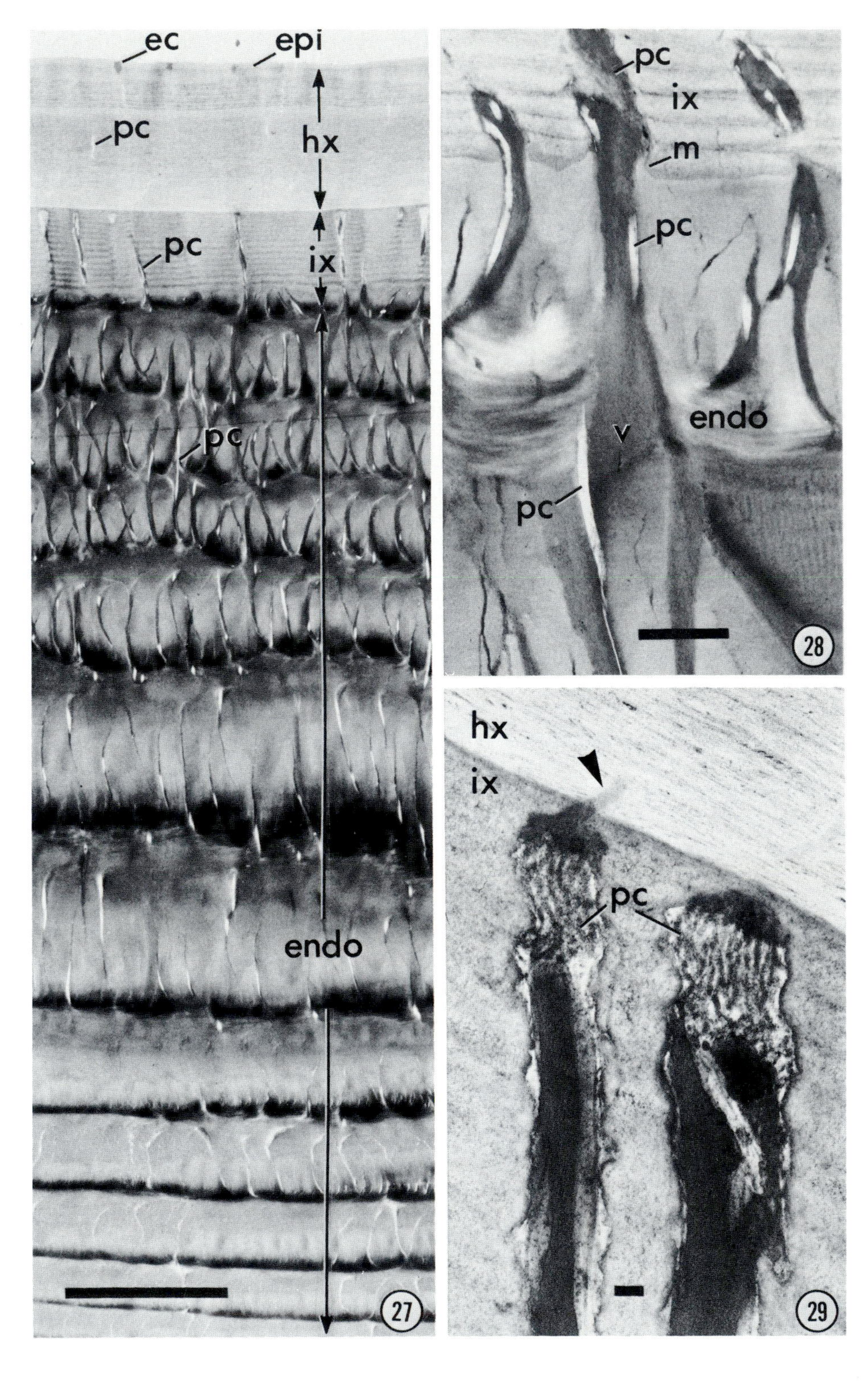
ec
epi
pc
hx
pc
ix
pc
endo
27
pc
ix
m
pc
endo
v
pc
28
hx
ix
pc
29

3. The Nonchitinous Cuticle

Although it is generally stated that the epicuticle of arthropods is nonchitinous, this is simply because it reacts negatively to tests for chitin. These tests may not be sensitive enough to detect very small amounts of chitin because the epicuticle (Richards, 1951), except for the wax layer, is always extremely thin (usually less than 1 μm in thickness).

Five distinct layers are generally recognized from combined studies by light and electron microscopy on fully formed cuticles. These layers are, in order of their deposition, the cuticulin layer, the inner epicuticle, the outer epicuticle, the wax layer, and the cement layer. The latter two layers are sometimes classified separately from the others and are known as *superficial layers* (see Chapter 12, section 2.3.3). Also present in the epicuticle is a transverse canal or filament system that connects the pore canals (where present) with the cuticulin layer (Figures 32, 33).

The epicuticle is apparently inextensible. Where an increase in the volume of the body is required (to accommodate postecdysial expansion, growth, or food intake), the epicuticle is formed initially in excess. For instance, the epicuticle of the unfed adult female cattle tick is deeply folded (Figure 31). During engorgement, the epicuticle unfolds, while the underlying procuticle stretches.

3.1. The Cuticulin Layer

The term "cuticulin" was first used by Wigglesworth (1933) to describe the chemical substance of the inner part of the epicuticle (i.e., all except wax and cement). The cuticulin *layer* was designated by Locke (1966) as the thin (approximately 10 nm, depending on stage, location, and species), electron-dense membrane that is the first layer of the cuticle to be secreted. The cuticulin layer is sometimes called the "outer epicuticle" (see Chapter 12, section 2.3.4), but I prefer to reserve the latter term for the superior layer (see section 3.3), which appears at a later stage in the formation of the epicuticle and may not be an integral part of the first-formed layer.

Locke's (1966) description of the structure of the cuticulin layer is still the most detailed for any single insect species. The cuticulin layer is probably universally present in arthropods. Its apparent absence over certain sense organs (Slifer, 1961) and the gut (Bertram and Bird, 1961) may have been failures to

←

Figure 27. Transverse section of the cuticle of the dorsal plate of the desert scorpion. The cuticle is subdivided into an epicuticle (epi), hyaline exocuticle (hx), inner exocuticle (ix), and endocuticle (endo). Epicuticular canals (ec) traverse the epicuticle and pore canals (pc) transverse the layers beneath. The canals have a varied structure in each layer. (Bar = 10 μm.)

Figure 28. Transverse section of scorpion dorsal plate cuticle showing the junction between endocuticle (endo) and inner exocuticle (ix). The two layers are separated by a membrane (m) which everts to form the lining of the pore canal (pc) in the inner exocuticle. Pore canals in the endocuticle lie within vertical columns (v) of microfibrils (not visible at this magnification). (Bar = 1 μm.)

Figure 29. Border of inner exocuticle (ix) and hyaline exocuticle (hx) of the scorpion dorsal plate, showing the complex structure of the pore canals (pc) at this interface. A thinner pore canal (arrow) continues distally into the hyaline exocuticle, passing out of the section plane. (Bar = 0.1 μm.)

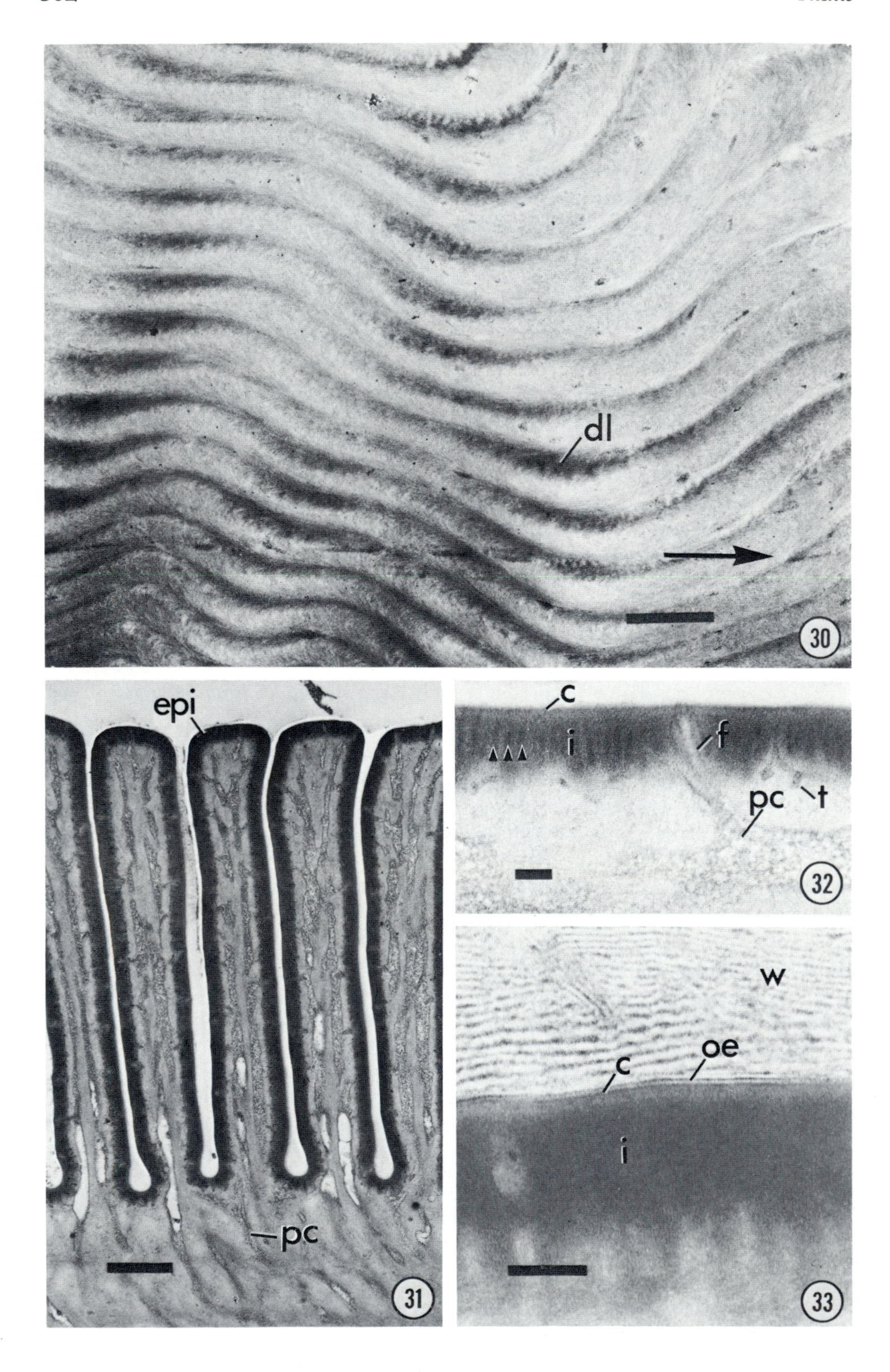
dl
30
epi
pc
31
c
i
f
pc
t
32
w
c
oe
i
33

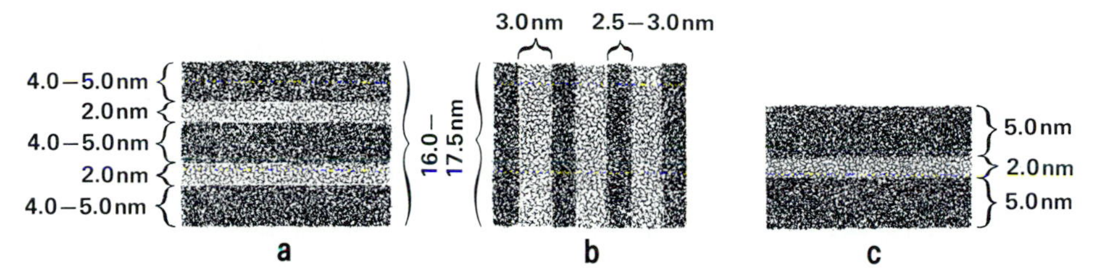

Figure 34. Diagram showing the changes in the structure of the cuticulin layer during its development (after Locke, 1966). (a) When first secreted. (b) Just before ecdysis, showing the porous phase. (c) Appearance between molts.

demonstrate it because of inadequacies in the embedding or staining procedures. For instance, in the third instar larva of *Lucilia cuprina,* 3 or 4 hr before the cuticle hardens and darkens to form the puparium, the cuticulin layer is visible and densely stained (Figure 37). In sections of the cuticle after pupariation the cuticulin layer cannot be visualized (Filshie, 1970b).

Apart from insects, the cuticulin layer has been demonstrated in a spider (Barth, 1970), in a scorpion (Filshie and Hadley, 1979), in ticks (Beadle, 1974; Filshie, 1976), and in mites (Wharton *et al.*, 1968; Brody, 1970; Kuo and McCully, 1978). In crustacea, layers apparently homologous to the cuticulin layer also exist in the epicuticle (Gharagozolou-van Ginneken and Bouligand, 1973; 1975; Filshie and Smith, 1980).

In *Calpodes,* cuticulin is secreted initially as a five-layered membrane (Figure 34a), which transforms to a layer of uniform density but with minute pores (Figure 34b) (Locke, 1966). These pores were thought to function in the selective uptake of digestion products from the exuvial space during breakdown of the old cuticle. Locke (1966; 1976) also demonstrated similar pores that were present at all times in the cuticulin layer of tracheolar linings. The appearance of the cuticulin layer between molts in *Calpodes* is shown in Figure 34c. It has been my experience that in most fully formed cuticles the trilaminar structure shown in this diagram is very often reduced in complexity to a single, moderately

←

Figure 30. Near-transverse section of the third instar larval cuticle of *Lucilia cuprina,* stained with barium permanganate and lead citrate. The discontinuous dense layers (dl) result from variations in section thickness caused by interaction between the fibrous structure of the procuticle and the microtome knife. A knife mark passes horizontally through the section about 2 cm from the bottom of the figure. The arrow indicates the direction of knife travel. (Bar = 1 μm.)

Figure 31. Transverse section of the outer part of the procuticle and epicuticle (epi) of the alloscutum of the pharate adult female cattle tick. The deep folding of the epicuticle is to allow for expansion during feeding. pc, pore canal. (Bar = 1 μm.)

Figure 32. Detail of the epicuticle of the alloscutum of the pharate adult female cattle tick. The wax layer has not yet been secreted. c, cuticulin layer; f, epicuticular filaments; i, inner epicuticle; pc, pore canal; t, tubules. The arrowheads indicate dense filaments traversing the inner epicuticle. The section is stained with uranyl acetate and lead citrate. (Bar = 0.1 μm.)

Figure 33. Epicuticle of the alloscutum of the adult female cattle tick after the wax layer (w) has been secreted. c, cuticulin layer; i, inner epicuticle; oe, outer epicuticle. The section is stained with barium permanganate and lead citrate. (Bar = 0.1 μm.)

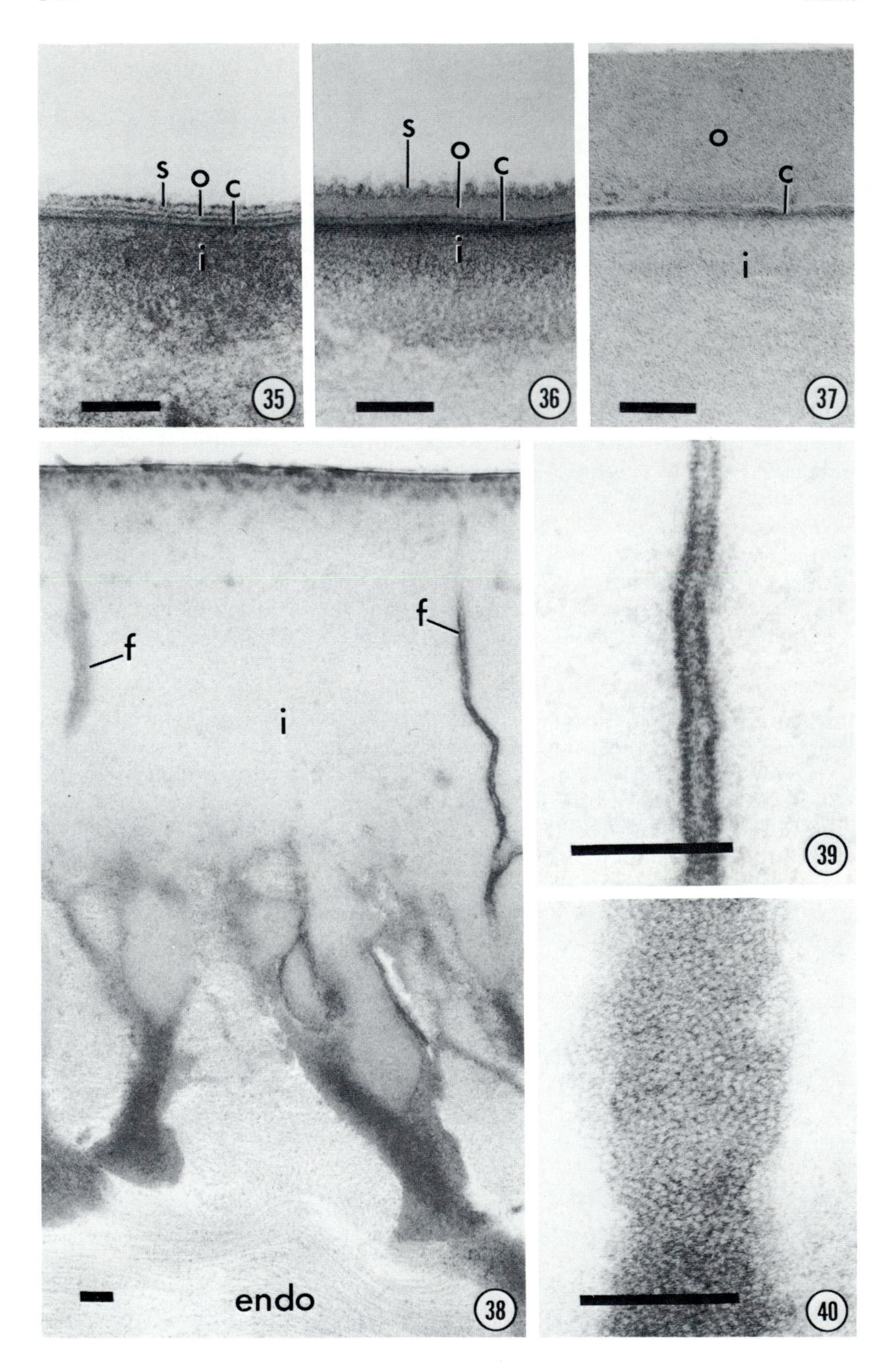
s
o
c
i
35
36
37
f
endo
38
39
40

electron-dense layer about 12 nm thick, even when very thin sections are examined. Zacharuk (1972) found that in wireworms (larvae of elaterid beetles) the trilaminar cuticulin membrane is preserved by glutaraldehyde fixation alone. Postecdysially secreted electron-dense material, which obliterates the trilaminar structure, is stabilized by osmium fixation but not by glutaraldehyde alone. He suggested, therefore, that the original cuticulin is a lipoprotein, which is impregnated with lipid after ecdysis.

3.2. The Inner Epicuticle

This layer is generally much thicker than the cuticulin layer and is situated between the cuticulin layer and the exocuticle (or procuticle in the case of untanned cuticle). It has alternatively been called the "dense layer" (Locke, 1961) or the "protein epicuticle" (Locke, 1969), the former because it sometimes appears electron dense and the latter because it was thought to be largely proteinaceous.

As the range of observations widens to different insect species and arthropod groups, it is becoming clear that this layer is quite varied in its appearance in the electron microscope in different species and in different areas of cuticle of a single species, which in turn probably indicates a complex and varied chemistry. It is thought, by different authors, to contain protein, lipid, lipoprotein, dihydroxphenols, and phenol oxidase (see Neville, 1975, Chapter 2 for references). Usually, the inner epicuticle is seen to have a uniformly fine granular substructure (Figures 35–38), but has been reported in some instances to be finely laminated, as in the spermathecal duct of *Periplaneta americana* (Gupta and Smith, 1969) and the larva of the cockroach *Blaberus trapezoideus* (Brück and Stockem, 1972a). Zacharuk (1972) recorded two kinds of lamination in the inner epicuticles of elaterid larvae. The lamination in *Limonius californicus* is caused by rows of tiny vesicular droplets, presumed to be lipoidal, while those in *Ctenicera destructor* were thought to be extra layers of cuticulin.

The appearance of the inner epicuticle of the alloscutum of the adult female cattle tick differs according to the developmental stage and the staining regime used on sections (Filshie, 1976). Vertically oriented filaments (as distinct from epicuticular filaments that connect to pore canals in the procuticle) were revealed with uranyl acetate and lead citrate staining (Figure 32). These have a density of about 800 per μm^2. When preecdysial cuticle was stained with potassium permanganate and lead citrate, the inner epicuticle was seen to be com-

←

Figures 35–37. Sections of the epicuticles of the first, second, and third instar larval cuticles of *Lucilia cuprina*, showing the progressive increase in thickness of the outer epicuticle (o). c, cuticulin layer; i, inner epicuticle; s, superficial layer. (Bars = 0.1 μm.)

Figures 38–40. Transverse sections of the epicuticle of the intersegmental cuticle of the desert scorpion. Epicuticular filaments (f) of this cuticle are flattened tubes. The one at left is sectioned across its greater diameter, the detail of a similar view being shown in Figure 40. The filament at right is sectioned across its smaller diameter, and a similar one is shown in detail in Figure 39. Small electron-lucent subunits are seen in the dense substance of the wall of the tube (Figure 40).
Figure 38: (Bar = 0.1 μm.)
Figures 39, 40: (Bars = 0.1 μm.)

posed of a closely packed array of vertically oriented, electron-lucent microfibers about 3.0 nm in diameter, embedded in an electron-dense matrix (Hackman and Filshie, 1982).

3.3. The Outer Epicuticle

In *Galleria* and *Calpodes*, Locke (1961) described an electron-lucent layer approximately 10 nm thick that seemed to be formed by delamination from the cuticulin layer just before ecdysis. Although this was originally thought to be the "lipid monolayer" postulated by Beament (1961), it was later shown in *Calpodes* to be resistant to extraction by lipid solvents (Filshie, 1970a) and is now generally known as the outer epicuticle. The layer can usually be demonstrated in intermolt cuticles, provided thin and accurate transverse sections are examined. Because it is commonly the outermost layer seen in electron microscope sections (wax and cement are usually dissolved during dehydration and embedding), its demonstration may depend on the presence of some superficial electron-dense (extraneous) deposit to reveal it in negative contrast (Figures 33, 35). Most authors have measured the thickness of the outer epicuticle to be about 10 nm, but in *L. cuprina* (Filshie, 1970b), the homologous layer differs in thickness in each of the three larval instars (Figures 35–37), being approximately 7.5 nm, 25–30 nm, and 250–300 nm in thickness in the first, second, and third instars respectively. The relatively thick outer epicuticle of the third instar cuticle and of the puparium (formed from the latter), survives as a coherent membrane in acid hydrolysates, and, in the puparium, is apparently composed of the same material as the epicuticular filaments (Filshie, 1970b).

Noirot *et al.* (1978) demonstrated, in freeze-fracture replicas, a preferred cleavage plane within the outer epicuticle of rectal pads of termites and cockroaches. This preferred cleavage plane is similar to that seen in the lipid bilayer membranes of animal cells, and the authors suggested that the outer epicuticle is lipoidal and is involved in the water impermeability properties of the cuticle. They also found that the structure and cleavage properties of the outer epicuticle were modified at the bases of epicuticular depressions. This observation was indicative of modified permeability properties over these areas of the surface. Flower and Walker (1979), by use of freeze fracturing, described an almost identical fracture plane in the rectal papillae of *Musca domestica*. In tracheae (Chapter 12, section 2.3.4 and Figures 15–17), the fracture plane follows the contours of the external face of the cuticulin layer.

3.4. The Wax Layer and Epicuticular Filaments

Although the wax layer is probably the most important layer of the cuticle with respect to the protection of terrestrial arthropods from water loss and desiccation, it is one of the components about which we know least from electron microscopy. As already mentioned, the techniques used for fixation and embedding for transmission microscopy are incapable of stabilizing saturated lipids, except possibly for the method described by Wigglesworth (1975a) (see section 2.4), so generally these components are removed during the dehydration and resin infiltration steps of the preparative procedures. Wigglesworth's method has so far been applied only to the epicuticle of *Rhodnius* (Wigglesworth, 1975b), where a discrete wax layer external to the outer epicuticle was not demonstrated in electron micrographs, although the myrcene partition-osmium staining procedure was successful in revealing lipid in the epicuticular channels (see Wigglesworth, 1975b, Figures 17–21). In ticks (Nathanson, 1967; Filshie, 1976; Hackman and Filshie, 1982), a well preserved "wax layer" has been detected in postecdysial cuticles. It is irregular in thickness, varying from 0.1 μm to 1 μm in *B. microplus* nymphs and adults (Figure 33), is generally poorly stained, and sometimes contains laminated electron-dense plates arranged in small crystallites that are not oriented in any preferred direction. The fact that the layer is preserved in electron microscope sections suggests that additional components other than saturated lipids comprise a significant proportion of the layer. Most chemical studies of tick cuticular lipids have been made of whole cuticles extracted with organic solvents (see review of Hackman and Filshie, 1982), so the loci of the lipids (superficial layers or permeating the rest of the cuticle) are unknown. Also, it is not known whether nonlipoidal substances are present in the "wax" layer.

Locke (1966) described a method purporting to demonstrate the fatty acid component of the cuticular wax of *Calpodes*. The cuticle was incubated in lead nitrate after fixation in glutaraldehyde and before osmium fixation. Dense deposits were seen confluent with the outer epicuticle and also within the spaces surrounding the epicuticular channels of the inner epicuticle. The above method has not been applied to any other cuticle.

In the scorpion, *H. arizonesis,* attempts to locate the lipids using scanning and transmission electron microscopy on cuticles before and after chemical treatments were only partially successful (Hadley and Filshie, 1979). No surface wax layers were preserved in embedded and sectioned specimens, although treatment with lipid solvents failed to remove completely superficial deposits seen by scanning microscopy of bulk specimens. These latter deposits were only removed by treatment with alkali and may have been components of the cement layer.

Some cuticles are specialized for the production of large quantities of wax, sometimes in the form of long filaments. The earlier examples of these are reviewed by Locke (1974). The tenebrionid beetle *Cryptoglossa verrucosa* secretes copious quantities of filamentous wax from tubercles distributed over the surface of its cuticle (Hadley, 1979). Under conditions of low humidity the filaments persist intact on the surface and light scattering gives a blue color to the cuticle. In a more humid environment the filaments are somehow incorporated into the surface wax layer and the color of the insect changes to black. Pope (1979) has described the complex structure of wax threads and the cuticular tubercles from which they arise in twenty species of coccinellid larvae.

Wax is secreted from the epidermal cells through the cuticle to the surface. The wax precursors may pass directly through the chitinous part of the cuticle even when pore canals are absent (Locke, 1974), but epicuticular filaments have always been demonstrated in cuticles known to secrete surface wax. The most likely route through the epicuticle is via the epicuticular filaments (see discussion by Locke and Krishnan, 1971).

Epicuticular filaments have a varied structure and distribution from one species to another and from different areas of the cuticle of a single species (Locke, 1961, 1974; Filshie and Hadley, 1979). Where pore canals are present in the underlying chitinous part of the cuticle, they usually connect to epicuticular filaments. In most published micrographs, the filaments are simple cylindrical rods or tubules, but may be more complex, as in *H. arizonensis* intersegmental cuticle, where they are in the form of flattened tubes with a finer substructure within the walls of the tubes (Figures 38–40, and see Filshie and Hadley, 1979).

3.5. The Cement Layer

Where present, this is the outermost layer of the cuticle, being superficial to the wax layer. By definition (Wigglesworth, 1933) it is a layer secreted by the dermal glands at ecdysis. Secretions from the ducts pour out over the surface and solidify to form a partial or complete layer. The cement has not been studied adequately by electron microscopy (see Neville, 1975), and in most published micrographs, layers have been identified as cement because of their location, rather than their origin as dermal gland secretions. The chemistry and function of the cement layer is poorly understood, but it is generally thought to act as a protective layer for the underlying wax.

4. Concluding Remarks

As with the application of most new techniques to the solution of old problems, the employment of electron microscopy to the study of cuticle structure in recent years has raised as many questions as it has answered. Certainly, a new dimension has been uncovered by the superiority of resolution of the electron microscope over the light microscope, but frequently the interpretation of electron micrographs is equivocal, to some extent because of the limitations of the instrument, but mainly as a result of the inadequacies of the techniques of preservation of the material under scrutiny (Filshie, 1980). The arthropod cuticle, being at once the skin and the skeleton, is very resistant, both chemically and physically. Although these properties are undoubtedly beneficial to the animal (one of the major factors leading to the evolutionary success of the Insecta—see Waterhouse and Norris, 1980), they are of little comfort to either the chemist wishing to isolate and characterize the intractable constituents of the cuticle, or the electron microscopist attempting to cleave into the horny material with his fragile glass (or expensive diamond!) knife and then, finally, after all the drastic treatments to which he has subjected his specimen, naively expecting its molecular structure to be revealed on the fluorescent screen of his trusty instrument.

It is tempting, at the conclusion of a review such as this, to make some predictions for the future, but, to paraphrase the words of Horridge and Blest

(1980), such prophecies have been generally ineffectual and their likely outcome will be laughter and consignment to the waste paper basket before the ink is dry. Suffice it to say that, because of the complexities of modern scientific methods, further significant advances in our knowledge of cuticle structure are more likely to come about from the combined efforts of microscopists, X-ray crystallographers, cell biologists, chemists, and others, than by the isolated labors of any one specialist.

References

Altner, H., 1975, Microfiber texture in a specialized plastic cuticle area within a sensillum field on cockroach maxillary palp as revealed by freeze fracturing, *Cell Tissue Res.* **165**:79–88.

Barth, F. G., 1970, Die Feinstruktur der Spinnerinteguments. II. Die räumliche Anordnung der Mikrofasern in der lamellierten Cuticula und ihre Beziehung zur Gestalt der Porenkanäle (*Cupiennius salei* Keys., adult, hautungsfern, Tarsus), *Z. Zellforsch.* **104**:87–106.

Barth, F. G., 1973, Microfiber reinforcement of an arthropod cuticle: Laminated composite material in biology, *Z. Zellforsch.* **144**:409–433.

Beadle, D. J., 1974, Fine structure of the integument of the ticks, *Boophilus decoloratus* Koch and *B. microplus* (Canastrini) (Acarina: Ixodidae), *Int. J. Insect Morphol. Embryol.* **3**:1–12.

Beament, J. W. L., 1961, The water relations of insects, *Biol. Rev. Cambridge Phil. Soc.* **36**:281–320.

Bertram, D. S., and Bird, R. G., 1961, Studies on mosquito-borne viruses and their vectors 1., *Trans. R. Soc. Trop. Med. Hyg.* **55**:404–423.

Biedermann, W., 1903, Geformta Sekrete, *Z. Allg. Physiol.* **2**:395–481.

Blackwell, J., and Weih, M. A., 1980, Structure of chitin-protein complexes: Ovipositor of the ichneumon fly *Megarhyssa*, *J. Mol. Biol.* **137**:49–60.

Bouligand, Y., 1965, Sur une architecture torsadée répandue dans de nombreuses cuticules d'arthropodes, *C. R. Hebd. Sceances Acad. Sci. Ser. D.* **261**:3665–3668.

Bouligand, Y., 1972, Twisted fibrous arrangements in biological material and cholesteric mesophases, *Tissue Cell*, **4**:189–217.

Bouligand, Y., 1978, Liquid crystalline order in biological molecules. In *Liquid Crystalline Order in Polymers*, edited by A. Blumstein, pp. 261–297, Academic Press, New York.

Brody, A. R., 1970, Observations on the fine structure of the developing soil mite *Oppia coloradensis* (Acarina: cryptostigmata), *Acarologia* **12**:421–431.

Brück, E., and Stockem, W., 1972a, Morphologische Untersuchungen an der Cuticula von Insekten. I. Die Feinstruktur der larvalen Cuticula von *Blaberus trapezoideus* Burm., *Z. Zellforsch.* **132**:403–416.

Brück, E., and Stockem, W., 1972b, Morphologische Untersuchungen an der Cuticula von Insekten. II. Die Feinstruktur der larvalen Cuticula von *Periplaneta americana* (L.), *Z. Zellforsch.* **132**: 417–430.

Caveney, S., 1970, Juvenile hormone and wound modelling of *Tenebrio* cuticle architecture, *J. Insect Physiol.* **16**:1087–1107.

Credland, P. F., 1978, An ultrastructural study of the larval integument of the midge, *Chironomus riparius* Meigen (Diptera: Chironomidae), *Cell Tissue Res.* **186**:327–335.

Dalingwater, J. E., 1975a, SEM observations on the cuticles of some decapod crustaceans, *Zool. J. Linn. Soc.* **56**:327–330.

Dalingwater, J. E., 1975b, Reality of cuticular laminae, *Cell Tissue Res.* **163**:411–413.

Dalingwater, J. E., 1977, Cuticular ultrastructure of a cretaceous decapod crustacean, *Geol. J.* **12**: 25–32.

Delachambre, J., 1971, La formation des canaux cuticulaires chez l'adulte de *Tenebrio molitor* L. Étude ultrastructurale et remarques histochemiques, *Tissue Cell* **3**:499–520.

Delachambre, J., 1975, Les variations de l'architecture dans la cuticule abdominale chez *Tenebrio molitor* L. (Ins. Col.), *Tissue Cell* **7**:669–676.

Dennell, R., 1973, The structure of the cuticle of the shore-crab *Carcinus maenas* (L.), *Zool. J. Linn. Soc.* **52**:159–163.

Dennell, R., 1974, The cuticle of the crabs *Cancer pagurus* L. and *Carcinus maenas* (L.), *Zool. J. Linn. Soc.* **54**:241–245.

Dennell, R., 1975, The structure of the cuticle of the scorpion *Pandinus imperator* (Koch), *Zool. J. Linn. Soc.* **56**:249–254.

Dennell, R., 1976a, The structure and lamination of some arthropod cuticles, *Zool. J. Linn. Soc.* **58**:159–164.

Dennell, R., 1976b, The fine structure of the cuticle of some Phasmida. In *The Insect Integument*, edited by H. R. Hepburn, pp. 177–192, Elsevier, Amsterdam.

Dennell, R., 1978, The laminae and pore canals of some arthropod cuticles, *Zool. J. Linn. Soc.* **64**:214–250.

Drach, P., 1953, Structure des lamelles cuticulaires chez les crustacés, *C. R. Hebd. Sceances Acad. Sci. Ser. D.* **237**:1772–1774.

Driest, E., and Müller, H. O., 1935, Elektronenmikroskopische Aufnahmen (Elektronenmikrogramme) von Chitinobjekten, *Z. wiss. Mikrosk.* **52**:53–57.

Filshie, B. K., 1970a, The resistance of epicuticular components of an insect to extraction with lipid solvents, *Tissue Cell* **2**:181–190.

Filshie, B. K., 1970b, The fine structure and deposition of the larval cuticle of the sheep blowfly (*Lucilia cuprina*), *Tissue Cell* **2**:479–489.

Filshie, B. K., 1976, The structure and deposition of the epicuticle of the adult female cattle tick (*Boophilus microplus*). In *The Insect Integument*, edited by H. R. Hepburn, pp. 193–206, Elsevier, Amsterdam.

Filshie, B. K., 1980, Insect cuticle through the electron microscope—distinguishing fact from artifact. In *Insect Biology in the Future, VBW 80*, edited by M. Locke and D. S. Smith, pp. 59–77, Academic Press, New York.

Filshie, B. K., and Hadley, N. F., 1979, Fine structure of the cuticle of the desert scorpion, *Hadrurus arizonensis*, *Tissue Cell* **11**:249–262.

Filshie, B. K., and Smith, D. S., 1980, A proposed solution to a fine-structural puzzle: The Organization of gill cuticle in a crayfish (*Panulirus*), *Tissue Cell* **12**:209–226.

Flower, N. E., and Walker, G. D., 1979, Rectal papillae in *Musca domestica*: The cuticle and lateral membranes, *J. Cell Sci.* **39**:167–186.

Fraenkel, G., and Rudall, K. M., 1947, The structure of insect cuticles, *Proc. R. Soc. London Ser. B.* **134**:111–143.

Gharagozolou-van Ginneken, I. D., 1974, Sur l'ultrastructure cuticulaire d'un crustacé copépode harpacticide: *Tisbe holothuriae* Humes, *Arch. Zool. Exp. gen.* **115**:411–422.

Gharagozolou-van Ginneken, I. D., 1976, Particularités morphologiques du tégument des Peltidiidae (Crustacés copepodes), *Arch. Zool. Exp. Gen.* **117**:411–422.

Gharagozolou-van Ginneken, I. D., and Bouligand, Y., 1973, Ultrastructures tegumentaires chez un crustacé copepode *Cletocamptus retrogressus*, *Tissue Cell* **5**:413–439.

Gharagozolou-van Ginneken, I. D., and Bouligand, Y., 1975, Studies on the fine structure of *Porcellidium*, Crustacea, Copepoda, *Cell Tissue Res.* **159**:399–412.

Gordon, H., and Winfree, A. T., 1978, A single spiral artifact in arthropod cuticle, *Tissue Cell* **10**:39–50.

Green, J. P., and Neff, M. R., 1972, A survey of the fine structure of the integument of the fiddler crab, *Tissue Cell* **4**:137–171.

Gubb, D., 1975, A direct visualization of helicoidal architecture in *Carcinus maenas* and *Halocynthia papillosa* by scanning electron microscopy, *Tissue Cell* **7**:19–32.

Gupta, B. L., and Smith, D. S., 1969, Fine structural organization of the spermatheca in the cockroach, *Periplaneta americana*, *Tissue Cell* **1**:295–324.

Hackman, R. H., 1971, The integument of arthropoda. In *Chemical Zoology*, edited by M. Florkin and B. T. Scheer, vol. 6, pp. 1–62, Academic Press, New York.

Hackman, R. H., 1974, The soluble cuticular proteins from three arthropod species: *Scylla serrata* (Decapoda: Portunidae), *Boophilus microplus* (Acarina: Ixodidae) and *Agrianome spinicollis* (Coleoptera: Cerambycidae), *Comp. Biochem. Physiol. B.* **49**:457–464.

Hackman, R. H., and Filshie, B. K., 1982, The tick cuticle. In *The Physiology of Ticks*, edited by F. D. Obenchain, in press, Pergamon Press, London.

Hadley, N. F., 1979, Wax secretion and color phases of the desert tenebrionid beetle *Cryptoglossa verrucosa* (Le Conte), *Science* **203**:367–369.

Hadley, N. F., and Filshie, B. K., 1979, Fine structure of the epicuticle of the desert scorpion, *Hadrurus arizonensis*, with reference to location of lipids, *Tissue Cell* **11**:263–275.

Hegdahl, T., Silness, J., and Gustavsen, F., 1977a, The structure and mineralization of the carapace of the crab (*Cancer pagurus* L.). 1. The endocuticle, *Zool. Scr.* **6**:89–99.

Hegdahl, T., Gustavsen, F., and Silness, J., 1977b, The structure and mineralization of the carapace of the crab (*Cancer pagurus* L.). 2. The exocuticle, *Zool. Scr.* **6**:101–105.

Horridge, A., and Blest, D., 1980, The compound eye. In *Insect Biology in the Future, VBW 80*, edited by M. Locke and D. S. Smith, pp. 705–733, Academic Press, New York.

Kayser-Wegmann, I., 1976, Ultrastructural differences between larval and pupal cuticle of *Pieris brassicae* (Lepidoptera), *Protoplasma* **90**:319–331.

Kennaugh, J., 1965, Pore canals in the cuticle of *Hypoderma bovis* (Diptera), *Nature* (London) **205**:207.

Kuo, J. S., and McCully, M. E., 1978, Fine structure of the integument and associated structures of *Caloglyphus mycophagus*, *Acarologia* **20**:572–589.

Livolant, F., Giraud, M. M., and Bouligand, Y., 1978, A goniometric effect observed in sections of twisted fibrous materials, *Biol. Cellulaire* **31**:159–168.

Locke, M., 1960, Cuticle and wax secretion in *Calpodes ethlius* (Lepidoptera, Hesperiidae), *Quart. J. Microsc. Sci.* **101**:333–338.

Locke, M., 1961, Pore canals and related structures in insect cuticles, *J. Biophys. Biochem. Cytol.* **10**:589–618.

Locke, M., 1966, The structure and formation of the cuticulin layer in the epicuticle of an insect, *Calpodes ethlius* (Lepidoptera, Hesperiidae), *J. Morphol.* **118**:461–494.

Locke, M., 1969, The structure of an epidermal cell during the development of the protein epicuticle and the uptake of molting fluid in an insect, *J. Morphol.* **127**:7–40.

Locke, M., 1974, The structure and formation of the integument of insects. In *The Physiology of Insecta*, edited by M. Rockstein, vol. 6, 2nd edition, pp. 123–213, Academic Press, New York.

Locke, M., 1976, Role of plasma membrane plaques and Golgi complex vesicles in cuticle deposition during the moult/intermoult cycle. In *The Insect Integument*, edited by H. R., Hepburn, pp. 237–258, Elsevier, Amsterdam.

Locke, M., and Krishnan, N., 1971, The distribution of phenoloxidases and polyphenols during cuticle formation, *Tissue Cell* **3**:103–126.

Meyer, K. H., and Pankow, G. W., 1935, Sur la constitution et la structure de la chitine, *Helv. Chim. Acta* **18**:589–598.

Mutvei, H., 1974, SEM studies on arthropod exoskeletons. 1. Decapod crustaceans, *Homarus gammarus* L. and *Carcinus maenas* (L.)., *Bull. Geol. Inst. Univ. Uppsala*: NS, **4**:73–80.

Mutvei, H., 1977, SEM studies on arthropod exoskeletons. 2. Horseshoe crab *Limulus polyphemus* (L.) in comparison with extinct eurypterids and recent scorpions, *Zool. Scr.* **6**:203–213.

Nathanson, M. E., 1967, Changes in the fine structure of the integument of the rabbit tick, *Haemaphysalis leporispalustris* (Acari: Ixodides: Ixodidae), which occur during feeding, *Ann. Entomol. Soc. Am.* **63**:1768–1774.

Neville, A. C., 1963, Growth and deposition of resilin and chitin in locust rubberlike cuticle, *J. Insect Physiol.* **9**:265–278.

Neville, A. C., 1970, Cuticle ultrastructure in relation to the whole insect, *Symp. R. Ent. Soc. London* **5**:17–39.

Neville, A. C., 1975, *Biology of the Arthropod Cuticle*, Springer-Verlag, Berlin.

Neville, A. C., and Berg, C. W., 1971, Cuticle ultrastructure of a Jurassic crustacean (*Eryma stricklandi*), *Paleontology* **14**:201–205.

Neville, A. C., and Caveney, S., 1969, Scarabaeid beetle exocuticle as an optical analogue of cholesteric liquid crystals, *Biol. Rev. Cambridge Phil. Soc.* **44**:531–562.

Neville, A. C., and Luke, B. M., 1969a, Molecular architecture of adult locust cuticle at the electron microscope level, *Tissue Cell* **1**:355–366.

Neville, A. C., and Luke, B. M., 1969b, A two-system model for chitin-protein complexes in insect cuticles, *Tissue Cell* **1**:689–707.

Neville, A. C., and Luke, B. M., 1971, Form optical activity in crustacean cuticle, *J. Insect Physiol.* **17**:519–526.

Neville, A. C., Thomas, M. G., and Zelazny, B., 1969, Pore canal shape related to molecular architecture of arthropod cuticle, *Tissue Cell* **1**:183–200.

Neville, A. C., Parry, D. A. D., and Woodhead-Galloway, J., 1976, The chitin crystallite in arthropod cuticle, *J. Cell Sci.* **21**:73–82.

Noirot, C., and Noirot-Timothée, C., 1969, La cuticule proctodéale des insectes. 1. Ultrastructure comparée, *Z. Zellforsch.* **101**:477–509.

Noirot, C., Noirot-Timothée, C., Smith, D. S., and Cayer, M., 1978, Cryofracture de la cuticule des insectes: mise en évidence d'un clivage dans l'épicuticle externe. Implications structurales et fonctionelles, *C. R. Hebd. Sceances Acad. Sci. Ser. D.* **287**:503–505.

Pace, A., 1972, Cholesteric liquid crystal-like structure of the cuticle of *Plusiotis gloriosa*, *Science* **176**:678–680.

Pope, R. D., 1979, Wax production by coccinellid larvae (Coleoptera), *Syst. Entomol.* **4**:171–196.

Richards, A. G., 1951, *The Integument of Arthropods*, University of Minnesota Press, Minneapolis.

Richards, A. G., 1967, Sclerotization and the localization of brown and black colors in insects, *Zool. Jb. Abt. Anat. Ontog. Tiere* **84**:25–62.

Richards, A. G., and Korda, F. H., 1948, Studies on arthropod cuticle. 2. Electron microscope studies of extracted cuticles, *Biol. Bull* **94**:212–235.

Rieder, N., 1972a, Ultrastruktur der Carapaxcuticula von *Triops cancriformis* Bosc. (Notostraca, Crustacea), *Z. Naturforsch.* **27B**:279.

Rieder, N., 1972b, Ultrastruktur und Polysaccharidanteile de Cuticula von *Triops cancriformis Bosc.* (Crustacea, Notostraca) während der Hautungsvorbereitung, *Z. Morphol. Okol. Tiere* **73**:361–380.

Rudall, K. M., 1965, Skeletal structures in insects. In *Aspects of Insect Biochemistry, Biochem. Soc. Symp.* **25**:83–92.

Rudall, K. M., 1967, Conformation in chitin-protein complexes. In *Conformation of Biopolymers*, edited by G. N. Ramachandran, vol. 2, pp. 751–765, Academic Press, London.

Rudall, K. M., 1969, Chitin and its association with other molecules, *J. Polym. Sci. Part C* **28**:83–102.

Rudall, K. M., 1976, Molecular structure in arthropod cuticles. In *The Insect Integument*, edited by H. R. Hepburn, pp. 21–41, Elsevier, Amsterdam.

Ruska, E., 1980, *The Early Development of Electron lenses and Electron Microscopy*, translated by T. Mulvey, S. Hirzel Verlag, Stuttgart.

Slifer, E. H., 1961, The fine structure of insect sense organs, *Int. Rev. Cytol.* **2**:125–159.

Waterhouse, D. F., and Norris, K. R., 1980, Insects and insect physiology in the scheme of things. In *Insect Biology in the Future, VBW 80*, edited by M. Locke and D. S. Smith, pp. 19–37, Academic Press, New York.

Wharton, G. W., Parrish, W., and Johnston, D. E., 1968, Observations on the fine structure of the cuticle of the spiny rat mite, *Laelaps echidnina* (Acari, mesostigmata), *Acarologia* **10**:206–214.

Wigglesworth, V. B., 1933, The physiology of the cuticle and of ecdysis in *Rhodnius prolixus* (Triatomidae, Hemiptera): With special reference to the function of the oenocytes and of the dermal glands, *Quart. J. Microsc. Sci.* **76**:269–318.

Wigglesworth, V. B., 1948, The insect cuticle, *Biol. Rev. Cambridge Phil. Soc.* **23**:408–451.

Wigglesworth, V. B., 1957, The physiology of insect cuticle, *Annu. Rev. Entomol.* **2**:37–54.

Wigglesworth, V. B., 1970, Structural lipids in the insect cuticle and the function of the oenocytes, *Tissue Cell* **2**:155–179.

Wigglesworth, V. B., 1971, Bound lipid in the tissues of mammal and insect: A new histochemical method, *J. Cell Sci.* **8**:709–725.

Wigglesworth, V. B., 1975a, Lipid staining for the electron microscope—A new method, *J. Cell Sci.* **19**:425–437.

Wigglesworth, V. B., 1975b, Distribution of lipid in lamellate endocuticle of *Rhodnius prolixus* (Hemiptera), *J. Cell Sci.* **19**:439–457.

Wigglesworth, V. B., 1975c, Incorporation of lipid into epicuticle of *Rhodnius* (Hemiptera), *J. Cell Sci.* **19**:459–485.

Zacharuk, R. Y., 1972, Fine structure of the cuticle, epidermis and fat body of larval Elateridae (Coleoptera) and changes associated with molting, *Can. J. Zool.* **50**:1463–1487.

Zelazny, B., and Neville, A. C., 1972, Quantitative studies on fibril orientation in beetle endocuticle, *J. Insect Physiol.* **18**:2095–2121.

11

The Structure and Development of Insect Connective Tissues

DOREEN E. ASHHURST

1. *Definition of Connective Tissue*

The term connective tissue is now used to denote the whole intercellular matrix, and thus it embraces all the material found between the cells; fibrous tissues, basement membranes, and other less structured components. Despite their variety, connective tissue matrices contain a limited number of components, the main ones being collagens, proteoglycans, glycoproteins, and elastin with its associated microfibrillar protein. All these constituents are found in the vertebrates, but some, for example elastin, have not yet been identified in the invertebrates. Another extracellular protein, resilin, is found in insects. This protein, which is peculiar to the invertebrates, will also be discussed in this chapter.

1.1. *Distribution of Connective Tissues in Insects*

The main skeletal tissue in an insect is provided by the cuticle, but the internal organs are all supported by collagenous connective tissues in exactly the same way as in other Metazoa. Thus, layers of fibrous tissue are found around the gut, the nervous system, the reproductive system, and so on. The epithelia lie on basement membranes, most of which are of typical structure; some modified basement membranes, peculiar to insects and other arthropods, will be discussed in section 4.

At the time when an earlier review of insect connective tissues was pub-

DOREEN E. ASHHURST • Department of Anatomy, St. George's Hospital Medical School, Cranmer Terrace, Tooting, London SW17 ORE, England.

lished (Ashhurst, 1968), there was still a widespread misconception that insects did not possess collagenous connective tissue. Thus, in that paper, many individual instances of the occurrence of connective tissues were described. Almost fifteen years later, there is no longer any dispute. It is, therefore, not necessary to list descriptions of, or to give references to, the connective tissues of the various organs. Instead, this chapter will describe the types of connective tissue which can be identified by their ultrastructure and discuss what is known of their constituents.

1.2. Constituents of Connective Tissues

Over the past two decades, much information has been accumulated about all aspects of the constituents of connective tissues. Inevitably, most of this information has derived from studies of avian and mammalian tissues, but as will be apparent later, the constituents of the extracellular matrices are very similar throughout the Animal Kingdom.

1.2.1. Collagens

At the present time, five different types of collagen have been characterized (see review by Prockop *et al.*, 1979), and several further collagenous polypeptide chains are under investigation. These have been isolated from vertebrate tissues, but as will be seen later, at least two of these collagens are present in insect tissues.

The collagen molecule is approximately 300 nm long and 1.5 nm in diameter. It consists of three polypeptide chains which have a central helical region, or domain, and short nonhelical terminal domains. The polypeptide chains are referred to as α-chains, and are coiled into a left-handed helix. Three α-chains are then coiled into a right-handed triple helix to form the collagen molecule. The nonhelical terminal peptides are involved in both intramolecular and intermolecular crosslinking. The triple-helical region is characterized by the occurrence firstly of glycine every third residue, secondly of many proline residues, some of which are hydroxylated to hydroxyproline, and thirdly of a few lysine residues, some of which are hydroxylated to form hydroxylysine, and which in turn may be glycosylated. Lysine and hydroxylysine also occur in the nonhelical terminal domains and are involved in the formation of the crosslinks, dihydroxylysinonorleucine and hydroxylysinonorleucine.

The collagen molecules are arranged very precisely to form the typically banded collagen fibril. The generally accepted model for aggregation is the modified quarter-stagger model of Hodge and Petruska (1963), which is illustrated in Figure 1A. This model reconciles the generally accepted periodicity of 67 nm, measured by X-ray diffraction of native hydrated collagen fibrils, and the length of the molecule, that is, approximately 300 nm. The gaps between the ends of the molecules are inherent in this model. It also allows for crosslinking between the terminal and the helical domains of adjacent molecules.

The distribution of the five types of collagen in vertebrate tissues is well known. Type I collagen forms thick fibrils and is widely distributed in adult tissues; it is the major collagen in bone and tendon. The molecule has two

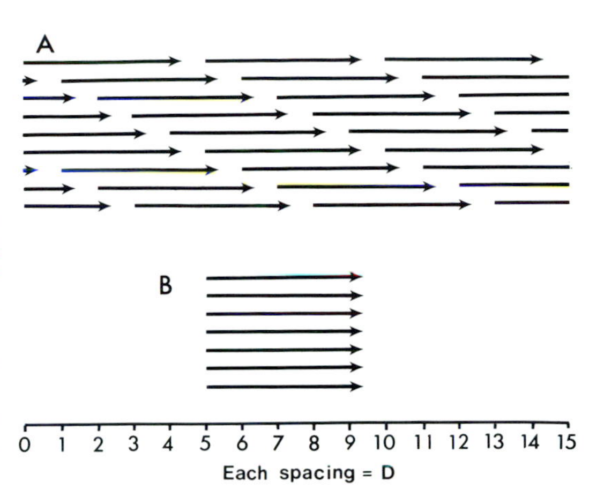

Figure 1. (A) A diagram to show the arrangement of the molecules in collagen fibrils according to the Hodge and Petruska (1963) model. The distance D is 67 nm, i.e., the banding periodicity. Each molecule is 4.4 D long. The "gaps" between the ends of the molecules are inherent in this model.
(B) A diagram to show the arrangement of the collagen molecules in a segment-long-spacing (SLS) crystallite.

similar $\alpha1$ chains and one dissimilar chain, denoted as $\alpha2$. Type II collagen has three similar α-chains, forms very thin, indistinctly banded fibrils, and is found in cartilage. Type III collagen also has three similar α-chains; it forms thin, banded fibrils, which are found as delicate networks in many organs, such as the spleen, and in embryonic tissues, especially skin. Type IV collagen is the nonfibrous basement membrane collagen, and it will be discussed in section 4. Type V is a more recently discovered molecule of widespread occurrence, but it is not fully characterized, although there are known to be two kinds of α-chain in the molecule.

1.2.2. Proteoglycans and Glycosaminoglycans

Glycosaminoglycans are long chain polymers formed of repeating, identical disaccharide units (see review by Muir and Hardingham, 1975). The disaccharides consist of an amino-sugar, glucosamine or galactosamine, and a uronic acid moiety, which may be either glucuronic or iduronic acid. The polymers include hyaluronate, chondroitin sulfate, dermatan sulfate, heparin and heparan sulfate, and keratan sulfate, in which the uronic acid is substituted by glucose, and all, with the exception of hyaluronate and heparin, have an obligatory association with protein. The polymers are attached to a polypeptide chain by a linkage region which consists of a trisaccharide unit (Gal-Gal-Xyl); the xylose residue is glycosidically linked to serine. This association of glycosaminoglycan polymers with a polypeptide forms the proteoglycan. Different glycosaminoglycan polymers may be attached to the same polypeptide to form a variety of proteoglycans; research so far suggests that the proteoglycans of each tissue are peculiar to it. Several proteoglycans may be attached to a single-stranded polymer of hyaluronate to form a very large aggregate.

Some properties of the glycosaminoglycans are important for the evaluation of the functions of connective tissue matrices. Firstly, the glycosaminoglycan polymers may be very long, especially in the case of hyaluronate, which may be greater than 1 μm in length (Fessler and Fessler, 1966), and thus they occupy a significant proportion of the volume of the matrix. Secondly, they are all polyanions, that is, each uronic acid moiety has a free carboxyl group, and, with the exception of hyaluronate, the amino-sugars are sulfated. Thirdly, the chondroitin, dermatan, and keratan sulfates are closely associated with the collagen fibrils, and are thought to influence fibril formation (Mathews, 1974). Evidence is slowly accumulating that these first two properties of the glycosaminoglycans make the matrix into a negatively charged sieve, which may retard the passage of large molecules, particularly those with negative charges (Laurent, 1977).

1.2.3. Glycoproteins

Knowledge of the glycoproteins of connective tissue is rudimentary; few molecules have been isolated or characterized. Recently, two glycoproteins, fibronectin and laminin, have been identified.

Fibronectin occurs in two forms, an insoluble form on cell surfaces and a soluble form in blood plasma and other body fluids. As a result, it has been the subject of many independent, unrelated studies, and it is only recently that it was realized that only one molecule is involved (see review by Ruoslahti *et al.*, 1981). Fibronectin is a dimer of two, large, similar polypeptide chains. Immunological evidence suggests that the molecules extracted from different sources, e.g., cell surfaces and plasma, are very closely related chemically, but some differences, which may be due to variations in the glycosylation of the molecules, are detected by gel electrophoresis. Fibronectin has binding sites for collagen, glycosaminoglycans, fibrinogen, actin, and cell surfaces. It is suggested that it forms a bridge between the cell and the matrix, but investigations of the significance of this molecule in normal developmental and healing processes have only just begun. It is noteworthy that fibronectin is not bound by many malignant cells.

The other glycoprotein, laminin, is a high molecular weight polypeptide which has two subunits, one twice the size of the other. The molecule contains about 15% by weight of carbohydrate. It is found in basement membranes, and its demonstration by immunofluorescence in the early mouse embryo suggests that it may have a role in morphogenesis (see review by Heathcote and Grant, 1981).

1.3. Functions of Connective Tissues

The most obvious function of any connective tissue is supportive. Collagen fibrils are inextensible and very strong; thus, the strength of any layer containing fibrils will be determined by the density of the collagen fibrils and their orientation with respect to the neighboring fibrils.

The physical and chemical properties of the glycosaminoglycans, mentioned previously, imply further functional properties. Because of the size of the polymers and of the large proteoglycan aggregates, they occupy a considerable volume within the matrix. Thus, it has been demonstrated by Laurent and his coworkers (Laurent, 1977), that large molecules may be retarded in their movement across the matrices by these molecular sieves. The charge on the penetrating molecules affects the retardation, since the molecules are moving between polyanions.

Recently it has become apparent that the matrix is essential for many morphogenetic movements and the subsequent differentiation of cells. Toole (1972), in a study of limb bud development in the chick, found that the presumptive chondroblasts move and become oriented in a matrix of hyaluronate; hyaluronidase is then produced to remove the hyaluronate before differentiation of the cells and the secretion of the cartilage matrix can occur. Other studies have shown, for example, that sulfated glycosaminoglycans are essential in the basement membrane of developing salivary glands for the formation of acini (Bernfield *et al.*, 1972), and that basement membranes are found in embryos prior to gastrulation (Hay, 1973). This field of research has expanded rapidly since this early work, but so far the studies have been confined to vertebrate tissues.

2. *Fibrous Connective Tissues*

2.1. *Morphological Studies*

2.1.1. *Connective Tissue Matrix*

The fibrous connective tissue matrices found in insects fall into two groups; those possessing large, clearly banded collagen fibrils and those with very thin, indistinctly banded fibrils. The latter are found in the Diptera, Coleoptera, Lepidoptera, and Protura (Baccetti, 1961; Ashhurst, 1968; François, 1972; Locke and Huie, 1972; de Biasi and Pilotto, 1976).

The collagen fibrils are randomly arranged in most insect matrices (Figure 2); in only a few structures, such as the sheath covering the transverse hypopharyngeal structure of *Blaberus craniifer* (Moulins, 1968) and the endosternite of *Campodea chardardi* (François, 1968), have parallel arrays of fibrils been reported. This random arrangement is appropriate, since most of the layers provide support around organs or under the epidermis, where deformation can occur in many directions.

In longitudinal sections, the large fibrils are clearly banded and have an obvious repeating periodicity. Many measurements of this periodicity have been made and these vary from just over 50 nm to 70 nm. Careful examination of the micrographs reveal, however, that the periodic banding pattern is in all instances identical. This pattern is shown in Figure 7, but it is much clearer in Figure 19A, which shows a reconstituted fibril made from extracted locust collagen. Bands I and XI are the most distinct and are separated from the others; these are readily recognized in sectioned fibrils, whereas bands III to X form the broad area in which the individual bands are less distinct. It will be argued later (see section 2.5.1), that *in vivo*, all these fibrils display the same periodicity of 67 nm, and that the variations measured on electron micrographs are due entirely to the preparative techniques.

In cross section, the larger fibrils may be up to 200 nm in overall diameter

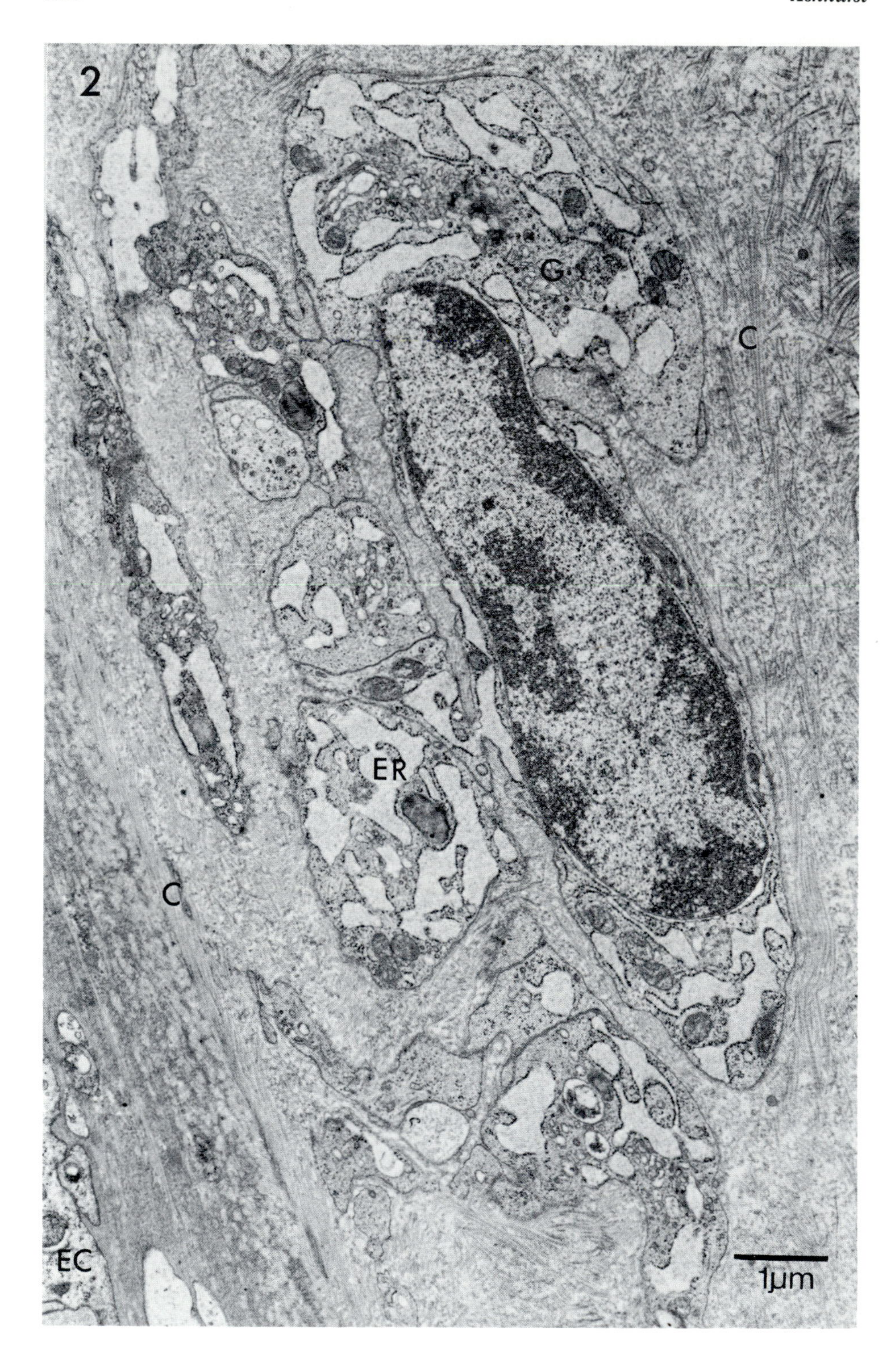
2
G
C
ER
C
EC
1μm

and may also be very irregular in shape (Figure 6). Such fibrils are frequently seen in the connective tissues of locusts, cockroaches, and other insects. These are not abnormal fibrils; similarly irregular fibrils have been observed in rat lung (Stephens *et al.*, 1971). It is not known whether the large fibrils result from the fusion of several smaller fibrils, or from the gradual accretion of molecules onto pre-existing fibrils. The gradual development of these large fibrils was observed in the ejaculatory duct of adult male locusts (*Locusta migratoria*) during the first three weeks after the adult molt (Ashhurst and Costin, 1974) and is illustrated in Figures 3 to 7.

The very thin fibrils found in coleopteran, dipteran, and lepidopteran connective tissues are often obscured by other matrix components, such as glycosaminoglycans and glycoproteins. That they are collagenous was demonstrated by Ashhurst and Richards (1964b); hydroxyproline was detected in hydrolysates of the nerve cord of adult wax-moths (*Galleria mellonella*). The lack of a readily distinguishable banding pattern along these fibrils (Figure 12) is probably unimportant in their identification. For banding to be clearly discernable, a collagen fibril must have a diameter of about 20 nm. The thin fibrils in developing tissues and cartilage of vertebrates do not exhibit clear banding. nor do those in the developing tissues of the locust (see Figures 13 and 15). Thus, the apparent lack of banding in these insect fibrils may simply be due to their small diameter. Because the amount of electron-dense stain bound must depend on the number of molecules in the fibril, a small fibril with few molecules will bind less than a large fibril with many molecules; hence, insufficient stain may be bound to make the individual bands discernable. It is noteworthy that Locke and Huie (1972), using techniques designed to enhance staining, were able to show a periodicity of about 66 nm in *Calpodes ethlius* fibrils. The indications of periodicities of around 15 and 20 nm suggested by many workers clearly do not represent a true collagen periodicity (see section 2.5.1) (Ashhurst, 1968).

The matrix also contains proteoglycans and glycoproteins (see section 2.2). These have no recognizable structure which can be distinguished in thin sections, but the amorphous material seen among the collagen fibrils (Figure 6, arrow) presumably consists of proteoglycans and glycoproteins. Occasionally a homogeneous, moderately dense body is seen; this is probably a lipid droplet.

The connective tissue around the ejaculatory duct of the adult male locust, which is illustrated in Figures 2 and 8, was described as a cartilage by Martoja and Bassot (1965) because the cells in the tissue are of one type and in their preparations appeared to be in lacunae. That this tissue is not cartilagenous was demonstrated by Ashhurst and Costin (1974). There are no other reports of cartilage-like connective tissues in insects.

←

Figure 2. A low-magnification electron micrograph of the connective tissue around the ejaculatory duct of a 6-day adult locust, *Locusta migratoria*. The cells are typical, active fibroblasts with dilated rough endoplasmic reticulum (ER) and small Golgi complexes (G). The matrix is packed with collagen fibrils (C) in varying orientations. A region of compact fibrils occurs adjacent to the basal region of the epidermal cells (EC).

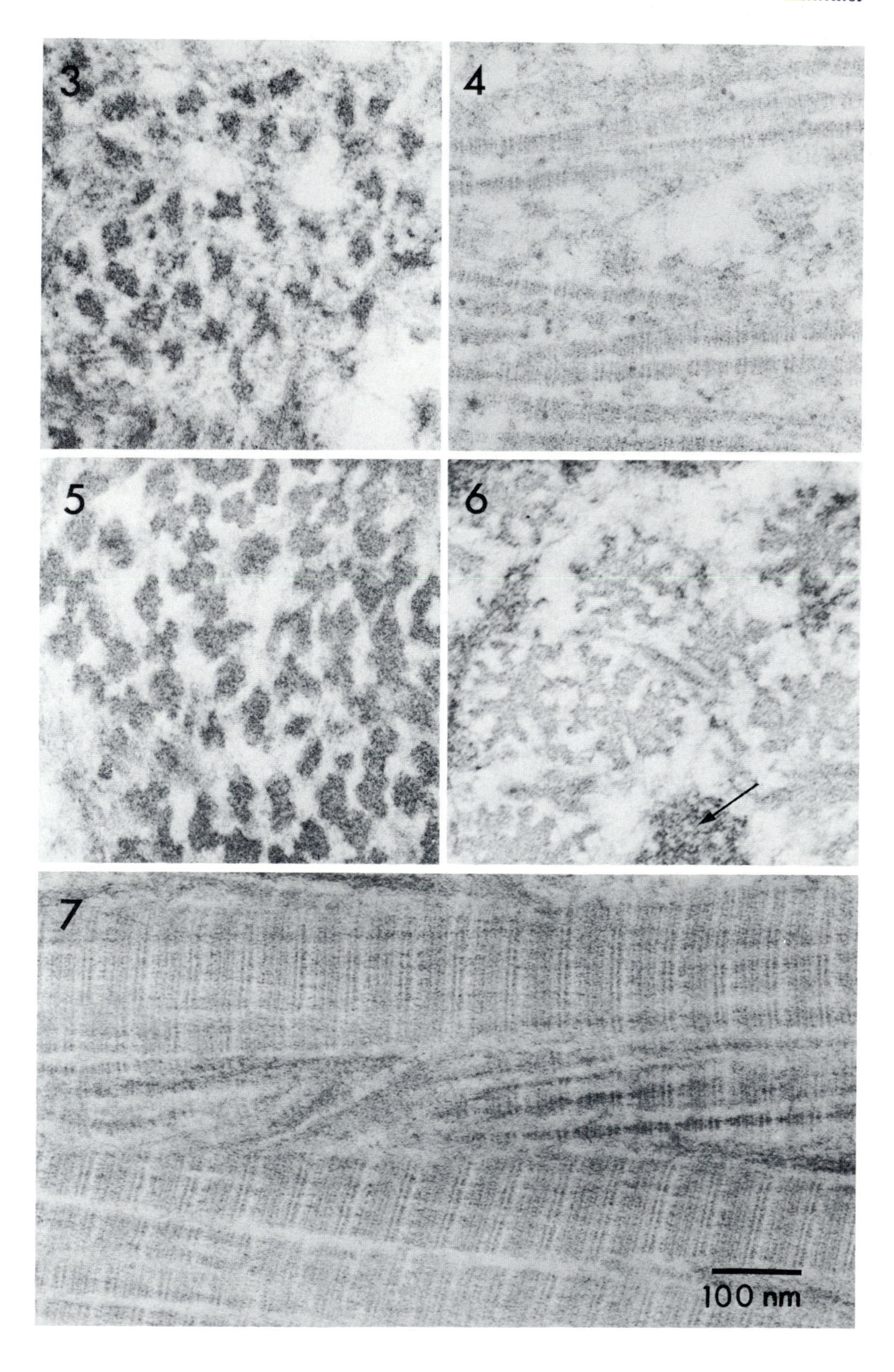
3
4
5
6
7
100 nm

2.1.2. Cells of Connective Tissue

In vertebrates, the fibrous connective tissues are usually populated by fibroblasts, or in the case of bone and cartilage, by osteocytes and chondrocytes, respectively. In insects, only a few connective tissues enclose cells within the matrix; in most instances the connective tissue lies on a layer of cells. The neural lamella around the nervous system is a good example of this latter situation. A result of this relationship between the cells and connective tissue matrix in insects is that many of the cells which produce connective tissue also have other functions.

A typical fibroblast is an irregular spindle-shaped cell. The cytoplasm of an active fibroblast (see Figures 2 and 8) contains a large amount of rough endoplasmic reticulum with irregularly dilated cisternae. The cisternae contain a moderately electron-dense amorphous material, which is thought to be procollagen. In areas where an *en face* view of the surface of a cisterna is visible, the ribosomes can be seen as spiral polyribosomes (Figure 9). The Golgi complexes occupy a much smaller volume of the cytoplasm than the rough endoplasmic reticulum and are typically a series of irregularly arranged vesicles. There are few lamellae arranged in parallel, and so it is impossible to distinguish the forming and maturing faces of these Golgi complexes (Figure 8). Secretory or storage granules are not observed in fibroblasts. The mitochondria are rather small and not especially numerous. A few lysosomes may be present.

The above description of a typical fibroblast applies equally to mammalian fibroblasts and to the insect fibroblasts from the connective tissue around the ejaculatory duct of the locust (Figure 2) (Ashhurst and Costin, 1974) the nervous system of the wax-moth, *Galleria mellonella* (Figure 11), the endosternites of *Thermobia domestica* (François, 1973), and the midgut of the cockroach, *Periplaneta americana* (François, 1978).

In most insects, the connective tissue is produced during a specific period in the life cycle. For example, in the locust most of the collagenous tissue around the ejaculatory duct is produced during the first 10 days after the adult molt (Ashhurst and Costin, 1974), and similarly, that in the dorsal mass on the adult nerve cord of the wax-moth is produced in the pupa from days 4 to 7 (Ashhurst, 1964; Ashhurst and Richards, 1964a). Thereafter, the cells cease active secretion and appear inactive; the rough endoplasmic reticulum becomes very reduced (Figures 10 and 12). In instances such as the neural lamella where the underlying

←

Figures 3–7. These figures show collagen fibrils in the connective tissue of the ejaculatory duct of adult male locusts at different ages and illustrate the development of the irregularly shaped fibrils typical of many mature insect connective tissues. All the micrographs are at the same magnification.

Figure 3. Transverse section of fibrils from a 0-day adult locust.

Figure 4. Longitudinal section of fibrils from a 0-day adult locust.

Figure 5. Transverse section of fibrils from a 6-day adult locust.

Figure 6. Transverse section of fibrils from a 25-day adult locust. The amorphous material (arrow) is presumably composed of glycosaminoglycans and glycoproteins.

Figure 7. Longitudinal section of fibrils from a 25-day adult locust.

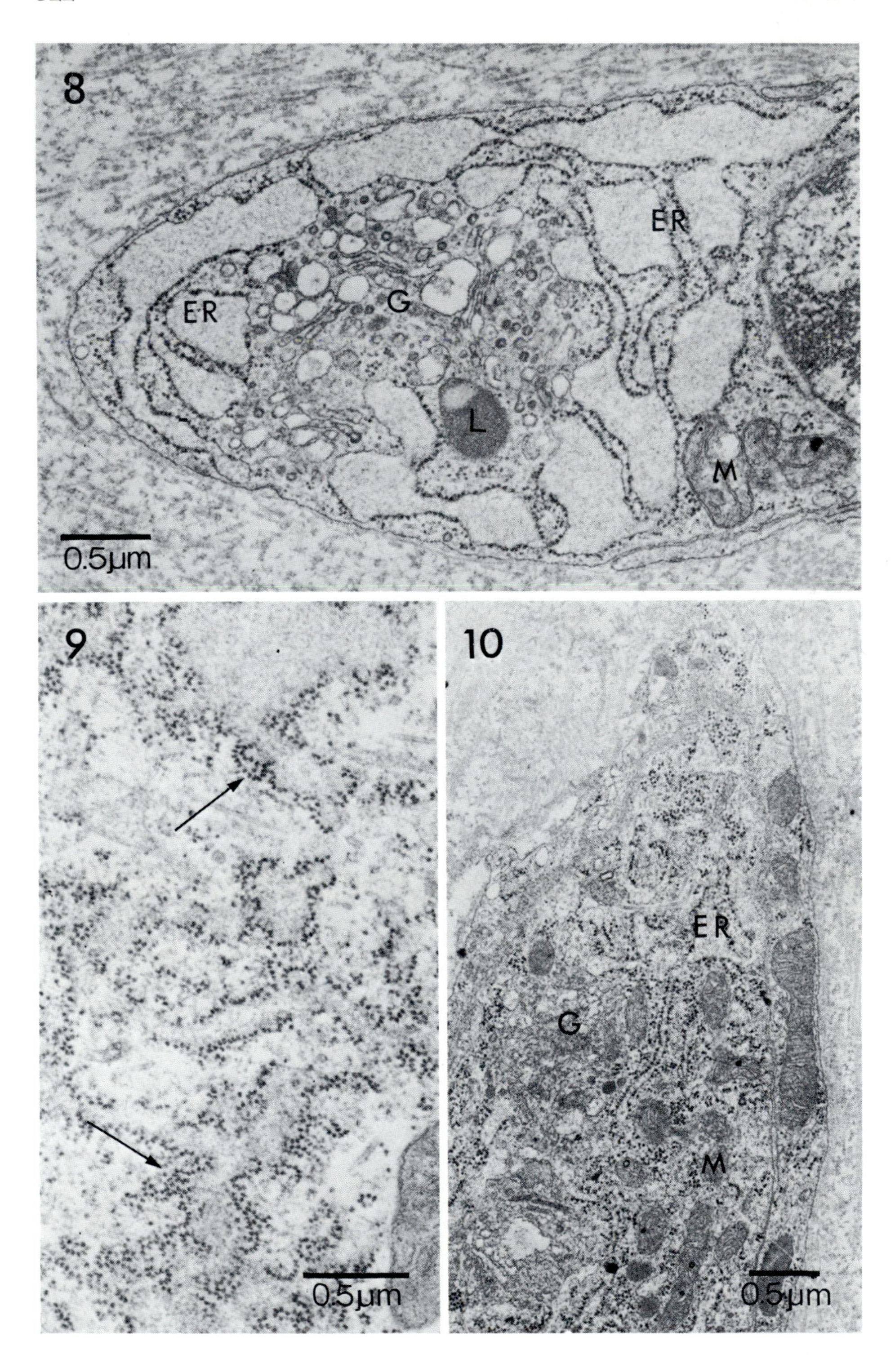
8
ER
G
ER
L
M
0.5µm
9
0.5µm
10
ER
G
M
0.5µm

cells produce the connective tissue, it may be again produced during specific periods, for example, just after a molt, and at this time the cells may acquire the ultrastructural characteristics of a fibroblast (see section 2.3).

2.2. *Histochemistry of Fibrous Connective Tissues*

Fibrous collagen is always associated with glycosaminoglycans; the only exception appears to be in the coelenterates (Katzman and Jeanloz, 1970). The early studies of connective tissues in grasshoppers and locusts (Baccetti, 1955; Ashhurst, 1959) were designed primarily to characterize the layers of connective tissue and to determine if collagen might be present; this aspect of the studies was rapidly superceded by electron microscopical evidence. Another conclusion from these studies was that the neural lamella of these insects posesses neutral polysaccharides but no glycosaminoglycans. Since these early studies, histochemical methods have been much improved, and it is now possible to demonstrate that insect connective tissues contain glycosaminoglycans.

The methods using the copper phthalocyanin dye, alcian blue 8GX, are the most reliable and sensitive of the tests for glycosaminoglycans available at the present time. This is a basic dye, with four cationic charges, and it is bound electrostatically by the anions along the glycosaminoglycan polymers. The dye may be used under differing conditions of pH and salt concentration (see Ashhurst, 1979b, for a detailed account of the staining properties of alcian blue), and it is possible to determine the presence of hyaluronate, chondroitin and dermatan sulfates, heparan sulfate, heparin and keratan sulfate. Table 1 shows the results of a typical histochemical study of the connective tissue around the ejaculatory duct of the adult male locust (Ashhurst and Costin, 1971a). From these studies, it was concluded that as the connective tissue develops in the young adult, chondroitin sulfates and perhaps, dermatan sulfate accumulate, and that later, keratan sulfate is also produced. The results with alcian blue were confirmed by enzyme digestions with hyaluronidase and neuraminidase, and with the high iron diamine test. The presence of neutral glycoproteins, which give a positive reaction with the periodic acid-Schiff (PAS) test, was also confirmed. A very similar pattern of histochemical reactions, and hence the occurrence of glycosaminoglycans and glycoproteins, is reported by François (1978) in the connective tissue surrounding the mesenteron of *Periplaneta americana*.

These studies were extended to the neural lamellae of *Locusta migratoria*, *Periplaneta americana*, and the stick insect, *Carasius morosus*. It was immedi-

←

Figure 8. Part of the cytoplasm of a fibroblast from the ejaculatory duct of a 6-day adult locust. The rough endoplasmic reticulum (ER) is very dilated, and an electron-dense material, which is thought to be procollagen, is present in the cisternae. The Golgi complex (G) consists of vesicles with only a few small lamellae. A lysosome (L) and mitochondria (M) are also present.

Figure 9. An area of endoplasmic reticulum in a 4-day adult locust fibroblast in which there is an *en face* view of the spiral polyribosomes (arrows) on the cisternal membranes.

Figure 10. Part of the cytoplasm of an inactive fibroblast from the ejaculatory duct of a 22-day adult locust. The cytoplasm is shrunken and the cisternae of the rough endoplasmic reticulum (ER) are no longer dilated. Golgi complexes (G) can be seen, and there are many mitochondria (M).

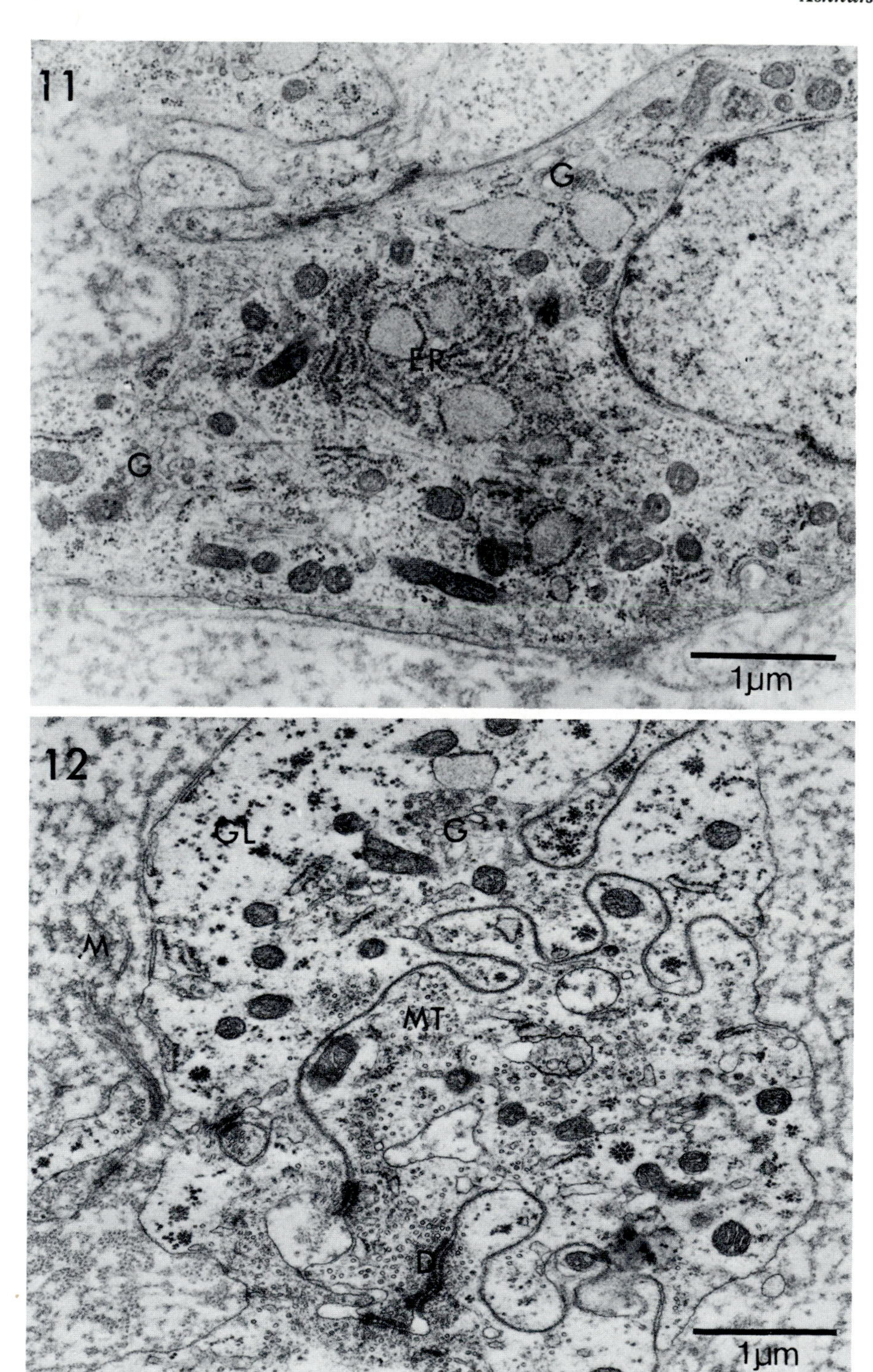
11
G
ER
G
1µm
12
GL
G
M
MT
D
1µm

Table 1. Histochemical Reactions of the Connective Tissue around the Ejaculatory Duct of the Adult Male Locust, *Locusta migratoria*[a]

Procedure	2-day	Adult 10-day	26-day
AB, pH2.5	1–3+	1–3+	1–3+
AB, pH1.0	1–4+	1–3+	2–4+
AB + $MgCl_2$, 0.05M	4+	4+	4+
AB + $MgCl_2$, 0.1M	3+	1–3+	4+
AB + $MgCl_2$, 0.2M	1–4+	2–4+	1–3+
AB + $MgCl_2$, 0.4M	1–3+	1–3+	1–3+
AB + $MgCl_2$, 0.6M	1+	0–1+	1–2+
AB + $MgCl_2$, 0.8M	1+	0–1+	1–2+
AB + $MgCl_2$, 1.0M	0	±	1–2+
Hyaluronidase-AB, pH2.5	+	±	±
Control-AB, pH2.5	2–3+	1–2+	2–3+
Hyaluronidase-AB, pH1.0	1+	1+	
Control-AB, pH1.0	1–2+	1–4+	1–4+
Neuraminidase-AB, pH2.5	1–3+	1–3+	1–3+
Control-AB, pH2.5	1–3+	1–3+	1–3+
PAS (Standard)	1–4+	1–4+	2–4+
PAS (Diastase control)	1–4+	1–4+	1–4+
PAS (Modified)	1–2+	1–2+	
HID	1–2+	1–2+	1–3+

[a]The intensity of the reactions, or staining, is as follows: no reaction or staining, 0; increasing intensity of reaction or staining, 1–5+ (1 = weak; 5 = very strong); weakly positive reaction or staining, ±.
Abbreviations: AB = alcian blue; PAS = periodic acid/Schiff test; HID = high iron diamine test.

ately apparent from the binding of alcian blue in the presence of magnesium chloride by the neural lamellae, that glycosaminoglycans are present; both chondroitin and keratan sulfates were identified (Ashhurst and Costin, 1971b,c). Thus, the earlier anomaly of fibrous connective tissue with no glycosaminoglycans was resolved, since they are undoubtedly present. The positive PAS reaction in these neural lamellae is due to the presence of glycoproteins.

In the early studies of Ashhurst and Richards (1964b), the dorsal connective

←

Figure 11. An active fibroblast in the dorsal connective tissue mass of a 7-day pupal wax-moth, *Galleria mellonella*. Many of the cisternae of the rough endoplasmic reticulum (ER) are dilated. The Golgi complexes (G) are very small.

Figure 12. The cytoplasm of several inactive fibroblasts in a 1-day adult wax-moth, *Galleria mellonella*. The cells are interdigitated and are held together by desmosomes (D). The rough endoplasmic reticulum (ER) is very reduced and the Golgi complexes (G) are very small. Most of the cytoplasm is filled by microtubules (MT), many of which are associated with desmosomes, and large areas of glycogen (GL). The matrix (M) contains only very thin fibrils. These are largely obscured by amorphous material, which presumably consists of glycosaminoglycans and glycoproteins.

tissue mass and neural lamella of the abdominal region of the nervous system of the adult wax-moth *Galleria mellonella*, was shown to possess glycosaminoglycans. Later investigations, again using alcian blue, confirmed this result and enabled the identification of chondroitin sulfate, but not of keratan sulfate (Ashhurst and Costin, 1971c). More recently, Dybowska and Dutkowsky (1977; 1979) have confirmed the presence of glycosaminoglycan in the neural lamella around the brain of *Galleria* using ruthenium red staining and enzyme extractions, but there appear to be more glycosaminoglycans in the larval neural lamella than in that of the adult. The presence of hyaluronate in the prepupal larva of *Musca domestica* was detected histochemically (Mustafa and Kamat, 1970).

The results obtained histochemically have been reinforced by biochemical studies; hyaluronate and sulfated glycosaminoglycans were identified in extracts from *Calliphora erythrocephala* and *Phormia regina* (Höglund, 1976a,b; Sharief *et al.*, 1973).

2.3. Development of Fibrous Connective Tissues

Very few studies have been designed to determine when and how the connective tissues are produced. Indeed, there has been much debate about the cells repsonsible for producing connective tissue. Some authors have suggested that hemocytes can produce connective tissues, but the evidence is inconclusive. Since this topic formed the subject of a recent paper (Ashhurst, 1979a), it will not be discussed further in this chapter.

The layers of connective tissue are usually produced after the organ they surround is almost fully developed. For example, the neural lamella develops around the nervous system of embryo locusts from the ninth day onwards (in conditions where hatching occurs on day 12) (Ashhurst, 1965), and during metamorphosis it does not re-form around the nervous system of lepidopteran pupae until the reorganization of the larval nervous system to the adult form is completed (Ashhurst, 1964; Ashhurst and Richards, 1964a; Pipa and Woolever, 1965; McLaughlin, 1974). Similarly the connective tissue around the ejaculatory duct of the adult male locust is produced for the most part after the adult molt (Ashhurst and Costin, 1974).

Studies of the developing neural lamella of locust embryos showed that Scharrer's (1939) suggestion that it is produced by the underlying glial cells is correct. In embryos of *Schistocerca gregaria* (Ashhurst, 1965), the perineurial cells form a thick layer from day 9 to 11 and the cytoplasm contains much rough endoplasmic reticulum with dialated cisternae and Golgi complexes (Figure 13). On day 9, the nervous system is surrounded by a basement membrane (Figure 14), but this rapidly thickens to form a thick amorphous layer; thin banded collagen fibrils are not seen until after hatching (Figure 15). By 12 days, the nymphal neural lamella is fully formed, the cytoplasm of the perineurial cells has contracted (Figure 15), and only a small amount of rough endoplasmic reticulum is present. The neural lamella continues to increase in thickness during the larval instars. Unfortunately, this process has not been studied, but it would seem appropriate for most growth to occur near the time of a molt. It would, therefore, be interesting to know if the perineurial cells display large amounts of dilated rough endoplasmic reticulum at this time. Osińska (1981) found that the perineurial cells of the first instar larva of *Galleria mellonella* have much rough endoplasmic reticulum, but she found much less in the second and third instars. As the larvae were not accurately aged, it is conceivable that the latter cells were in a quiescent, intermolt stage. In the recently molted adult tick, *Boophilus microplus,* the perineurial cells do contain large dilated cisternae of rough endoplasmic reticulum (Binnington and Lane, 1980).

The development of the connective tissues around the adult nerve cord of moths has received much more attention. An early study of the nerve cord during the pupal stage of the wax-moth, *Galleria mellonella,* (Ashhurst and Richards, 1964a) showed that in the prepupa the larval neural lamella starts to split and adipohemocytes (sometimes referred to as granular hemocytes) penetrate the neural lamella. In conditions in which the pupal stage lasts for 7 days, the neural lamella is broken down by the adipohemocytes within 24–36 hr of molting. During this time, the reorganization of the larval central nervous system to the adult form is initiated. When this is complete, at around 2 days, the perineurial cells are swollen, and in addition, their number has increased on the dorsal side of the abdominal interganglionic connectives. During the next 4 days, these cells produce the fibrous connective tissue of the neural lamella and also that of the dorsal connective tissue mass. During the time they are actively synthesizing collagen and glycosaminoglycans, etc., the cytoplasm contains large, almost spherical cisternae of rough endoplasmic reticulum and small Golgi complexes (Figure 11) (Ashhurst, 1964; Ashhurst and Costin, 1976). The cells contain large amounts of glycogen which are labile to fixatives; hence the large empty spaces seen in the cytoplasm. In the dorsal mass, the cells are pushed apart as the connective tissue is formed, and thus the cells decrease in size. In the adult, the cells are small, with little rough endoplasmic reticulum, and are held together by numerous desmosomes with many associated microtubules (Ashhurst, 1970). Similar changes in the perineurial cells during metamorphosis in *Manduca sexta* and *Pieris brassicae* were reported by McLaughlin (1974) and Ali (1973) respectively.

The ejaculatory duct of the male reproductive system of *Locusta migratoria* has been the subject of extensive experimental study (Ashhurst and Costin, 1974). The duct is seen as a small invagination of the epidermis in the second larval instar, and a few mesodermal cells are associated with the basal surface of the cells. These cells remain *in situ* and divide during the subsequent larval instars, so that a sheath of cells is formed around the duct. In the fourth instar, a layer of connective tissue with some thin collagen fibrils (Figure 16) is present between the epidermal and surrounding cells. A small amount of connective tissue is present between the cells, which are differentiating (Figure 17). Connective tissue continues to be formed throughout the fifth instar, but the process is rapidly accelerated during the first 10 days after the adult molt (Figure 2). During this period, the cells are characterized by the large amounts of rough endoplasmic reticulum with very dilated cisternae in the cytoplasm (Figures 2, 8); the Golgi complexes are small. A few small, dense granules, presumably lysosomes, may also be found in the cytoplasm. The area occupied by the matrix

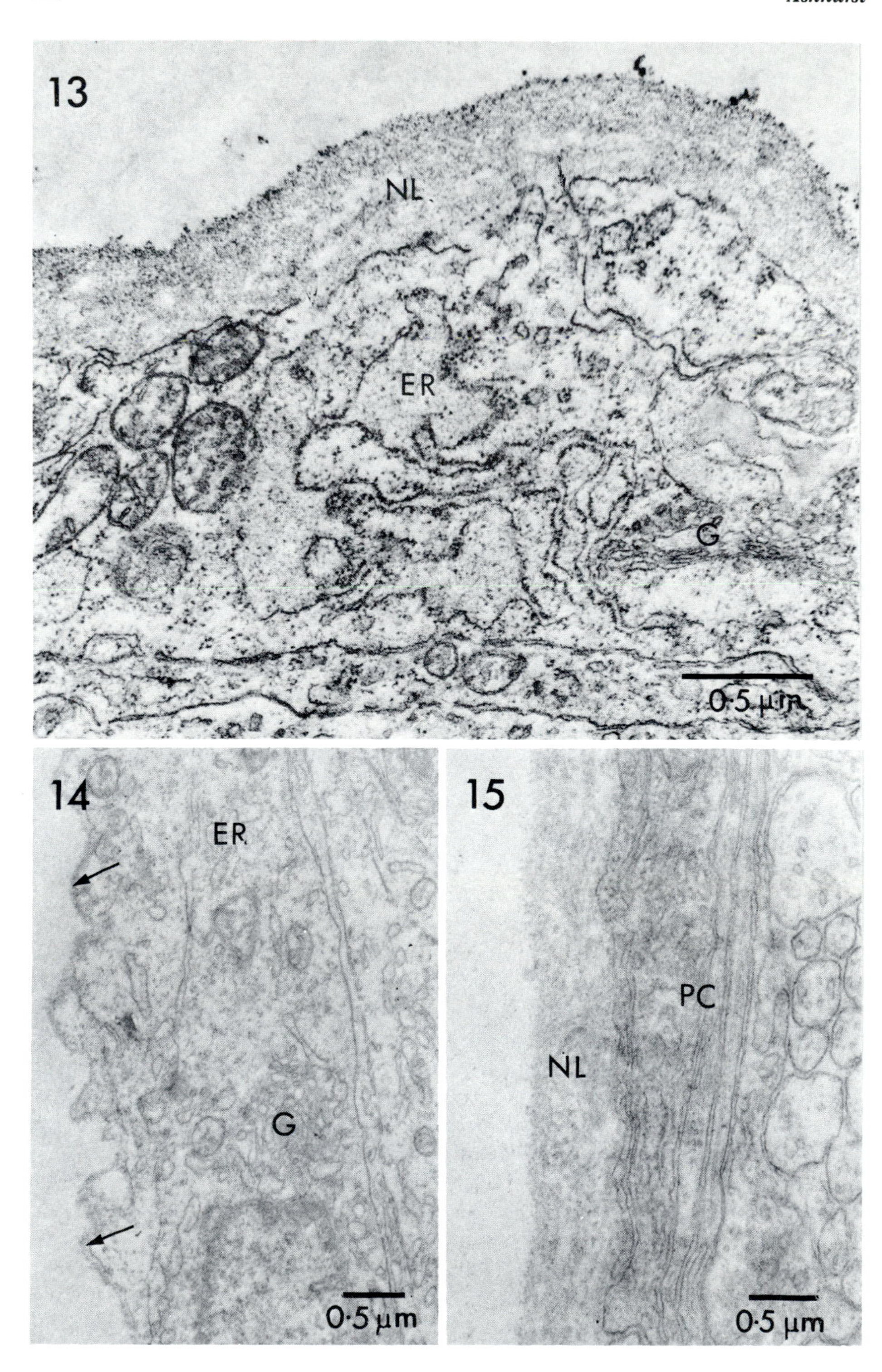
13
NL
ER
G
0·5 µm
14
ER
G
0·5 µm
15
PC
NL
0·5 µm

rapidly increases at the expense of the cells. The collagen fibrils, which in the 1-day adult are of small diameter and indistinctly banded, increase in diameter during the succeeding days, as illustrated in Figures 3 to 7. By 10 days, the production of the matrix is slowing down, and at the onset of sexual maturity at about 21 days, the cells are much reduced in size and have little rough endoplasmic reticulum, but they possess large lysosomes (Figure 10). Large collagen fibrils of very irregular cross section are now present in the matrix (Figures 6,7).

In the latter examples of a developing connective tissue, the matrix is laid down over a defined period in the life cycle, after which the cells become quiescent. It is not known whether these cells can be stimulated into further synthetic activity.

2.4. Biosynthetic and Secretory Pathways of Fibrous Collagen

The synthetic pathways of collagen in a variety of vertebrate connective tissue cells, fibroblasts, chondroblasts, osteoblasts, and odontoblasts (see review by Ross, 1975) have been investigated using electron autoradiography. Differing conclusions were reached; Ross and Benditt (1965), working on guinea pig dermal fibroblasts, suggested that the newly synthesized collagen passes straight from the rough endoplasmic reticulum to the matrix, whereas similar studies of osteoblasts and odontoblasts indicated that the collagen passes from the rough endoplasmic reticulum to the Golgi complex before leaving the cell (Weinstock and Leblond, 1974). In chondroblasts, Salpeter (1968) was unable to make an unequivocal decision about the pathway.

Two series of experiments using the developing connective tissues described in the previous section, the dorsal connective tissue mass of the pupal nerve cord of *Galleria* and the ejaculatory duct of the adult male *Locusta*, were designed to elucidate this pathway in insect fibroblasts (Ashhurst and Costin, 1976). The cells were exposed to tritiated proline, either by injection, or a pulse in organ culture, followed by a cold chase. ^{3}H-proline is usually chosen as the radioactive label for collagen studies since proline and hydroxyproline comprise about 20% of the total amino acids. The experiments were analyzed by the methods of Williams (1973), in which the location of the silver grains is determined by placing a circle over the silver grain and recording the tissue compartment(s) within it. The radius of the circle is calculated from the data

←

Figure 13. A perineurial cell and the developing neural lamella around the central nervous system of an 11-day locust embryo. The perineurial cell has dilated rough endoplasmic reticulum (ER) with amorphous material in the cisternae and a small Golgi complex (G). The neural lamella (NL) looks amorphous; no fibrils are visible.

Figure 14. A perineurial cell in a 9-day locust embryo. The developing neural lamella (arrows) appears similar to a basement membrane. The cytoplasm contains some rough endoplasmic reticulum (ER) and a Golgi complex (G).

Figure 15. The neural lamella (NL) and perineurial cells (PC) in a 2-day first instar locust nymph. The neural lamella contains many thin fibrils. The perineurial cells are much reduced in size.

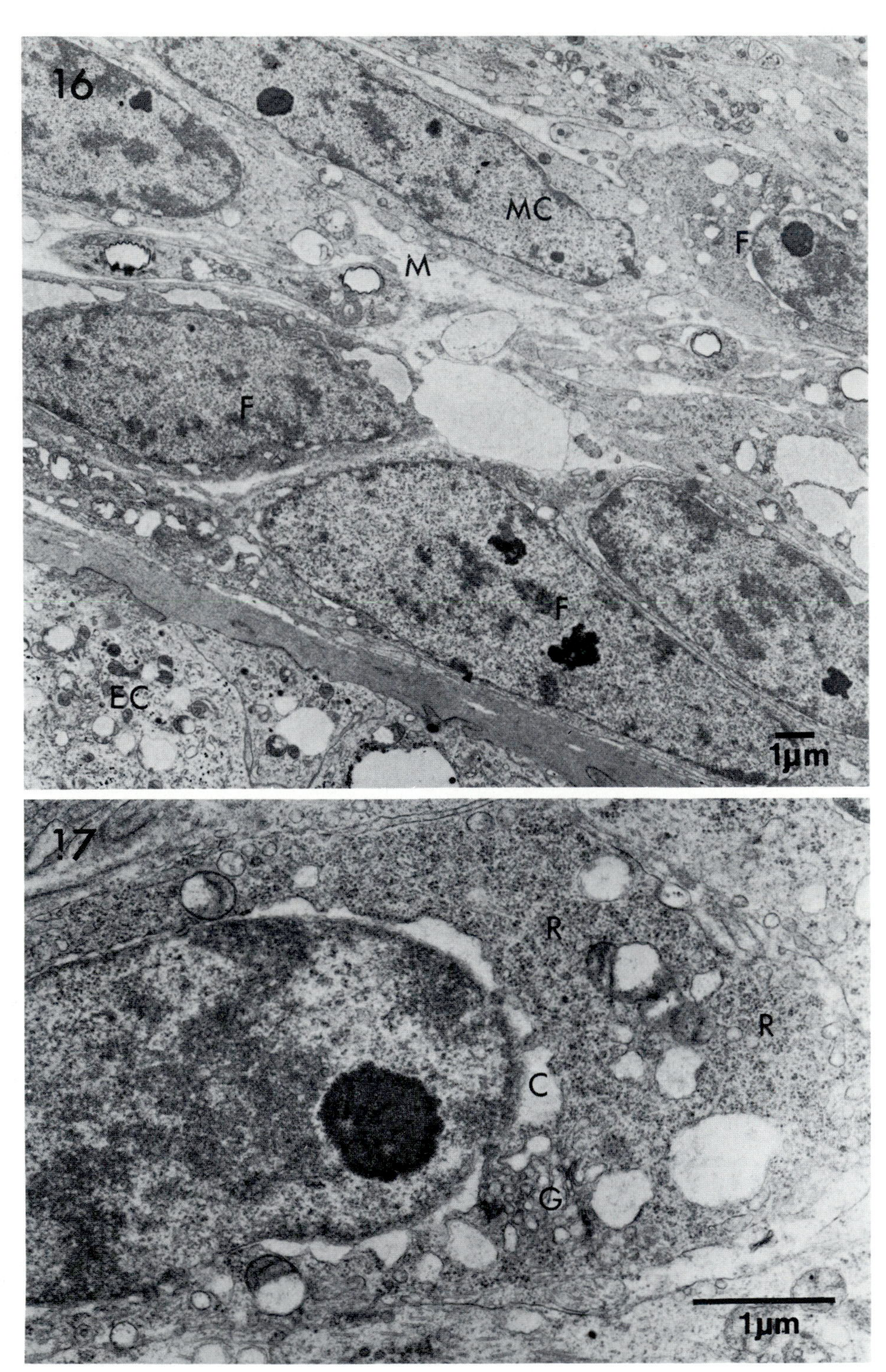
16
MC
F
M
F
F
EC
1µm
17
R
R
C
G
1µm

of Salpeter *et al.*, (1969) on the scatter of silver grains around a tritium source. The relative area of each tissue compartment is determined by placing a grid of the same sized circles over the electron micrographs and recording the compartment(s) within each circle. This method avoids the problems involved in scoring grains which fall over two compartments. The relative amounts of radioactivity over each tissue compartment can be calculated from the data (see Table 2). The circle diameter chosen for this study was equivalent to 280 nm on the prints; since most of the organelles are small, a larger circle would enclose two compartments too frequently. The results of the experiments are shown in Figure 18. The interpretation of this type of experiment is based on the premise that if the radioactively labeled substance is passing from one tissue compartment to another, the relative number of grains per unit area over compartment 1 will rise to a maximum and then start to fall before that over compartment 2 reaches a maximum, and that the subsequent relative numbers over this second compartment will fall later than in the preceding compartment, and so on. In none of these experiments did the results of the relative numbers of grains over the rough endoplasmic reticulum and Golgi complexes show this relationship. Hence, the conclusion reached was that in the locust and wax moth fibroblasts the collagen molecules are synthesised in the rough endoplasmic reticulum, but that for the most part, they then pass straight to the matrix bypassing the Golgi complexes. Statistical analysis provided further evidence for this interpretation of the results. The passage of a small amount of the newly synthesized collagen through the Golgi complex is not excluded. A problem inherent in these studies is that the relative area occupied by the Golgi complex in these tissues is very small, as are the individual complexes, and this leads to problems of interpretation which are discussed in detail by Ashhurst and Costin (1976).

In a similar study of collagen synthesis in the firebrat, *Thermobia domestica*, François (1980) suggested that the collagen molecules do pass through the Golgi complex. While his method of analysis is similar to that of Ashhurst and Costin (1976), he used a grid of much larger circles for the stereological analysis and grain location. This circle had a diameter equivalent to 493 nm on the prints, i.e., almost 0.5 μm. Thus, since most Golgi complexes are about 0.5 μm by 1.0 μm in size, very few circles would fall just over a Golgi complex, while the majority would fall over two adjoining compartments, such as Golgi complex-cytoplasm, or Golgi complex-rough endoplasmic reticulum. Thus, the figures cited for both the area and the grain counts over the Golgi complexes seem to be somewhat higher than might be expected. In addition, the decrease

←

Figure 16. A low-magnification micrograph of the developing connective tissue around the ejaculatory duct of a 0-day fourth instar locust nymph. There is a layer of connective tissue along the base of the epidermal cells (EC). There are differentiating fibroblasts (F) and a few muscle cells (MC) with small areas of matrix (M) separating them.

Figure 17. A differentiating fibroblast from the area of connective tissue in Figure 16. The outer nuclear membrane is enlarging and budding off cisternae (C) into the cytoplasm, which is full of ribosomes (R). A Golgi complex (G) is also present.

Table 2. Data from the Stereological and Grain Analyses and the Calculations of the Relative Number of Grains per Unit Area over Each Compartment for an Experiment on the Uptake of ^{3}H-Proline by Fibroblasts of *Locusta migratoria*

	Experimental times														
	15 min			30 min			1 hr			2 hr			4 hr		
Compartment	Area count	Grain count	Relative no. of grains/ unit area	Area count	Grain count	Relative no. of grains/ unit area	Area count	Grain count	Relative no. of grains/ unit area	Area count	Grain count	Relative no. of grains/ unit area	Area count	Grain count	Relative no. of grains/ unit area
Matrix	2759	18	0.14	2295	24	0.24	2320	48	0.41	3194	108	0.52	2838	124	0.77
Rough endoplasmic reticulum	211	53	5.37	215	40	4.27	134	21	3.12	167	41	3.78	96	19	3.50
Golgi complexes	123	16	2.78	92	9	2.25	71	14	3.92	77	5	1.00	74	7	1.67
Pooled counts for all other components[a]	4471	267	1.28	3554	195	1.26	2988	194	1.29	3741	312	1.28	2857	182	1.13
Total	7564	354	—	6156	268	—	5513	277	—	7179	466	—	5865	332	—

[a]For the complete data, see Ashhurst and Costin (1976).

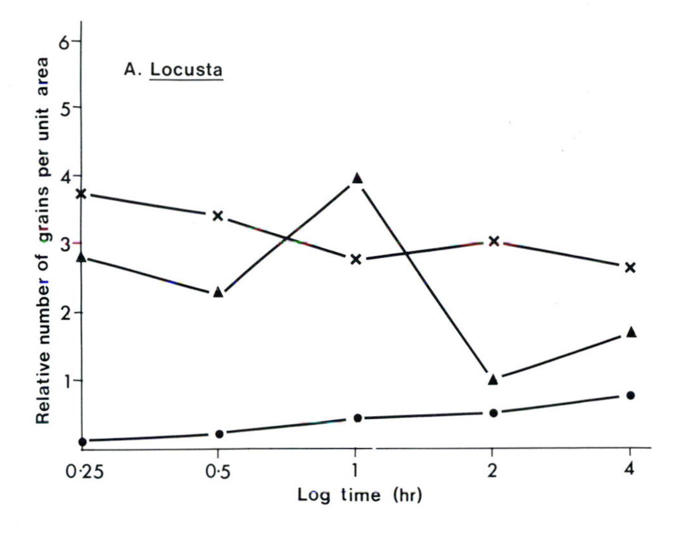

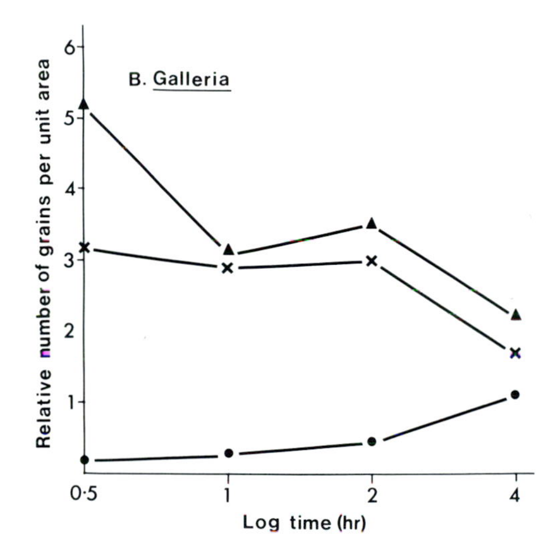

Figure 18. Graphs of the relative number of silver grains per unit area over the rough endoplasmic reticulum, **x**, Golgi complexes, ▲, and matrix, ●, against log time; the data was obtained from experiments on the uptake of ^{3}H-proline by *Locusta* and *Galleria* fibroblasts. (A) *Locusta* fibroblasts, data given in Table 2. (B) *Galleria* fibroblasts, data in Ashhurst and Costin (1976).

in the number of grains per unit area (which is the relative ^{3}H-concentration) over the Golgi complexes does not follow the decrease over the rough endoplasmic reticulum, since it is lower at 8 hr (the figure given in his table for the ^{3}H-concentration over the Golgi complex is incorrectly calculated from the data given). Thus, it is suggested that François does not provide convincing evidence that the Golgi complex is part of the obligatory pathway for all collagen molecules synthesized by the firebrat cells.

2.5. Collagen Molecule

Collagen molecules can be characterized by their morphological, biochemical, and physical properties. The two former groups of properties are more readily determined, especially if the amount of collagen available for study is small. Any biochemical study on insects is restricted by the very small amount of material present in any one insect, and this is especially difficult for studies of collagen. In an early biochemical study, Harper *et al.* (1967) used whole cockroaches, *Blaberus craniifer* and *Leucophaea maderae,* and thus did not know the precise source of their collagen, whereas in more recent studies, Ashhurst and Bailey (1978; 1980) used the ejaculatory duct of the adult male *Locusta migratoria,* and François *et al.* (1980) used the mesenteron of adult *Periplaneta americana.*

2.5.1. Morphological Studies

It is possible to make collagen molecules in solution reaggregate in several ways to form different types of fibrils. If a solution of collagen in weak acetic acid is dialyzed against tap water or a weakly alkaline solution, such as 0.02 M disodium hydrogen phosphate (approx. pH 9.0) for periods up to 2 or 3 days, native type fibrils will be formed, i.e., the molecules will align and aggregate as in Figure 1A. If the same collagen solution is dialyzed against ATP in acid solution, the molecules aggregate as shown in Figure 1B, and form segment-long-spacing (SLS) crystallites. These synthetic fibrils and SLS crystallites may be examined with the electron microscope after either negative or positive staining. Reconstituted fibrils display the periodic banding pattern more clearly than sectioned fibrils, since they are not obscured by the other matrix substances. The SLS crystallites demonstrate the length of the molecule and the distribution of charged, or polar, amino acids along its length. In positively stained preparations, the heavy metals are bound by the polar amino acids.

Insect collagens, in common with other invertebrate collagens, are very insoluble. It is thus difficult to obtain collagen solutions without pretreatment. In order to obtain soluble locust collagen, newly molted adult male locusts, *Locusta migratoria,* were fed for 5 to 6 days on wheat seedlings sprayed with β-aminoproprionitrile, an agent that reduces the amount of molecular cross-linking. After 6 days, during which it is known that active collagen synthesis occurs, the ejaculatory ducts were removed and a solution of collagen in dilute acetic acid was obtained (Ashhurst and Bailey, 1978; 1980). The SLS crystallites and reconstituted fibrils made from this solution are shown in Figures 19 and 20.

In Figure 20, an SLS crystallite made from locust collagen is shown alongside a crystallite made from rat tail collagen, which is known to be type I collagen, i.e., $\alpha 1(\mathrm{I})\alpha 2$. The exact correspondence of the banding pattern along these crystallites is at once apparent. Similarly, the reconstituted locust fibril in Figure 19 is shown with a reconstituted rat tail tendon fibril, and again the exact correspondene of the periodic banding pattern of the two fibrils is obvious.

The conclusions which can be drawn from these experiments are as follows. The SLS crystallites demonstrate firstly that the distribution of the polar amino acids along the length of the locust collagen molecule is very similar to that of mammalian type I collagen, and secondly that the lengths of the molecules must be the same, i.e., approximately 295 nm, since the same number and distribution of amino acid residues must be present to create the same banding pattern. The reconstituted fibrils show that the molecular organization within the fibrils is the same. Thus, it follows that the periodicity of the locust collagen must be the same as that of mammalian type I collagen, that is, 67 nm*. The variations in periodicity measured on electron micrographs must, therefore, be due to the preparative procedures.

A collagen from the mesenteron of the cockroach, *Periplaneta americana*, has recently been used for similar studies. This collagen was subjected to pepsin digestion in order to solubilize it before SLS crystallites and reconstituted fibrils were made (François *et al.*, 1980). The SLS crystallites and reconstituted fibrils made from this collagen are identical to those made from the locust collagen. It should be noted here that the same is true of other invertebrate collagens, such as those of the sea anemone, *Actinia equina*, the liver fluke, *Fasciola hepatica*, and the snail, *Helix pomatia* (Nordwig *et al.*, 1970). Thus, it appears from the morphological studies that invertebrate collagens are very similar to mammalian type I collagen and display the same banding periodicity of 67 nm. The collagen of the thin fibrils of the Diptera, Coleoptera, and Lepidoptera has not yet been used for this type of study, but the indications from work such as that of Locke and Huie (1972) on *Calpodes ethlius* collagen, are that it will be shown to be similar to the locust and cockroach collagens.

2.5.2. *Biochemistry*

It is not pertinent in the context of a chapter on the ultrastructure of connective tissue to devote much space to biochemical studies, but a brief review must be made. Since the early studies of insect connective tissues, workers have endeavored to prove the presence of collagen biochemically. At first, crude chromatographic identification of hydroxyproline was the criterion used to prove the presence of collagen (Ashhurst, 1959; Ashhurst and Richards, 1964b). Later, Harper and his co-workers (1967) attempted a first complete amino acid analysis of collagens extracted from the cockroaches, *Blaberus craniifer* and *Leucophaea maderae*, and concluded that their extracts contained collagens with typical amino acid compositions.

It is now possible to characterize collagens very precisely, and it was shown

*The figure of 67 nm for the periodicity of the banding of collagen fibrils is that obtained from X-ray diffraction studies of native, hydrated rat tail tendon collagen. When water is removed from collagen, as is essential for electron microscopy, shrinkage is inevitable. In the author's experience, measurements of the length of SLS crystallites taken from micrographs can vary from 180 nm to 230 nm in similar preparations made at different times. For this reason, the magnifications and measurements of collagen fibril periodicity are deduced from the measurement of 67 nm above, and for SLS crystallites from the length between bands 6 and 22, of 79 nm, which is calculated from data on the number of amino acids known to be present. For further details of these calculations see Ashhurst and Bailey (1980).

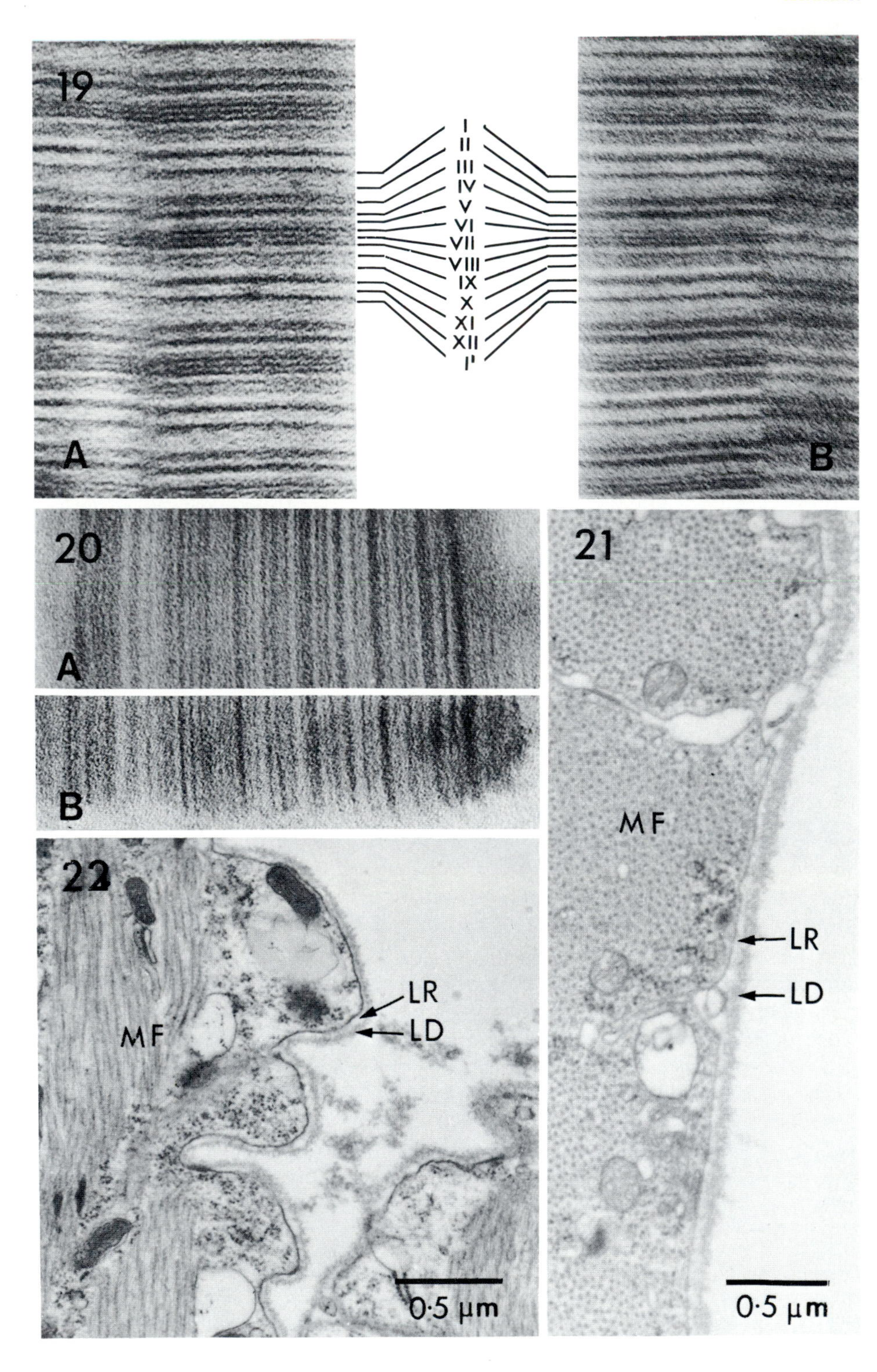
19
I
II
III
IV
V
VI
VII
VIII
IX
X
XI
XII
I'
A
B
20
A
B
21
MF
LR
LD
0·5 μm
22
MF
LR
LD
0·5 μm

by Ashhurst and Bailey (1978; 1980) that the fibrous collagen extracted from the ejaculatory duct of the locust has the following properties. Firstly, there is only one type of α-chain, and so the molecule must be an $\alpha 1$ trimer; secondly, the amino acid composition is that of a typical collagen, but the hydroxylysine content is somewhat elevated; thirdly, the peptides produced by cleaving the α-chains with cyanogen bromide are unique; fourthly, the collagen crosslinks are dihydroxylysinonorleucine and hydroxylysinonorleucine; and finally, the α-chains are precipitated preferentially in 4.0 M sodium chloride.

A comparison of these properties with those of mammalian α-chains immediately indicates a resemblance to the α-chains of the newly characterized type I trimer (Uitto, 1979). Both collagens precipitate at 4.0 M salt concentration and both have a high hydroxylysine content. The differences in cyanogen bromide peptides are of minimal significance as they indicate only the distribution of the small number of methionine residues.

A more recent examination of the collagen of the mesenteron of *Periplaneta americana* (François *et al.*, 1980) indicates that this collagen is essentially similar to that of the locust. The amino acid analysis reveals a high level of hydroxylysine, it precipitates at 4.0 M salt concentration, and it has only one type of α-chain. Nevertheless, the authors suggest that this collagen is more similar to vertebrate type II collagen, i.e., the cartilage collagen, because they consider that its amino acid composition is more similar to that of type II than that of type I collagen. Their additional results show that its elution on CM-cellulose chromatography is later than type II collagen. A further point is that their comparison with normal type I ($\alpha 1(\mathrm{I})\alpha 2$) collagen is obviously not as valid as comparison with type I trimer. The SLS crystallite is similar to that of type I collagen whereas the banding pattern of type II SLS crystallites differs from that of type I (Stark *et al.*, 1972). The evidence for the similarity of cockroach collagen to type II collagen is, therefore, rather equivocal.

A minor component of the collagen extracted from the locust tissue was precipitated at 2.4 M salt concentration. This minor component has similarities with vertebrate type IV collagen and will be discussed later.

2.5.3. Conclusions

The fibrous collagen molecules of the insects so far studied reveal similarities with the vertebrate type I molecule and more particularly with the type I

Figure 19. (A) A reconstituted fibril of locust collagen. (B) A reconstituted fibril of rat tail tendon collagen. The nomenclature of the bands within the periodic pattern is taken from Burns and Gross (1974). Both fibrils are positively stained with uranyl acetate and phosphotungstic acid. Calculated magnification: × 266,000.

Figure 20. (A) SLS crystallite made from rat tail tendon collagen. (B) SLS crystallite made from locust collagen. Both crystallites are positively stained with uranyl acetate and phosphotungstic acid. Calculated magnification: × 240,000.

Figure 21. The basement membrane of a muscle fiber (MF) of fourth instar nymph of *Rhodnius prolixus*, 5 days after feeding, showing the lamina rara (LR) and lamina densa (LD).

Figure 22. Muscle fibers (MF) of the ventral diaphragm of a 6-day pupa, *Galleria mellonella*, with basement membranes showing the lamina rara (LR) and lamina densa (LD).

trimer. This is also true of a number of other invertebrate collagens. It appears, therefore, that a collagen molecule similar to the vertebrate type I trimer evolved in the primitive metazoan phyla and that this molecule has been conserved with little change during evolution.

3. Basement Membranes

In vertebrates, a basement membrane is produced by most cells, with the notable exception of fibroblasts and blood cells. With the light microscope, the basement membrane appears as a dense line adjacent to the basal region of the epithelial cell, which is especially clear after the PAS reaction. The electron microscope reveals a more complex structure, which is shown diagrammatically in Figure 23. Immediately adjacent to the cell membrane there is an electron-lucent layer, the lamina rara, or lamina lucida, which is about 50 nm thick. The electron-dense lamina densa appears amorphous, but very thin filaments can be discerned within the matrix. This layer is of variable thickness, but it is not usually more than 100 nm thick. Below the lamina densa, there may be a feltwork of thin collagen fibrils, which is sometimes referred to as the reticular layer. This layer is produced by the cells in the underlying connective tissue and not by the epithelial cells. The term "basement membrane" is normally used to denote the laminas rara and densa (Heathcote and Grant, 1981) but some authors prefer to restrict the term to the lamina densa (Kefalides *et al.*, 1979).

The collagen of the basement membrane is located in the lamina densa, and it has been designated as type IV. It was originally considered that the type IV collagen molecule consists of three identical α-chains; now the evidence increasingly suggests that there is a group of type IV collagens. At least two genetically distinct polypeptides have been isolated from mammalian lens capsule and glomerular basement membranes (Heathcote and Grant, 1981). The collagen molecules do not form typical banded fibrils, but instead form a random network within the lamina densa.

The glycoproteins of basement membranes are only now being characterized. Laminin has been located by immunofluorescence, and it appears to be localized in the lamina rara (Foidart *et al.*, 1980). Fibronectin has also been located by immunofluorescence in the region of the basement membrane of several cells, but its exact location remains uncertain (Heathcote and Grant,

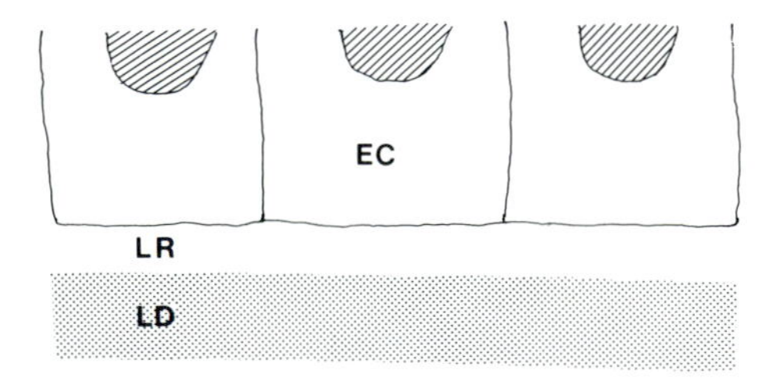

Figure 23. A diagrammatic representation of a typical basement membrane. The lamina rara (LR) separates the lamina densa (LD) from the basal cell membrane of the epithelial cells (EC).

1981). Until recently, proteoglycans had been detected only in embryonic basement membranes, such as that of the developing mouse submandibular gland (Bernfield *et al.*, 1972), but in the glomerular basement membrane of adult rats a glycosaminoglycan has now been detected that was identified by enzymatic digestion as heparan sulfate, the glycosaminoglycan associated with cell surfaces (Kanwar and Farquhar, 1979). Much work remains to be done before the constituents of basement membranes are as well characterized as those of the rest of the intercellular matrix.

3.1. Structure and Location of Basement Membranes

Basement membranes are found in insects under epithelia and surrounding muscle fibers and fat body cells. The morphology illustrated in the micrographs is exactly the same as that found in vertebrate tissues, but the lamina densa of the *Galleria* muscle fiber (Figure 21) is considerably thinner than that of the *Rhodnius* muscle fiber (Figure 22). The occurrence of basement membranes in the typical form described here is not as frequent as might be expected, since very often they merge into the adjacent connective tissues. For example, the first connective tissue found around the developing nervous system in *Schistocerca* embryos looks like a basement membrane (Ashhurst, 1965), but this is rapidly obliterated as the neural lamella develops (Figures 13, 14). Evidence that the epithelial cells which form the ejaculatory duct of the male locust produce a basement membrane came from the biochemical analyses (see section 3.2); a basement membrane was never observed in electron micrographs (Ashhurst and Costin, 1974).

3.2. Composition of Basement Membranes

The first indication of the presence of a basement membrane collagen in insects came from analyses of the collagen of the locust ejaculatory duct. During the fractionation procedures, a minor collagenous component was identified in the 2.4 M salt precipitate. Insufficient material was available to pursue a detailed characterization, but its possible amino acid composition was estimated and found to be similar to that of mammalian type IV collagen (Ashhurst and Bailey, 1980). A similar collagen has recently been isolated from locust muscle, and preliminary results confirm its similarity to the mammalian type IV collagens (unpublished observations).

Insect basement membranes are PAS-positive, but there are few reports of histochemical evidence for the presence of glycosaminoglycans. One such report is that of Francois (1978), who found that the basement membrane of the epithelial cells of the mesenteron of adult *Periplaneta* binds alcian blue under varying conditions, which suggests the presence of sulfated glycosaminoglycans. Another report (Dutkowski, 1977) also suggests that glycosaminoglycans may be present in the basement membrane of the fat body of *Galleria*, since it binds ruthenium red. No attempts to locate laminin or fibronectin have yet been published. Doubtless, the insect basement membrane will prove to be very similar to that of the mammals.

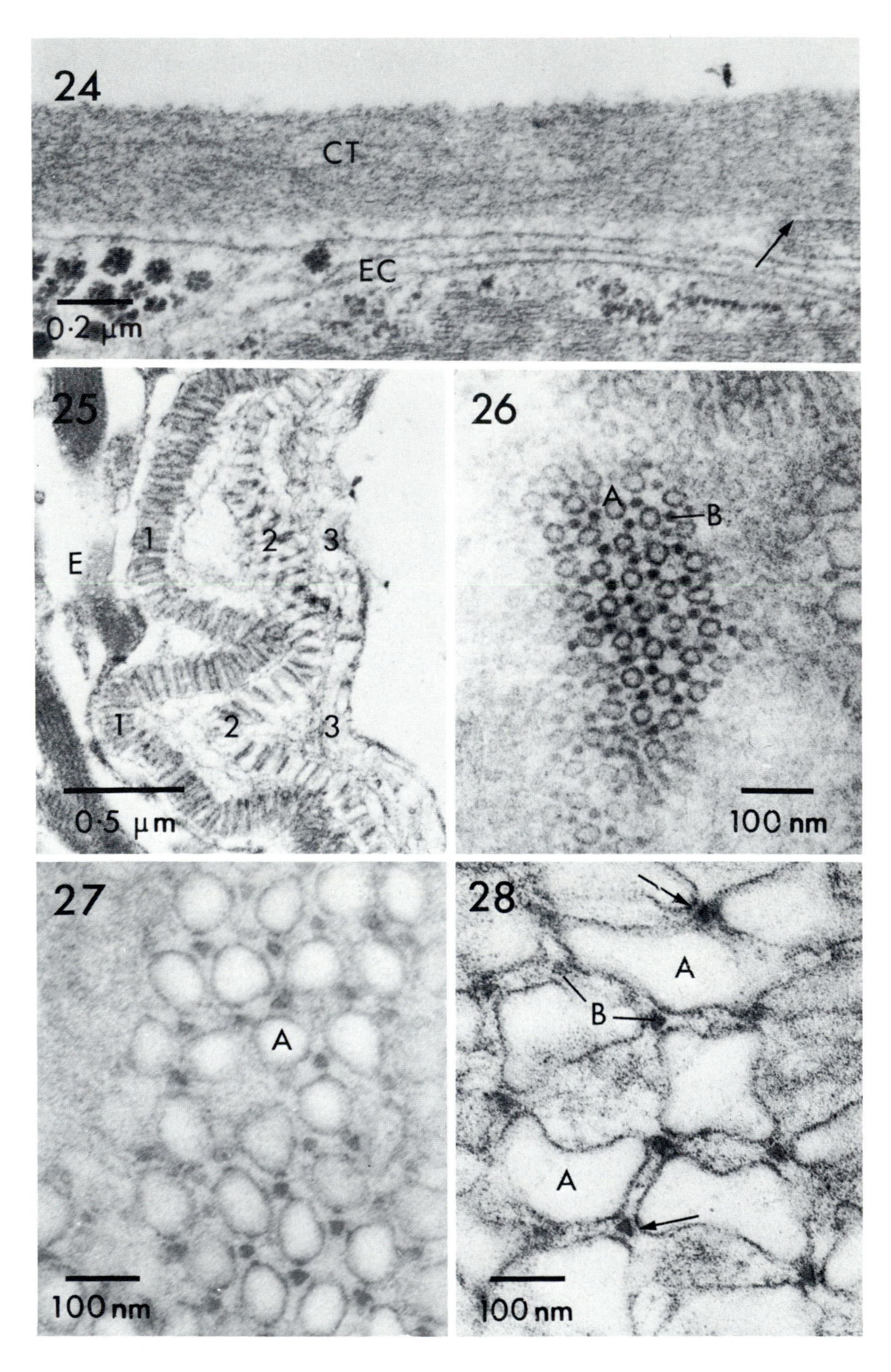
24
CT
EC
0·2 μm
25
E
1
2
3
1
2
3
0·5 μm
26
A
B
100 nm
27
A
100 nm
28
A
B
B
A
100 nm

4. Other Connective Tissues

Insects possess some structures within the intercellular matrix that are either unique or rarely seen in other animals. They will be considered separately, since there is no biochemical evidence to relate them to structures in other animals.

4.1. Thick, Nonfibrous Layers of Connective Tissue

In many instances, insect tissues are surrounded by a precisely delimited layer of dense amorphous connective tissue. Such layers may be seen as developmental stages in the formation of a fibrous connective tissue (see section 2.3), but often the nonfibrous tissue persists. An obvious example is the subepidermal layer of *Rhodnius prolixus* (Figure 24) (Wigglesworth, 1973). In some areas very thin fibrils are seen, but no banding or other detail can be resolved. These layers do not display the structure of a basement membrane because there is no lamina rara separating the electron-dense layer from the cell membrane, and this latter layer is many times thicker than the typical lamina densa. Nothing is known about the composition of these layers.

An increase in the thickness of the lamina densa occurs in sites such as the mammalian glomerulus, but here it is interpreted as the fused laminae of the endothelial and epithelial basement membranes, and it is known to contain type IV collagen. It is, therefore, suggested that until more is known of the chemical composition of these layers, they should be considered as a separate category of connective tissue structure.

4.2. Specialized Connective Tissue Associated with Midgut Epithelium

The epithelium of the midgut of some insects of the orders Coleoptera, Diptera, Heteroptera and Siphonaptera lies on a highly specialized layer of connective tissue which differs from a true basement membrane. These layers

Figure 24. The connective tissue layer (CT) under the epidermal cells (EC) of a newly molted, fifth instar *Rhodnius* nymph. Thin filaments can be seen, but no periodic banding can be resolved. At this stage the epidermal cells are reorganizing and moving away from the connective tissue, but in one region (arrow) it is closely applied to the adjacent cell.

Figure 25. The gridlike layers (1, 2, 3) under the midgut epithelium (E) of a third instar larva of *Oryctes nasicornis* in transverse section.

Figure 26. A tangential section of layer 1. The larger cylindrical units (A) have less dense cores and are separated by smaller uniformly electron dense units (B); both types of unit are arranged in regular lattices. In places (arrows), there are indications of bridges joining the units.

Figure 27. A tangential section of layer 2. In this layer, the units with less dense cores (A) are larger than in layer 1 and the lattices are less regular.

Figure 28. A tangential section of layer 3. The less electron-dense units (A) are much larger and irregular in shape. The small dense units (B) are triangular in some places and an increased density in the space between the units (arrows) suggests that they may be joined. (Figures 25 to 28 are reproduced courtesy of Dr. J. François.)

were first noticed by Bertram and Bird (1961) in *Aedes aegypti,* and have since been described in detail by Terzakis (1967) and Reinhardt and Hecker (1973). In longitudinal sections, three or four thin layers consisting of alternating dark and light bands are seen, but tangential sections reveal an array of cylindrical units with less electron dense centers joined to form a network, which is expanded and distorted after a blood meal. Similar structures have been observed in several fleas, but here there is an array of large-diameter cylindrical units separated by a network of smaller-diameter units (Richards and Richards, 1968; Reinhardt *et al.*, 1972), and the individual units do not appear to be joined in any way.

In a survey of thirteen species of Coleoptera, Holter (1970) found gridlike layers under the midgut epithelia of six species. He described a single layer which in longitudinal section appears as a ribbonlike structure transected by light and dark bands. In tangential sections, an array of two types of units is seen; small, electron-dense units occur around larger less dense units. These cylindrical units appear to be held together by bridges. Studies of the midgut of the larva of the beetle *Oryctes* (Hess and Pinnock, 1975; Bayon and François, 1976), revealed three layers under this epithelium (Figure 25). The innermost layer is like that described by Holter in the other beetles, but while the two outer layers are made up of similar small units, the less electron-dense units increase in size in the middle and again in the outer layer (Figures 26, 27, 28).

Yet another type of modified connective tissue layer was found under the midgut epithelial cells of two species of Heteroptera, *Nepa cinerea* and *Ranatra linearis* (Gouranton, 1970). Here, a typical basement membrane-like layer (lamina densa) is separated from the cell membrane firstly by an electron-lucent layer of about 35 nm and secondly by a discontinuous layer about 90 nm thick. Tangential sections show that this layer is composed of a series of polygonal plaques, separated by gaps of approximately 30 nm which are crossed by a series of filaments less than 5 nm in diameter. Digestion with collagenase and pronase suggest that both the dense plaques and the outer layer contain collagen.

The composition of these connective tissue structures is unknown. The only evidence for the presence of collagenous protein is from Gouranton's work on *Ranatra,* but the layers in *Oryctes* appear to be resistant to collagenase digestion (Bayon and François, 1976). Holter (1970) attempted a histochemical analysis on *Harpalus rufipes* and found that the layers are PAS positive and give a strong reaction with tests for tyrosine. The tests for glycosaminoglycans gave equivocal results. Similarly, the layers in *Ranatra* are PAS positive, but the tests for glycosaminoglycans gave no clear results (Gouranton, 1970). *Oryctes* again appears different, since Bayon and François (1976) give histochemical evidence for the presence of sulfated glycosaminoglycans. Thus, while glycoproteins are clearly present, the other constituents of the layers have yet to be positively determined.

The function of these connective tissue networks is unknown. It may be significant that mosquitos and fleas feed only occasionally and their midgut becomes very distended; the network may allow stretching and provide extra support for the midgut in these insects. It is much more difficult to ascribe a possible function to the structures in the Coleoptera and Heteroptera; indeed, Gouranton considers it impossible on the present evidence.

Similar structures have not yet been reported elsewhere in insects, except around the ovarioles of *Aedes aegypti* (Bertram and Bird, 1961). Recently, however, modified connective tissue structures have been described under the epithelium of the midgut of two crustaceans (Factor, 1981), but again their significance is unknown. It may be that the epithelium of the midgut of many arthropods is supported by specialized connective tissue structures, but until more evidence is accummulated about their distribution, structure, and composition, it will be impossible to deduce their physiological significance.

5. Elastic Fibers

The first suggestion that a second type of fiber might be present in certain insect connective tissues was made by Baccetti and Bigliardi (1969). They found bundles of thin fibrils in the wall of the dorsal vessel of the orthopteran, *Aiolopus strepens,* which were removed by elastase digestion, and thus they suggested that the bundles of fibrils might be a form of elastic fiber. Later, Locke and Huie (1972) found similar fibrils in the neural lamella and in the pericardial connective tissue and its attachment strands of *Calpodes ethlius* larvae. When a reaction for peroxidase is performed on these tissues, or when sections of the tissues fixed only in glutaraldehyde are stained with phosphotungstic acid, a layer of darkly stained fibril bundles is observed peripherally in the neural lamella and as discrete bundles in the connective tissue associated with the heart (Figure 29). Subsequently, Locke and Huie (1975) found that the fibers stained with tannic acid if it is added to the glutaraldehyde fixative. Two types of fibril can be distinguished in the bundles; thick fibrils about 40 nm in diameter are separated by a space of about 70 nm in which microfibrils approximately 6 nm thick occur. The identification of these composite fibril bundles as a form of elastic fiber is based on the known elasticity of the tissues concerned. Similar fibers are present in the neural lamella of the abdominal nerve cord of adult female locusts (Figure 30) (Locke and Huie, 1975); the abdomen of the female locust is extended to about three times its normal length during oviposition. More recently, François (1978) noticed similar fibers in the connective tissue of the mesenteron of *Periplaneta americana.*

The relationship of these fibers to vertebrate elastic fibers is unknown. Morphologically, vertebrate fibers consist of an electron-lucent amorphous material which forms the core of the fiber and a peripheral layer of microfilaments. The electron-lucent material is the protein elastin.

Fibers, identified as elastic fibers on the basis of their staining with spirit blue, have been located in animals from all the invertebrate phyla (Elder, 1973), but the protein elastin has not been isolated from any invertebrate, despite an exhaustive study of fourteen species (Sage and Gray, 1979). Thus, if these fibers in insects and other invertebrates have elastic properties, the "elastic" protein must be unique to the invertebrates. It would be interesting to discover if the

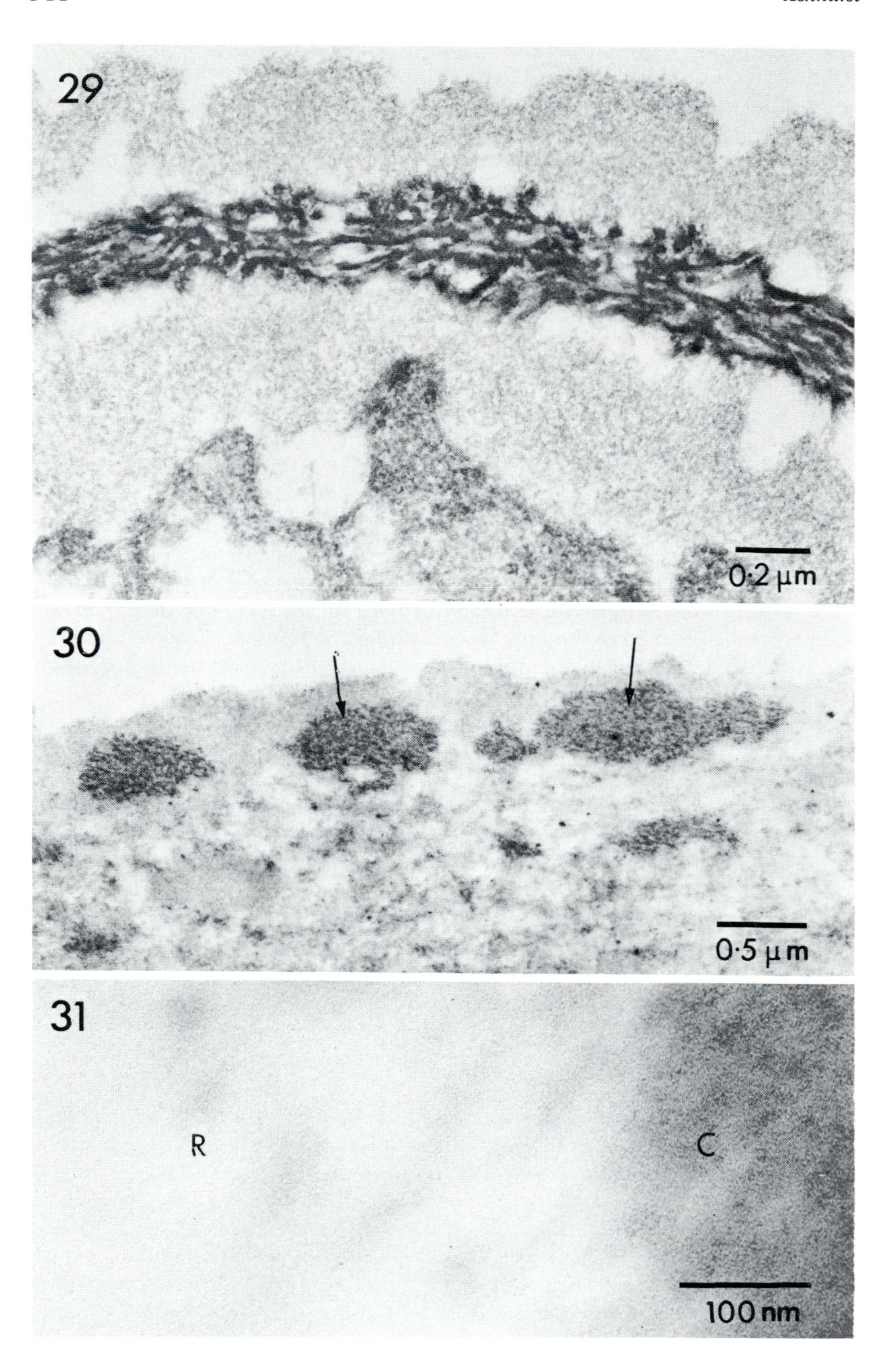
29
0·2 μm
30
0·5 μm
31
R
C
100 nm

invertebrate fibers contain a protein similar to the microfibrillar protein of the mammalian elastic fiber.

6. Resilin

Resilin is a perfectly elastic protein which is unique to the invertebrates. It was first identified by Weis-Fogh in 1959 (see review by Neville, 1975), and it consists of a randomly coiled polypeptide chain in which dityrosine and tertyrosine cross links maintain the tertiary structure. It is present in all insects, and it has two main functions. It stores energy in flying and jumping insects, or it may replace muscles. Thus, it is found in the wing hinges of locusts and attached to an apodeme in the jumping mechanism of fleas. No muscles occur in the proboscis of butterflies; it is retracted by resilin. Apart from its presence in arthropods, resilin is associated with the jaw cuticle of *Eoperipatus weldoni* (Onychophora), and more recently it has been shown to form the retraction mechanism for the jaws of an annelid, *Perinereis* (Sasikala and Sundara Rajulu, 1978).

Ultrastructurally, resilin is disappointing (Figure 31); it appears as an electron-lucent area with no apparent structure. It follows that if seemingly homogeneous areas are encountered in electron micrographs, it may be useful to test for the presence of resilin. The tertiary cross links of resilin cause it to exhibit autofluorescence, so its presence can be readily demonstrated (see Neville, 1975).

7. Conclusions

This present survey of the ultrastructure of insect connective tissue underlines the many similarities which exist in the connective tissues of all animals. The components, that is, the collagens, proteoglycans, and glycoproteins of which the different connective tissues are composed, have similar chemical and physical properties throughout the animal kingdom. The identification and cataloguing of the insect connective tissue is nearing completion; one doubts that new categories will be discovered. The future problems will be the detailed characterization of the components, which is necessary before their full biological significance can be appreciated.

One problem which cannot be overcome is that because insects are so small, it is very difficult to obtain material for exhaustive biochemical investigations,

←

Figure 29. The pericardial sheath of *Calpodes ethlius*. An elastic fiber in longitudinal section is seen within an amorphous matrix; this fiber is stained with phosphotungstic acid. (Micrograph courtesy of Dr. M. Locke.)

Figure 30. Transverse sections of elastic fibers (arrows) in the neural lamella of a mature adult female locust. A peroxidase reaction has been performed on this tissue.

Figure 31. Resilin (R) in the pleural arch of the flea, *Xenopsylla cheopis*. The resilin is unstained and grades into the densely stained cuticle (C). (Micrograph courtesy of Dr. A. C. Neville.)

and histochemical techniques which depend on staining reactions have inherent limitations. The recent development of immunohistochemical techniques, however, may provide a breakthrough in the precise characterization of the connective tissue matrices of insects and other invertebrates. Antibodies have now been produced to all the mammalian collagens and to fibronectin and laminin. The antibodies have resulted in the precise localization of these macromolecules in various tissues and have enabled investigation of the sequential deposition of the different collagens and glycoproteins during developmental and healing processes. There is, however, one major problem; in many instances the antigenic sites on the molecules are species-specific, and thus many antibodies will not cross-react with the macromolecules of species other than those closely related. Thus, the potential information to be gained by using these techniques on insect connective tissues may be limited by the availability of antibodies which will cross-react with the insect proteins. Unfortunately, it will be impossible to obtain sufficient amounts of purified insect connective tissue proteins to immunize and raise antibodies in even a small mammal. Let us hope that suitable vertebrate antibodies will be found.

While the characterization of the collagens is progressing, little is known about the "elastic" fibers. Baccetti and Bigliardi (1969) identified the fibers in *Aiolopus strepens* as elastic because they were liable to digestion with elastase. It is pertinent to comment here that no commercial sample is pure. Often only about 25% of the activity of a sample of a connective tissue enzyme is attributable to the specific enzyme and the remaining 75% of the activity is due to other contaminating proteolytic enzymes. Hence, the results of such extractions should be interpreted with due caution. Perhaps when a suitable source of insect elastic fibers is found, biochemical analysis will be possible. Since it seems unlikely that these fibers possess the same proteins as vertebrate elastic fibers (Sage and Gray, 1979), immunological techniques may not be applicable to their study.

The connective tissue layers under the midgut epithelium of many insects would seem to be very suitable candidates for immunological study. At least, it should be possible to ascertain whether collagenous proteins are present. Similarly it should be possible to determine the type of collagen found in thick, nonfibrous layers and subsequently to deduce their relationship to basement membranes.

The localization of fibronectin and laminin is another problem which might be solved by immunological investigation. Fibronectin has been located in sponges using antimammalian fibronectin antibody (Labat-Robert *et al.*, 1980), so hopefully similar antibodies may cross-react with insect fibronectin. The study of laminin is less developed, but its distribution in insect and other invertebrates should be amenable to immunological study.

There are many aspects of insect connective tissue research awaiting investigation, but the future appears promising, especially if it proves possible to apply the new methods developed for work on vertebrates to insect tissues.

References

Ali, F. A., 1973, Post-embryonic changes in the central nervous system and perilemma of *Pieris brassicae* (L), (Lepidoptera: Pieridae), *Trans. R. Ent. Soc. London.* **124**:463–498.

Ashhurst, D. E., 1959, The connective tissue sheath of the locust nervous system: a histochemical study, *Quart. J. Microsc. Sci.* **100**:401–412.

Ashhurst, D. E., 1964, Fibrillogenesis in the wax-moth, *Galleria mellonella, Quart. J. Microsc. Sci.* **105**:391–403.

Ashhurst, D. E., 1965, The connective tissue sheath of the locust nervous system: its development in the embryo, *Quart. J. Microsc. Sci.* **106**:61–73.

Ashhurst, D. E., 1968, The connective tissue of insects, *Annu. Rev. Entomol.* **13**:45–74.

Ashhurst, D. E., 1970, An insect desmosome, *J. Cell Biol.* **46**:421–425.

Ashhurst, D. E., 1979a, Haemocytes and connective tissue: a critical assessment. In *Insect Hemocytes: Development, Form, Functions, and Techniques*, edited by A. P. Gupta, pp. 319–330, Cambridge University Press, Cambridge.

Ashhurst, D. E., 1979b, Histochemical methods for hemocytes. In *Insect Hemocytes: Development, Form, Functions, and Techniques*, edited by A. P. Gupta, pp. 581–599, Cambridge University Press, Cambridge.

Ashhurst, D. E., and Bailey, A. J., 1978, Insect collagen. In *Coll. Int. CNRS*, **287**:79.

Ashhurst, D. E., and Bailey, A. J., 1980, Insect collagen, morphological and biochemical characterization, *Eur. J. Biochem.* **103**:75–83.

Ashhurst, D. E., and Costin, N. M., 1971a, Insect mucosubstances. I. The mucosubstances of developing connective tissue in the locust, *Locusta migratoria, Histochem. J.* **3**:279–295.

Ashhurst, D. E., and Costin, N. M., 1971b, Insect mucosubstances. II. The mucosubstances of the central nervous system, *Histochem. J.* **3**:297–310.

Ashhurst, D. E., and Costin, N. M., 1971c, Insect mucosubstances. III. Some mucosubstances of the nervous systems of the wax-moth (*Galleria mellonella*) and the stick insect (*Carausius morosus*), *Histochem. J.* **3**:379–387.

Ashhurst, D. E., and Costin, N. M., 1974, The development of a collagenous tissue in the locust, *Locusta migratoria, Tissue Cell* **6**:279–300.

Ashhurst, D. E., and Costin, N. M., 1976, The secretion of collagen by insects: uptake of ^{3}H-proline by collagen-synthesizing cells in *Locusta migratoria* and *Galleria mellonella, J. Cell Sci.* **20**:377–403.

Ashhurst, D. E., and Richards, A. G., 1964a, A study of the changes occurring in the connective tissue associated with the central nervous system during the pupal stage of the wax-moth, *Galleria mellonella*, L., *J. Morphol.* **114**:225–236.

Ashhurst, D. E., and Richards, A. G., 1964b, The histochemistry of the connective tissue associated with the central nervous system of the pupa of the wax-moth, *Galleria mellonella*, L., *J. Morphol.* **114**:237–246.

Baccetti, B., 1955, Ricerche sulla fine struttura del perilemma nel sistema nervoso degli insetti, *Redia* **40**:197–212.

Baccetti, B., 1961, Indagini comparative sulla ultrastruttura della fibrilla collagene nei diversi ordini degli insetti, *Redia* **46**:1–7.

Baccetti, B., and Bigliardi, E., 1969, Studies on the fine structure of the dorsal vessel of Arthopods. I. The "heart" of an orthopteran, *Z. Zellforsch.* **99**:13–24.

Bayon, C., and François, J., 1976, Ultrastructure de la lame basale du mesenteron chez la larve d'*Oryctes nasicornis* L. (Coleoptera: Scarabaeidae), *Int. J. Insect Morphol. Embryol.* **5**:205–217.

Bernfield, M. R., Banerjee, S. D., and Cohn, R. H., 1972, Dependence of salivary epithelial morphology and branching morphogenesis upon acid mucopolysaccharide-protein (proteoglycan) at the epithelial surface, *J. Cell Biol.* **52**:674–689.

Bertram, D. S., and Bird, R. G., 1961, Studies on mosquito-borne viruses in their vectors. I. The normal fine structure of the midgut epithelium of the adult female *Aedes aegypti* (L.) and the functional significance of its modification following a blood meal, *Trans. R. Soc. Trop. Med. Hyg.* **55**:404–423.

Binnington, K. C., and Lane, N. J., 1980, Perineurial and glial cells in the tick *Boophilus microplus* (Acarina: Ixodidae): freeze-fracture and tracer studies, *J. Neurocytol.* **9**:343–362.

Burns, R. R., and Gross, J., 1974, High resolution analysis of the modified quarter-stagger model of the collagen fibril, *Biopolymers* **13**:931–941.

de Biasi, S., and Pilotto, F., 1976, Ultrastructural study of collagenous structures in some Diptera, *J. Submicrosc. Cytol.* **8**:337–345.

Dutkowski, A. B., 1977, The ultrastructure and ultracytochemistry of the basement membrane of the *Galleria mellonella* fat body, *Cell Tissue Res.* **176**:417–429.

Dybowska, H. E., and Dutkowski, A. B., 1977, Ruthenium red staining of the neural lamella of the brain of *Galleria mellonella, Cell Tissue Res.* **176:**275–284.

Dybowska, H. E., and Dutkowski, A. B., 1979, Developmental changes in the fine structure and some histochemical properties of the neural lamella of *Galleria mellonella* (L.) brain, *J. Submicrosc. Cytol.* **11:**25–37.

Elder, H. Y., 1973, Distribution and functions of elastic fibers in the invertebrates, *Biol. Bull.* **144:** 43–63.

Factor, J. R., 1981, Unusually complex basement membranes in the midgut of two decapod crustaceans, the stone crab (*Menippe mercenaria*) and the lobster (*Homarus americanus*), *Anat. Rec.* **200:**253–258.

Fessler, J. H., and Fessler, L. I., 1966, Electron microscopic visualization of the polysaccharide hyaluronic acid, *Proc. Nat. Acad. Sci. USA* **56:**141–147.

Foidart, J. M., Bere, E. W., Yaar, M., Rennard, S. I., Gullino, M., Martin, G. R., and Katz, S. I., 1980, Distribution and immunoelectron microscopic localization of laminin, a non-collagenous basement membrane glycoprotein, *Lab. Invest.* **42:**336–342.

François, J., 1968, Nature conjonctive du "tentorium" des Diploures (Insectes, Apterygotes). Étude ultrastructurale, *C. R. Acad. Sc. Paris* **267:**1976–1978.

François, J., 1972, L'endosquelette céphaliques des insectes aptérygotes. II. Protoures et Thysanoures, *Arch. Anat. Microsc.* **61:**279–300.

François, J., 1973, Sur la prèsence de fibroblastes caractéristiques chez le Thysanoure *Thermobia domestica, C. R. Acad. Sc. Paris* **277:**2505–2507.

François, J., 1978, The ultrastructure and histochemistry of the mesenteric connective tissue of the cockroach *Periplaneta americana* L. (Insecta, Dictyoptera), *Cell Tissue Res.* **189:**91–107.

François, J., 1980, Secretion of collagen by insects: autoradiographic study of L-proline ^{3}H-5 incorporation by the firebrat *Thermobia domestica, J. Insect Physiol.* **26:**125–133.

François, J., Herbage, D., and Junqua, S., 1980, Cockroach collagen: isolation, biochemical and biophysical characterization, *Eur. J. Biochem.* **112:**389–396.

Gouranton, J., 1970, Étude d'une lame basale présentant une structure d'un type nouveau, *J. Microsc.* (Paris) **9:**1029–1040.

Hay, E. D., 1973, Origin and role of collagen in the embryo, *Amer. Zool.* **13:**1085–1107.

Harper, E., Seifter, S., and Scharrer, B., 1967, Electron microscopic and biochemical characterization of collagen in blattarian insects, *J. Cell Biol.* **33:**385–394.

Heathcote, J. G., and Grant, M. E., 1981, The molecular organization of basement membranes, *Int. Rev. Connective Tissue Res.* **9:**191–264.

Hess, R. T., and Pinnock, D. E., 1975, The ultrastructure of a complex basal lamina in the midgut of larvae of *Oryctes rhinoceros* L. *Z. Morphol. Ökol. Tiere.* **80:**277–285.

Hodge, A. J., and Petruska, J. A., 1963, Recent studies with the electron microscope on ordered aggregates of the tropocollagen macromolecule. In *Aspects of Protein Structure,* edited by G. N. Ramachandran, pp. 289–300, Academic Press, London.

Höglund, L., 1976a, Changes in acid mucopolysaccharides during the development of the blowfly, *Calliphora erythrocephala, J. Insect Physiol.* **22:**917–923.

Höglund, L., 1976b, The comparative biochemistry of invertebrate mucopolysaccharides. V. Insecta (*Calliphora erythrocephala*), *Comp. Biochem. Physiol.* **53:**9–14.

Holter, P., 1970, Regular grid-like substructures in the midgut epithelial basement membrane of some Coleoptera, *Z. Zellforsch.* **110:**373–385.

Kanwar, Y. S., and Farquhar, M. G., 1979, Presence of heparan sulfate in the glomerular basement membrane, *Proc. Nat. Acad. Sci. USA* **76:**1303–1307.

Katzman, L. R., and Jeanloz, R. W., 1970, The carbohydrate chemistry of invertebrate connective tissue. In *Chemistry and Molecular Biology of the Intercellular Matrix,* edited by E. A. Balazs, vol. 1., pp. 217–227, Academic Press, London.

Kefalides, N. A., Alper, R., and Clark, C. C., 1979, Biochemistry and metabolism of basement membranes, *Int. Rev. Cytol.* **61:**167–228.

Labat-Robert, J., Robert, L., Garrone, R., Auger, C., and Lethias, C., 1980, Immunofluorescence localization of a fibronectin-like protein in sponges: its role in cell aggregation, *Abstracts of VII European Symposium on Connective Tissue Research,* Prague, p. 192.

Laurent, T. C., 1977, Interaction between proteins and glycosaminoglycans, *Fed. Proc.* **36:** 24–27.

Locke, M., and Huie, P., 1972, The fiber components of insect connective tissue, *Tissue Cell*, **4:** 601–612.

Locke, M., and Huie, P., 1975, Staining of the elastic fibers in insect connective tissue after tannic acid/glutaraldehyde fixation. *Tissue Cell* **7:**211–216.

McLaughlin, B. J., 1974, Fine structural changes in a lepidopteran nervous system during metamorphosis, *J. Cell Sci.* **14:**369–387.

Martoja, R., and Bassot, J. M., 1965, Existence d'un tissue conjonctif de type cartilagineux chez certain insectes orthoptères, *C. R. Acad. Sc. Paris* **261:**2954–2957.

Mathews, M. B., 1974, *Connective Tissue: Macromolecular Structure and Evolution*, Springer Verlag, Berlin, p. 318.

Moulins, M., 1968, Étude ultrastructurale d'une formation de soutien épidermo-conjonctive inédite chez les Insectes, *Z. Zellforsch.* **91:**112–134.

Muir, H., and Hardingham, T. E., 1975, Structure of proteoglycans, *MTP Int. Rev. Sci. Biochem.*, series One, volume 5, edited by W. J. Whelan, pp. 153–222, Butterworths, London.

Mustafa, M., and Kamat, D. N., 1970, Mucopolysaccharide histochemistry of *Musca domestica*. I. A report on the occurrence of a new type of KOH-labile alcianophilia, *Histochemistry* **21:** 54–63.

Neville, A. C., 1975, *Biology of Arthropod Cuticle*, Springer Verlag, Berlin, p. 448.

Nordwig, A., Rogall, E., and Hayduk, U., 1970, The isolation and characterization of collagen from three invertebrate tissues. In *Chemistry and Molecular Biology of the Intercellular Matrix*, vol., 1, edited by A. E. Balazs, pp. 27–41, Academic Press, London.

Osińska, H. E., 1981, Ultrastructural study of the postembryonic development of the neural lamella of *Galleria mellonella* L. (Lepidoptera), *Cell Tissue Res.* **217:**425–433.

Pipa, R. L., and Woolever, P. S., 1965, Insect neurometamorphosis. II. The fine structure of perineurial connective tissue, adipohaemocytes, and the shortening ventral nerve cord of a moth, *Galleria mellonella* (L), *Z. Zellforsch.* **68:**80–101.

Prockop, D. J., Kivirikko, K. I., Tuderman, L., and Guzman, N. A., 1979, The biosynthesis of collagen and its disorders, *N. Eng. J. Med.* **301:**13–23.

Reinhardt, C., and Hecker, H., 1973, Structure and function of the basal lamina and of the cell junctions in the midgut epithelium (stomach) of female *Aedes aegypti* L. (Insecta, Diptera), *Acta Tropica* **30:**213–236.

Reinhardt, C., Schulz, U., Hecker, H., and Freyvogel, T. A., 1972, Zur Ultrastruktur des Mitteldarmepithels bei Flohen (Insecta, Siphonaptera), *Rev. Suisse Zool.* **79:**1130–1137.

Richards, A. G., and Richards, P. A., 1968, Flea *Ctenophthalmus*: heterogeneous, hexagonally organized layer in the midgut, *Science* **160:**423–424.

Ross, R., 1975, Connective tissue cells, cell proliferation and synthesis of extracellular matrix—a review, *Phil. Trans. R. Soc. London Ser. B.* **271:**247–259.

Ross, R., and Benditt, E. P., 1965, Wound healing and collagen formation. V. Quantitative electron microscope radioautographic observations of proline-H^3 utilization by fibroblasts, *J. Cell Biol.* **27:**83–106.

Ruoslahti, E., Engvall, E., and Hayman, E. G., 1981, Fibronectin: current concepts of its structure and functions, *Coll. Res.* **1:**95–128.

Sage, H., and Gray, W. R., 1979, Studies on the evolution of elastin. I. Phylogenetic distribution, *Comp. Biochem. Physiol.* **64B:**313–327.

Salpeter, M. M., 1968, H^3-Proline incorporation into cartilage: electron microscope autoradiographic observations, *J. Morphol.* **124:**387–422.

Salpeter, M. M., Bachmann, L., and Salpeter, E. E., 1969, Resolution in electron microscope radioautography, *J. Cell Biol.* **41:**1–20.

Sasikala, K., and Sundara Rajulu, G., 1978, Occurrence of resilin-like protein in an annelid *Perinereis* sp. *Nat. Acad. Sci. Letters*, (India) **1:**193–194.

Scharrer, B. C. J., 1939, The differentiation between neuroglia and connective tissue sheath in the cockroach (*Periplaneta americana*), *J. Comp. Neurol.* **70:**77–88.

Sharief, F. S., Perdue, J. M., and Dobrogosz, W. J., 1973, Biochemical, histochemical and autoradiographic evidence for the existence and formation of acid mucopolysaccharides in larvae of *Phormia regina*, *Insect Biochem.* **3:**243–262.

Stark, M., Miller, E. J., and Kühn, K., 1972, Comparative electron microscope studies on the collagens extracted from cartilage, bone and skin, *Eur. J. Biochem.* **27:**192–196.

Stephens, R. J., Freeman, G., and Evans, M. J., 1971, Ultrastructural changes in connective tissue of rats exposed to NO_2, *Arch. Intern. Med.* **127**:873–883.

Terzakis, J. A., 1967, Substructures in an epithelial basal lamina (basement membrane), *J. Cell Biol.* **35**:273–278.

Toole, B. P., 1972, Hyaluronate turnover during chondrogenesis in the developing chick limb and axial skeleton, *Develop. Biol.* **29**:321–329.

Uitto, J., 1979, Collagen polymorphism: isolation and partial characterization of $\alpha 1$(I)-trimer molecules in normal human skin, *Arch. Biochem. Biophys.* **192**:371–379.

Weinstock, M., and Leblond, C. P., 1974, Formation of collagen, *Fed. Proc.* **33**:1205–1218.

Wigglesworth, V. B., 1973, Haemocytes and basement membrane formation in *Rhodnius*, *J. Insect Physiol.* **19**:831–844.

Williams, M. A., 1973, Electron microscopic autoradiography: its application to protein biosynthesis. In *Techniques in Protein Biosynthesis*, edited by P. N. Campbell and J. R. Sargent, vol. 3, pp. 125–190, Academic Press, London and New York.

12

The Structure and Development of the Tracheal System

CHARLES NOIROT AND
CÉCILE NOIROT-TIMOTHÉE

1. Introduction

The tracheal system is unique in the animal kingdom, as gas exchanges occur directly between the tissues and the air in the tracheal tubes without any noticeable contribution by the blood. The tubes are inward invaginations of the body surface, being continuous with the integument at the spiracular openings.

The tracheae were clearly identified as the respiratory system of insects by Malpighi (1669), Swammerdam (1737) (Figure 1), and Réaumur (1737). However, Lyonet (1760), who pictured with unsurpassed accuracy the tracheal system of the larva of *Cossus*, was probably the first to recognize the tripartite composition of a tracheal tube (Figure 2), namely the cuticular intima with its taenidial thread, the surrounding epithelium (described as an amorphous thin layer), and the connective sheath or tunica propria. Subsequent observations confirmed the widespread occurrence of this organization in insects and added considerable information on the structure of the tracheal system. However, the smallest branches, or tracheoles, were at the limit of the resolution of the light microscope, and thus their detailed morphology and their relationships with other cells and tissues were not clarified until the advent of the electron microscope. The classical textbook by Richards appeared in 1951 and gave both a good resume of the classical data and the first observations showing the possibilities of the new instrument.

CHARLES NOIROT AND CÉCILE NOIROT-TIMOTHÉE • Laboratory of Zoology, University of Dijon, 6, Boulevard Gabriel, 21100 Dijon, France.

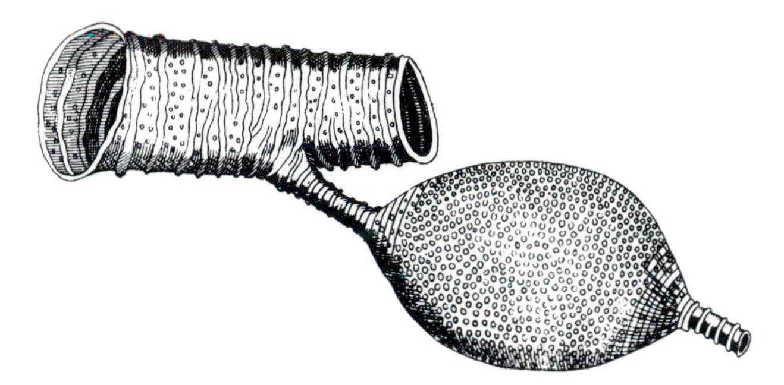

Figure 1. A tracheal trunk and an air-sac of the horned beetle as seen by Swammerdam (1737). The variable pattern of the taenidial fold and the punctate ornamentation of the air sac are shown.

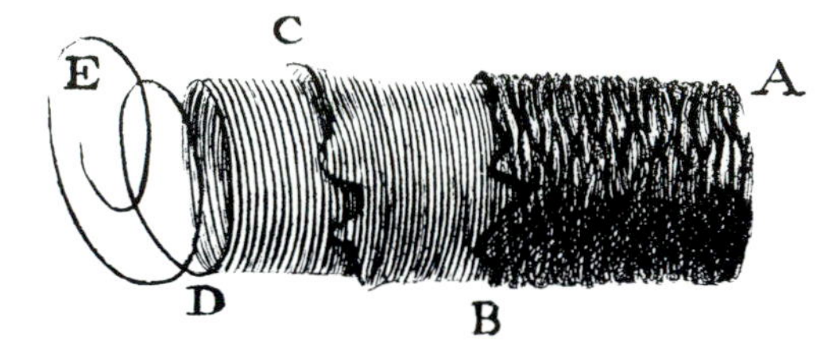

Figure 2. The components of a trachea of the willow caterpillar, as observed by Lyonet (1760). From the description given by Lyonet, AB can be interpreted as the connective sheath, BC as the tracheal epithelium, and CD as the intima, with the taenidium uncoiled in DE.

2. *General Organization*

2.1. Tracheae and Tracheoles

The distinction between a small trachea and a tracheole remains controversial. The tracheoles were at first believed to be devoid of taenidia, but the pioneering work of Richards and Anderson (1942) and especially Richards and Korda (1950) with the electron microscope showed well-structured taenidia in all the tracheoles examined (Figures 3–5). As pointed out in section 3.1, it is no longer possible to say that tracheoles are characterized by their behavior during the molt, i.e., renewal of the intima in the tracheae *versus* its persistence in tracheoles. No sharp differences seem evident in the structure of the cuticle, except perhaps in the buckling frequency of the taenidium (section 2.3.1). The relationship between the tracheal epithelium and the cuticular intima may offer a better criterion. In the smallest tracheae, the cuticular tube is surrounded by a unique tracheal cell that is rolled up around the intima and forms a hollow cylinder, with the two cell edges facing and linked together in a *mestracheon*, as defined by Edwards *et al.* (1958). A mestracheon is shown in Figure 8, and at higher magnification in Figure 11. In the tracheoles, in contrast, no mestracheon is observed (Figures 9, 10, 21). The plasma membrane adjacent to the intima is a part of the plasma membrane of the tracheal end cells that deeply infolds at the point of entry of the tracheolar trunk into the cell. This distinction was clearly made by Smith (1968), and although not always accepted by subsequent reviewers (e.g., Locke, 1974), it is completely consistent with our own observa-

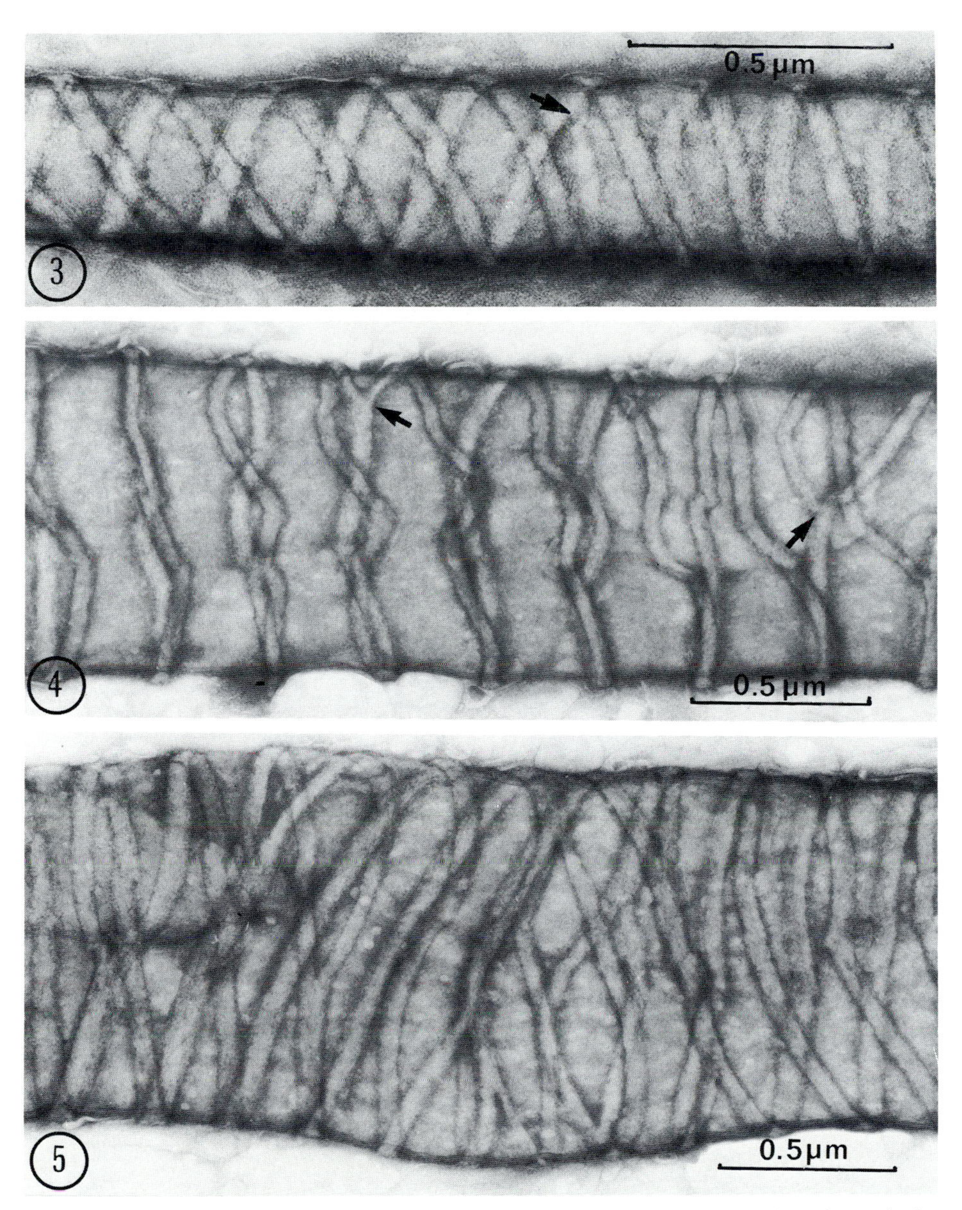

Figures 3–5. Tracheal tubes in the hindgut of the firebrat, *Thermobia domestica,* observed after dilaceration of the organ and negative staining with sodium phosphotungstate.

Figure 3. A tracheole with the taenidium in a simple helix (right), and a division (arrow) giving rise to a double helix (left).

Figure 4. In this small trachea, the taenidium is annular (left), then a simple helix (center), and finally a double helix (right). The arrows point to branchings of the taenidium.

Figure 5. A small trachea with a complex taenidial pattern.

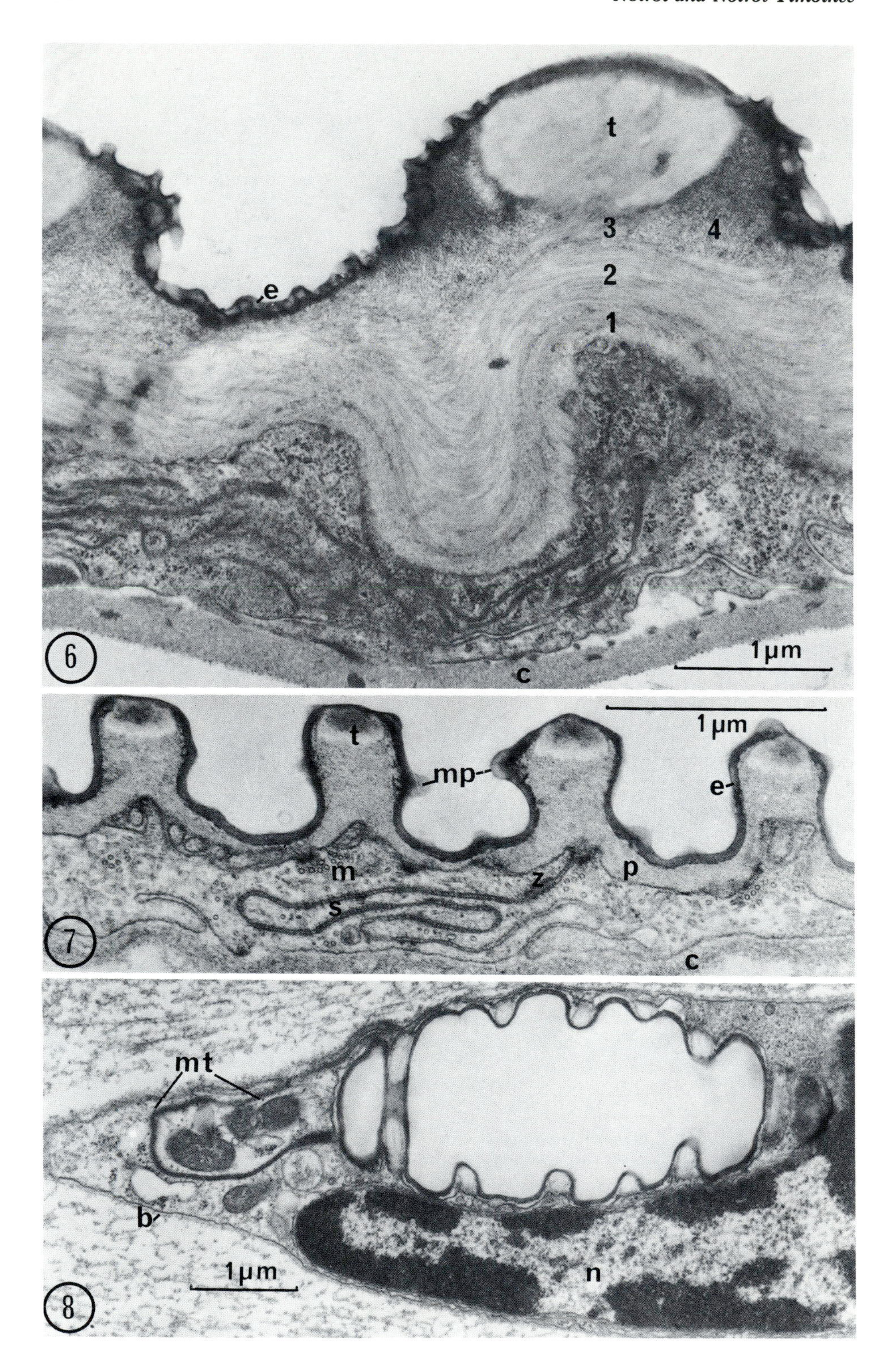
t
3
4
2
e
1
6
1 µm
c
1 µm
t
mp
e
m
z
p
s
7
c
mt
b
1µm
n
8

tions on several species and tissues. With this exception, the structures of the smallest tracheae and the tracheoles are so similar that there is no convincing evidence of a real *functional* difference between the two. However, from a morphological point of view, the distinction remains useful.

2.2. Anastomoses between the Tracheal Tubes

In most insects, the segmental tracheae arising from each spiracle are joined, both longitudinally and transversely, to the adjacent ones. Anastomoses occur through specialized structures called the *nodes*. During the molt, the network thus realized breaks at the nodes so that the old intima may be shed. Additionally, the nodes seem to play an important function in the morphogenesis of the tracheal system (see Whitten, 1972, for a review). The complete absence (as far as we know) of any ultrastructural study of such structures is to be regretted.

Although blind endings are usual in tracheoles, a network of anastomosing tracheoles was sometimes described by the light microscopists, especially in the light organs (reviewed in Richards, 1951). So far, no convincing evidence for such anastomoses has been given (see Peterson and Buck, 1968). However, it must be admitted that the electron microscope is not well adapted to the solution of this problem.

2.3. Relationships with the Aerated Organs and Cells

The tracheal supply of a tissue is in proportion to its oxygen demand, and the relationships between the cells and the tracheoblasts vary accordingly. In most cases, the tracheoblasts and their digitations form a network apposed at the basal surface of epithelia. Formerly it was reported that the basement lamina of the tracheoblasts coalesce with that of the epithelial cells, thus producing a "fenestrated membrane" or "peritoneal membrane." This was believed to be the functional equivalent of connective tissue in other animals, at a time when

←

Figure 6. A longitudinal section of a trachea in the silkworm *Antheraea pernyi*. The tracheal cells are surrounded by a relatively thick connective sheath (c). The four fibrillar layers of the procuticle (1 to 4) are seen, as well as the taenidium (t) and the epicuticle (e). After Beaulaton, 1964. Courtesy of Jacques Beaulaton.

Figure 7. A longitudinal section of a trachea around the hindgut of the carabid Coleoptera *Amara eurynota*. The very flat tracheal cells are joined by a zonula adherens (z) and a septate junction (s) and are surrounded by a relatively thick (0.1 μm) connective sheath (c). In register with the taenidial folds, microtubules (m) are cut transversally. The cuticular intima is composed of a continuous procuticle (p) and an epicuticle (e) which protrudes as micropapillae (mp) and is thinner in the taenidial folds, where the taenidium (t) appears of a variable electron density.

Figure 8. An oblique section of a small trachea in the musculoconnective envelope in the hindgut of the firebrat *Thermobia domestica*. The tracheal epithelium is formed by only one cell, which has a flattened nucleus (n). The mestracheon (mt) unites the outer plasma membrane with that facing the cuticle. The cell is surrounded by a thin basement lamina (b) clearly distinct from the connective fibers of the gut envelope. The intima is composed of a dark epicuticle and a pale taenidium.

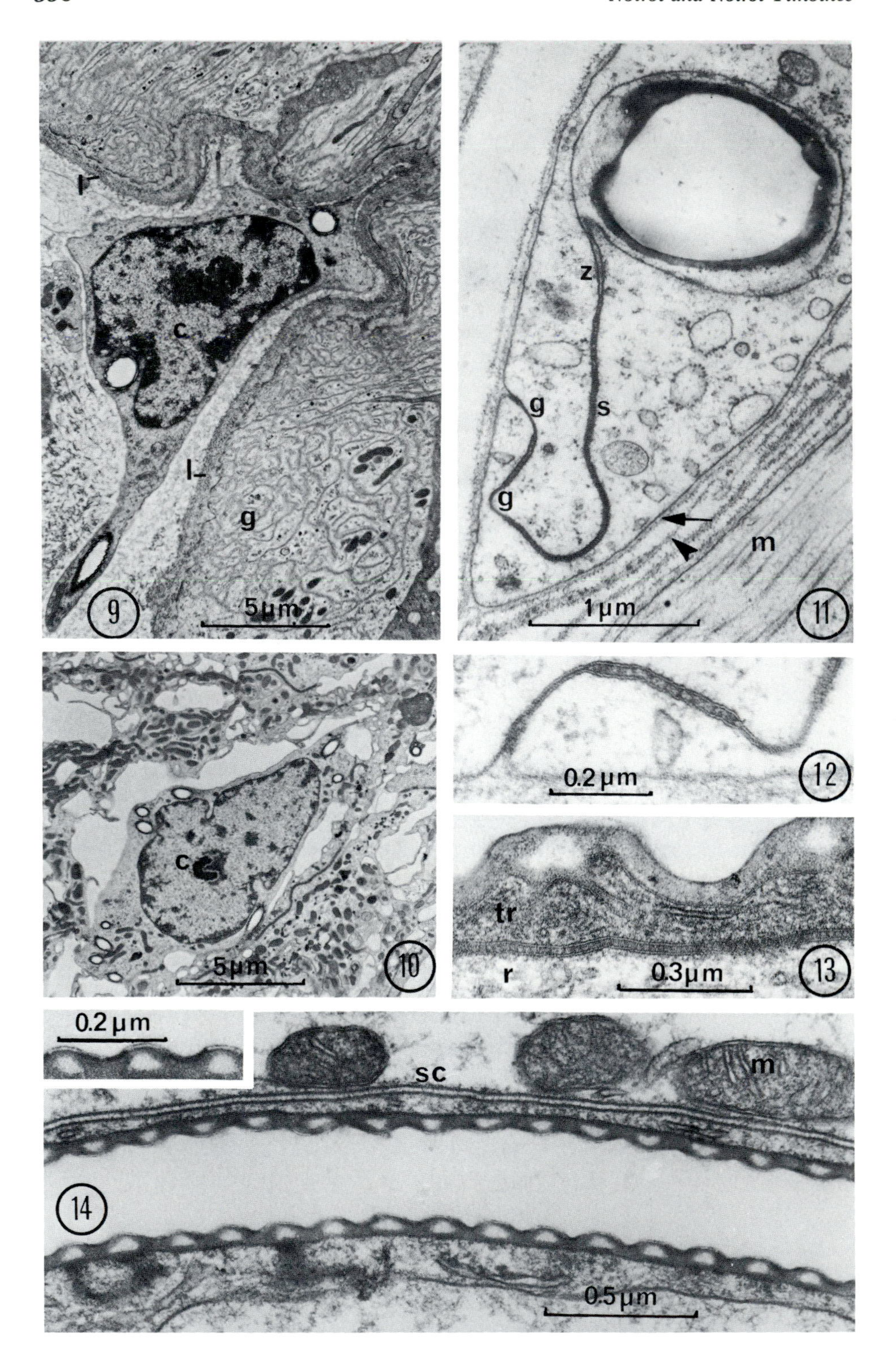
l
c
l
g
5 µm
9
z
g
s
g
m
1 µm
11
0.2 µm
12
c
5 µm
10
tr
r
0.3 µm
13
0.2 µm
sc
m
14
0.5 µm

the existence of a real connective tissue with collagen fibrils was not recognized in insects (see Edwards *et al.*, 1958). In light of ultrastructural studies, the so-called peritoneal membrane includes also muscle cells and a network of fibrillar material where structured collagen fibrils are frequently visible. However, the *supporting function* of the tracheae for most internal organs is evident, and the tracheal network plays an essential role in maintaining the position of the gut, of the reproductive organs, etc. In such cases, the tracheolar cells may be very close to the aerated cells, but always separated by at least a basement lamina (Figure 9).

A closer association is seen in more compact and/or more metabolically active organs, where the tracheoles penetrate the tissue and are located in the intercellular spaces. In most cases, the basement lamina disappears and the tracheoblasts are separated from the cells by only a small intercellular gap that may be no more than 20 nm. Furthermore, in the case of very large and active cells, the tracheoles extend into deep cell invaginations, and it was previously believed that the tracheoles became "intracellular." Use of the electron microscope clearly demonstrated the extracellular situation of such tracheoles, showing at the same time the importance of the tracheolar investment, as observed in the fibrillar flight muscles (Smith, 1961; 1964; 1968), the rectal papillae (Gupta and Berridge, 1966; Oschman and Wall, 1969; Noirot and Noirot-Timothée, 1976), and the light organs (see section 3.2).

In the *tracheal gills* observed in many aquatic insects, the tracheal supply is adapted for the gas exchanges with the surrounding water. In all the examples studied a common organization was observed, regardless of the position and morphology of the gill: numerous tracheoles, generally running parallel to the long axis of the gill, are arranged regularly under the thin cuticle covering the organ. These tracheoles indent the epidermal cells and thus lie at a very short distance from the gill surface. This was observed in Trichoptera (Wichard, 1973), Diptera Blepharoceridae (Komnick and Wichard, 1975), Plecoptera (Wichard and Komnick, 1974), Ephemeroptera (Wichard, 1979a), and Odonata.

Figure 9. A section showing a tracheal end cell (c) at the base of the midgut epithelium (g) in *Blaberus craniifer.* The cell is in close contact with, but does not cross, the midgut basement lamina (l). No mestracheon is visible.

Figure 10. A section of a rectal pad of the American cockroach. A tracheal end cell (c) is seen in an intercellular space between the rectal cells.

Figure 11. A section of a small trachea in the hindgut envelope of *Kalotermes flavicollis.* The basement lamina, in close contact with the tracheal cell (arrow) is quite distinct from the connective lamellae (arrowhead), especially between the tracheal cell and a muscle (m). In the mestracheon are a zonula adherens (z) and a septate junction (s) interrupted by two gap junctions (g).

Figure 12. A higher magnification of the gap junctions of Figure 11.

Figure 13. A section of a trachea entering the rectal pad of the American cockroach. The flattened tracheal cell (tr) is joined to a basal rectal cell (r) by a septate junction.

Figure 14. A longitudinal section of a tracheole inside the rectal pad of the American cockroach. On the lower side, the tracheoblast is free in the intercellular space, but on the upper side it is joined to a rectal cell by a long scalariform junction (sc) with associated mitochondria (m). The cuticular intima, seen at higher magnification in the inset, is made of a dense material (interpreted as the epicuticle) interrupted by the electron-lucent taenidium; the very thin dense line at the surface may be either part of the outer epicuticle or a superficial layer.

In the latter order, the arrangement of the tracheoles is essentially the same in both Zygoptera (Wichard, 1979b), where the gills are external as in the preceding cases, and Anisoptera (Greven and Rudolph, 1973; Saini, 1977).

The compound eye of several Lepidoptera, in the dark-adapted state, exhibits the phenomenon of eye glow when illuminated. The glow is light reflected by the tracheae just underlying the rhabdom. In diurnal butterflies this reflecting layer, or *tapetum lucidum,* consists of a specialized region of the tracheae just below (proximal to) the rhabdom. At this level, under each ommatidium, a trachea divides in two or three branches lying close together under the rhabdom. The taenidium of these branches is greatly extended and forms a series of 20–40 platelets perpendicular to the ommatidial axis. These platelets, alternating with air spaces, function as an interference filter (Miller and Bernard, 1968; Ribi, 1979).

3. Fine Structure

3.1. Basement Lamina and Connective Envelope

The tracheal epithelium is always surrounded by a fibrillar sheath, except in some tracheoles closely associated with specialized cells (section 2.3). Although described in a few cases, it seems loosely structured, with a microfibrillar texture and a feltlike appearance. Under a light microscope, it shows a strong positive reaction to the PAS test and to paraldehyde-fuchsin after oxidation. Its thickness seems variable; for example, in the prothoracic gland of the larvae of *Antheraea pernyi,* it ranges from 300 nm for the larger tracheae to 20 nm for the tracheoles (Beaulaton, 1964). From our observations on several tissues from a variety of insect species, the thickness appears to be related not only to the size of the tube, but also to its location, being maximum where the trachea is free in the hemolymph or runs at the surface of an organ, and minimum where embedded in a musculoconnective tissue layer. In the latter case, it is reduced to a distinct but fluffy layer, about 15 nm thick, closely apposed to the cell membrane and clearly separated from the connective layers of similar appearance (Figures 8, 11). In our opinion, this 15-nm layer is the true *basement lamina* and probably is secreted by the tracheal cells. Where the envelope is thicker, this enlargement is caused by the deposition of additional material, probably by other cells, which according to the studies of Wigglesworth (1973; 1979) could be some varieties of hemocytes. It is usually difficult to separate the basement lamina from the other layers, and our interpretation remains speculative. However, it is significant that in the intercellular spaces where the envelope (if present) is obviously produced by the tracheal cells, it corresponds to our basement lamina *stricto sensu.*

3.2. Tracheal Cells and Tracheoblasts

The tracheal cells, elongated along the axis of the trachea, have a variable thickness, almost in proportion to the tracheal diameter, rarely exceeding a few micrometers in the large trunks and being as thin as 0.1 or 0.2 μm in the smallest, except in the nuclear region. The nucleus is flattened and elongated. The tracheoblasts, or tracheal end cells, are often called stellate cells, owing to the long fingerlike projections radiating from the nuclear region and containing the ramifications of the tracheoles. The variously shaped nuclei are frequently indented by the tracheolar ramifications (Figure 9). The tracheal cells are joined by the same type of junctional complex as the epidermal cells, namely a belt desmosome (zonula adherens) at the apical (cuticular) side, followed by a pleated septate junction (see Noirot-Timothée and Noirot, 1980). However, the septate junction in most cases extends the length of the cell contact. Macular gap junctions are regularly observed. The same type of junctional complex, and occasional gap junctions, exist along the mestracheon (Figures 11, 12). One may speculate about the function of such gap junctions between two ends of the *same* cell. Nevertheless, such "autocellular" or "reflexive" gap junctions are present in other tissues, in vertebrates and in invertebrates (Larsen, 1977).

The tracheal cells rarely come into direct contact with other cell types, usually being separated from them by the basement lamina (section 3.1). Where this lamina is not present, junctional specializations are not usually observed even where the association is very close, as in the fibrillar flight muscles (Smith, 1961; 1964). The rectal papillae or rectal pads of certain species are an exception.

In the cockroaches we observed various types of junctions, septate (Figure 13), scalariform (Figure 14), and spot desmosomes (Noirot and Noirot-Timothée, 1976). These junctions, especially the septate, may play a role in the compartmentalization of the extracellular spaces. Indeed, the tracheal trunks entering the pads branch within a complex "intercellular sinus" (Gupta and Berridge, 1966; Oschman and Wall, 1969) which can communicate with the hemolymph only at the points of tracheal penetration. The junctions between tracheal and and rectal cells may determine both the relative isolation of the intercellular sinus from the hemolymph and the occurrence of subcompartments within the pads. Although the precise interpretation remains controversial, the pattern of the tracheation and the junctions linking the tracheal cells with the other cells must be taken into account to understand the functioning of these organs.

The basal plasma membrane of tracheal cells may be attached to the basal lamina by hemidesmosomes, which are inconspicuous and few in number. No special device was observed for the attachment of the apical plasma membrane to the cuticle. The plasma membrane shows a very simple topography, with very few infoldings on the basal side and occasionally short irregular microvilli on the apical side (Figure 7). Basal infoldings and apical microvilli are absent in tracheoblasts.

The cytoplasmic organelles, both in tracheal cells and tracheoblasts, are not indicative of a marked specialization nor of an important metabolic activity, except during the secretion of the cuticle (section 4.2). Microtubules are usually observed with various orientations, although mainly in two perpendicular directions: parallel to the long axis of the cell and/or to the taenidial fold (Figure 7). However, they are neither numerous nor organized in bundles, which may be surprising for such elongated structures, often running free in the body cavity. The cuticular intima is believed to give the necessary mechanical strength. No information exists about other cytoskeletal components such as microfilaments and intermediary filaments.

The cytological structure of tracheal and tracheolar cells is in agreement with the common view that attributes only a passive role to these cells in gas transport, but the luminescent organs of fireflies provide a noticeable exception. From comparative histological data, special differentiations, the *tracheal end organs,* were observed in all firefly species able to produce sharp flashes (Buck, 1948). With the electron microscope, they were beautifully analyzed by Smith (1963) in *Photuris* (adult males). Additional data were given for the same genus by Hanna *et al.* (1976) and Ghiradella (1977; 1978), and for adults belonging to other genera by Peterson and Buck (1968) and Ghiradella (1978). In contrast, in *Photuris* larvae, which glow continuously, the tracheal end organs are not present (Peterson, 1970). When present, the tracheal end organs have the following structure: The tracheal trunks enter the light organ through cylinders surrounded by photogenic cells or photocytes. Each trachea branches, putting out lateral tracheal *twigs,* each of which branches further into several tracheoles. These are ensheathed by a *tracheolar cell,* itself enfolded by the so-called *tracheal end cell,* which also surrounds the tracheal twig. (The term "tracheal end cell" is used here to denote a peculiar tracheal cell and not, as usual, a tracheolar cell.) The tracheolar cell has unusual specializations. The plasma membrane facing the intima is folded regularly in a honeycomb fashion. Mitochondria are especially numerous in the region surrounded by the tracheal end cell, and they protrude into the folds of the plasma membrane. The tracheal end cell, which has a mestracheon, is less differentiated, only containing more and larger mitochondria than ordinary tracheal cells. The most important feature is the presence of naked nerve endings between the tracheolar and the tracheal end cells (i.e., without direct contact with the photocytes). Although the significance of these differentiations is not yet completely understood, it seems to relate to the flashing signals, and an active control of the oxygen supply for the photogenic reactions by the tracheal and tracheolar cells may be supposed.

3.3. Cuticular Intima

3.3.1. Geometry of the Taenidium

Although in continuity with the peripheral cuticle, the tracheal intima is thrown into regular folds reinforced by a thread, the taenidium. As soon as the electron microscope was used, the extension of the taenidium into the tracheoles was recognized (Richards and Anderson, 1942; Richards and Korda, 1950; Santos *et al.*, 1954). Most commonly, the taenidium is helical and the microscopists of the 17th and 18th centuries could unwind the taenidium after pulling apart the two halves of a broken trachea (Figure 2). However, the geometry of the taenidium is highly variable, as well documented by Swammerdam (1737) (Figure 1), and best demonstrated by the early electron microscopists who recognized in the small tracheae and the tracheoles either an annular or a helical pattern, with the helix being either simple or multiple. This geometry was clearly demonstrated by negative staining (Smith, 1968), and Figures 3–5 illustrate several arrangements observed by us. As observed by Richards and Korda (1950) and Santos *et al.* (1954), the taenidial pattern may vary in tracheal tubes of the same organ, and even along the same tube. At the present time, we have no idea of the significance of such variations. This problem deserves more detailed research.

Is there a difference in the taenidial pattern between the small tracheae and the tracheoles? In our opinion, the only difference is in the number of taenidial folds per unit length, or "buckling frequency" (Locke, 1958), which is higher in tracheoles (from 4 to 8 per micrometer). In their pioneering observations, Richards and Korda (1950) observed in several cases an abrupt change in taenidia, with or without a simultaneous change in tube diameter, and justifiably interpreted it as the junction of trachea and tracheole. If these conclusions can be verified and generalized, they may have important implications both for the formation of the tracheoles *vs.* the tracheae and for the determination of the buckling frequency.

Apart from the taenidium, the tracheal intima has other localized thickenings, frequently in the form of micropapillae or tubercules (Figure 7) (Richards and Korda, 1950; Santos *et al.*, 1954; Edwards *et al.*, 1958; Bordereau, 1975), seen clearly using the freeze-fracture technique (Figure 15). These micropapillae are observed mainly on the intertaenidial surfaces, and seem to be absent in tracheoles.

In the *air sacs*, which are dilatations of various tracheal trunks observed in many species, especially the good flyers, the arrangement of the taenidium is modified to various degrees, as recognized by Swammerdam (1737) (Figure 1). From studies with the scanning electron microscope (SEM) (Faucheux and Sellier, 1971; Faucheux, 1972; 1974), three principal types may be distinguished. In the "taenidial sacs," a taenidium is clearly present, and the taenidial folds are regularly arranged. In the "recticulate sacs," the taenidial folds are anastomosed in a more or less regular and complex network. In the "punctate sacs," the taenidium is no longer recognizable, and the intima only has glomerular protrusions of various arrangements. Intermediate cases were described. According to Faucheux (1972; 1974), the taenidial sacs, mainly observed in the abdomen, can sustain very large variations in volume, and apart from their function in ventilation have an important role in the maintenance of the abdominal volume. The mainly thoracic reticulate and punctate sacs undergo only limited variations of volume and seem specially adapted for the ventilation of the flight muscles.

Other dilations of the tracheae do not relate to a respiratory function. In the cockroach *Gromphadorhina portentosa*, the tracheae from the second abdominal spiracle are modified into a sound-producing system. The dilated tracheae have a normal taenidium, but their communication with the spiracular trachea is greatly narrowed, forming a whistle. At this place, the taenidium disappears, giving way to a more rigid cuticle with a paved ornamentation (Baudet, 1974).

Very little information appears to be available on the cuticle of the tracheal dilatations forming the tympanic cavity when present in auditory organs. From

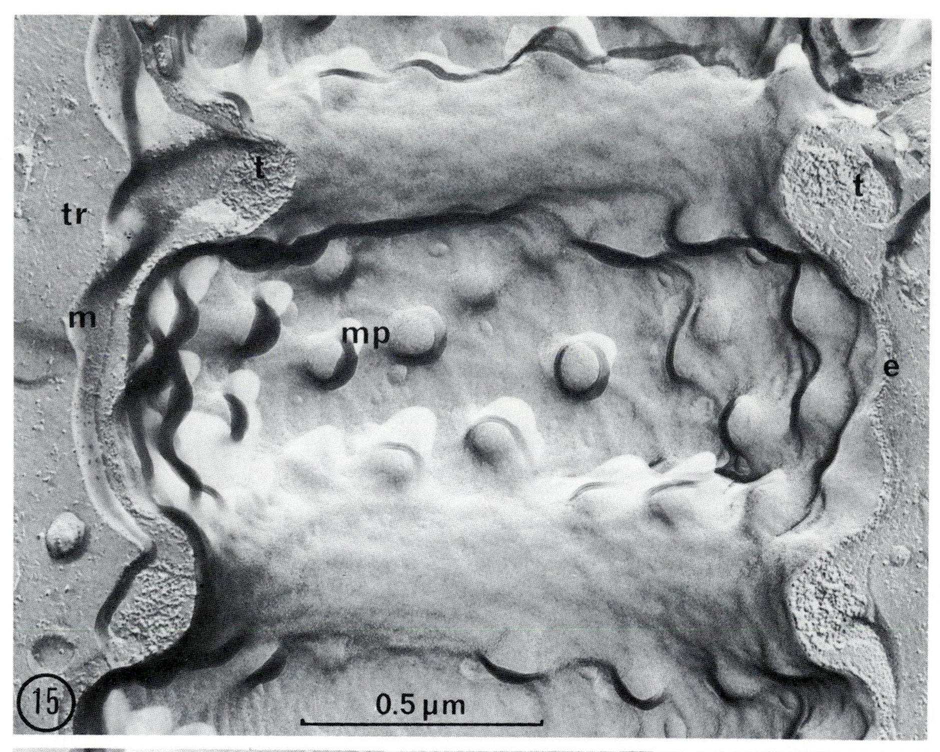
t
t
tr
m
mp
e
15
0.5 µm

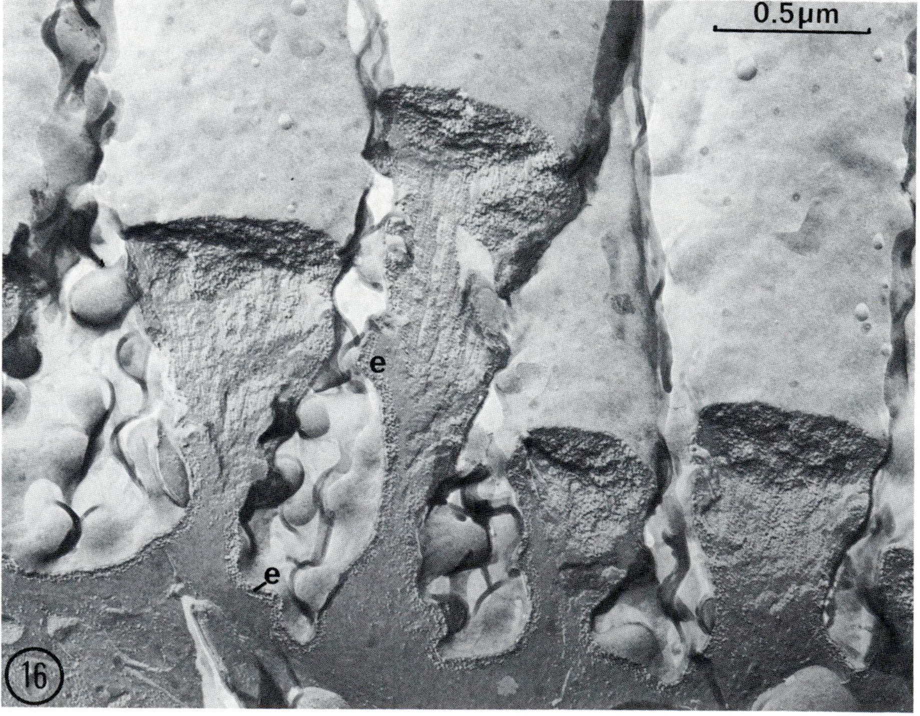
0.5µm
e
e
16

the published micrographs of Ghiradella (1971) and Ball and Cowan (1978), a taenidial fold is present, but its pattern was not specified.

In the tracheal end organ of the fireflies (see section 3.2) the taenidium seems absent in the tracheal twig. In the tracheoles inserted on this twig, the taenidium is stiffened by a series of bars running from one taenidial fold to the next (Ghiradella, 1977). This stiffening was interpreted by Ghiradella (1977; 1978) as a device to keep the tracheoles open during the osmotic pressure changes that were supposed to occur during the light emission.

This review emphasizes the various ways in which the tracheal intima is reinforced, obviously to avoid the collapse or dilatation of the tubes by the internal pressure. No doubt additional research may show greater variability. As an example, the cuticle of the *atrium* (the chamber lying just behind the spiracle) remains poorly understood at the ultrastructural level, although SEM views show a great complexity of cuticular patterns in Diptera (Faucheux, 1974).

3.3.2. Terminology

Are the components of the tracheal cuticle homologous with those of the integument? In most studies the answer is yes, and although some difficulties are encountered it seems better to follow this practice for a sake of clarity. However, a great deal of confusion remains about the terminology used for the different layers of the cuticle. In the present review we distinguish, starting from the luminal surface, the *superficial layers* (e.g., cement, wax), the *epicuticle*, itself subdivided into an *outer epicuticle* (named cuticulin by Locke, 1958; 1966) and an *inner epicuticle* (or protein epicuticle of Locke, 1969), and the deeper layer, named *procuticle*, in which an *exocuticle* and an *endocuticle* can sometimes be distinguished. The *taenidium*, a reinforcement of the taenidial folds, is characteristic of the tracheal cuticle and deserves special attention. The possible occurrence of a *subcuticle* is briefly examined.

3.3.3. Superficial Layers

No *cement layer* is observed at the tracheal surface, and this is consistent with the absence of dermal glands in the tracheal epithelium. The so-called peritracheal glands observed in some species seem to perform other functions (see Whitten, 1972). Whether a *wax layer* is present is not so clear. It is generally believed to be absent, and indeed is not visible in most published electron micro-

---

Figure 15. A freeze-fracture replica of a small trachea in a *Tenebrio* pupa. On both sides, the cuticle was cross fractured, and the taenidium (t) and the granular epicuticle (e) are clearly visible above the plasma membrane (m) of the tracheal cell (tr). In the center, the fracture followed the surface of the intima, in spite of its complex relief, especially pointing up the micropapillae (mp) between the taenidial folds.

Figure 16. A freeze-fracture replica of a larger trachea in a *Tenebrio* pupa. Four taenidial folds were obliquely fractured, showing the fibrillar nature of the taenidium, at least in its basal part, in contrast to the granular structure of the epicuticle (e). In the upper part, the fracture followed the surface of the cuticle, as in the preceding figure.

graphs, although some exceptions were noted in *Calpodes* by Locke (1966) and in termites by Bordereau (1975). In several instances (e.g., Figure 14) we observed a distinct superficial layer in small tracheae and tracheoles of the American cockroach. Moreover, it must be emphasized that tissue processing probably alters and even dissolves this layer if present. A thick wax layer is out of the question, but it is perhaps premature to suppose a complete absence of wax. This problem is especially important in relation to the permeability of the tracheal cuticle. Freeze-fracture techniques could be very useful, because material is processed without any treatment with a lipid solvent. We obtained some views suggestive of a superficial layer (e.g., Figure 17). However, the interpretation of the replicas remains controversial (see section 3.3.4).

3.3.4. Epicuticle

According to Locke (1958; 1966), an outer epicuticle or "cuticulin" layer, about 20 nm in thickness, is present over the whole surface of the tracheal and the tracheolar intima, as well as the peripheral cuticle. There is some discrepancy with regard to the thickness of this layer. It is stated to be only 8.5 nm on the tracheoles described in Locke's 1958 paper. The constant occurrence of this layer seems confirmed by all available observations, but its detailed structure is questionable. When newly secreted, the outer epicuticle appears pentalaminar (two electron-lucent layers sandwiched between three electron-dense ones) in thin sections of tracheae, tracheoles, and integument. In the integument, the outermost layer becomes denser and further separated from the others, giving rise to a superficial layer, tentatively identified as a wax monolayer. This separation also occurs in tracheae, but in the fully formed cuticle a superficial layer is generally not visible, and the outer epicuticle appears only as a dense line at the surface (Locke, 1966).

It is often difficult to separate this outer epicuticle from the *inner epicuticle* which, as in the integument, appears homogeneous and electron dense. This inner epicuticle, always present in tracheae, was believed to be absent in tracheoles (Edwards *et al.*, 1958; Locke, 1958). However, we observed numerous cases where it was clearly visible (Figure 14), and the absence of an inner epicuticle cannot be considered as a general characteristic of tracheoles. The thickness of the inner epicuticle is always minimal at the apex of the taenidial folds, where it is not well separated from the taenidium (Figures 7, 13). When present in tracheoles, it may be absent at this place (Figure 14). The maximum thickness observed in the intertaenidial regions rarely exceeds 0.1 μm, except at the level of the tubercules or micropapillae.

In contrast with the cuticle of the integument and the gut, no epicuticular filaments (wax canal filaments) were observed in the tracheal cuticle, except perhaps in the large trunk of *Sarcophaga* larvae (Whitten, 1976). These filaments are believed to be the site of wax transport to the surface, and thier absence is consistent with the probable absence of a wax layer above the epicuticle.

The freeze-fracture technique can give additional information. When the tracheal cuticle is cross fractured, the epicuticle appears as a granular layer (Figures 15–17) regardless of the direction (longitudinal or transverse) of the

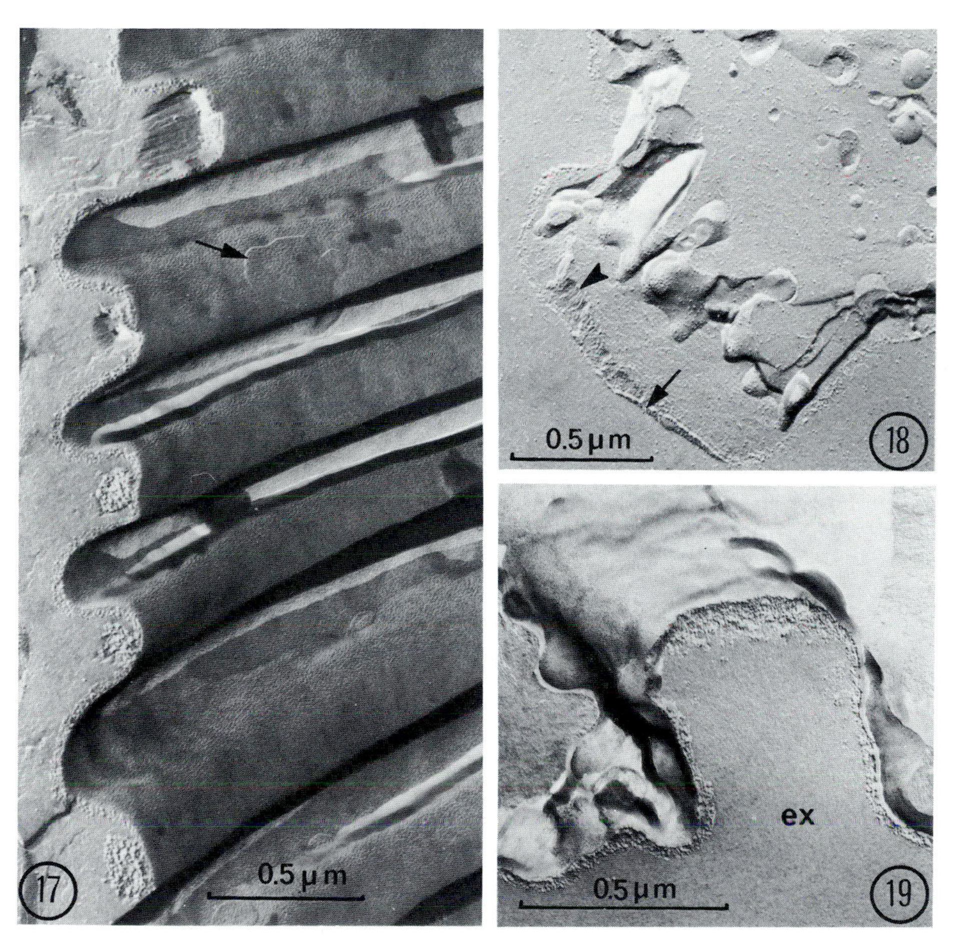

Figure 17. A freeze-fracture replica of a trachea around the hindgut of the firebrat, *Thermobia domestica*. As in Figure 15, the intima was cross fractured, and (right) the fracture followed the surface, which is seen to be smooth, marked only by the taenidial folds. The arrow indicates a patch of material that may be a remnant of a superficial layer. In the upper part, the base of a taenidial thread is obliquely fractured, showing its fibrillar nature. In the other taenidial threads these fibrils appear in cross section.

Figure 18. A freeze-fracture replica of a tracheal trunk in a pharate adult of *Tenebrio*. The secretion of the new cuticle was still in progress. The epicuticle is continuous and, at the arrow, it is possible to distinguish an outer and an inner epicuticle. The taenidial fold is well formed, although not very regular, and the first taenidial fibrils had already been secreted (arrowhead). Note the irregular, short microvilli under the taenidial fold. There is no tendency for the fracture plane to follow the cuticular surface, in contrast to the fully secreted intima (Figures 15–17).

Figure 19. An electron micrograph from a pharate adult *Tenebrio*. A freeze fracture of the old (pupal) cuticle of a tracheal trunk, where (owing to the digestion by the molting fluid) the intima is reduced to the epicuticle and a remnant of taenidial material. ex: exuvial space.

fracture plane; i.e., the granularity is not caused by cross-sectioned fibrils. It is not possible to differentiate an outer and an inner epicuticle on the replicas. However, a remarkable feature of the replicas is the extensive areas where the fracture plane follows what seems to be the *surface* of the intima. Indeed, the very high frequency of views such as Figures 15–17 suggest a strong tendency for tissues to fracture along this level, in spite of the very convoluted relief (taenidial folds, tubercules, etc.) followed by the fracture. The views can be interpreted in two ways. One is that the fracture occurred at the interface between the cuticle and the glycerol filling the tubes, and that the true surface of the intima is thus revealed. Another possibility is a cleavage *inside* the outer epicuticle, which is considered, by analogy with a plasma membrane, to include a lipid bilayer. For the rectal cuticle, we gave evidence (Noirot *et al.*, 1978) for the latter interpretation, and similar conclusions were reached by Flower and Walker (1979) on the same type of cuticle. So far, it is not possible to decide whether the same interpretation can be applied to the tracheal cuticle, and the question must be reexamined with the technique of deep etching. However, whatever the interpretation, these fractures indicate a strong hydrophobic region, either at the surface itself or into the outermost layer. This may have important implications for the permeability of this type of cuticle. Accordingly, it must be noted that newly secreted epicuticle has different characteristics, and there is no preferential fracture plane along its surface (Figure 18).

Unfortunately, the precise chemical composition of the epicuticle, especially of the tracheae, remains unknown. It is generally believed that chitin is lacking, although Locke (1957; 1974) supposed it to be present in the inner epicuticle of the tracheae in the form of micelles oriented in the axis of the tube, mainly on the basis of an axial birefringence which may well be caused by the procuticle (see below). Longitudinal fibers were not seen in the epicuticle on freeze-fracture replicas, whereas fibers were easily observed in the taenidium and the lamellated peripheral cuticle. The only evidence of chitin in the tracheal epicuticle is that the inner epicuticle is digested by a chitinase (Beaulaton, 1969), but experiments with the inhibitor of chitin synthesis, diflubenzuron (B. Mauchamp, personal communication), do not support the same assumption (see section 3.3.5). Furthermore, the strong resistance to concentrated acids may be caused by a tanning of proteins or lipoproteins, as stated by Wigglesworth (1971) and many others. A good indication is given by the cytochemical localization of phenoloxydase in the tracheal epicuticle (Locke and Krishnan, 1971), and the tanning is consistent with the refractoriness obtained with the histological stains. However, the epicuticle does not usually react with the argentaffin or silver methenamine tests, which are indicative of polyphenols (Bordereau, 1975).

3.3.5. *Taenidium*

The taenidial thread, or more simply the taenidium, so distinctive of the tracheal system, is difficult to homologize with the usual cuticular layers, although somewhat similar thickenings may be observed in the ducts of salivary glands (Richards, 1951) or in some rigid cuticles, as in the dorsal pulvillus of the footpads of the fly (Whitten, 1969). The size of the taenidium is proportionate to that of the tubes, from about 50 nm in tracheoles (Figure 14) to several micrometers in the larger trunks (Figure 6). The taenidium may be absent in very small tracheoles (the taenidial fold being reduced to the buckled epicuticle), but according to our observations, this is certainly not a general feature of the tracheoles, and the results obtained after negative staining (Figures 3–5) give clear indications of a real thickening of the cuticle along the taenidial folds. In conventional sections, the taenidium usually appears of variable electron density (Figures 6, 7). A fibrous structure is visible only in sections where the taenidium is cut along its long axis, that is, in the same direction as the microfibrils, and this is confirmed by the freeze-fracture technique (Figures 16, 17). This fibrous structure is more evident in the basal part of the taenidium, and it is not certain that the fibrous structure extends throughout the taenidium; differences in electron density, frequently observed between the center and the periphery of the taenidium (Figures 6, 7), may be indicative of a heterogeneous composition and/or structure. Apically, the taenidium is in close contact with the epicuticle, without any clear-cut boundary (Figures 6, 7). Basally and laterally, its border is more sharply delineated (Figures 6, 7). The chemical composition of the taenidium remains subject to speculation. The claim for a chitinous component is based mainly on the fibrous structure and the birefringence observed after mild extraction with potassium hydroxide (Locke, 1957). Indeed, Beaulaton (1969) observed a dissolution of the taenidium by a chitinase on thin sections of tracheae of *Antheraea*, but this result was not obtained by Bordereau (1975) using the same technique on a termite. However, the susceptibility of the taenidium to concentrated potassium hydroxyde (Locke, 1957) and proteolytic enzymes (Beaulaton, 1969; Bordereau, 1975) is in marked contrast to the resistance of the chitinous layers of the integumental cuticle. This susceptibility is suggesstive of a high protein content. The elastic protein, resilin, characterized by its blue fluorescence in ultraviolet light, was not detected by Bordereau (1975). This relative fragility of taenidia is compatible with the presence of chitin. For example, the peritrophic membrane of the midgut, which contains chitin, is no more resistant. The molecular architecture may be different: for example, perhaps the protein chains are not closely associated with the chitin fibers and are not arranged in parallel (such a hypothesis could explain why the fibrillar structure is poorly visible in the electron micrographs); the proteins might be in a more hydrated state and thus more accessible to the reagents. The problem was recently reexamined by B. Mauchamp (unpublished observations), who studied the effect of diflubenzuron, an inhibitor of cuticle formation believed to block the chitin synthesis. In the last larval instar of *Pieris brassicae*, the application of the inhibitor profoundly affects the formation of the pupal cuticle, and in the tracheae, the new intima is often reduced to an epicuticle separated from the tracheal cells by a very loose material; usually no taenidial folds are visible, but in some cases irregular spots of electron-lucent material, reminiscent of the taenidium, may be observed under the epicuticle (Figure 26). The disruption of the taenidium by this inhibitor is a strong argument for the presence of chitin, provided that the specificity of diflubenzuron as an inhibitor of chitin synthesis is confirmed.

The taenidium was often homologized with the exocuticle, based on its mechanical strength and the frequent red coloration after Mallory triple stain. This view, which implies a tanning, is in agreement with the cytochemical localization of phenoloxidases and polyphenols (Locke and Krishnan, 1971). However, the taenidium is far more easily destroyed by alkali solutions, proteolytic enzymes, and the molting fluid than is a normally tanned cuticle. Considering the absence of pore canals (section 3.3.6), a tanning of the tracheal cuticle, if it occurs, may well be different from that observed in the integument. While awaiting precise chemical analysis, it seems better to consider the taenidium as an original structure, adapted for keeping the tracheae open without hindering their flexibility.

3.3.6. Procuticle

The presence of a layer similar to the procuticle (endocuticle) is not universal in tracheae, contrary to the earlier interpretations of histological sections, and was frequently believed by the electron microscopists to occur only rarely (Locke, 1957; Edwards *et al.*, 1958; Smith, 1968). This discrepancy seems to be due, at least in part, to histologists using the largest tracheal trunks and electron microscopists the smaller ones (Whitten, 1972), but the possibility of real differences among species (or perhaps orders) can not be dismissed. From the published micrographs and our own unpublished observations, the absence of a procuticle in the tracheoles and the small tracheae seem obvious. In contrast, the large trunks of Lepidoptera possess a well developed procuticle, briefly mentioned by Locke (1957) and well studied by Beaulaton (1964) in *Antheraea* larvae, in which four components may be recognized (Figure 6): (1) A thin basal layer of randomly orientated fibrils. (2) A layer 0.5 to 1 μm thick made of fibrils oriented along the axis of the tube. (3) A hypotaenidial layer, which is discontinuous, since it is only present at the base of the taenidium, and is made of fibrils oriented along the axis of the taenidium. In these three layers, the fibrils are about 5 nm in diameter. (4) A fourth layer is formed by larger fibers (8–10 nm) irregularly interwoven and found in the lateral portions of the taenidial folds on both sides of the third layer. The procuticle becomes thinner in smaller tracheae, and disappears in the tracheoles. An essentially similar organization was observed in the larvae of the cabbage butterfly, *Pieris brassicae*, and the experiments with diflubenzuron clearly indicate the chitinous nature of the several components (B. Mauchamp, personal communication). It is now not certain whether or not such a complex procuticle is a characteristic of Lepidoptera. In *Amara eurynota* (Coleoptera), we observed a distinct procuticle in the tracheae aerating the hindgut but with a looser organization (Figure 7). In the imagoes of termites (Bordereau, 1975), no real procuticle was observed, but under the taenidium, and only there, another material was present that was either granular or fibrillar depending on the species. This material should perhaps be compared with the hypotaenidial layer described by Beaulaton (1964). The case of *Sarcophaga* deserves special attention. Whitten described the cuticle of the large tracheal trunks first of the larvae (1969), and later of the pupae and adults (1976). In larvae, the taenidium is well developed but does not protrude into the lumen of the trachea. Under the epicuticle, the intertaenidial areas are filled by a homogeneous material of low electron density, interpreted by Whitten to be a homogeneous exocuticle. More basally, a fibrillar endocuticle is very distinct and has a regular thickness of about 1 μm. In pupae, the picture is similar except for a reduced amount of endocuticle. In adults, the taenidial folds protrude far into the lumen as usual, and under them lies a thick (more than 1 μm), regular homogeneous layer where no fibrils are evident, interpreted by Whitten as an exocuticle. No fibrillar endocuticle is present.

All these examples indicate large variations in the structure of the tracheal cuticle, according to the species, the stage of development and, most probably, the size and location of the tracheae. More comparative studies are badly needed, especially for the endocuticular and exocuticular layers. So far, no laminated endocuticle has been clearly described in tracheae, and in the few examples where a fibrillar endocuticle was shown, the orientation of the fibers was only briefly examined. No pore canals were ever described, except in the pupae of *Sarcophaga* (Whitten, 1976), but the filaments pictured may best be compared to epicuticular filaments.

3.3.7. Subcuticle

The only mention of this layer is by Bordereau (1975), who studied the tracheae of the termite queen (see section 6.2). The subcuticle is characterized by its coloration with aldehyde-fuchsin after oxidation, and is usually rich in acid mucopolysaccharides. Using several histochemical tests, Sharief *et al.*, (1973) localized such mucosubstances in the tracheae of *Phormia* larvae, mainly in the intertaenidial region. In our opinion, this problem deserves more attention.

4. The Molt Cycle and the Formation of the New Cuticle

4.1. Do the Tracheoles Molt?

At each molt, the tracheal intima is shed and replaced by a new one. The tracheal exuvium remains in continuity with the integumental exuvium and is expelled at the spiracles. In the classical textbooks (Smith, 1968; Wigglesworth, 1971; Locke, 1974; Neville, 1975), the cuticle of the tracheoles is believed to persist from one instar to the next, i.e., not to molt. This seems to be founded essentially on the well-known work of Wigglesworth (1954) on *Rhodnius prolixus*, where the process of molting is admirably described. There is no doubt about the validity of these observations, but the generalization must be questioned. Indeed, Keister (1948) studied in great detail the evolution of the tracheal system in the fungus gnat *Sciara coprophila* from hatching to the adult. During the pupal and the imaginal molts, the tracheolar intima was not shed, as observed later by Wigglesworth. However, Keister noticed that during the larval molts the whole tracheal intima was eliminated, including that of the tracheoles. Keister's paper is frequently cited to provide evidence concerning the persistence of tracheolar intima during metamorphosis, but her description of its loss during the larval molts is often overlooked.

We had the opportunity to make some ultrastructural observations of the termite *Kalotermes flavicollis* during its larval molts and obtained convincing evidence of the replacement of the intima in even the finest tracheoles in the rectal pads. The electron micrograph (Figure 21) taken of a larva before ecdysis clearly shows a tracheole in cross section, with the old cuticle in the lumen and the new one being secreted. Such observations are possible only over a short time, because after apolysis the tracheolar intima tends to retract (probably by its own elasticity) and to be displaced in more proximal and larger tubes, leaving the tracheolar lumen empty. In Figure 22, taken from an animal at a slightly later stage than in Figure 21, a small trachea is seen to contain the exuvia of several tracheoles retracted in it. Similar observations were made in the rectal pads of the cockroach *Blaberus craniifer*. These facts can by no means be generalized. Most probably, tracheoles do or do not molt depending on the species, the developmental instar, and even the organ. Again, more comparative research is needed.

4.2. Formation of the New Cuticle

Although studied in only a few cases, the process of formation seems essentially similar to that observed in the integument. The first step is the separation of the intima from the cells (*apolysis*). The exuvial space is filled with molting fluid, which is able to digest some parts of the old cuticle. In the integument, an *ecdysial membrane* is generally formed just after the apolysis, but in tracheae its presence seems more or less apparent, depending on the species. It is well delineated in *Rhodnius* (Locke, 1958; Wigglesworth, 1973), and its possible involvement in the formation of the tracheal pattern was discussed by Whitten (1972). However, the ecdysial membrane was not clearly visible in *Antheraea* (Beaulaton, 1968). In our unpublished observations, it was clearly seen in the termite *Kalotermes* (Figures 22, 23) whereas its existence may be questioned in *Blaberus* and *Tenebrio*. Besides, when present it seems to be quickly digested by the molting fluid (Locke, 1958), whereas in the integument it persists and is shed with the exuvium. Possibly, the "ecdysial membrane" of the tracheae may be different and deserves more precise investigation.

As in all types of cuticle so far examined, the secretion of the new cuticle begins with the formation of *plaques* at the apical cell surface (Locke, 1966), where the cytoplasm adjacent to the membrane appears denser and a granular extracellular material is observed at the opposite side. In the epidermis, the plaques are located at the apex of microvilli, whereas in tracheae they commonly lie on only small elevations of the plasma membrane (Locke, 1966; Beaulaton, 1968; Wigglesworth, 1973; and our unpublished observations). At each plaque, a patch of outer epicuticle is secreted, then the patches progressively extend their borders and eventually coalesce. At the time of secretion, the outer epicuticle appears pentalaminar, formed by three electron-dense layers separated by two electron-lucent ones (Locke, 1966; Beaulaton, 1968). The middle dense layer is usually less distinct, sometimes barely visible (Figures 20, 23). The total thickness is approximately 15 nm. Later, the pentalaminar structure is obscured, and the outer epicuticle appears as a dense line, often difficult to distinguish from the inner epicuticle, (Figure 24). In the integument, the outermost layer of the outer epicuticle separates from the other two to become a superficial layer (Locke, 1966). What happens in the tracheae, where a superficial layer seems generally absent? The secretion of the inner epicuticle immediately follows the completion of the outer epicuticle. The taenidial folds are well formed by then (see below), and the inner epicuticle appears thicker at the top of the folds (Beaulaton, 1968). Thus, the outer part of the taenidium seems of epicuticular nature (cf. section 3.3.5). Indeed, the formation of the taenidium was only described by Beaulaton (1968) in *Antheraea* and *Bombyx*, where the different components are produced shortly before the ecdysis, whereas the continuous procuticle is only completed after the ecdysis. During this time, the old cuticle is partially digested. This digestion begins very early, and only the epicuticle and sometimes the outer part of the taenidium (possibly epicuticular in nature) remain (Figure 19) and are shed at ecdysis. Limited information seems available about the way the molting fluid and the products of this digestion may be resorbed. In the integument, the role of muscle attachment zones was well evidenced by Lensky *et al.* (1970), and such structures obviously do not exist in the tracheal system. In a further paper, Lensky and Rakover (1972) showed the absorption of macromolecules in the tracheae and spiracles during ecdysis in the honeybee, without precise details. Possibly, the cuticle is sufficiently permeable, at least before ecdysis, to allow the reabsorption. Also it must be remembered that, in certain conditions, some parts of the tracheal system may be temporarily filled by fluid. In some aquatic insects the newly differentiated tracheal system is likewise fluid-filled, and the replacement of this fluid by gas occurs under water, by movement of fluid and gas across the cuticle.

The modifications of the tracheal cells during the molt cycle were not followed with the same precision as in the epidermal cells, but seem essentially similar (Locke, 1966; Beaulaton, 1968), with the development of the granular endoplasmic reticulum and the activation of the Golgi apparatus. The realization of the taenidial pattern, i.e., the final product of cellular activity, is an interesting problem of morphogenesis. The involvement of physical forces was for a long time favored (see Richards, 1951), and this was elaborated by Locke (1958, 1974) as the "buckling hypothesis." The newly formed cuticular tubes of the tracheae (with only an outer epicuticle) can be considered as thin-walled, elastic cylinders. The buckling, giving rise to the taenidial folds, could be produced by the increase in surface area against the restraint of the epidermal cells. Using reasonable assumptions, Locke developed a simple mathematical expression of buckling frequency taking into account the radius of the tracheal tubes, the wall thickness (i.e., of the outer epicuticle), and a factor (Poisson's ratio) characteristic of the material. The measurements made on the tracheae of *Rhodnius* and *Tenebrio* conformed to the formula.

Several objections were made to the buckling hypothesis, especially by Whitten (1972) and Wigglesworth (1973). It seemed difficult to explain the

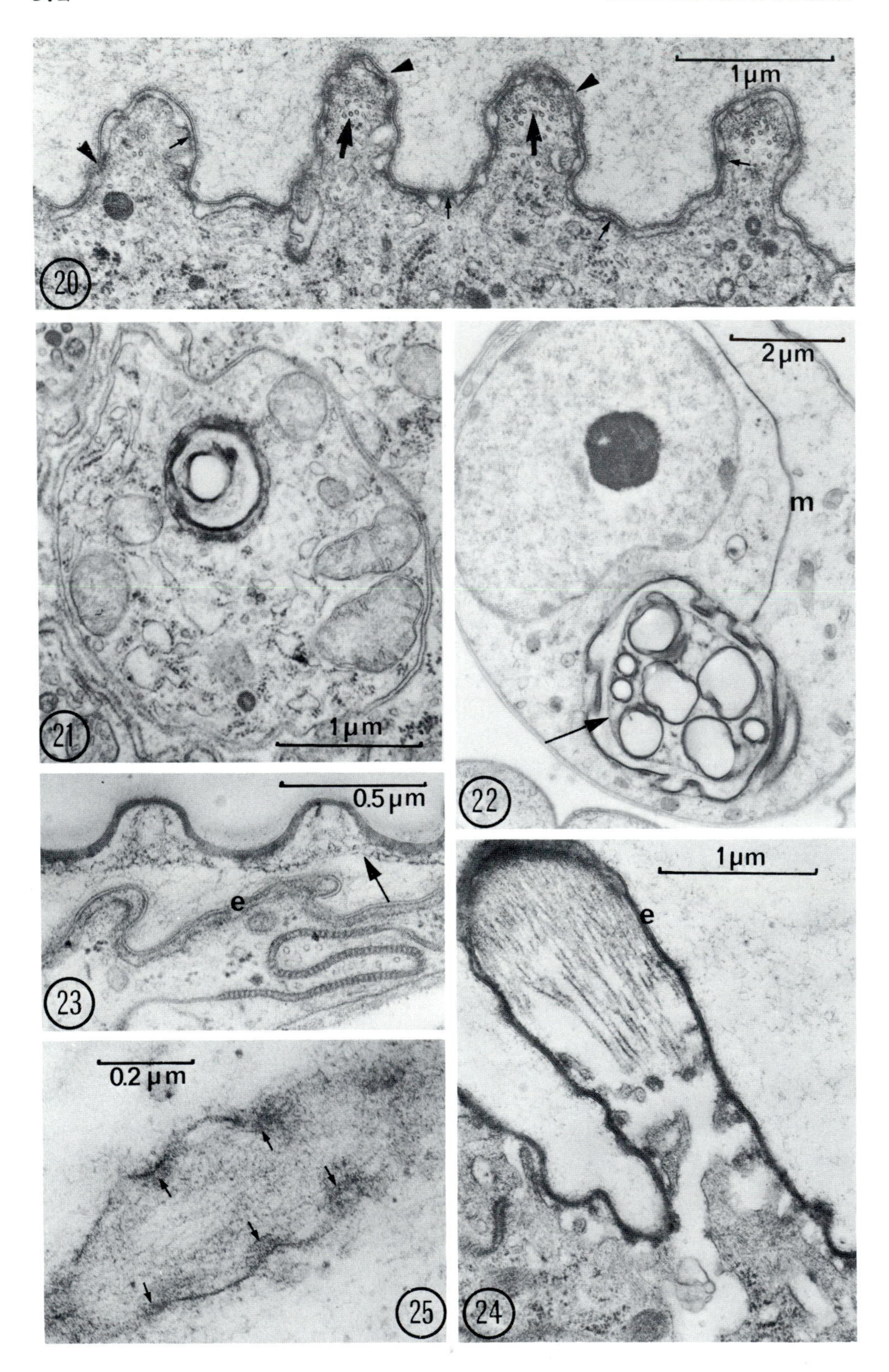
1 μm
20
21
1 μm
2 μm
m
22
0.5 μm
e
23
1 μm
e
0.2 μm
25
24

very regular and sometimes very complex taenidial pattern (e.g., the seven taenidia in series observed by Santos *et al.*, 1954) by simple physical forces. The secretion of the taenidium (with its several components) in a precise location in the taenidial folds can hardly be determined solely by previous cuticular buckling. In most cases, the taenidial folds are formed *before* the completion of the outer epicuticle (Wigglesworth, 1973; and our unpublished observations) (Figure 20); although at that time the folds often appear crumpled, they most probably result from the growth of the plasma membrane, i.e., an active process controlled by the epidermal cells. The deposition of the taenidium occurs, at least in *Rhodnius*, at the apex of newly formed microvilli, which are only present under the taenidial folds (Wigglesworth, 1973) (see also Figure 24), and this again involves the plasma membrane. The physical forces were supposed to comprise only an axial compression and a small tangential shearing force (to account for the taenidia of helicoidal form). This seems an oversimplification. In the wind discs of *Galleria*, a set of tracheoles is formed early in the pupal development at the base of the disc and have a very convoluted arrangement. Afterward, they extend into the growing disc to take up their final position, but during their formation they are surely subject to strong lateral compression. Nevertheless, their taenidia are perfectly normal (Hasskarl *et al.*, 1973). Without dismissing the role of physical forces, an active control of the tracheal cells on the patterning of the cuticle seems highly probable. The importance of the plasma membrane is evident, and a careful examination of the cytoskeleton could be very informative. In our preparations, microtubules were frequently observed running parallel to the taenidial folds, especially at the time of their formation (Figure 20), and may well affect the patterning. Moreover, the other elements of the cytoskeleton (microfilaments and intermediary filaments) ought to be taken into account, but this field remains completely unexplored with regard to the tracheal system.

Figure 20. A section of a trachea in the hindgut wall of *Blaberus* at the beginning of cuticle secretion. The trilaminar outer epicuticle still shows some discontinuities (arrowheads), but the taenidial folds are well formed and filled by cytoplasmic protrusions containing cross sectioned microtubules (large arrows). The plasma membrane supports numerous plaques (small arrows) on very small elevations.

Figure 21. A cross section of a tracheoblast and a tracheole in a rectal pad of a molting termite *Kalotermes*. Inside the new cuticular tube the apolysed intima is clearly visible. Note the absence of mestracheon.

Figure 22. A preparation similar to that seen in Figure 21, but the stage of development is slightly more advanced. In this small trachea [note the mestracheon, (m)] the exuviae of the tracheolar ramifications are ensheathed by a common "ecdysial membrane" (arrow).

Figure 23. In this section of a trachea in the hindgut of a molting *Kalotermes*, the partially digested intima is separated from the new one by an "ecdysial membrane" (arrow). The new cuticle still consists only of the outer epicuticle (e), and the taenidial folds are formed but appear crumpled.

Figure 24. An oblique section of a trachea of a molting *Blaberus craniifer* during the secretion of the taenidium. The newly secreted fibrils of the taenidium are clearly seen under the epicuticle (e). The plasma membrane forms irregular microvilli, especially under the taenidial thread.

Figue 25. An oblique section of a new tracheole at the beginning of its formation in a rectal pad of *Kalotermes*. The cavity is partly filled with fine longitudinal fibrils and the plasma membrane plaques (arrows) are already formed.

4.3. Formation of New Tracheoles

There is a controversy as to whether new tracheoles are of intracellular or extracellular origin. Several observations were made on embryos (Shafiq, 1963; Hillman and Lesnik, 1970) or on insects during their postembryonic development (Locke, 1958; 1966; Beaulaton, 1968; Matsuura and Tashiro, 1976). These studies left the problem incompletely resolved because it is impossible to identify on the electron micrographs the very first steps of the formation; a new tracheole is recognized only when its lumen is completely formed and shows cuticle secretion. This lumen is bound by a unit membrane, which in some cases is connected, and in other cases not connected, by a mestracheon with the peripheral plasma membrane of the tracheoblast; both situations may be encountered in the same cell (Figure a, Pl. VI in Beaulaton, 1968)! In the first instance, a rolling up of the tracheoblast can be postulated, and so it was described by Hillman and Lesnik (1970) in *Drosophila* embryo. The second situation has led to the hypothesis of an intracellular origin, either from the endoplasmic reticulum (Edwards *et al.*, 1958) or by fused Golgi vacuoles (Locke, 1966). Whatever it may be, the secretion of the cuticle is exactly the same as that observed during the molt of the tracheae (where the secreting membrane is surely the plasma membrane) (Locke, 1966; Beaulaton, 1968; Hillman and Lesnik, 1970). Thus, the membrane lining the lumen of the new tracheole shows the same secretory behavior as the apical plasma membrane of the tracheal cell, and a continuity with the plasma membrane of the tracheoblast is evident in favorable sections. When such a connection is not observed, its formation by a fingerlike infolding can be supposed. So in our opinion, an intracellular origin is not supported by any definitive evidence and an extracellular origin seems to be the best hypothesis. However, this morphogenesis implies a large increase of the surface area of the plasma membrane and also a regional differentiation in relation to the cuticle secretion.

The lumen of the new tracheole is filled with longitudinally arranged fibrils (Figure 25), which are still present when the epicuticle is completed (Locke, 1966; Beaulaton, 1968; Matsuura and Tashiro, 1976). The dissolution of this fibrillar material and its absorption again relate to the problem of the cuticular permeability.

5. Postembryonic Development

As emphasized by Whitten (1972), the modification of the tracheal system during the insect development remains a neglected field of research. This is particularly true at the ultrastructural level. The few available studies nevertheless give a good indication of the interest in the tracheal system for developmental biology.

During the larval–pupal development of *Galleria mellonella*, tracheoles migrate into lucanae of the wing imaginal discs. This migration was followed *in vivo* and *in vitro* by Hasskarl *et al.* (1973). Before the migration, the tracheal mass, at the base of the wing disc, is formed of tightly packed tracheolar cells each containing a coiled tracheole with many microtubules aligned parallel to it, and the microtubules follow the course of the uncoiling tracheole during migration. Colchicine and vinblastine inhibit the migration and disrupt the microtubules, supporting the hypothesis that microtubules are required for tracheole migration. The importance of microtubules in tracheole migration was also shown by Wigglesworth in a quite different situation. In his first series of experiments (1954), he observed a migration of tracheoles towards a region of the epidermis deprived of its oxygen supply, and later demonstrated the role of the epidermal cells (1959). They give rise to fine cytoplasmic processes, which become attached to neighboring tracheoles and draw them into the oxygen-deficient zone. Finally, the ultrastructural analysis (1977) revealed both numerous microtubules and microfilaments in the cytoplasmic strands. Wigglesworth suggested that the microtubules are there to resist tension and that the contractile force may be provided by the microfilaments. In contrast to the wing disks studied above, the epidermal, not the tracheolar, cells are active in the migration. Nevertheless, Wigglesworth emphasized the regular occurrence of microtubules in the tracheolar cytoplasm. The experiments of Wigglesworth suggested a direct relationship between the demand for oxygen of a tissue and its tracheal supply, and indicated that the tracheal pattern may be, at least in part, determined by the target organs, a view that was supported by the experiments of Pihan (1971). However, some observations point the opposite way. Studying the flight muscles in various flightless Coleoptera, Smith (1964) described several cases where strongly reduced muscles possess a very extensive tracheolar supply far in excess of their oxygen requirement. Indeed, such muscles are reduced because their development was arrested soon after the tracheolar invasion, which occurs, as in normal flight muscles, during the pupal stage. The tracheal development follows a programed pattern, which anticipates the needs of a fully mature tissue and is preserved in case of regressive evolution. A similar situation is observed in the queen of the higher termites: in the young adult (alate), the organs that will increase during the physogastric growth (ovaries, mid-gut, fat body) are supplied with many tracheae, in proportion not to their present size but to their importance in the fully mature queen (Bordereau, 1971b).

6. Modifications in Adults

As a cuticular structure, the tracheal system is completed at the imaginal molt. However, some later modifications may occur in special cases.

6.1. Degeneration of Tracheal Epithelium

Several examples may be found in the literature of small tracheae and tracheoles reduced to just cuticle, without any cell lining. Gupta and Berridge (1966) frequently observed such naked tracheoles in rectal papillae of adult *Calliphora*, especially in older flies (two days or more after emergence), whereas in newly emerged adults the cytoplasm was well developed. Similar observations were made by Kümmel and Zerbst-Boroffka (1974) in the rectal papillae of the honeybee. In our comparative studies on the rectum, we observed naked tracheoles and small tracheae in imagoes of *Chrysopa carnea* (Neuroptera), *Ctenocephalides canis* (Siphonaptera), and *Apis mellifera* (Hymenoptera) (Figure 27). However, in *Amara eurynota* (Coleoptera), several species of cockroaches (Dictyoptera), *Kalotermes flavicollis* (Isoptera), and *Carausius morosus* (Cheleutoptera), we did not find any tracheole without surrounding cytoplasm. Apart from the rectum, we suspect some cases of tracheolar cell degeneration in the flight muscles. In several micrographs of the flightless Coleoptera published by Smith (1964, Figures 101, 105–107) the tracheolar cuticles seem to be in direct contact with the muscle cells and appear to be intracellular. Likewise, the reported "intramitochondrial" tracheoles in flight muscle from the hornet (Afzelius and Gonnert, 1972) are most probably tracheoles whose secreting cells disappeared after completion of the cuticle. A similar situation may be expected in other organs penetrated by tracheae.

A different type of degeneration of tracheal cells in adults occurs in the auditory organs of some insects. In *Teleogryllus* (Orthoptera), the tympanum

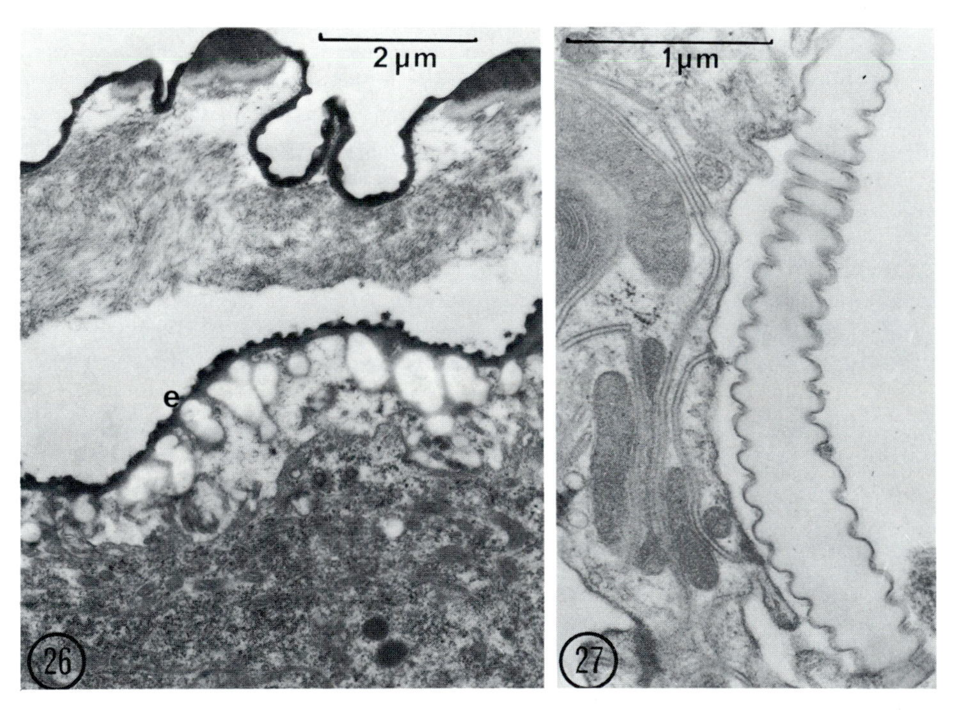

Figure 26. A section of a trachea in a pharate pupa of *Pieris brassicae* that was treated by diflubenzuron in the middle of the last instar. Under the partially digested larval intima, the newly formed cuticle seems profoundly modified. While the epicuticle (e) appears nearly normal, it is separated from the tracheal cells by an irregular meshwork tentatively interpreted as disorganized taenidial material. Unpublished micrograph kindly provided by Bernard Mauchamp.

Figure 27. A section showing a "naked" tracheole in the rectal pad of an *Apis mellifera* worker. In the dilated intercellular spaces between the rectal cells, only the cuticle of the tracheole is apparent with no trace of surrounding tracheolar cytoplasm.

is made by the apposition of two cuticular layers; one is a modification of the peripheral cuticle, the other a part of the tracheal dilatation forming the tympanal cavity. These two cuticles are secreted respectively by the epidermis and the tracheal epithelium, which at the time of formation are sandwiched between the two cuticles. Once the imaginal molt is completed, the two epithelia disappear completely leaving a purely cuticular tympanum (Ball and Cowan, 1978). In a Noctuid moth the tracheal epithelium diasppears along the tympanum (except at the attachment site of the sensillae) but persists in the rest of the tympanal cavity (Ghiradella, 1971).

6.2. *Tracheal Growth in the Adult*

The queen of the higher termites undergoes a tremendous enlargement of its abdomem (physogastry) during the development of the colony. The tracheal system of the female alate has a classical structure, both at the anatomical and the ultrastructural level (Bordereau, 1971a, 1975). During the imaginal growth, the general organization is maintained, but the length of the tracheae may be increased by 15 to 20 times and their diameter by 5 to 10 times in extreme cases of physogastry (Bordereau, 1971b). According to the detailed study of Bordereau (1975), this is caused by an activation of the tracheal epithelium (involving cell multiplication and renewed secretory activity), which produces new cuticular material giving rise to an exceptionally thick (up to 35 μm) and partially sclerotized cuticle. The taenidial folds are preserved but greatly en-

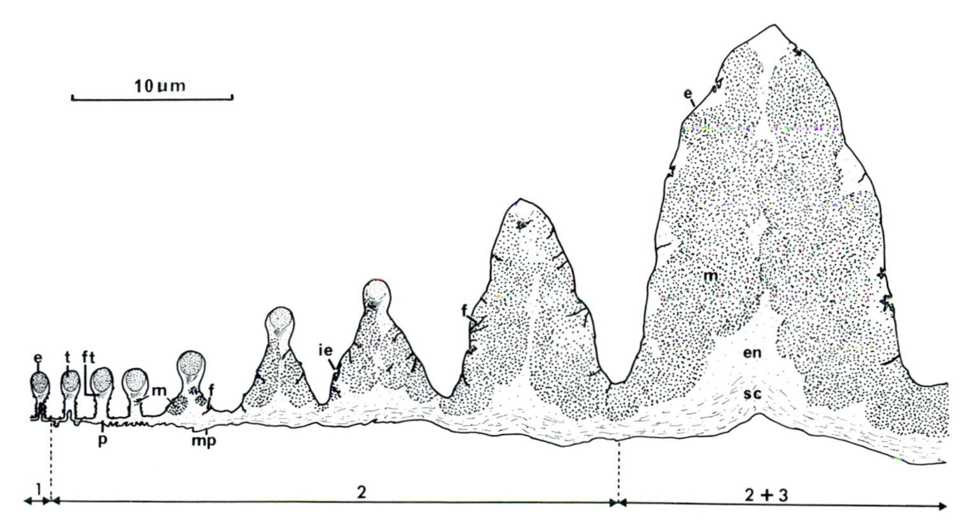

Figure 28. A diagram summarizing the modifications of the tracheal intima during the physogastry in a queen of *Macrotermes bellicosus* (formerly *Bellicositermes natalensis*). (1) In the female alate, the cuticle is composed of an epicuticle (e), the taenidium (t) and some subtaenidial fibrils (ft). (2 and 3) The progressive enlargement is accompanied by the growth of the epicuticle with epicuticular filaments (f) and epicuticular invaginations (ie), the secretion of a procuticle (p) above the plasma membrane (mp), differentiating in mesocuticle (m), endocuticle (en), and subcuticle (sc). The taenidium progressively disappears, and the taenidial fold becomes composed of mainly mesocuticular material. After Bordereau (1975).

larged, and thus the successive folds become wider spaced. The new cuticular material shows the same histochemical reactions as a mesocuticle, but appears poorly structured under the electron microscope and not birefringent in polarized light. It contains chitin, as indicated by a strong reaction with the chitosan test, but no resilin. However, most surprising seems the true growth of the epicuticle. These modifications are summarized in Figure 28.

So far, the termite queen seems to be a unique example, perhaps because of a lack of information on similar cases of imaginal enlargement.

7. *Conclusion*

Until the present time, the ultrastructural research on the tracheal system has been essentially descriptive. Although many gaps remain, as underlined in this review, it now seems necessary to combine the electron microscope techniques with biochemical, biophysical, and physiological methods. The physicochemical composition of the different cuticular layers ought to be clarified, in relation to the permeability and the mechanical properties of the tracheal intima. The development of the tracheal system offers a promising field of investigation, both at the cellular (i.e., the role of the cytoskeleton) and the organismic level. We hope that this review can promote a renewed interest in the tracheal system.

ACKNOWLEDGMENTS

We are grateful to Dr. Bernard Mauchamp who has communicated to us various unpublished results, especially concerning experimental studies with diflubenzuron. We also acknowledge the kind help of Dr. and Mrs. Collin R. Green for correction of the manuscript language.

References

Afzelius, B., and Gonnert, N., 1972, Intramitochondrial tracheoles in flight muscle from the hornet *Vespa crabro*, *J. Submicrosc. Cytol.* **4**:1–6.

Ball, E. E., and Cowan, A. N., 1978, Ultrastructural study of the development of the auditory tympana in the cricket *Teleogryllus commodus* (Walker), *J. Embryol. Exp. Morphol.* **46**:75–87.

Baudet, J. L., 1974, Morphologie et ultrastructure du dispositif trachéen permettant l'émission sonore chez *Gromphadorhina portentosa* Schaum (Dictyoptera, Blaberidae), *C. R. Acad. Sci. Paris* **279(D)**:1175–1178.

Beaulaton, J., 1964, Les ultrastructures des trachées et leurs ramifications dans la glande prothoracique du ver à soie Tussor (*Antheraea pernyi* Guer. Lepidoptère, Attacide), *J. Microsc.* (Paris) **3**:91–104.

Beaulaton, J., 1968, Modifications ultrastructurales des trachées et trachéoles chez les vers à soie en période de mue, *J. Microsc.* (Paris) **7**:621–646.

Beaulaton, J., 1969, Etude cytochimique en microscopie électronique de l'intima cuticulaire de trachées d'Insects: Action de chitinases et de quelques proteases sur coupes ultrafines d'épon, *C. R. Acad. Sci.* **269(D)**:2388–2391.

Bordereau, C., 1971a, Dimorphisme sexuel du système trachéen chez les imagos ailés de *Bellicositermes natalensis* Haviland (Isoptera, Termitidae). Rapports avec la physogastrie de la Reine, *Arch. Zool. Exp. gén.* **112**:33–54.

Bordereau, C., 1971b, Le système trachéen de la Reine physogastre et du Roi chez *Bellicositermes natalensis* Haviland (Isoptera, Termitidae), *Arch. Zool. Exp. gén.* **112**:747–760.

Bordereau, C., 1975, Croissance des trachées au cours de l'évolution de la physogastrie chez la reine des Termites supérieurs (Isoptera, Termitidae), *Int. J. Insect Morphol. Embryol.* **4**:431–465.

Buck, J., 1948, The anatomy and physiology of the light organ in fireflies, *Ann. N.Y. Acad. Sci.* **49**:397–482.

Edwards, G. A., Ruska, H. and de Harven, E., 1958, The fine structure of insect tracheoblasts, tracheae and tracheoles, *Arch. Biol. Belg.* **69**:351–369.

Faucheux, M. J., 1972, Relations entre l'ultrastructure de l'intima cuticulaire et les fonctions des sacs aériens chez les Insectes, *C. R. Acad. Sci. Paris* **274(D)**:1518–1521.

Faucheux, M. J., 1974, Recherches sur l'appareil respiratoire des Diptères adultes. III. Evolution du système trachéen, *Ann. Soc. ent. Fr.* **10**:99–121.

Faucheux, M. J., and Sellier, R., 1971, L'ultrastructure de l'intima cuticulaire des sacs aériens chez les Insectes, *C. R. Acad. Sci. Paris* **272(D)**:2197–2200.

Flower, N. E., and Walker, G. D., 1979, Rectal papillae in *Musca domestica:* The cuticle and lateral membranes, *J. Cell Sci.* **39**:167–186.

Ghiradella, H., 1971, Fine structure of the Noctuid moth ear. I. The transducer area and connections to the tympanic membrane in *Feltia subgothica* Haworth, *J. Morphol.* **134**:21–46.

Ghiradella, H., 1977, Fine structure of the tracheoles of the lantern of a photurid firefly, *J. Morphol.* **153**:187–204.

Ghiradella, H., 1978, Reinforced tracheoles in three firefly lanterns: Further reflections on specialized tracheoles, *J. Morphol.* **157**:281–299.

Greven, H., and Rudolph, R., 1973, Histologie und Feinstruktur der larvalen Kiemenkammer von *Aeshna cyanea* Müller (Odonata, Anisoptera), *Z. Morphol. Ökol. Tiere* **76**:209–226.

Gupta, B. L., and Berridge, M. J., 1966, Fine structural organization of the rectum in the blowfly, *Calliphora erythrocephala* (Meig.) with special reference to connective tissue, tracheae and neurosecretory innervation in the rectal papillae, *J. Morphol.* **120**:23–82.

Hanna, C. H., Hopkins, T. A., and Buck, J., 1976, Peroxisomes of the firefly lantern, *J. Ultrastruct. Res.* **57**:150–162.

Hasskarl, E., Oberlander, H., and Stephens, R. E., 1973, Microtubules and tracheole migration in wing disks of *Galleria mellonella, Develop. Biol.* **33**:334–343.

Hillman, R., and Lesnik, L. H., 1970, Cuticle formation in the embryo of *Drosophila melanogaster, J. Morphol.* **131**:383–396.

Keister, M. L., 1948, The morphogenesis of the tracheal system of *Sciara, J. Morphol.* **83**:373–423.

Komnick, H., and Wichard, W., 1975, Vergleichende Cytologie der Analpapillen, Abdominalschläuche und Tracheenkiemen aquatischer Mückenlarven (Diptera, Nematocera), *Z. Morphol. Ökol. Tiere* **81**:323–341.

Kummel, G., and Zerbst-Boroffka, I., 1974, Electron microscopic and physiological studies on the rectal pads in *Apis mellifica, Cytobiologie* **9**:432–459.

Larsen, W. J., 1977, Structural diversity of gap junctions: A review, *Tissue Cell* **9**:373–394.

Lensky, Y., Cohen, C., and Schneiderman, H. A., 1970, The origin, distribution and fate of the molting fluid proteins of the *Cecropia* silkworm, *Biol. Bull.* **139**:277–295.

Lensky, Y., and Rakover, Y., 1972, Resorption of moulting fluid proteins during the ecdysis of the honey bee, *Comp. Biochem. Physiol.* **41**:521–531.

Locke, M., 1957, The structure of insect tracheae, *Quart. J. Microsc. Sci.* **98**:487–492.

Locke, M., 1958, The formation of tracheae and tracheoles in *Rhodnius prolixus, Quart. J. Microsc. Sci.* **99**:29–46.

Locke, M., 1966, The structure and formation of the cuticulin layer in the epicuticle of an insect, *Calpodes ethlius* (Lepidoptera, Hesperiidae), *J. Morphol.* **118**:461–494.

Locke, M., 1969, The structure of an epidermal cell during the development of the protein epicuticle and the uptake of molting fluid in an insect, *J. Morphol.* **127**:7–39.

Locke, M., 1974, The structure and formation of the integument in insects. In *The Physiology of Insecta,* edited by M. Rockstein, vol. 6, pp. 123–213, Academic Press, New York.

Locke, M., and Krishnan, N., 1971, The distribution of phenoloxidases and polyphenols during cuticle formation, *Tissue Cell* **3**:103–126.

Lyonet, P., 1760, *Traité Anatomique de la Chenille qui Ronge le Bois de Saule*, De Hondt, La Haye.

Malpighi, M., 1669, Dissertatio Epistolica de Bombyce, Martyn, Londini.

Matsuura, S., and Tashiro, Y., 1976, Ultrastructural changes in the posterior silk gland cells in the early larval instars of the silkworm, *Bombyx mori*, *J. Insect Physiol.* **22**:967–979.

Miller, W. H., and Bernard, G. D., 1968, Butterfly glow, *J. Ultrastruct. Res.* **24**:286–294.

Neville, A. C., 1975, *Biology of the Arthropod Cuticle*, Springer-Verlag, Berlin.

Noirot, C., and Noirot-Timothée, C., 1976, Fine structure of the rectum in cockroaches (Dictyoptera): General organization and intercellular junctions, *Tissue Cell* **8**:345–368.

Noirot, C., Noirot-Timothée, C., Smith, D. S., and Cayer, M. L., 1978, Cryofracture de la cuticule des Insectes, mise en évidence d'un plan de clivage dans l'épicuticule externe: Implications structurales et fonctionnelles, *C. R. Acad. Sci. Paris* **287(D)**:503–505.

Noirot-Timothée, C., and Noirot, C., 1980, Septate and scalariform junctions in arthropods, *Int. Rev. Cytol.* **63**:97–140.

Oschman, J. L., and Wall, B. J., 1969, The structure of the rectal pads of *Periplaneta americana* L. with regard to fluid transport, *J. Morphol.* **127**:475–510.

Peterson, M. K., 1970, The fine structure of the larval firefly light organ, *J. Morphol.* **131**:103–116.

Peterson, M. K., and Buck, J., 1968, Light organ fine structure in certain asiatic fireflies, *Biol. Bull.* **135**:335–348.

Pihan, J. C., 1971, Mise en évidence d'un facteur tissulaire intervenant au cours de la morphogenèse du système trachéen chez les Insectes Diptères, *J. Embryol. Exp. Morphol.* **26**:497–521.

Réaumur, R. A. F. de., 1734–1742, *Mémoires pour Servir à l'Histoire des Insectes*, Impremerie Royale, Paris.

Ribi, W. A., 1979, Structural differences in the tracheal tapetum of diurnal butterflies, *Z. Naturforsch. C.* **34**:284–287.

Richards, A. G., 1951, *The Integument of Arthropods*, University of Minnesota Press, Minneapolis.

Richards, A. G., and Anderson, T. E., 1942, Electron micrographs of insect tracheae, *J. N.Y. Entomol. Soc.* **50**:147–167.

Richards, A. G., and Korda, F. H., 1950, Studies on arthropod cuticle. IV. An electron microscope survey of the intima of arthropod tracheae, *Ann. Entomol. Soc. Am.* **43**:49–71.

Saini, R. S., 1977, Ultrastructural observations on the tracheal gills of *Aeshna* (Anisoptera, Odonata) larva, *J. Submicrosc. Cytol.* **9**:347–353.

Santos, H. L. S., Edwards, G. A., Santos, P. S., and Sawaya, P., 1954, Electron microscopy of insect tracheal structures, *An. Acad. Brasileira Cienc.* **26**:309–315.

Shafiq, S. A., 1963, Electron microscopy of the development of tracheoles in *Drosophila melanogaster*, *Quart. J. Microsc. Sci.* **104**:135–140.

Sharief, F. S., Perdue, J. M., and Dobrogosz, W. J., 1973, Biochemical, histochemical and autoradiographic evidence for the existence and formation of acid mucopolysaccharides in larvae of *Phormia regina*, *Insect Biochem.* **3**:243–262.

Smith, D. S., 1961, The structure of insect fibrillar flight muscle: A study made with special reference to the membrane systems of the fiber, *J. Biophys Biochem. Cytol.* **10**:123–158.

Smith, D. S., 1963, The organization and innervation of the luminescent organ in a firefly, *Photuris pennsylvanica* (Coleoptera), *J. Cell Biol.* **16**:323–359.

Smith, D. S., 1964, The structure and development of flightless Coleoptera: A light and electron microscopic study of the wings, thoracic exoskeleton and rudimentary flight musculature, *J. Morphol.* **114**:107–183.

Smith, D. S., 1968, *Insect Cells*, Oliver and Boyd, Edinburgh.

Swammerdam, J. J. 1737, *Bybel der Natuur*, Severin, Leyden.

Whitten, J. M., 1969, Coordinated development in the fly foot: Sequential cuticle secretion, *J. Morphol.* **127**:73–104.

Whitten, J. M., 1972, Comparative anatomy of the tracheal system, *Annu. Rev. Entomol.* **17**:373–402.

Whitten, J. M., 1976, Stage specific larval, pupal, and adult cuticles in the tracheal system of *Sarcophaga bullata*, *J. Morphol.* **150**:369–398.

Wichard, W., 1973, Zur Morphogenese des respiratorischen Epithels der Tracheenkiemen bei Larven der *Limnephilini* Kol. (Insecta, Trichoptera), *Z. Zellforsch.* **144**:585–592.

Wichard, W., 1979a, Structure and function of the respiratory epithelium in the tracheal gills of mayfly larvae, *Proc. 2nd Int. Conf. Ephemeroptera*, Krakow 1975, pp. 307–309.

Wichard, W., 1979b, Zur Feinstruktur der abdominalen Trachenkiemen von Larven der klein-libellen-Art *Epallage fatime* (Odonata, Zygoptera, Euphaeidae), *Entomol. Gen.* **5**:129–134.

Wichard, W., and Komnick, H., 1974, Structure and function of the respiratory epithelium in the tracheal gills of stonefly larvae, *J. Insect Physiol.* **20**:2397–2406.

Wigglesworth, V. B., 1954, Growth and regeneration in the tracheal system of an Insect, *Rhodnius prolixus* (Hemiptera), *Quart. J. Microsc. Sci.* **95**:115–137.

Wigglesworth, V. B., 1959, The role of epidermal cells in the "migration" of tracheoles in *Rhodnius prolixus* (Hemiptera), *J. Exp. Biol.* **36**:632–640.

Wigglesworth, V. B., 1971, *The Principles of Insect Physiology*, 7th ed., Methuen and Co., London.

Wigglesworth, V. B., 1973, The role of the epidermal cells in moulding the surface pattern of the cuticle in *Rhodnius* (Hemiptera), *J. Cell Sci.* **12**:683–705.

Wigglesworth, V. B., 1977, Structural changes in the epidermal cells of *Rhodnius* during tracheole capture, *J. Cell Sci.* **26**:161–174.

Wigglesworth, V. B., 1979, Secretory activities of plasmatocytes and oenocytoids during the moulting cycle in an Insect (*Rhodnius*), *Tissue Cell* **11**:69–78.

13

Structural and Functional Analysis of Balbiani Ring Genes in the Salivary Glands of *Chironomus tentans*

BERTIL DANEHOLT

1. *Introduction*

During recent years, much attention has been devoted to studies of the regulation of gene activity in eukaryotic cells. Individual eukaryotic genes have been characterized by the new recombinant DNA technology and defined sequences have been tested for functions in both *in vitro* (Wasylyk *et al.*, 1980) and *in vivo* systems (Grosschedl and Birnstiel, 1980). Furthermore, the rapidly expanding knowledge of chromosome structure suggests that the chromosomal fiber and its organization into higher order structures is probably important for the regulation of gene activity (Finch and Klug, 1976; Sedat and Manuelidis, 1978, Suau *et al.*, 1979). Certain results even indicate that a change in chromosome structure could, in fact, be the primary event upon gene activation in eukaryotes (e.g., Foe, 1978). The structure of inactive chromatin is rather well understood (Kornberg, 1977; Chambon, 1978; Felsenfeld, 1978; Klug *et al.*, 1980), while less information is available on the structure of the fiber in the active state (Mathis *et al.*, 1980). It has been demonstrated in micrococcal digestion experiments that the active fiber maintains the 200 base pair (bp) subunit structure characteristic for inactive chromatin (Bellard *et al.*, 1978). Other biochemical data show, however, that the chromosome fiber is altered upon transcription; an increased sensitivity to DNase I has, for example, been noted (Weintraub

BERTIL DANEHOLT • Department of Medical Cell Genetics, Medical Nobel Institute, Karolinska Institutet, S-10401 Stockholm, Sweden.

and Groudine, 1976). The main obstacle in the biochemical investigations of transcriptionally active chromatin is that this fraction of chromatin is a minor one and cannot easily be isolated from total chromatin (for discussion, see Mathis *et al.*, 1980). Electron microscopy offers here a distinct advantage since active genes can be directly studied by this method. It has even been feasible to investigate two specific active genes, the silk fibroin gene in the spinning glands of *Bombyx mori* (McKnight *et al.*, 1976) and the Balbiani ring 75 S RNA genes in the salivary glands of *Chironomus tentans* (Lamb and Daneholt, 1979).

In this review I will sum up the present ultrastructural information on the Balbiani ring genes in *Chironomus tentans*. These genes have been studied in the electron microscope after having been spread on a surface according to Miller (Miller and Bakken, 1972), but also *in situ* by conventional electron microscopy techniques, including serial sectioning and reconstruction. A detailed picture of the active 75 S RNA genes has emerged, and the results can be related to the information available on the 75 S RNA genes in their inactive state. With this basic knowledge at hand, the nature of the gene activation process can be discussed.

2. Balbiani Rings and Production of 75 S RNA

The *Chironomus* salivary glands contain polytene chromosomes, which owing to their large size and detailed morphology have proven most useful as an experimental material (for reviews, see Case and Daneholt, 1977; Grossbach, 1977). Some regions are more or less expanded and have been designated puffs. These sites are active in RNA synthesis and are likely to correspond to active genes. Three puffs on chromosome IV are particularly conspicuous and have been named Balbiani rings (BR1, BR2, and BR3). The morphology of a Balbiani ring is shown in Figure 1. These giant puffs can be isolated by micromanipulation (Edström, 1964), which proved important in early biochemical studies of RNA synthesis in the BRs (for review, see Case and Daneholt, 1977). It was shown that 75 S RNA molecules are generated in BR2 (Daneholt, 1972) and in BR1 (Egyházi, 1975), while the BR3 product is still unknown. Although similar in size, the BR1 and BR2 RNA molecules are different, as demonstrated by *in situ* hybridization (Lambert, 1972). The 75 S RNA transcription products are released into nuclear sap and are translocated into cytoplasm (Daneholt and Hosick, 1973; Egyházi, 1976). There, BR1 and BR2 75 S RNA molecules enter polysomes (Daneholt *et al.*, 1977; Wieslander and Daneholt, 1977) and are likely to encode the message for giant salivary polypeptides (Grossbach, 1977). According to recent studies (Hertner *et al.*, 1980; Rydlander and Edström, 1980), these polypeptides are of exceptional sizes (about 10^6 daltons) and match the extraordinary sizes of the mRNA molecules (about 37 kilo base pairs (kb) according to Case and Daneholt, 1978). It should be added that the secretory polypeptides are utilized by the larvae to build tubes, in which the larvae reside (for review, see Grossbach, 1977). In conclusion, the transfer of genetic information has been followed from the BR1 and BR2 genes to the corresponding protein

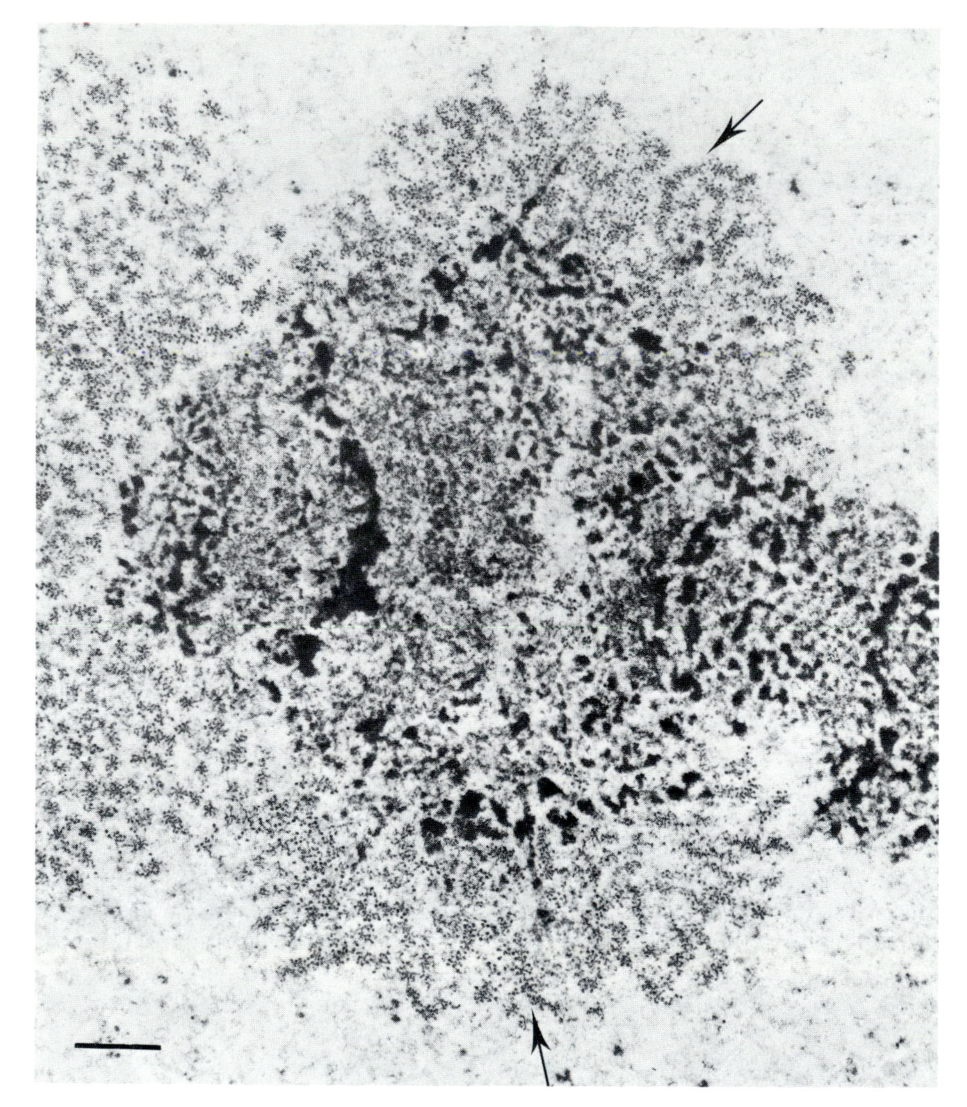

Figure 1. Electron micrograph of a Balbiani ring on chromosome IV in the salivary glands of *Chironomus tentans*. Two peripheral lampbrushlike loops have been indicated by arrows. The bar represents 1 μm.

products, and the functional significance of the proteins is known. Therefore, the BRs should now be suitable for studies of the regulation of gene activity.

3. *Active Balbiani Ring Genes as Observed in Miller Spreads*

The active BR genes have recently been visualized in the electron microscope by Lamb and Daneholt (1979), essentially following the Miller method

(Miller and Bakken, 1972). *Chironomus* salivary glands were treated in a non-ionic detergent solution, and single chromosomes were released when the glands were forced through a fine pipette opening. Individual fourth chromosomes were transferred to a low ionic strength medium at pH 9, allowed to expand, and subsequently sedimented onto a grid and analyzed in the electron microscope (Figure 2). Most of the material consisted of beaded fibers characteristic of chromatin inactive in transcription (Olins and Olins, 1974; Oudet *et al.*, 1975; Woodcock *et al.*, 1976); the beads are likely to correspond to the chromatin subunits, the nucleosomes (for review, see Kornberg, 1977). Some of the chromosomal fibers, however, exhibited features typical of genes active in transcription, i.e., a lateral fiber gradient corresponding to the growing ribonucleoprotein (RNP) fibers, and a densely staining granule at the base of each lateral fiber corresponding to an RNA polymerase (e.g., Miller *et al.*, 1970) (Figure 2). Most of the active genes recorded are likely to be of the 75 S RNA type, as more than 85% of the RNA synthesis on chromosome IV corresponds to 75 S RNA (Daneholt *et al.*, 1969; unpublished data). Moreover, the length of the most prevalent active gene amounted to 7.7 μm, which agrees well with the predicted value for a 75 S RNA gene, assuming that a 75 S RNA gene exhibits a similar DNA compaction as other nonribosomal genes (DNA compaction 1.1–1.9; for examples, see Laird *et al.*, 1976; McKnight *et al.*, 1978). Finally, the active gene was closely packed with RNP fibers (123 fibers per gene as an average, or about one fiber per 300 base pairs). Such a high fiber density is to be expected from a 75 S RNA gene producing an adundant mRNA. We therefore conclude that the recorded active genes are likely to represent the BR 75 S RNA genes. In Miller spreads, we have, however, no means of distinguishing between active 75 S RNA genes in BR1 and BR2.

The Miller method provided us with information on the arrangement of the active 75 S RNA genes. In the electron micrographs it appeared as if the active genes were clustered (Figure 2), but the large number of active genes had to be attributed to the polytene nature of the chromosomes. Sometimes the chromosomal fiber could be followed for a considerable distance (up to 15 μm) on one or both sides of the active gene, and then in no case was a second active 75 S RNA gene recorded along the same fiber. Furthermore, we could calculate the average number of active 75 S RNA genes per chromosomal fiber in a BR. The amount of RNA per active gene was estimated from the number of lateral fibers (123 per gene) and the size of primary transcript (37 kb). Knowing the amount of RNA per BR2 (10.8 pg according to Edström *et al.*, 1978), we could then establish that there are about 8600 active genes in BR2. This figure corresponds approximately to the number of chromatids in a salivary gland cell in *Chironomus tentans* (Daneholt and Edström, 1967). We therefore conclude that there is only one active 75 S RNA gene per chromatid in BR2. Since BR1 contains only 3.7 pg RNA (Edström *et al.*, 1978), the same conclusion should also be applicable to BR1.

3.1. The Transcription Process in a Balbiani Ring

The RNA polymerases were distributed along the entire 75 S RNA genes. Sometimes gaps lacking RNP fibers could be seen. However, there were no seg-

ments of the gene that consistently lacked RNA polymerases with associated RNP fibers. It seems likely, therefore, that the RNA polymerases move along the entire active gene from the initiation to the termination site. Although the fibers appeared to increase in length along the gene, a strict length gradient was not established. Stronger evidence for a continuous transcription can be derived from biochemical analysis of nascent RNA in BR2. When rapidly labeled RNA was studied in agarose gels, a smooth distribution of radioactivity up to the peak at 75 S was observed (e.g., Figure 1 in Daneholt *et al.*, 1978a). This population of RNA molecules probably represents growing RNA molecules (for discussion, see Daneholt, 1975), and the gradual increase in activity with no peaks smaller than 75 S RNA (except for 4–5S RNA) suggests a continuous transcription along the unit. It can, therefore, be concluded that the RNA polymerases are probably not only traversing the whole unit but are also synthetically active along the whole unit.

The available biochemical and electron microscopic information on the transcription process in BR2 made it possible to characterize the process as to relevant quantitative parameters. The initiation frequency was estimated to be six RNA polymerases per min per transcription unit, as derived from the average number of fibers per unit (123) and the transcription time (20 min at 18°C according to Egyházi, 1975). The elongation rate amounted to 31 nucleotides per sec as calculated from the length of the unit (37 kb) and the transcription time. This knowledge of the transcription process in the BR2 system forms the basis for further analysis of the regulation of the gene activity. In this context it is of importance that the RNA synthesis in BR2 can be modified by galactose (Beermann, 1973; Nelson *et al.*, 1978). At different levels of RNA production, the distribution of the RNA polymerases, the length and distribution of the RNP fibers, and the transcription time can now be investigated at the 75 S RNA genes. It remains to be seen if such an approach will help us understand the mechanisms involved in regulation of RNA synthesis on the chromosomal level (initiation, pretermination, posttranscriptional degradation, etc.).

3.2. The Chromosomal Fiber Within and Outside the Active Genes

The chromosomal fibers of the *Chironomus* polytene chromosomes were regularly beaded, as observed in Figures 2 and 3. The bead, or nucleosome, frequency, 28 per μm fiber, was similar to that reported for chromatin of other sources; for example, embryonic chromatin in *Drosophila melanogaster* also has a bead frequency of 28 beads per μm when spread according to Miller (Laird *et al.*, 1976). It was, however, evident that the chromosomal fiber within the active genes had been modified. The fiber segments between the RNA poly-

←

Figure 2. Balbiani ring region of chromosome IV, spread by the Miller method. Several active 75 S RNA genes can be seen as well as a number of chromosomal fibers not engaged in transcription. The two arrows denote active regions with incompletely unfolded RNP fibers. The bar is equal to 1 μm. For further information, see Lamb and Daneholt (1979).

merases were usually thin (about 5 nm) and contained no or very few beads (the average number of beads amounted to 4–5 per μm chromosomal axis) (Figure 3). The few beads were also less regularly distributed than in inactive chromatin. Our results can be related to other studies of the beaded structure of nonribosomal genes showing various degrees of transcriptional activity. The moderately active genes in *Drosophila* and *Oncopeltus* (Laird *et al.*, 1976) harbor somewhat fewer beads than inactive chromatin (19 *versus* 34 beads per μm chromatin in *Oncopeltus,* and 18 *versus* 28 beads in *Drosophila*). No or very few beads were recorded in the highly active nonribosomal genes studied in, e.g., embryos of *Drosophila melanogaster* (McKnight and Miller, 1976) and the silk gland of *Bombyx mori* (McKnight *et. al.*, 1976). The 75 S RNA genes having an RNP fiber density between the moderately and highly active genes also show an intermediate bead frequency (4–5 beads per μm chromatin). When the active genes become increasingly loaded with RNA polymerases, the number of beads seems

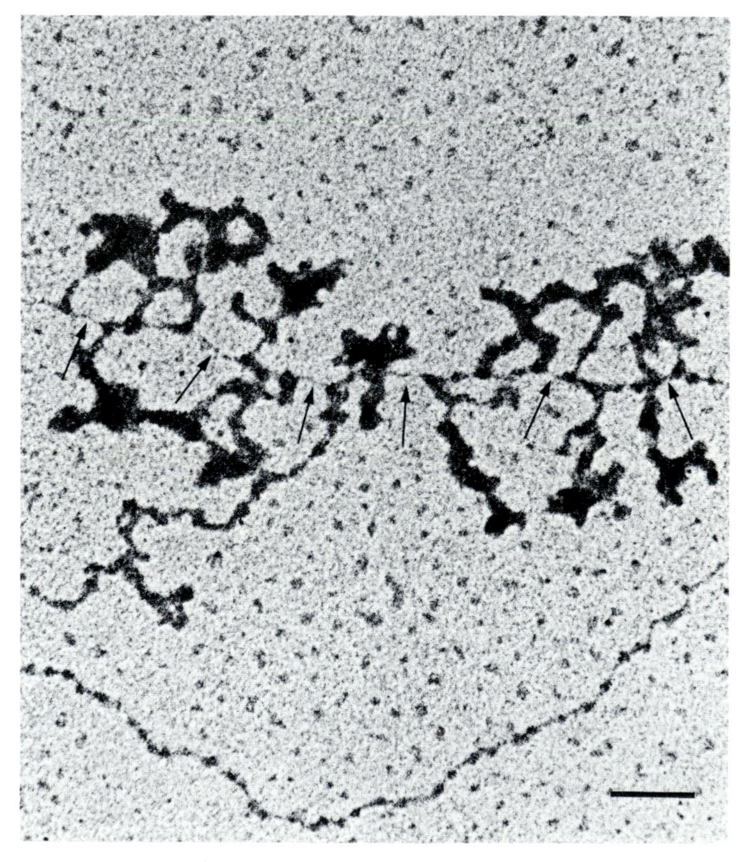

Figure 3. The morphology of the chromosomal axis in transcriptionally active and inactive regions. The active part (above) represents a segment of a 75 S RNA gene. Nonbeaded portions of the axis have been indicated by arrows. The inactive fiber (below) exhibits a bead-on-a-string structure reflecting the organization of the fiber into defined subunits, the nucleosomes. The bar equals 100 nm.

to diminish gradually. Such a relationship has also been noted by Scheer (1978) in a study of the transcriptional activity during amphibian oogenesis. No or very few beads were recorded on the active genes showing maximal activity in mid-oogenic stages, while beads were apparent during stages of lower activities.

We were able to determine that the loss of beads on the active gene was accompanied with an extension of the chromosomal fiber. This was done by comparing the DNA compaction for active and inactive chromatin (DNA compaction is defined as μm DNA per μm chromosomal fiber). Knowing that DNA in *Chironomus* polytene chromosomes is organized into nucleosomes with a repeat of 189 bp (Andersson *et al.*, 1980) we could estimate that the DNA compaction for inactive chromatin is 1.8 when spread according to the Miller method. Since the B form length of the active gene is 12.6 μm (Case and Daneholt, 1978) and the length of the active gene in spread preparations is 7.7 μm, we can arrive directly at a value of 1.6 for the DNA compaction within the active 75 S RNA genes. The data therefore suggest that the fiber is more extended in the transcriptionally active gene than in adjacent inactive chromatin. This result agrees with studies of other nonribosomal genes. The DNA compaction values for active nonribosomal chromatin range from about 1.1 for the highly active silk fibroin gene (McKnight *et al.*, 1976) to 1.6–1.9 for genes with low activity recorded in *Drosophila* (Laird and Chooi, 1976) and *Oncopeltus* (Foe *et al.*, 1976) embryos. It has been suggested by McKnight *et al.* (1978) that the degree of chromatin extension in active nonribosomal genes is directly proportional to the density of RNA polymerase molecules on the chromatin. Our data for the 75 S RNA gene support this concept, since the 75 S RNA gene exhibits an intermediate fiber density (16 fibers per μm chromatin) as well as an intermediate DNA compaction (1.6).

The properties of the chromosomal fiber within the active 75 S RNA genes suggest that the structure of the fiber is modified upon transcription to an extended and nonbeaded configuration. Whatever the alteration of the chromatin may be, micrococcal nuclease experiments on active chromatin in other systems suggest that the organization of the fiber into subunits corresponding to about 200 bp of DNA, is still maintained (for review, see Mathis *et al.*, 1980). Since the histone proteins are responsible for the subunit organization of the chromosomal fiber, they are likely to remain bound to the DNA, although evidently in a modified arrangement. Finally, the fact that the bead frequency is related to the transcriptional activity indicates that the structural change is probably rapidly reversible in nonribosomal genes.

Interestingly, it was noted that a nonbeaded segment of chromatin preceded the first putative RNA polymerase molecule on the active gene. In Figure 4, three examples of the nonbeaded region are displayed. The length of the region was measured to be 0.15–0.20 μm. To establish whether this initial segment is, in fact, outside the transcribed sequence, we determined the start of RNA synthesis from the length of the growing RNP fibers according to the method of Laird *et al.* (1976). It was found that the RNA synthesis begins at about the position of the first polymerase, and the smooth region proximal to this polymerase is therefore probably not transcribed. A similar nonbeaded segment has also been recorded in the SV40 minichromosome (Saragosti *et al.*,

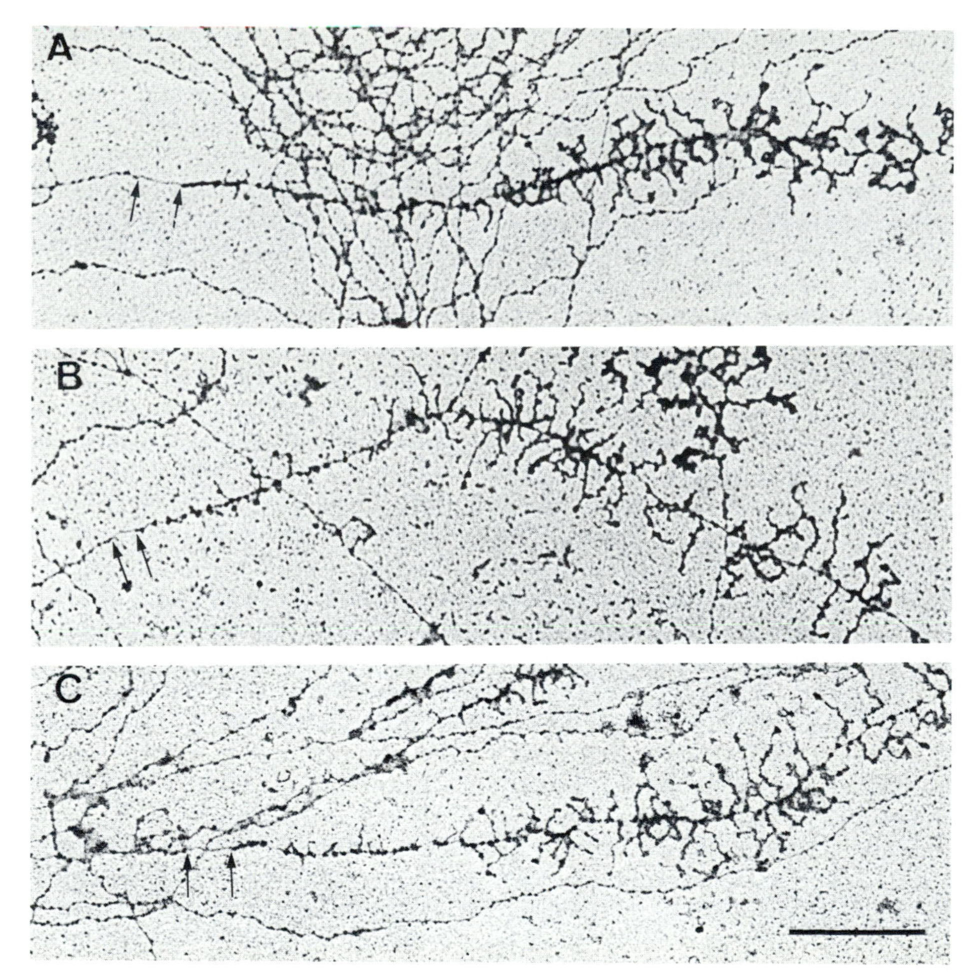

Figure 4. Three examples of the presence of a nonbeaded segment preceding the active 75 S RNA genes in Balbiani rings. The smooth segments have been indicated by arrows. The bar equals 0.5 μm. For further details, consult Lamb and Daneholt (1979).

1980). The significance of the nonbeaded segment can only be a matter of speculation. The idea that the segment might correspond to a promoter region must be considered (cf. modulator elements, as demonstrated by Grosschedl and Birnstiel, 1980). Another possibility is that a fiber segment somewhat longer than the actual active gene is changed upon the activation of the gene.

4. The in Situ Structure of Active Balbiani Ring Genes

Several electron microscopy studies on sectioned material (Beermann and Bahr, 1954; Stevens and Swift, 1966; Vasquez-Nin and Bernhard, 1971; Daneholt,

1975) have shown that Balbiani rings contain characteristic lampbrushlike loop structures consisting of a deoxyribonucleoprotein axis and RNP granules attached to this axis by RNP stalks (Figure 1) (for cytochemistry, consult Stevens and Swift, 1966). Daneholt (1975) also proposed that the loops in BRs probably correspond to active 75 S RNA genes. This hypothesis was recently tested when the loops observed *in situ* were compared with the morphology of the spread active 75 S RNA genes (Lamb and Daneholt, 1979). Most RNP fibers are extended in spread preparations, which is probably due to unfolding during preparation for electron microscopy. Often, however, we observed tightly packed fibers resembling in size and conformation the stalked granules in sections (examples indicated in Figure 2 by arrows; cf. also Figure 3). On the basis of these striking similarities, we interpreted the stalked granules linked to the loop axis in the sectioned material as growing RNP transcripts and concluded that the lampbrushlike loops correspond to active 75 S RNA genes, although in thin sections normally only short segments of the genes can be revealed. Furthermore, it seems likely that each loop comprises just one active 75 S RNA gene.

In order to characterize in further detail the active 75 S RNA genes as they appear within the cell, we have recently reconstructed longer segments of the loops from serial sections through the BRs (Andersson *et al.*, 1980). Loop segments optimal for reconstruction were located in the periphery of the BRs. In Figure 1 two such segments have been marked by arrows. Apart from being clearly separated from other segments, the peripheral ones displayed the same morphology as loop segments observed in more central regions of the BRs. Usually our reconstructions made from serial sections comprised one quarter to one half of the loops. We failed to reconstruct entire individual loops because of the interpretation problems when several loops were located close together. Nevertheless, from a series of segment reconstructions we have arrived at a tentative picture of a complete loop (see section 4.1).

4.1. The Growing RNP Fibers

Proximal and distal segments of active 75 S RNA genes have been identified and the nascent RNP fibers (or particles) studied. An example of a putative distal segment is displayed in Figure 5. Three consecutive sections are exhibited in the figure (A, B, and C) and the reconstruction, representing a projected picture of the loop segment is also displayed (D). The RNP particles can be recognized, and the approximate position of the chromosomal axis has been indicated. Each of the RNP particles is built from two portions, a 2-nm fiber stalk and a peripheral globular part. As can be observed in Figure 5 B and D the globular part increases in size along the loop axis, while the stalk component maintains its length. Proximal segments as well as transitional regions between the proximal and distal segment have been analyzed in an analogous way. On the basis of a number of such reconstructions covering various parts of the loop, we can present a tentative structure of the entire 75 S RNA gene *in situ*. The complete loop, derived from five overlapping segments, is shown in Figure 6. When our data from the reconstruction work and the Miller spreads are com-

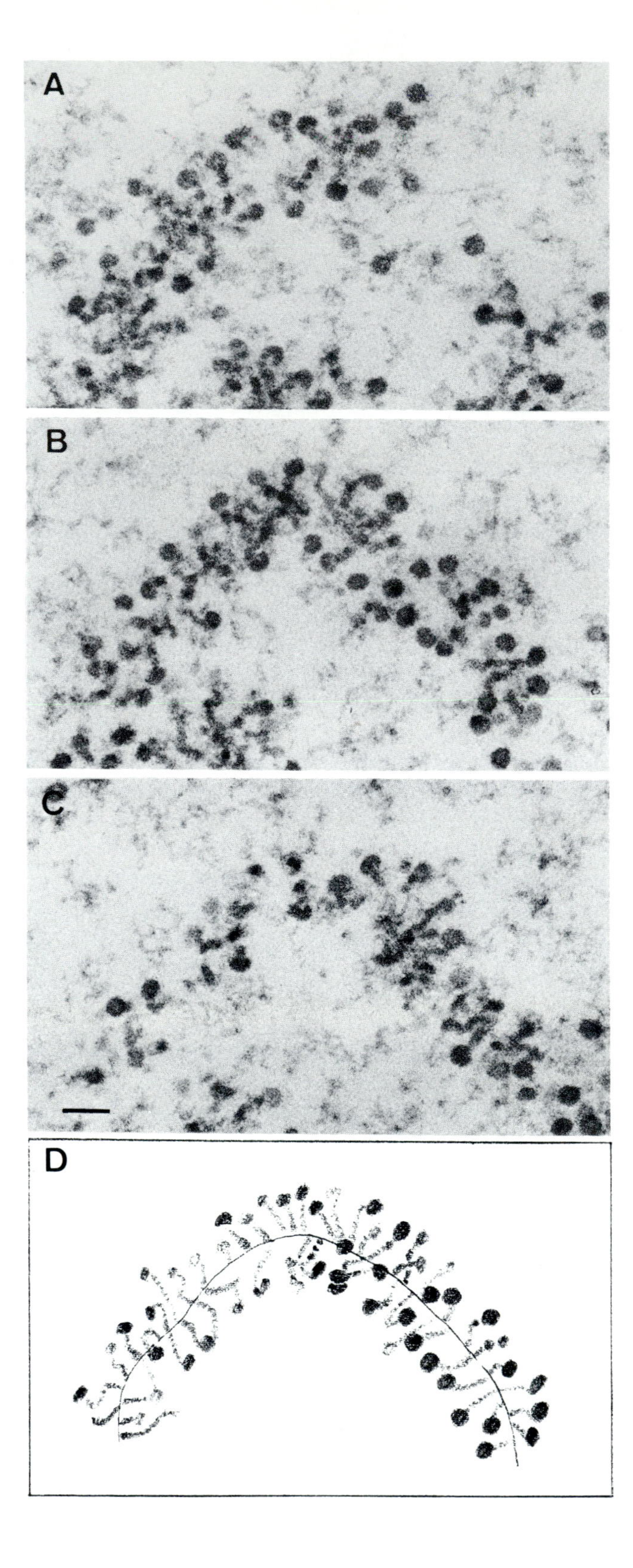
A
B
C
D

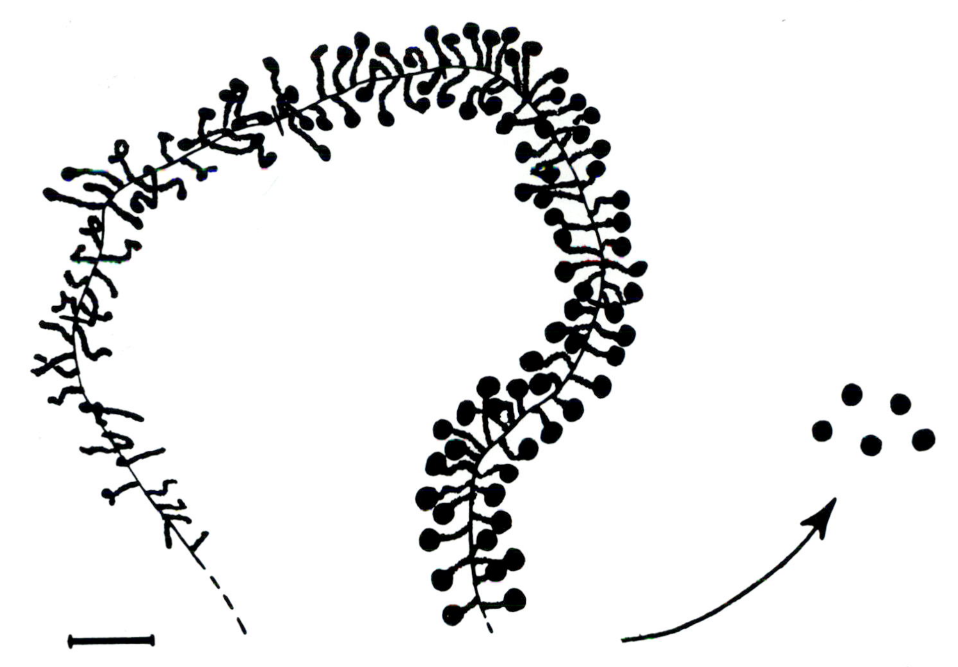

Figure 6. The structure of a complete Balbiani ring loop as derived from a series of reconstructions of 75 S RNA gene segments (cf. Figure 5). The proximal portion of the gene is characterized by growing 20-nm-thick RNP fibers, while the distal portion harbors RNP particles, each consisting of a 20-nm stalk and a peripheral globular portion. The globular part is increasing in size along the gene. The released RNP product is an RNP granule with a diameter of 50 nm. The bar equals 200 nm.

bined, we can draw the following conclusions. The RNP fibers increase in length along the gene as demonstrated in the Miller spreads. The *in situ* picture reveals the formation of the transcription product. In the first quarter of the gene, a 20-nm RNP fiber is formed. The further increase in length of the RNP product corresponds to a continuous packing of the peripheral part of the fiber into a globular structure. Evidently the transcription process and the packing of the product proceed in parallel. Towards the end of the gene, the globular part of the RNP particles is about 50 nm in diameter. In the nuclear sap, abundant granules of this size are recorded, each of them presumably containing a 75 S RNA molecule (for discussion, see Case and Daneholt, 1977). The stalked granules are, therefore, likely to be released and appear in the nuclear sap as granules devoid of stalks. The translocation of the BR products from the nucleus to the cytoplasm through the nuclear pores has also been visualized in the electron microscope (e.g., Stevens and Swift, 1966; Case and Daneholt, 1977).

Figure 5. Reconstruction of a segment of an active 75 S RNA gene. A, B, and C display electron micrographs from three consecutive sections through a Balbiani ring, and D shows the projected picture of the gene segment as reconstructed from the three electron micrographs. The bar equals 100 nm. For further information, see Andersson *et al.* (1980).

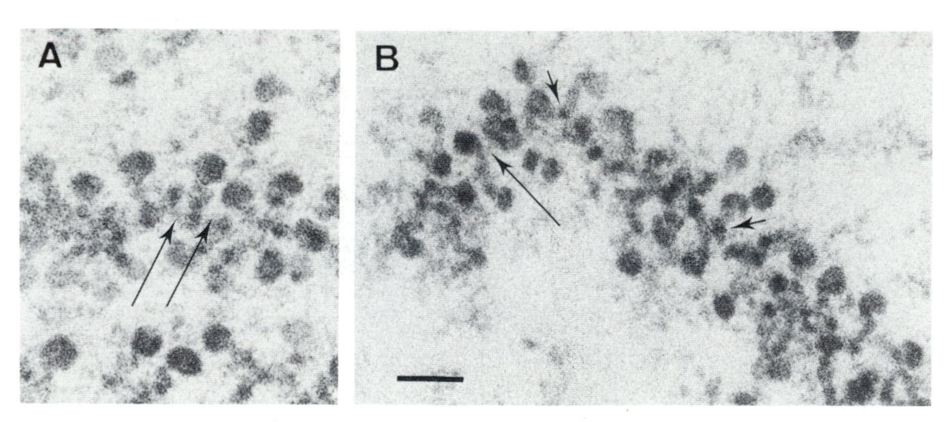

Figure 7. The thin chromosomal axis of active 75 S RNA genes *in situ*. Long arrows indicate the position of the axis. In B, putative RNA polymerases have been denoted by short arrows. The bar represents 100 nm. For further details, see Anderson *et al.* (1980).

4.2. The Chromosomal Axis

The electron micrographs of the active gene *in situ* allowed us to study the chromosomal fiber. The axis itself could usually not be seen due to the high density of RNA polymerases with associated RNP fibers. However, sometimes the RNA polymerases were clearly separated from one another. In such gaps we observed a fine axial fiber, as shown in Figure 7. It should be noted that the diameter of the fiber corresponds to that of the chromosomal axis recorded in spread preparations (about 5 nm; cf. Figure 3), but is thinner than that expected from a nucleofilament, i.e., a fiber consisting of a string of closely packed nucleosomes and with a diameter of 10 nm.

Although it is difficult to discern the chromosomal fiber in the *in situ* preparations, we can still determine the DNA compaction for the chromosomal axis from the information available on the structure of the gene *in situ* and in spread preparations. We know from the spread genes that the RNP fiber density is 10 fibers per μm DNA (Lamb and Daneholt, 1979), while *in situ* there are 36 RNP fibers per μm chromosomal axis (Andersson *et al.*, 1980; Olins *et al.*, 1980). Those figures give a DNA compaction of 3.6 for the chromosomal fiber *in situ*. This value might be too high, since the course of the loop axis is uncertain to some extent and the estimated length of the axis *in situ* represents the minimum length. In any case, the DNA compaction of the fiber is lower than that of a nucleofilament (5–7 according to Chambon, 1978). We conclude that both the observation in the electron microscope of a thin fiber and the calculation of the low DNA compaction for the chromosomal axis suggest that the active gene *in situ* is present as an extended nucleofilament. This result is in good agreement with our Miller type study (see section 3.2), although the axis is probably artificially stretched to some extent in the Miller method (DNA compaction of the active gene *in situ* is 3.6, and in Miller spreads it is 1.6).

When considering the structure and compaction of the fiber in the active 75 S RNA genes, it should be realized that the 75 S RNA genes are unusually

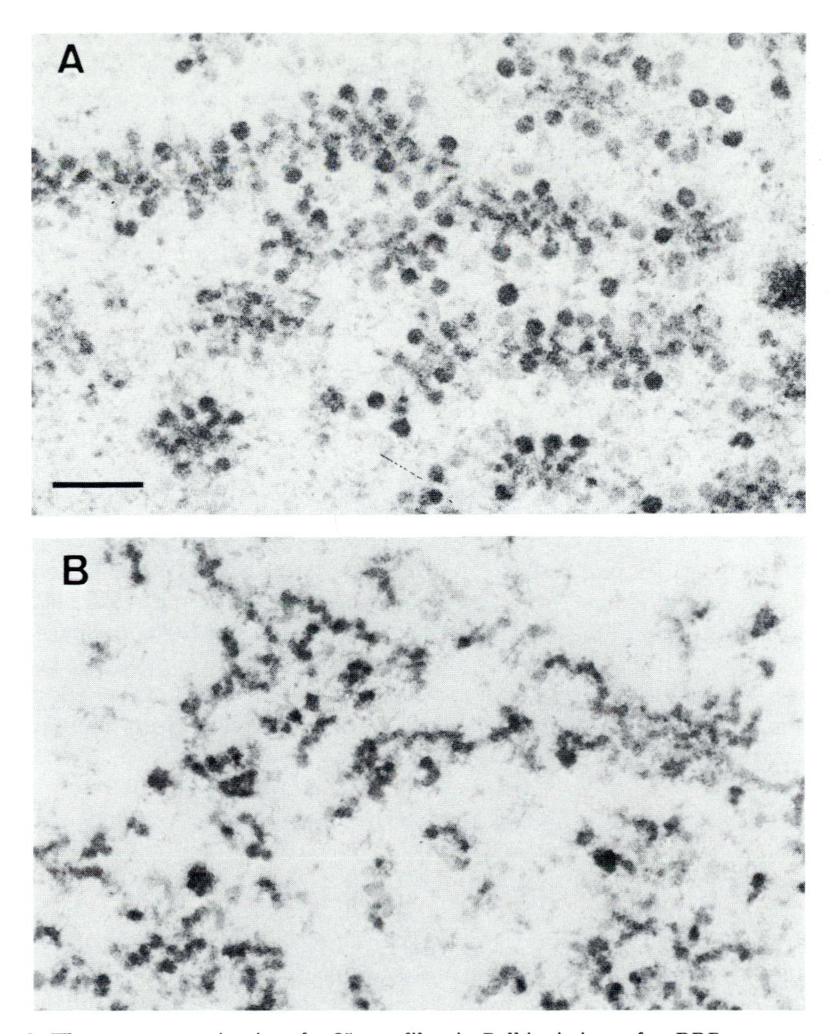

Figure 8. The appearance *in situ* of a 25-nm fiber in Balbiani rings after DRB treatment. The Balbiani ring control (A) exhibits active 75 S RNA genes with stalked RNP granules attached to the axis. The Balbiani rings treated for 30 min in DRB (B) is characterized by the appearance of 25-nm fibers, while the RNP particles have been completed and released from the Balbiani ring. The bar denotes 200 nm.

active ones. In fact, as discussed above, the chromosomal fiber of less active genes is less extended and displays a larger number of nucleosomes (e.g., Laird and Chooi, 1976; Scheer, 1978). Perhaps the nucleofilament might even transiently attain a supercoiled state during transcription, provided that the RNA polymerases are far enough apart. A recent electron microscopy study of DRB-treated glands hints at such a possibility (our unpublished data). DRB (5,6-dichlororibofuranosylbenzimidazole) is known to cause an early premature termination; the RNA polymerases already on the template when DRB is added, will, however, complete their run (for review, see Sehgal and Tham, 1978). The effect is that a gradually increasing part of the gene will lack growing RNP fibers. In our study of the BR 75 S RNA genes it was possible to observe the appearance of a 25-nm-thick fiber along with the disappearance of the growing RNP particles on the template (Figure 8). The fiber most likely corresponds to a rapidly re-formed supercoiled nucleofilament. Often the 25-nm fiber seemed to constitute a loop with a length of less than 0.5 μm and could be seen joining chromatin packed into a still higher order structure (not shown). The ultrastructural information on the 75 S RNA genes suggests that perhaps the active chromosomal fiber attains a series of conformational states during transcription; it can be assumed that an RNA polymerase opens up the higher order structures to permit transcription, but if proper time is allotted, it is conceivable that the nucleosomes, the nucleofilament, and even the supercoiled nucleofilament are re-formed prior to the passage of the next polymerase.

5. Balbiani Ring Genes in the Inactive State

The transcriptional activity in BR2 can be repressed by galactose treatment (Beermann, 1973; Nelson *et al.*, 1978); the giant puff regresses and a distinct band morphology is restored (Daneholt *et al.*, 1978b). After a galactose treatment the band pattern of the BR2 region in salivary gland cells is found to be similar to the pattern of the corresponding chromosomal region in Malpighian tubule cells. In the latter tissue, no or very little transcriptional activity takes place in the BR2 region. The sequences complementary to BR2 RNA are confined to a single band, the 3B10 band, which could be demonstrated by cytological hybridization (Derksen *et al.*, 1980). Microspectrophotometric measurements on Feulgen-stained chromosomes were carried out by Derksen *et al.* (1980), and they could determine the amount of DNA in the BR2 band relative to that of the whole chromosome set (0.205%). Since in *Chironomus* essentially all DNA is replicating to the same extent (Beermann, 1962), it was possible to estimate from the obtained percent figure and the haploid DNA amount (0.25 pg) that the band on the chromatid level, i.e., the BR2 chromomere, contains 470 kb of DNA. This amount of DNA exceeds by far the amount corresponding to a single 75 S RNA gene (37 kb according to Case and Daneholt, 1978). There are several possibilities to account for this discrepancy. First, there might be as many as four 75 S RNA genes in a BR2 chromomere (Wieslander, 1979), although only one is likely to be active (Lamb and Daneholt, 1979). It is also quite feasible that there might be genes other than the 75 S RNA gene(s). If not, we are facing another example of the DNA excess paradox, but here it is on the level of the cytological unit, the chromomere (for discussion, see Derksen *et al.*, 1980).

The structure of the chromosomal fiber in the BR2 chromomere has not yet been amenable for detailed ultrastructural and biochemical analysis, but it can be assumed that the general information available on chromomeres in polytene chromosomes most likely are applicable to the BR2 chromomere. The polytene chromosome fiber is likely to consist of nucleosomes, since the fiber displays a beads-on-a-string morphology in Miller spreads (Lamb and Daneholt,

1979). Moreover, when *Chironomus* polytene chromosomes were digested with micrococcal nuclease, a repeating structure with a repeat of 189 bp of DNA was observed (Andersson *et al.*, 1980). Assuming that the polytene chromosome fibers are built like chromatin fibers in diploid cells (for review, see, e.g., Kornberg, 1977), we can state that the nucleosomes are closely associated, to constitute a nucleofilament with a diameter of 10 nm. This fiber is likely to be supercoiled *in vivo* and form a 25-nm fiber (Finch and Klug, 1976; Suau *et al.*, 1979; Thoma *et al.*, 1979). For polytene chromosomes there is direct electron microscopic evidence for the presence of 25-nm fibers (Ris and Korenberg, 1979; our unpublished data). The DNA compaction in the 25-nm fibers has been estimated to be about 40 (Finch and Klug, 1976; Suau *et al.*, 1979). This value is considerably below the compaction of DNA in a BR2 chromomere, which has been determined from the DNA content of the chromomere and the thickness of the BR2 band to be 380 (Derksen *et al.*, 1980). To explain the structure of the BR2 chromomere one has to invoke still another higher order structure than the 25-nm fiber.

Several studies suggest that tightly packed loops might represent this higher order structure. Upon removal of protein from *Drosophila* polytene chromosomes, Sorsa *et al.* (1970) could show that the DNA unfolds into side loops. The recorded chromosome morphology resembles in a striking manner the lateral loops observed in meiotic chromosomes of *Chironomus* spermatocytes after spreading of the chromosomes on a hypotonic solution (Keyl, 1975). More recently, loops have also been revealed in human mitotic chromosomes after depletion of the histone proteins (Laemmli *et al.*, 1978) or after chelation of the divalent cations (Marsden and Laemmli, 1979). It seems, therefore, as if the organization of the chromosomal fiber into packed structural loops is a general feature of chromosomes, but the functional significance of the loops is unclear. An obvious possibility is that the structural loops correspond to the open transcriptional loops formed upon gene activation. Hopefully, the BR2 system might offer the possiblility to analyze a specific chromomere in terms of structural as well as transcriptional loops.

6. *The Gene Activation Process in Balbiani Rings*

Studies on polytene chromosomes have suggested that the genes are switched on in a two-step process (for discussion, see, e.g., Ashburner, 1977). A striking example is a study by Berendes (1968). He investigated puff induction in *Drosophila* and observed that the puffing elicited by, e.g., ecdysone, takes place also when the RNA synthesis is inhibited. The puffing step, i.e., the uncoiling of the tightly packed chromosomal fiber, can evidently be uncoupled from the transcriptional process. On the basis of our information on the BR2 chromomere, the revelation of a 25-nm fiber loop, and the structure of the active 75 S RNA gene, we want to propose that the two steps proceed according to the scheme in Figure 9. First, the puffing is initiated by a release of a tightly packed 25-nm fiber into an open loop-like configuration. This organizational change of the 25-nm fiber is accompanied by a drop in DNA

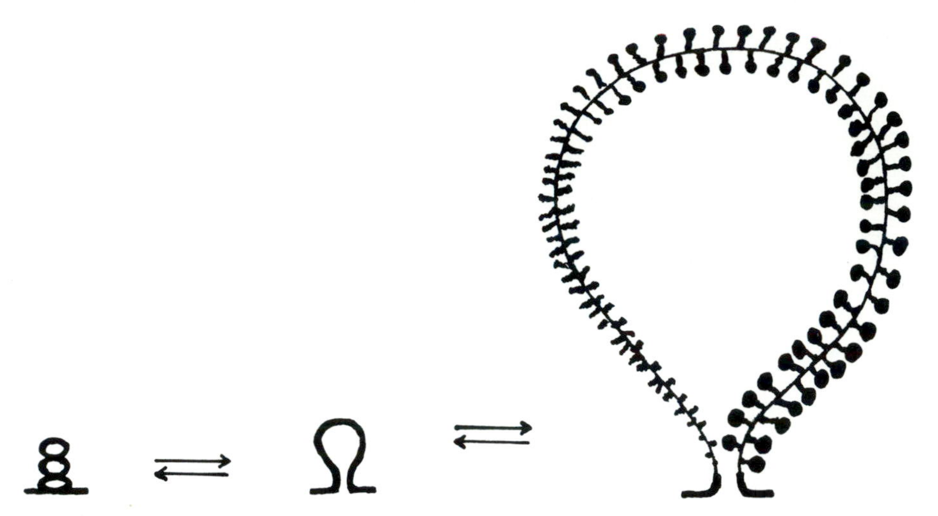

Figure 9. Model for induction of gene activity at the 75 S RNA genes in Balbiani rings. The inactive gene resides in a highly compacted state (here tentatively shown as a twisted 25-nm fiber loop; cf. Marsden and Laemmli, 1979). Upon activation, the gene is relaxed into an open 25-nm fiber loop, which is subsequently used as substrate for the transcription process. As a result of transcription, the 25-nm fiber is converted into a thin fiber, corresponding to an extended nucleofilament. If the transcriptional modification of the 25-nm fiber is rapidly reversible, the size of the loop might be proportional to the number of RNA polymerases on the template, i.e., the loop size would be related to the transcriptional activity. It is indicated in the model that the configurational changes might be reversible.

compaction from 380 to 40. Subsequently the RNA synthesis takes place, and the RNA polymerases will convert the 25-nm fiber into an extended nucleofilament. As a consequence of this modification of the template, the DNA compaction will decrease from 40 to 3.6. If our model is correct, the 25-nm fiber plays a key role in the gene activation process. First, the 25-nm fiber has to be transformed from a compacted state to an open configuration, when the gene is turned on. Furthermore, at least the first RNA polymerase has to recognize the promoter as a constituent part of the 25-nm fiber in order to initiate transcription.

ACKNOWLEDGMENTS

The research was supported by the Swedish Cancer Society, the Swedish Natural Science Research Council, Gunvor and Josef Anér Foundation, Magnus Bergvall Foundation, and Karolinska Institutet (Reservationsanslaget). I am indebted to Mrs. Vivi Jacobson for typing the manuscript and to Miss Birgitta Björkroth for preparing the illustrations.

References

Andersson, K., Björkroth, B., and Daneholt, B., 1980, The *in situ* structure of the active 75 S RNA genes in Balbiani rings of *Chironomus tentans*, *Exp. Cell Res.* **130**:313–326.

Ashburner, M., 1977, Happy Birthday-puffs. In *Chromosomes Today*, vol. 6, edited by A. de la Chapelle and M. Sorsa, pp. 213–222, Elsevier, Amsterdam.

Beermann, W., 1962, Riesenchromosomen. In *Protoplasmatologia*, vol. 6 D, pp. 1–165, Springer-Verlag, Vienna.

Beermann, W., 1973, Directed changes in the pattern of Balbiani ring puffing in *Chironomus*: Effects of a sugar treatment, *Chromosoma* **41**:297–326.

Beermann, W., and Bahr, G. F., 1954, The submicroscopic structure of the Balbiani-ring, *Exp. Cell Res.* **6**:195–201.

Bellard, M., Gannon, M., and Chambon, P., 1978, Nucleosome structure. III. The structure and transcriptional activity of the chromatin containing the ovalbumin and globin genes in chick oviduct nuclei, *Cold Spring Harbor Symp.* **42**:779–791.

Berendes, H., 1968, Factors involved in the expression of gene activity in polytene chromosomes, *Chromosoma* **24**:418–437.

Case, S. T., and Daneholt, B., 1977, Cellular and molecular aspects of genetic expression in *Chironomus* salivary glands, *Int. Rev. Biochem.* **15**:45–77.

Case, S. T., and Daneholt, B., 1978, The size of the transcription unit in Balbiani ring 2 of *Chironomus tentans* as derived from analysis of the primary transcript and 75 S RNA, *J. Mol. Biol.* **124**:223–241.

Chambon, P., 1978, Summary: The molecular biology of the eukaryotic genome is coming of age, *Cold Spring Harbor Symp.* **42**:1209–1234.

Daneholt, B., 1972, Giant RNA transcript in a Balbiani ring, *Nature New Biol.* **240**:229–232.

Daneholt, B., 1975, Transcription in polytene chromosomes, *Cell* **4**:1–9.

Daneholt, B., and Edström, J.-E., 1967, The content of deoxyribonucleic acid in individual polytene chromosomes of *Chironomus tentans*, *Cytogenetics* **6**:350–356.

Daneholt, B., and Hosick, H., 1973, Evidence for transport of 75 S RNA from a discrete chromosome region *via* nuclear sap to cytoplasm in *Chironomus tentans*, *Proc. Nat. Acad. Sci. USA* **70**: 442–446.

Daneholt, B., Edström, J.-E., Egyházi, E., Lambert, B., and Ringborg, U., 1969, RNA synthesis in a Balbiani ring in *Chironomus tentans* salivary gland cells, *Chromosoma* **28**:418–429.

Daneholt, B., Andersson, K., and Fagerlind, M., 1977, Large-sized polysomes in *Chironomus* salivary glands and their relation to Balbiani ring 75 S RNA, *J. Cell Biol.* **73**:149–160.

Daneholt, B., Case, S. T., Lamb, M. M., Nelson, L., and Wieslander, L., 1978a, The 75 S RNA transcription unit in Balbiani ring 2 and its relation to chromosome structure, *Philos. Trans. R. Soc. London Ser.* B **283**:383–389.

Daneholt, B., Case, S. T., Derksen, J., Lamb, M. M., Nelson, L. G., and Wieslander, L., 1978b, The size and chromosomal location of the 75 S RNA transcription unit in Balbiani ring 2, *Cold Spring Harbor Symp.* **42**:867–876.

Derksen, J., Wieslander, L., van der Ploeg, M., and Daneholt, B., 1980, Identification of the Balbiani ring 2 chromomere and determination of the content and compaction of its DNA, *Chromosoma* **81**:65–84.

Edström, J.-E., 1964, Microextraction and microelectrophoresis for determination and analysis of nucleic acids in isolated cellular units, *Meth. Cell Physiol.* **1**:417–447.

Edström, J.-E., Lindgren, S., Lönn, U., and Rydlander, L., 1978, Balbiani ring RNA content and half-life in nucleus and cytoplasm of *Chironomus tentans* salivary gland cells, *Chromosoma* **66**:33–44.

Egyházi, E., 1975, Inhibition of Balbiani ring RNA synthesis at the initiation level, *Proc. Nat. Acad. Sci. USA* **73**:947–950.

Egyházi, E., 1976, Quantitation of turnover and export to the cytoplasm of Hn RNA transcribed in the Balbiani rings, *Cell* **7**:507–515.

Felsenfeld, G., 1978, Chromatin, *Nature* **271**:115–122.

Finch, J. T., and Klug, A., 1976, Solendoidal model for superstructure in chromatin, *Proc. Nat. Acad. Sci. USA* **73**:1897–1901.

Foe, V. E., 1978, Modulation of ribosomal RNA synthesis in *Oncopeltus fasciatus*: An electron microscope study of the relationship between changes in chromatin structure and transcriptional activity, *Cold Spring Harbor Symp.* **42**:723–740.

Foe, V. E., Wilkinson, L. E., and Laird, C. D., 1976, Comparative organization of active transcription units in *Oncopeltus fasciatus*, *Cell* **9**:131–146.

Grossbach, U., 1977, The salivary gland of *Chironomus* (Diptera): A model system for the study of cell differentiation. In *Biochemical Differentiation in Insect Glands*, edited by W. Beermann, pp. 147-196, Springer-Verlag, Berlin.

Grosschedl, R., and Birnstiel, M. L., 1980, Identification of regulatory sequences in the prelude sequences of an H2A histone gene by the study of specific deletion mutants *in vivo*, *Proc. Nat. Acad. Sci. USA* **77**:1432-1436.

Hertner, T., Meyer, B., Eppenberger, H. M., and Mähr, R., 1980, The secretion proteins in *Chironomus tentans* salivary glands: Electrophoretic characterization and molecular weight estimation, *Wilhelm Roux's Arch. Dev. Biol.* **189**:69-72.

Keyl, H.-G., 1975, Lampbrush chromosomes in spermatocytes of *Chironomus*, *Chromosoma* **51**:75-91.

Klug, A., Rhodes, D., Smith, J., Finch, J. T., and Thomas, J. O., 1980, A low resolution structure for the histone core of the nucleosome, *Nature* **287**:509-516.

Kornberg, R. D., 1977, Structure of chromatin, *Annu. Rev. Biochem.* **46**:931-954.

Laemmli, U. K., Cheng, S. M., Adolph, K. W., Paulson, J. R., Brown, J. A., Baumbach, W. R., 1978, Metaphase chromosome structure: The role of nonhistone proteins, *Cold Spring Harbor Symp.* **42**:351-360.

Laird, C. D., and Chooi, W. Y., 1976, Morphology of transcription units in *Drosophila melanogaster*, *Chromosoma* **58**:193-218.

Laird, C. D., Wilkinson, L. E., Foe, V. E., and Chooi, W. Y., 1976, Analysis of chromatin associated fiber arrays, *Chromosoma* **58**:169-190.

Lamb, M. M., and Daneholt, B., 1979, Characterization of active transcription units in Balbiani rings of *Chironomus tentans*, *Cell* **17**:835-848.

Lambert, B., 1972, Repeated DNA sequences in a Balbiani ring, *J. Mol. Biol.* **72**:65-75.

Marsden, M. P. F., and Laemmli, U. K., 1979, Metaphase chromosome structure: Evidence for a radial loop model, *Cell* **17**:849-858.

Mathis, D., Oudet, P., and Chambon, P., 1980, Structure of transcribing chromatin, *Prog. Nucleic Acid Res. Mol. Biol.* **24**:1-55.

McKnight, S. L., and Miller, O. L., Jr., 1976, Ultrastructural patterns of RNA synthesis during early embryogenesis of *Drosophila melanogaster*, *Cell* **8**:305-319.

McKnight, S. L., Sullivan, N. L., and Miller O. L., Jr., 1976, Visualization of the silk fibroin transcription unit and nascent silk fibroin molecules on polyribosomes of *Bombyx mori*, *Prog. Nucleic Acid Res. Mol. Biol.* **19**:313-318.

McKnight, S. L., Bustin, M., and Miller, O. L., Jr., 1978, Electron microscopic analysis of chromosome metabolism in the *Drosophila melanogaster* embryo, *Cold Spring Harbor Symp.* **42**: 741-754.

Miller, O. L., Jr., and Bakken, A. H., 1972, Morphological studies of transcription, *Acta Endocrinol.* **168**:155-177.

Miller, O. L., Jr., Beatty, B. R., Hamkalo, B. A., and Thomas, C. A., Jr., 1970, Electron microscopic visualization of transcription, *Cold Spring Harbor Symp. Quant. Biol.* **35**:505-512.

Nelson, L. G., Derksen, J., Lamb, M. M., Wieslander, L., and Daneholt, B., 1978, Suppression of transcription in Balbiani ring 2 and the effect on chromosome structure. In *FEBS 11th Meeting*, edited by B. F. C. Clark, H. Klenow and J. Zeuthen, vol. 43, pp. 279-286, Pergamon Press, Oxford.

Olins, A. L., and Olins, D. E., 1974, Spheroid chromatin units (*v*-bodies), *Science* **183**:330-332.

Olins, A. L., Olins, D. E., and Franke, W. W., 1980, Stereoelectron microscopy of nucleoli, Balbiani rings and endoplasmic reticulum in *Chironomus* salivary glands, *Eur. J. Cell Biol.* **22**:714-723.

Oudet, P., Gross-Bellard, M., and Chambon, P., 1975, Electron microscopic and biochemical evidence that chromatin structure is a repeating unit, *Cell* **4**:281-300.

Ris, H., and Korenberg, J., 1979, Chromosome structure and levels of chromosome organization. In *Cell Biology*, vol. 2, edited by L. Goldstein and D. M., Prescott, pp. 267-361, Academic Press, New York.

Rydlander, L., and Edström, J.-E., 1980, Large sized nascent protein as dominating component during protein synthesis in *Chironomus* salivary glands, *Chromosoma* **81**:85-99.

Saragosti, S., Moyne, G., and Yaniv, M., 1980, Absence of nucleosomes in a fraction of SV 40 chromatin between the origin of replication and the region coding for the late leader RNA, *Cell* **20**:65-73.

Scheer, U., 1978, Changes of nucleosome frequency in nucleolar and nonnucleolar chromatin as a function of transcription: An electron microscopic study, *Cell* **13**:535–549.

Sedat, J., and Manuelidis, L., 1978, A direct approach to the structure of eukaryotic chromosomes, *Cold Spring Harbor Symp.* **42**:331–350.

Sehgal, P. B., and Tham, I., 1978, Halogenated benzimidazole ribosides. Novel inhibitors of RNA synthesis, *Biochem. Pharmacol.* **27**:2475–2485.

Sorsa, V., Sorsa, M., Virrankoski, V., and Pusa, K., 1970, An electron microscopy study of alkali-treated salivary gland chromosomes of *Drosophila*, *Ann. Acad. Sci. Fenn.* (Ser. A, IV, Biol.) **166**:1–10.

Stevens, B. J., and Swift, H., 1966, RNA transport from nucleus to cytoplasm in *Chironomus* salivary glands, *J. Cell Biol.* **31**:55–78.

Suau, P., Bradbury, E. M., and Baldwin, J. P., 1979, Higher-order structures of chromatin in solution, *Eur. J. Biochem.* **97**:593–602.

Thoma, F., Koller, Th., and Klug, A., 1979, Involvement of histone H 1 in the organization of the nucleosome and of the salt-dependent superstructures of chromatin, *J. Cell Biol.* **83**:403–427.

Vasquez-Nin, G., and Bernhard, W., 1971, Comparative ultrastructural study of perichromatin and Balbiani ring granules, *J. Ultrastruct. Res.* **36**:842–861.

Wasylyk, B., Kédinger, C., Corden, J., Brison, O., and Chambon, P., 1980, Specific *in vitro* initiation of transcription on conalbumin and ovalbumin genes and comparison with adenovirus-2 early and late genes, *Nature* **285**:367–373.

Weintraub, H., and Groudine, M., 1976, Chromosomal subunits in active genes have an altered conformation, *Science* **193**:848–856.

Wieslander, L., 1979, Number and structure of Balbiani ring 75 S RNA transcription units in *Chironomus tentans*, *J. Mol. Biol.* **134**:347–367.

Wieslander, L., and Daneholt, B., 1977, Demonstration of Balbiani ring RNA sequences in polysomes, *J. Cell Biol.* **73**:260–264.

Woodcock, C. L. F., Safer, J. P., and Stanchfield, J. E., 1976, Structural repeating units in chromatin. 1. Evidence for their general occurrence. *Exp. Cell Res.* **97**:101–110.

14

Insect Intercellular Junctions: Their Structure and Development

NANCY J. LANE

1. Historical Introduction

Since the advent of the electron microscope, many fine structural parameters of insect tissues have been elucidated, not least of which are the features of the junctions by which their cells are associated. Although the Schleiden and Schwann cell theory holds that cells are independent units, it has gradually become apparent that they are in fact joined together in a variety of ways, depending, it seems, on the function of the tissue under investigation. By the 1960s, junctions such as septate "desmosomes" were being reported as existing in insect systems, although few niceties of fine structure were understood owing to technical shortcomings. With the arrival of thinner sections and *en bloc* staining, distinctions could be made between subtly different junctional varieties so that, for example, gap junctions could be distinguished from other close membrane appositions. Moreover, tracer studies made it possible to, in effect, negatively stain junctions, thereby determining the fine details of the intercellular elaborations specific for each type. Hitherto undiscovered junctional differences were also revealed by these methods, so that the septate junctions could be subdivided into various types. After freeze fracture became available in the 1970s, the intramembranous modifications peculiar to different junctions could also be analyzed, and some junctional types, such as tight junctions, were demonstrated unequivocally for the first time. These intramembranous features, together with the greater understanding already obtained for the characteristic intercellular structures, enabled investigators to construct three-dimensional models of cell-to-cell associations from which it was hoped

NANCY J. LANE • Agricultural Research Council, Unit of Invertebrate Chemistry and Physiology, Department of Zoology, University of Cambridge, Downing Street, Cambridge CB2 3EJ, England.

to be able to comprehend more clearly the functional significance of each different junctional type.

However, in most cases, the precise physiological roles of the different insect intercellular junctions is still unclear. In some instances, correlated electrophysiological or permeability studies have led to what appear to be very reasonable assumptions, but frequently the coexistence of several junctional types along the same intercellular border has created confusion as to which structure is actually responsible for the observed effect. In spite of the occasional controversy, however, it has gradually emerged that the intercellular junctions in insect tissues play important physiological roles, including cell-to-cell coupling or communication, the production of partial or complete permeability barriers, the maintenance of tissue integrity, the enhancement of ionic fluxes, and the passage of nervous impulses from cell to cell. This last function takes place *via* the synaptic junctions, such as the axoaxonic synapse or neuromuscular junctions. This highly specialized area has not been included in this chapter due to limitations of space. Studies on developing junctional systems in embryonic tissues have just recently been initiated, since the mature structures had first to be elucidated. It is to be hoped that the analysis of the stages in assembly of different junctions may throw some light on their functions, which should become apparent concurrent with junctional maturity. One fact emerging from a survey of the junctional studies carried out over the last decade is that the recent developments and improvements in techniques and instrumentation have permitted the evolution of a whole new area of understanding in the realm of the structure of insect junctional complexes.

2. *Pleated Septate Junctions*

2.1. Location

Septate "desmosomes" were originally described in two species of hydra in 1959 by Wood. In fact, they were first seen, although not recognized as such, in insect compound eyes (Fernández-Morán, 1958). Numerous publications reporting their presence in the tissues of many invertebrates have appeared subsequently. Septate junctions tend to lie in circumferential belts around lateral cell borders on the outer or luminal surface of epithelia, and they have been subdivided into various categories as further investigations have revealed subtle differences in structure among different types. Within the Insecta, the chief classes are the pleated septate and the smooth septate junctions (Flower and Filshie, 1975). The latter were first described as continuous junctions, or *zonula continua* by Noirot and Noirot-Timothée (1967) and were thought to be a form of tight junction because of their intramembranous structure in freeze-fracture replicas. However, they have since been shown to be a variation of the basic septate type and will be considered separately (see section 3).

The pleated septate junctions occur widely in insect tissues of ectodermal origin and are found, for example, in the esophagus and rectum of the gut tract, although not in the midgut, which is endodermal. They also exist in such tissues as the salivary glands, testis, ovary, compound eye, Malpighian tubules, certain portions of the tracheal and neuroglial systems, sensory organs, and epidermal cells (see references in Lane and Skaer, 1980; Noirot-Timothée and Noirot, 1980). They are usually said to be homocellular in that they are found joining cells of the same type rather than of different type; however, examples of the latter are known, such as the heterocellular axoglial septate junctions in dipteran compound eye (Chi *et al.*, 1979; Lane, 1981a,c,d). They are said to be autocellular when they occur between membranes of the same cell, such as in the mestracheon of the tracheae and tracheoles. They may also be tricellular, when the borders of three cells are juxtaposed; here the structure becomes yet further specialized (Noirot-Timothée and Noirot, 1980). Pleated septate junctions also coexist, in insect tissues, with a range of other junctional types, including gap junctions, scalariform junctions, retinular junctions, tight junctions, and, occasionally, desmosomes or hemidesmosomes.

2.2. *Structure*

2.2.1. *Thin Sections*

In thin sections, pleated septate junctions exhibit a regular intercellular cleft of about 15–20 nm; across this run cross striations or septa, which produce a ladderlike appearance (Figure 1). The septa may be either numerous or infrequent, and are either regularly or unevenly spaced. After infiltration with electron-opaque tracers such as lanthanum, the spaces between the septa become stained; tangential sections of such preparations reveal the unstained septa to be undulating and ribbonlike against a dense background, producing a honeycomb appearance when deeply pleated (Figure 2); the extent of pleating, like the septal spacing, is variable depending on the tissue being investigated.

2.2.2. *Freeze-Fracture Replicas*

After freeze cleaving, the junction-bearing membrane faces display undulating rows of particles on the PF (protoplasmic or inner membrane half) and pits on the complementary EF (extracellular or outer membrane half) (Figure 3). These rows are separated from one another by varying distances, while the particles, about 8 nm in diameter, exhibit a variable center-to-center spacing of 10–20 nm (Noirot-Timothée *et al.*, 1978). The rows may run together in broad bands (insert, Figure 3), or they may wander separately across the membrane faces (Figure 3); this differs in different tissues and also may be variable within a single junctional belt.

2.2.3. *Model*

Analyses of tracer-infiltrated specimens together with freeze-fracture replicas make possible the production of three-dimensional models that attempt to describe the actual junctional configurations (for example, Caveney and Podgorski, 1975; Flower and Filshie, 1975; Lane and Skaer, 1980) (Figure 6A).

These models tend to assume that the intercellular septal ribbons insert into the membrane at the points where the intramembranous particles (IMPs) are seen to emerge in replicas, but this correlation has yet to be proven unequivocally (Noirot-Timothée *et al.*, 1978).

2.3. Development

The development of pleated septate junctions has not been investigated extensively as yet. Studies have been made on embryonic tissues in the nervous system of the blowfly, *Calliphora erythrocephala* (Lane and Swales, 1978a), and the locust *Schistocerca gregaria* (Lane and Swales, 1981). The other analyses that have been made deal with noninsect tissues, such as sea urchin embryos (Gilula, 1973) and regenerating *Hydra attenuata* (Wood and Kuda, 1980).

During junctional assembly in the locust (Lane and Swales, 1981), the junctional IMPs, having been inserted into the presumptive junctional membranes, become aligned in rows that at first meander across the membrane face (Figure 4). These rows then become lined up approximately parallel to one another, to give rise, ultimately, to the multiple rows that are typical of fully formed pleated septate junctions (as in Figure 3). In thin sections, the intercellular septa become organized in ladderlike arrays (Figure 5) at about the same time as the IMP alignments form.

In holometabolous insects, such as the blowfly, *Calliphora erhthrocephala*, the fate of the septate junctions has been followed during metamorphosis in the outer glial layer that surrounds the nervous tissue. The cells in this layer become disassociated during pupation as the larval ganglia become transformed into the adult form (Lane and Swales, 1978b). In so doing, the intercellular junctions seem to disappear, but apparently not by internalization (see Larsen, 1977). It has been suggested that junctional particles may disperse over the membrane faces, possibly to be subsequently reutilized to form the adult junctions, which reassemble in late pupation (Lane and Swales, 1978b, 1980). However, the pleated septate junctions have not yet been followed during metamorphosis as closely as the smooth septate or gap junctions (Lane *et al.*, 1980).

2.4. Physiological Role

The function of pleated septate junctions appears almost certainly to be, in part, adhesive, occurring as they do along the luminal borders of so many epithelial types. Some investigators (Noirot-Timothée *et al.*, 1978; Green *et al.*, 1979; Noirot-Timothée and Noirot, 1980) consider that they form the basis of permeability barriers in insect tissues. They may form a partial barrier to paracellular flow, but tracer penetration throughout their length frequently occurs (as in Figure 2), suggesting that they are unlikely to form a completely restrictive seal (see references in Lane and Skaer, 1980), especially since they occur in so many tissues in which a barrier has not been demonstrated. "Blistering" experiments (Green *et al.*, 1980) are considered to support their capacity to function as limiting junctions, like the tight junctions in vertebrates (di Bona, 1972), but recently authentic tight junctions have been found in insects, both

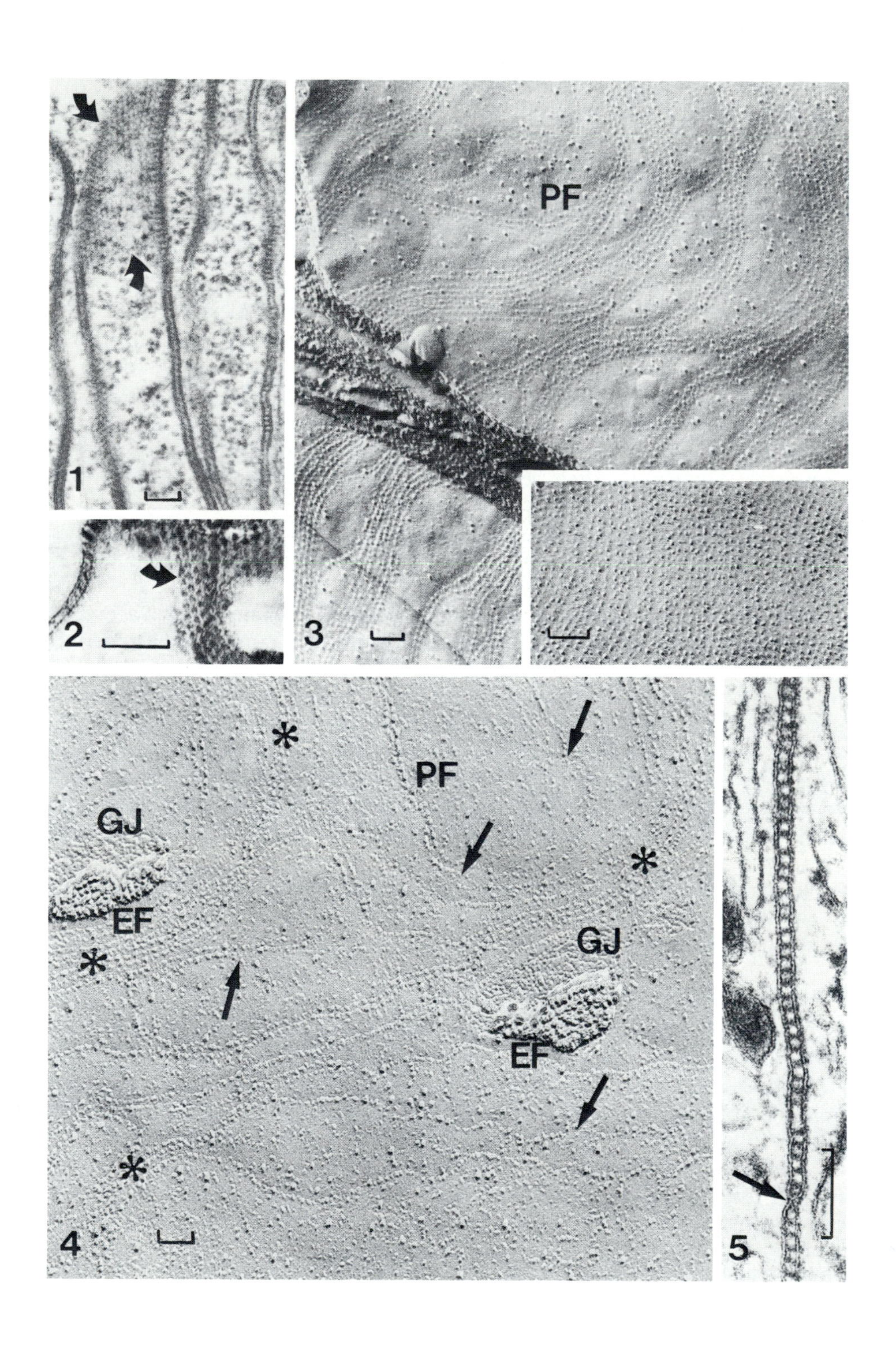
1
2
PF
3
PF
GJ
EF
GJ
EF
4
5

of the simple variety (Lane *et al.*, 1977; Lane and Swales, 1978a,b; 1979) and of the complex type (Lane, 1981a), which seem more likely to act as effective intercellular occlusions (see section 6). Nevertheless, the septate junctions may serve to slow down the inward penetration of molecules, and perhaps their matrix, if charged, may bind certain ions and in so doing, prevent their further entry (see Skaer *et al.*, 1979).

3. *Continuous Junctions*

3.1. Location

Continuous, or smooth septate, junctions, referred to as *zonula continua* by Noirot and Noirot-Timothée (1967), are found in circumferential belts around the luminal borders of tissues which are, on the whole, of endodermal origin. These include the midgut, the hepatopancreas, and parts of the Malpighian tubules (Dallai, 1976; Skaer *et al.*, 1979). Hence, the gut undergoes a transformation, possessing pleated septate junctions in the esophagus, smooth septate junctions in the proventiculus and midgut, and pleated septate junctions again in the rectum (Skaer *et al.*, 1980). Although they are chiefly homocellular, continuous junctions are occasionally heterocellular (Flower and Filshie, 1975).

Along the lateral borders of the midgut, the smooth septate junctions are distributed with highest frequency at the luminal border of microvilli and gradually decrease in density along the border toward the basal lamina. Where they tend to be less prevalent, they begin to coexist with gap junctions (see Skaer *et al.*, 1979; Lane and Skaer, 1980).

Figures 1–5 illustrate features of pleated septate junctions. (The bars equal 0.1 μm.)

Figure 1. Thin sections through pleated septate junctions in the mestracheon around a tracheole in the midgut of the blood-sucking bug *Rhodnius prolixus*. The curved arrows indicate the honeycomb appearance of tangentially sectioned septa.

Figure 2. Lanthanum infiltration into the pleated septate junctions of the sheath around the testis of the cockroach *Periplaneta americana*. The undulating, ribbonlike septa can be seen in *en face* view (curved arrow).

Figure 3. A freeze-fracture replica of the pleated septate junctions in the rectum of a cockroach displaying the P face (PF) with many irregular bands of aligned particle rows. These rows may be more highly ordered, as in the cockroach testis (insert).

Figure 4. Developing pleated septate junctions in the esophagous of a late embryo of the locust, *Schistocerca gregaria*. This P face (PF) view shows some recently organized particle rows (*) and many blindly terminating particle alignments (arrows). These are rows in the midst of formation. At this stage, fully mature clusters of gap junctional pits (GJ) are evident, with their complementary E face (EF) particles adhering to them in patches.

Figure 5. A thin section through a pleated septate junction that has just formed in a 13–14-day-old locust embryo. The septa are not always regularly spaced, and, in some cases, punctate tight junctional appositions may coexist (at arrow).

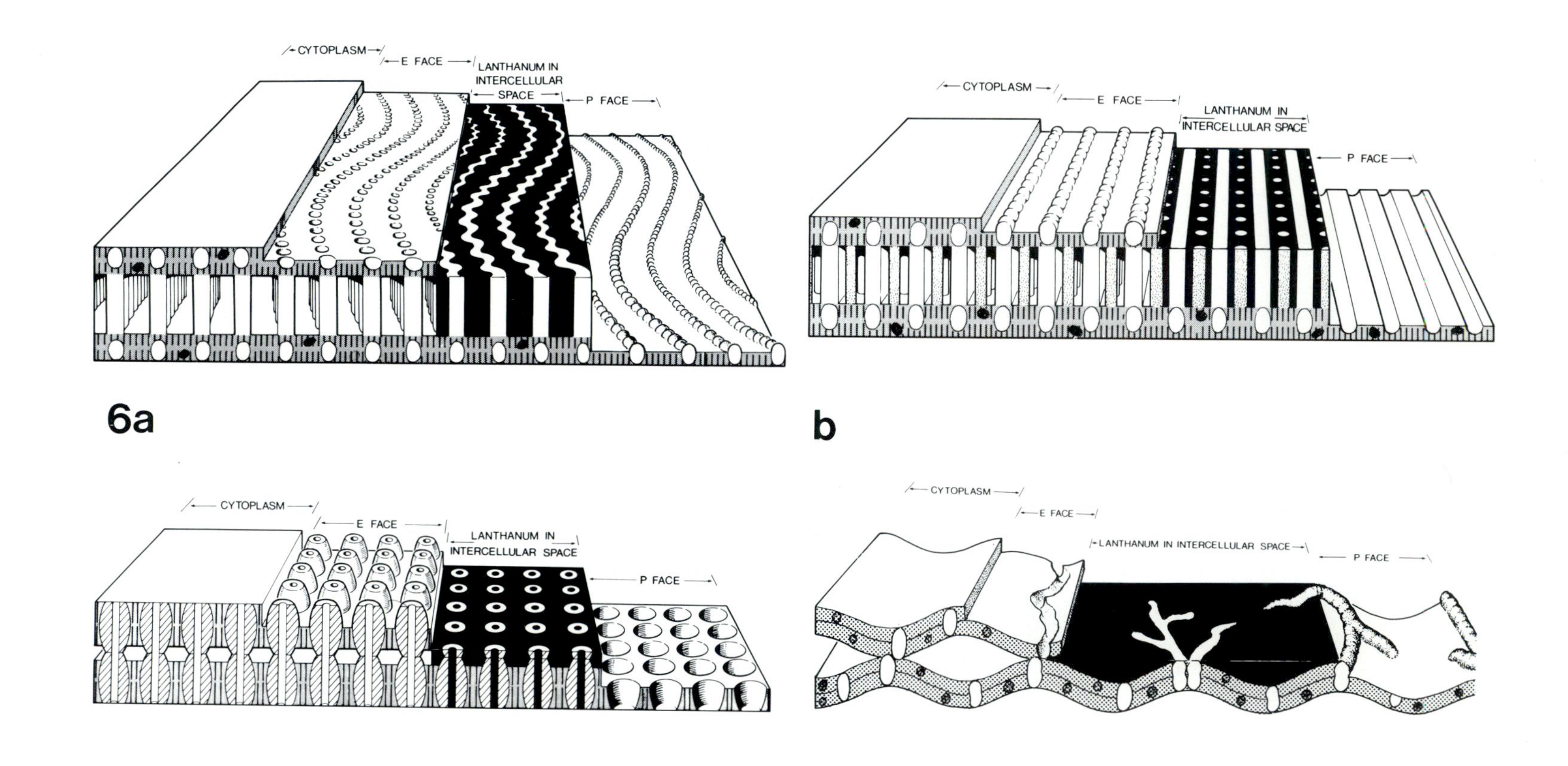

Figure 6. Models of insect intercellular junctions constructed from thin-sectioned, lanthanum-infiltrated preparations and freeze-fracture replicas. (A) Pleated septate junctions. (B) Smooth septate (continuous) junctions. (C) Gap junctions. (D) Tight junctions. See text for explanations. (Reproduced from Lane and Skaer, 1980; Lane, 1981b.)

3.2. Structure

3.2.1. Thin Sections

In cross section, smooth septate junctions exhibit a constant 15–20-nm intercellular cleft, which may exhibit no septa or may merely give an impression of cross striations. After tracer infiltration however, septa are definitely apparent, and examination of tangential sections reveals unstained, gently undulating intercellular sheets or septa (Figure 7). Between these septa occur columns that appear *en face* as round profiles (Figure 7); they may also be missing altogether. The septa, as in all arthropods, appear to have a particulate substructure (Lane and Harrison, 1978) of several subunits.

3.2.2. Freeze-Fracture Replicas

After freeze cleaving, the intramembranous particles forming the junctions are revealed lying in continuous ridges, which are particularly prominent at the apical or luminal border of the cells (Figure 8). These ridges are found in the P face with EF grooves in fixed tissues; whereas in unfixed tissues, the ridges shift over to the E face (Flower and Filshie, 1975; Lane and Skaer, 1980). The midgut in different insect groups may exhibit different patterns of intramembranous ridges (as in Figure 12), although in general, the freeze-fracture images are remarkably similar.

3.2.3. Model

Three-dimensional models can be constructed by correlating the possible insertion of the intercellular septa with the intramembranous ridges visible in freeze-fracture replicas. The two appear to coincide, and so it may be that the septa insert into the membranes in the region of the ridges (as in Figure 6B) (Flower and Filshie, 1975; Lane and Skaer, 1980). However, not all investigators agree about this correlation (Noirot-Timothée *et al.*, 1978), and, apart from particles, there are no specific intramembranous structures relating to the interseptal columns seen in thin sections.

3.3. Development

There have been no studies on the development of continuous junctions, although there have been suggestions of turnover in adult tissues, such as in the midgut of insects (Lane, 1979a). In embryonic midgut of the moth *Manduca sexta*, the first signs of junctional formation are individual particles becoming aligned into short rows of three to four IMPs (Figure 9). These subsequently become more extensive as further particles are added, and ultimately these rows fuse into ridges and become aligned in parallel to form the complex arrays typical of the mature junction (as in Figure 8) (Lane, unpublished data; Lane and Swales, 1981).

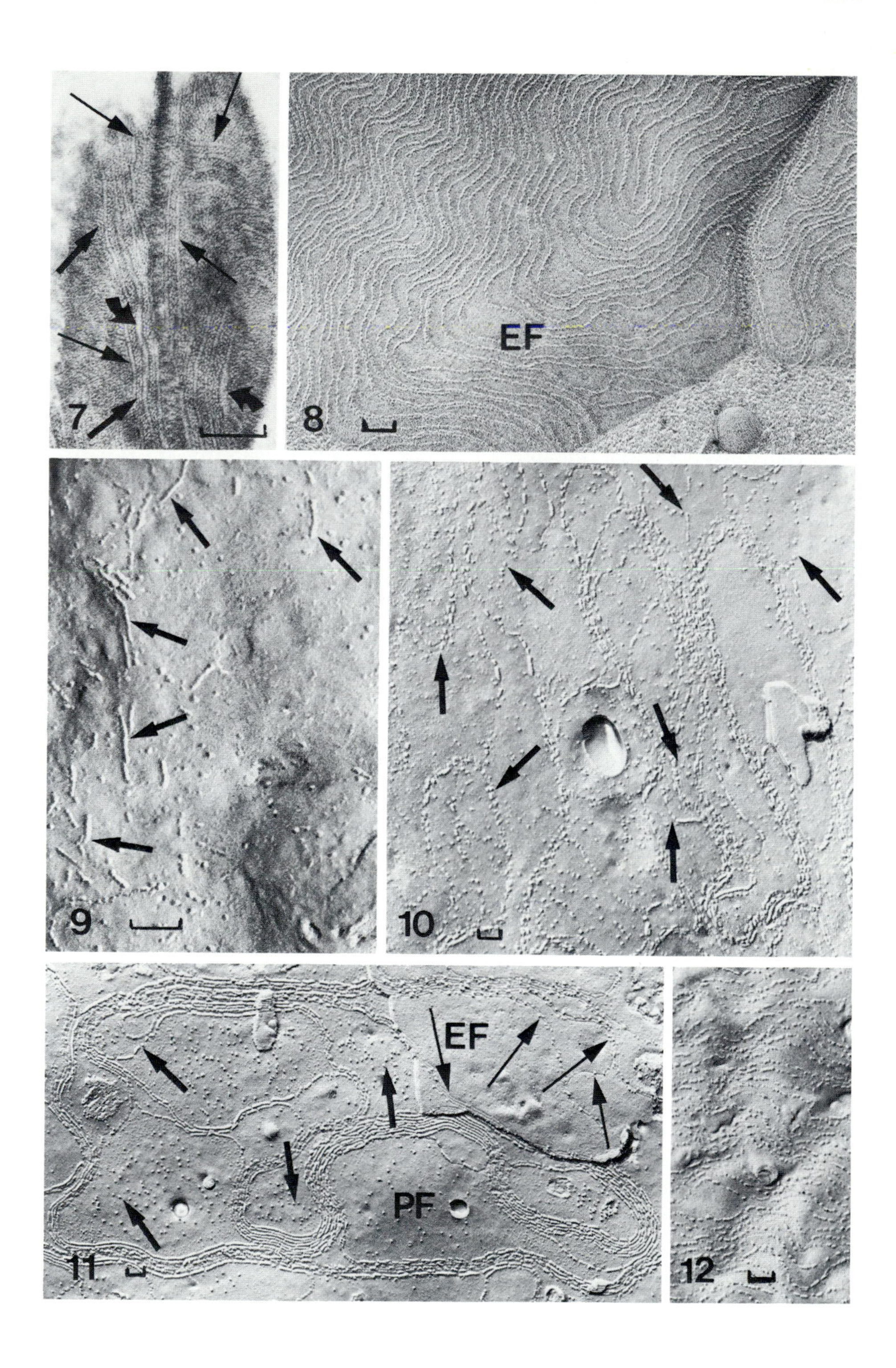
7
8
EF
9
10
11
EF
PF
12

In pupal tissues of holometabolous insects, the continuous junctions appear to undergo turnover (Lane *et al.*, 1980). The larval gut is much transformed when it becomes that of the adult, and the cells appear to partly dissociate, becoming much less columnar, and then reassociate, re-forming their smooth septate junctions in so doing. The junctional ridges appear to disaggregate in early pupal stages when the gut cells are separating (Figure 10); in midpupal stages they begin to reaggregate (Figure 11) by realignment of particles [of which some may be larval ones being reutilized (Lane *et al.*, 1980)] into ridges until the typical mature structures are re-formed again, at the time of emergence of the adult moth.

3.4. Physiological Role

Continuous junctions are usually found in tissues that are undergoing cell turnover, so they may be implicated in cell renewal. However, since they also occur in Malpighian tubules (Dallai, 1976; Skaer *et al.*, 1979), which do not regenerate, this need not necessarily be so. Certainly they must perform some adhesive role, maintaining coherence and integrity in the tissues in which they are found, since, for example, the pupal gut becomes very frail when the junctions are dissociating.

They have also been said to be involved in the formation of permeability barriers to paracellular flow, as is thought for the pleated septate junctions. There has been a report of "blistering" when the midgut of crayfish is exposed to hypertonic solutions (Mills *et al.*, 1976), suggesting a parallel with vertebrate

Figures 7–12 represent different views of smooth septate junctions. Figures 8–12 are freeze-fracture replicas. (The bars equal 0.1 μm.)

Figure 7. Thin section cut tangentially through a smooth septate junction of insect midgut after infiltration with lanthanum. The relatively straight, pale septa are evident (thin arrows), sometimes in double arrays (curved arrows). Between the septa lie columns, which appear as nonopaque dots because they were cut in cross section as they lie in the background of dense tracer (thick arrows).

Figure 8. Alignments of intramembranous particles in the midgut of *Rhodnius prolixus* after cryoprotection with glycerol, but no prefixation. The arrays of beadlike ridges occur on the E face (EF) under these circumstances, and the aligned particles seem to be close-packed laterally, although they do not always lose their particulate appearance.

Figure 9. A 3-day embryo from the moth, *Manduca sexta*, in which the smooth septate junctions of the midgut are forming from intramembranous particles. These become aligned into short rows and gradually increase in length by particle accretion (arrows).

Figure 10. The midgut from a 1-day pupa of *Manduca sexta*. The smooth septate junctions, normally orderly in their linear arrays (as in Figure 8), are undergoing disruption and disassembly (arrows) by transformation of the ridges into separate particles.

Figure 11. The midgut from a metamorphosing moth in which reaggregating smooth septate junctions can be seen; these are found in the lata pupa, before emergence of the adult. Here the junctional ridges are re-forming by addition of particles to the ridges that have already assembled (large arrows). The tissue is fixed, and so the IMPs and ridges fracture onto the P face (PF) with the complementary grooves (arrows) appearing on the E face (EF).

Figure 12. Smooth septate junctions have very different appearances in the adult state of different insects. These E face ridges from the midgut of *Rhodnius prolixus* are less continuous than in other insects (compare with Figure 8).

occluding junctions. However, no such work has been done on insects, and other reports show that tracers are able to move through these junctions under some circumstances (e.g., Figure 7) (Skaer *et al.*, 1980), as are viral particles (see Houk, 1977). Since their matrix may be a mucosubstance, this may cause partial inhibition to the passage of certain ions and molecules, perhaps perferentially if the matrix is charged (see Skaer *et al.*, 1979; Lane and Skaer, 1980).

4. Desmosomes and Hemidesmosomes

4.1. Location

Desmosomes (*maculae, fasciae,* or *zonulae adhaerentes*) are found rather ubiquitously in most insect tissues (Figure 13, insert) in the form of spots or bands, or as "intermediate" junctions. There are certain exceptions; for example, there are none in the glial cells in the CNS of the cockroach and the locust. However, this contrasts with the situation in the moth, where many are present between the glial membranes in the nerve cord (Lane, 1972). Hemidesmosomes occur in the CNS of locusts and cockroaches, where they are found between glial cells and extracellular space. Usually desmosomes are found in the macular or plaquelike conformation, rather than in the more extensive zonular form. They are normally homocellular, but they may be heterocellular or autocellular. They tend to coexist with a variety of other junctional types, including septate junctions, gap junctions, and tight junctions (see Lane and Skaer, 1980).

4.2. Structure

In thin sections, desmosomes exhibit a 20-nm or greater intercellular space which contains cross fibrillae (insert, Figure 13); they also feature dense striations on the cytoplasmic side, with associated microtubules (Ashhurst, 1970) which lie in parallel with, rather than inserting into, the membranes. Their fine structure in insects has not been elucidated by tracer impregnation or freeze fracture, since the former does not infiltrate sufficiently to produce the appropriate negatively stained appearance, while the latter rarely produce any recognizable desmosomal freeze-cleave profile in replicas (see Lane *et al.*, 1977; Lane and Skaer, 1980).

4.3. Development

Given the lack of a constant and characteristic freeze-fracture profile for desmosomes in insect tissues, it is not possible to follow changes in their outline during development. Any analysis of their changes during development must therefore be by thin-section criteria, which in the embryonic or pupal tissue of several insects involves the alignment of the adjacent plasma membranes before the fibrillae insert into the plasmalemma (Lane and Swales, 1978a; Lane and Skaer, 1980).

4.4. Physiological Role

The function of the desmosome must be to maintain tissue integrity in adult systems. In the case of the hemidesmosomes, they presumably maintain the physical association between the cell membrane and the basal lamina or the extracellular matrix.

5. Gap Junctions

5.1. Location

Gap junctions, also called nexuses, electrotonic junctions, and *maculae communicantes*, are fairly ubiquitous. They have been found between the cells of most insect tissues, inexcitable as well as excitable (see references in Lane and Skaer, 1980), except for such cells as circulating blood cells, striated muscle, and the majority of neurones. They occur most commonly between homologous cells but are sometimes located between processes of the same cell, or of different cell types forming autocellular or heterocellular associations. Since they are so universal, they frequently co-exist with one or more of the many other insect junctional types.

5.2. Structure

5.2.1. Thin Sections

In cross sections of *en bloc* stained tissues, gap junctions are characterized by a reduced intercellular cleft of 2 to 4 nm which may exhibit a cross-striated appearance (Figure 13). After infiltration with tracers the reduced cleft is still apparent, and tangential sections reveal the presence in the junctional areas of unstained particles lying against a densely stained background (Figure 14); in the center of each particle there may be a stained pore or channel (insert, Figure 14). The particles may be closely packed (Figure 14), but in insect tissues they usually are loosely aggregated (Figure 14, insert).

5.2.2. Freeze-Fracture Replicas

After freeze cleaving, the gap junctions take the form of different sized clusters of intramembranous particles (Figure 15), which, in insects, are 12–14 nm in diameter, larger than the normal particle populations, and which fracture onto the E face (Flower, 1972). These EF particles are complementary with P face pits (Figure 15, insert), and when the fracture plane undergoes transition

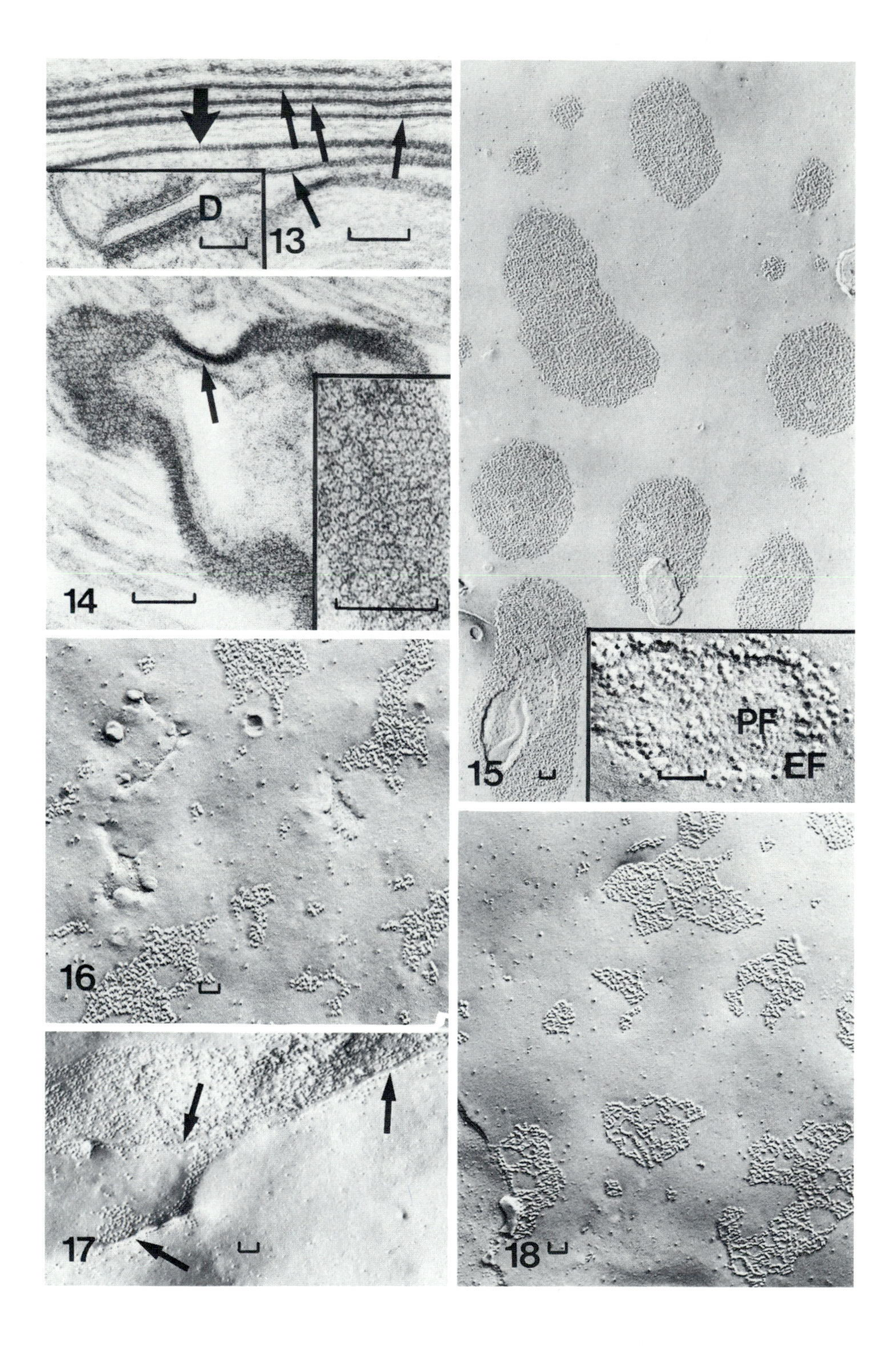
D
13
14
15
PF
EF
16
17
18

from EF to PF, the reduced intercellular cleft of the junction can be visualized. The particles in the plaques may be very irregularly and loosely aggregated in comparison with those of vertebrate tissues where the particles are usually packed in regular hexagonal arrays in fixed preparations (Gilula, 1978). Recent studies on rapidly frozen, nonfixed and noncryoprotected vertebrate tissue, however, reveal a looser packing (Raviola *et al.*, 1980). This may more accurately reflect the degree of cell-to-cell coupling (Peracchia, 1980) in that the cells are chemically untreated, and the more loosely packed the IMPs, the more likely are the cells to be actively exchanging ions and molecules. The results on rapidly frozen, unfixed insect tissues will be awaited with considerable interest.

5.2.3. Model

From the distribution of the tracer in the intercellular cleft and the intramembranous features viewed in freeze-fracture replicas, three-dimensional models of insect gap junctions may be constructed (Figure 6C) (Lane and Skaer, 1980). The junctional particles can be presumed to be completely transmembrane, with each junctional particle meeting another across the intercellular cleft so that their central channels are aligned and in continuity. Clearly, tracers can permeate around the points where these individual particles are apposed, and they also often stain the internal cores of the particles (Zampighi *et al.*, 1980). It is this central channel that is thought to be the site of exchange of ions and molecules between the coupled cells. As yet, there is no biochemical evidence on the substructure of the particles, as to whether, like vertebrates (Casper *et al.*, 1977), they are each composed of six hexamers that surround the central pore.

←

Figures 13–18 represent some of the different forms of gap junctions in insect tissues. (The bars equal 0.1 μm.)

Figures 13 and 14. Thin sections through the compound eye of the blowfly, *Calliphora erythrocephala*, which demonstrate lanthanum-impregnated gap junctions in transverse section (at arrows). In some regions, cross striations (thick arrow) are apparent. Gap junctions may also coexist with desmosomes (D in Figure 13, insert). In tangential sections (Figure 14), *en face* views of the particles that compose the junctions appear as unstained entities against an electron-opaque background. At higher magnification a stained central channel is visible within each particle (Figure 14, insert).

Figures 15–18. Freeze-fracture replicas from the central nervous system of *Manduca sexta* reveal the varying forms of glial gap junctional plaques at different stages in the life cycle. These gap junctional particles cleave onto the E face (EF), leaving complementary pits in the P face (PF) (insert to Figure 15).

Figure 15. Fully developed macular plaques demonstrating the broad size range of junctions found in insect tissues.

Figure 16. Early stages in the formation of the larval junctions where the junctional clusters are aggregating together.

Figure 17. An early pupal stage when the junctions become disaggregated and their particles disperse owing to a streaming out of the junctional particles in rows (arrows) from the maculae.

Figure 18. After junctional dispersal, the particles undergo a second round of reaggregation into macular arrays in the late pupal stage.

5.3. Development

5.3.1. Embryonic Stages

In insects, the development of gap junctions has been followed in embryonic tissues in only a few species, including the blowfly (Lane and Swales, 1978a), the moth *Manduca* (Lane and Swales, 1979), and the locust (Swales *et al.*, 1981). In the grasshopper CNS, electrical recordings have shown early dye and ionic coupling of neuroblasts with other cells, while later, dye, but not electrical, uncoupling developed (Goodman and Spitzer, 1979). In this system and in locust testis (Eley and Shelton, 1976), it may be that differentiation is the result of gap junctional uncoupling or loss with the concomitant disappearance of cell-to-cell communication, as has been suggested for a number of vertebrate tissues (see references in Lane and Skaer, 1980).

The actual mechanics of gap junctional development in insects has been followed in freeze-fracture replicas made at different stages of development. It seems that junctional assembly proceeds by translateral migration of 13-nm EF intramembranous particles into first linear alignments and then loose clusters (Figure 16) that gradually coalesce to produce the macular EF plaques of 13-nm particles (as in Figure 15) (Lane, 1978a; Lane and Swales, 1978a; 1979) that characterize mature junctions. The same sequence of events occurs in the assembly of gap junctions during the last stages of pupal metamorphosis in holometabolous insects (Figure 18) (Lane and Swales, 1978b; 1980) but is rather different from those in vertebrate tissues where there are special formation plaques and larger precursor particles (for example, Decker, 1976).

5.3.2. Pupal Stages

There appear to be junctional transformations in insect metamorphosis in that, upon entering pupation, the gap junctions between the glial and perineurial cells of the insect CNS are disrupted and the cells are separated. By freeze-fracture criteria, the junctional particles are seen to disperse over the glial membrane faces (Figure 17) (Lane and Swales, 1978b; 1980), but no internalization appears to occur such as may be found in certain vertebrate tissues (Larsen, 1977). As a result of this 13-nm particle dispersal, when the junctions come to re-form in late pupal stages (Figure 18) it seems possible that the dispersed particles may be reutilized in the assembly of the adult gap junctions (Lane and Swales, 1978b; 1980).

There also appears to be some turnover in the gap junctions in adult arthropod tissues that undergo cellular turnover, such as the midgut (Lane, 1978b). This may involve the insertion of new cells and the consequent need to reassemble communicating junctions between the membranes of the old cells and these new ones. Freeze-fracture studies reveal images consistent with this hypothesis (Lane, 1978b).

5.3.3. Experimental Advantages over Junctions of Vertebrates

Two features of the arthropod gap junctional particles in freeze fracture are that they cleave onto the E face and measure about 13 nm in diameter. In this respect they are very different from the tight junctional particles, which fracture, normally, on the P face and measure 8–10 nm in diameter. This distinction does not occur in vertebrates, where both tight and gap junctional particles are on the P face and around 8–10 nm in diameter. Hence stages in the development of these two junctional types in arthropods, even if concurrent, may be distinguished one from another. This is particularly striking in the case of arachnids. In the spiders, the tight junctions are very complex and extensive (Lane and Chandler, 1980), and stages in their development, which can be seen occurring in the same membranes as the gap junctions, are very clear-cut and distinct (Lane, 1980; 1981e). In fact, this evidence supports the contention that gap and tight junctional particles do not derive from one and the same precursor particle as has been suggested for vertebrate tissues (for example, Elias and Friend, 1976; Porvaznik *et al.*, 1979). In insect tissues, the simple tight junctions are sufficiently difficult to find that the studies made thus far on concurrent development of tight and gap junctions in embryonic and pupal tissues (Lane and Swales, 1978a,b; 1979) have not yet permitted simultaneous examination of the various stages of formation of both junctional types. Investigations have not yet been carried out on the assembly of the more complex tight junctions in insect compound eye (Lane, 1981a).

5.4. Physiological Role

Gap junctions are thought to represent low-resistance pathways between cells, through which ions and small molecules may be exchanged. Although they were referred to in early studies in excitable arthropod tissues as electrotonic junctions, they also have been found to couple cells in nonexcitable systems.

In excitable cells, such as nerve cells, giant axons are sometimes coupled electrically; one of the most striking examples of this is that of the crustacean CNS giant fibers (Peracchia, 1973a,b). The other arthropod system that has been much investigated is that of an insect, the dipteran salivary gland. It was originally believed that septate junctions were responsible for the coupling observed electrophysiologically in insect salivary gland cells (Loewenstein and Kanno, 1964), but it now seems that they coexist with gap junctions, and it is the latter that actually carry out the coupling. Studies on nonexcitable cells have shown that the coupled cells are associated both ionically and metabolically and will transfer molecules of molecular weight up to about 1000–1200 daltons (Simpson *et al.*, 1977) depending on the $Ca2^{+}$ concentration. Second messengers such as cyclic AMP are thought to be exchanged between coupled mammalian cells (Lawrence *et al.*, 1978), so that they will respond in an integrated way to hormonal stimulation. The suggestion that developmental events are coordinated by cyclic nucleotides diffusing via gap junctional channels in insect epidermis has also been put forward (Caveney, 1978). Moreover, the channels may be asymmetric, exhibiting directional permselectivity (Flagg-Newton and Loewenstein, 1980), and their effective diameter may be modulated by various factors (Loewenstein, 1977). These features may be important in differentiation in terms of the kinds of regulatory molecules that can be exchanged.

6. Tight Junctions

6.1. Location

Tight junctions, or *zonulae occludentes*, are found only in certain insect tissues—those that exhibit permeability barriers. These tissues usually are epithelial or epithelium-like cell layers, and the occluding junctions occur along their lateral borders, which have very extensive interdigitations. The junctions may coexist in intimate spatial association with gap junctions as well as with septate junctions and desmosomes (Figure 5). They are found between the borders of the glial cells that ensheath the insect CNS, forming the perineurium, in both adult (Lane and Treherne, 1971; 1972; Lane, 1972; 1978a; 1981b; Lane *et al.*, 1977) and embryonic tissues (Lane and Swales, 1978a,b; 1979). They are also found in the insect testis (Toshimori *et al.*, 1979; Lane and Skaer, 1980), as well as between the lateral cell borders of the rectal pads of cockroaches (Lane 1979b; 1979c), in the esophagus (Skaer *et al.*, 1980; Lane 1981b), and along the lateral mestracheon borders of tracheal cells (Lane and Skaer, 1980). These last are autocellular, in that the mestracheal junctions are joining together processes from the same cell (as occurs also in the tight junctions between Schwann cells around vertebrate axons [Schnapp and Mugnaini, 1975; Wiley and Ellisman, 1980; Shinowara *et al.*, 1980]). Tight junctions, however, are usually homocellular and are only rarely heterocellular in insects, as for example in the CNS, when outer perineural and inner glial cells are linked (Lane and Swales, 1979).

Tight junctions of a rather more complex type, distinguishable as complex only by freeze-fracture criteria (see section 6.2.2) are found between glial cells in the compound eye of dipterans, between the outer retina and the inner neuropile (Lane, 1981a). In this situation, the cell orientations are rather different than the usual perineurial sheath arrangements, and these geometrical considerations may account for their difference in complexity.

6.2. Structure

6.2.1. Thin Sections

In thin sections, tight junctions are seen as punctate appositions between adjacent plasma membranes (Figure 19). The membranes appear to fuse at these points, so that the intercellular space is thereby obliterated. If treated with stains such as uranyl calcium, the outer leaflet of the component membranes is heavily stained. However, the stain does not seem to gain access to those parts of the membranes where the junctional fusion has occurred so that in *en face* sections those fused regions appear as electron-lucent strands or ridges lying in a densely stained background. In tracer-infiltrated sections the same phenomenon can occur if the junctional strands are discontinuous, allowing tracer to leak around them (see Lane, 1981b). In preparations incubated in physiological saline plus tracer, the dense tracer cannot penetrate beyond the points of occlusion or membrane fusion, as seen in cross section (Figure 20).

6.2.2. Freeze-Fracture Replicas

For the most part, the intramembranous components of insect tight junctions are very simple moniliform ridges that lie in the PF in discontinuous alignments, usually unbranched. These ridges are composed of 8–10-nm IMPs that cleave onto the P face, leaving complementary grooves on the E face. The more complex variety of junction is recognizable only in freeze-cleaved preparations, and thus far is found only in the dipteran compound eye (Lane, 1981a). These exhibit a rather more reticular arrangement of junctional grooves (Figures 21, 22) and ridges, composed likewise of 8–10-nm particles closely aligned in rows which may fuse into ridges (Figure 23). These bear a great resemblance to vertebrate tight junctions (Claude and Goodenough, 1973) in terms of their reticular patterns and the frequency with which they are observed. The latter feature has led to more numerous observations of E–P fracture face transitions, so that the coincidence of EF grooves and PF ridges (Figures 21, 23) indicates the complementary nature of the two.

6.2.3. Models

The relatively recent literature on insect tight junctions, particularly those found in the dipteran compound eye, has produced enough information from several lines of evidence to construct models of insect tight junctions (Figure 6D) (Lane and Skaer, 1980; Lane, 1981b). Thin sections have revealed punctate appositions, and tracer infiltration through leaky regions has led to an understanding of the discontinuous ridges seen in freeze-fracture replicas. The complementary nature of the PF ridges with respect to the EF grooves is incorporated in the model, as is the fact that the ridges appear to be, like the tight junctions of arachnids (Lane and Chandler, 1980; Lane *et al.*, 1981), offset with respect to the grooves (Figure 21). This feature is comparable to that observed in vertebrate epithelial *zonulae occludentes* (Bullivant, 1978).

6.3. Development

In embryonic insect CNS, although tight junctional development has been followed in the perineurium of the blow fly, *Calliphora erythrocephala* (Lane and Swales, 1978a), the moth *Manduca sexta* (Lane and Swales, 1979), and the locust *Schistocerca gregaria* (Swales *et al.*, 1981), the process is not entirely clear-cut because of the relative paucity of junctional ridges and grooves. The advent of the permeability barrier is heralded by the exclusion of exogenous tracers. The process leading up to junctional maturity seems to involve a gradual

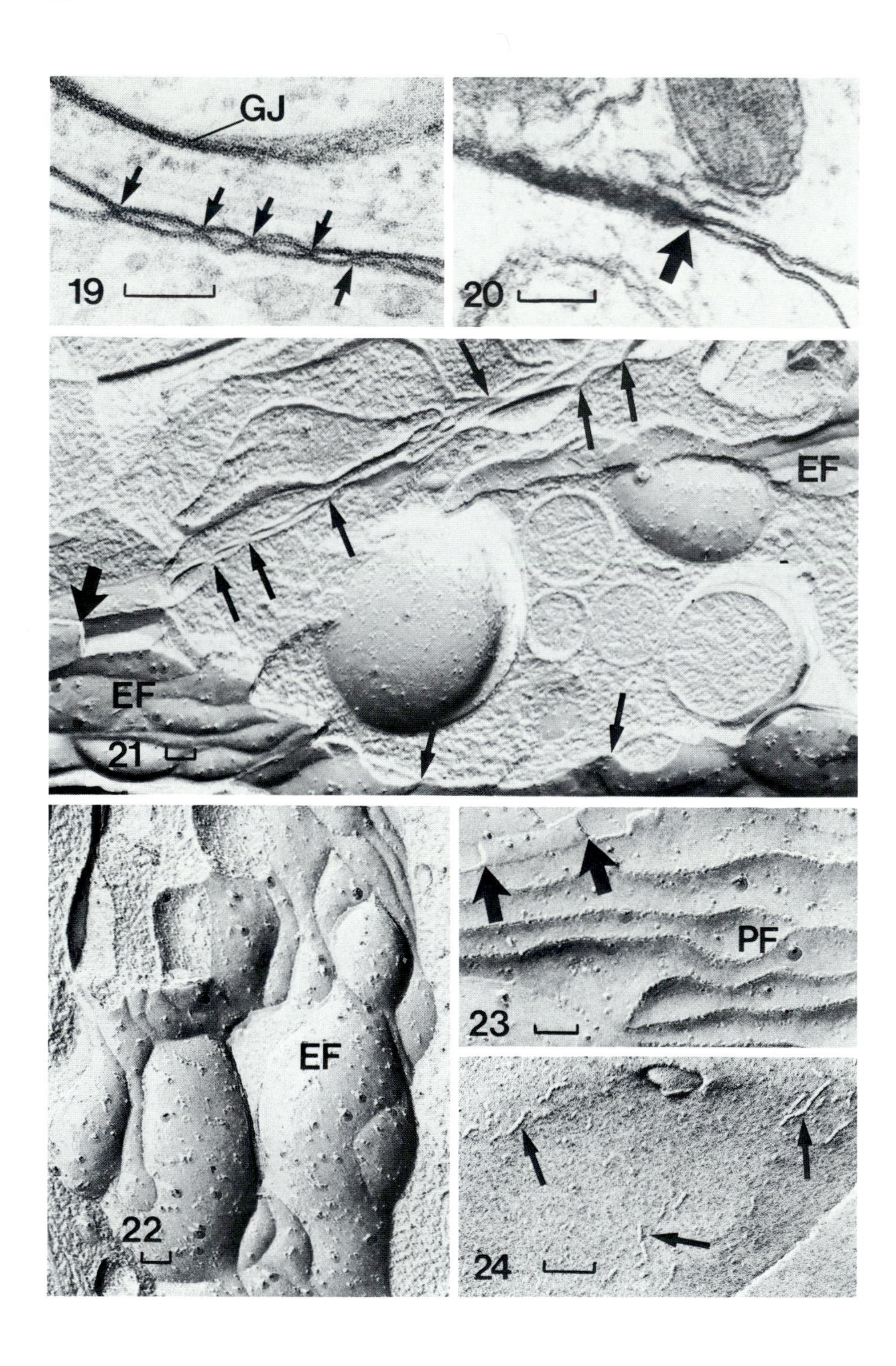
GJ
19
20
EF
EF
21
22
EF
PF
23
24

filtering out of tracer as the junctions form. This appears to be related to the gradual alignment of individual intramembranous PF particles into short rows (Figure 24), which then fuse into longer ridges to form the tight junctional structures.

In holometabolous insects, as they enter metamorphosis there is a breakdown of the blood–brain barrier involving junctional ridge disaggregation and consequent entry of exogenous tracers (Lane and Swales, 1978b). After the pupal CNS reorganization, the perineurial and glial cells become reestablished in the adult configuration and their intercellular junctions reform. When these occluding junctions have reassembled, the barrier to exogenous molecules reappears (Lane and Swales, 1978b). This could involve reutilization of the dispersed tight junctional ridge components, as occurs in the semiconservative process in vertebrate tissues (Dermietzel *et al.*, 1977; Polak-Charcon and Ben-Shaul, 1979).

6.4. Physiological Role

The tight junctions in insects form, as they do in vertebrates (Claude and Goodenough, 1973), diffusion or permeability barriers (Lane, 1981b) by the fusion of the strands or ridges in adjacent cells, to seal or obliterate the intercellular cleft with respect to the leakage of ions and molecules. In some cases, there are a large number of discontinuous ridges, in which case the junctions are said to be highly "leaky." In insects, a finite leak to K^+ ions has been observed in the perineurial barrier (Pichon *et al.*, 1971) to the CNS. This ionic leak can be enhanced by urea treatment (Treherne *et al.*, 1973), but ions such as La^{3+} or molecules are still unable to enter, unlike urea-treated vertebrate *zonulae occludentes* (Rapaport *et al.*, 1971).

Tight junctions are considered, in vertebrates, to separate the IMP population of the apical cell borders from that of the lateral cell borders (DiCamilli *et al.*, 1974). In insects, however, either septate junctions fulfill this role or the IMP populations being kept separate are those of the apical and lateral borders together, as distinct from the basal border (Lane, 1981b). Variations in tight junctional structure have also been implicated in tissue expansion. The num-

←

Figures 19–24 illustrate different features of insect tight junctions. (The bars equal 0.1 μm.)
Figure 19. A young larva of *Manduca sexta* showing the perineurium, which possesses both gap (GJ) and tight junctions (arrows). The punctate appositions of the latter form the morphological basis of the blood–brain permeability barrier.
Figure 20. A thin section through the CNS of the cockroach, *Periplaneta americana* after incubation with 10 nM lanthanum in physiological saline. The tracer stops at a tight junctional apposition (arrow), such as occur in the perineurial layer ensheathing nervous tissues.
Figures 21–23. Freeze-fracture replicas from the compound eye of *Calliphora erythrocephala.* As in *Musca domestica*, it exhibits glial layers that form tight junctional punctate appositions, here seen in cross fracture (thin arrows). The E face (EF) possesses junctional grooves or furrows (Figures 21 and 22), and the P face (PF) features raised particle rows or ridges (Figure 23). These ridges are coincident over fracture face transitions with the EF grooves (thick arrows).
Figure 24. Tight junctional PF ridges assembling in the embryonic perineurium of the moth, *Manduca sexta*, CNS. The IMPs are becoming aligned into ridgelike arrays (arrows).

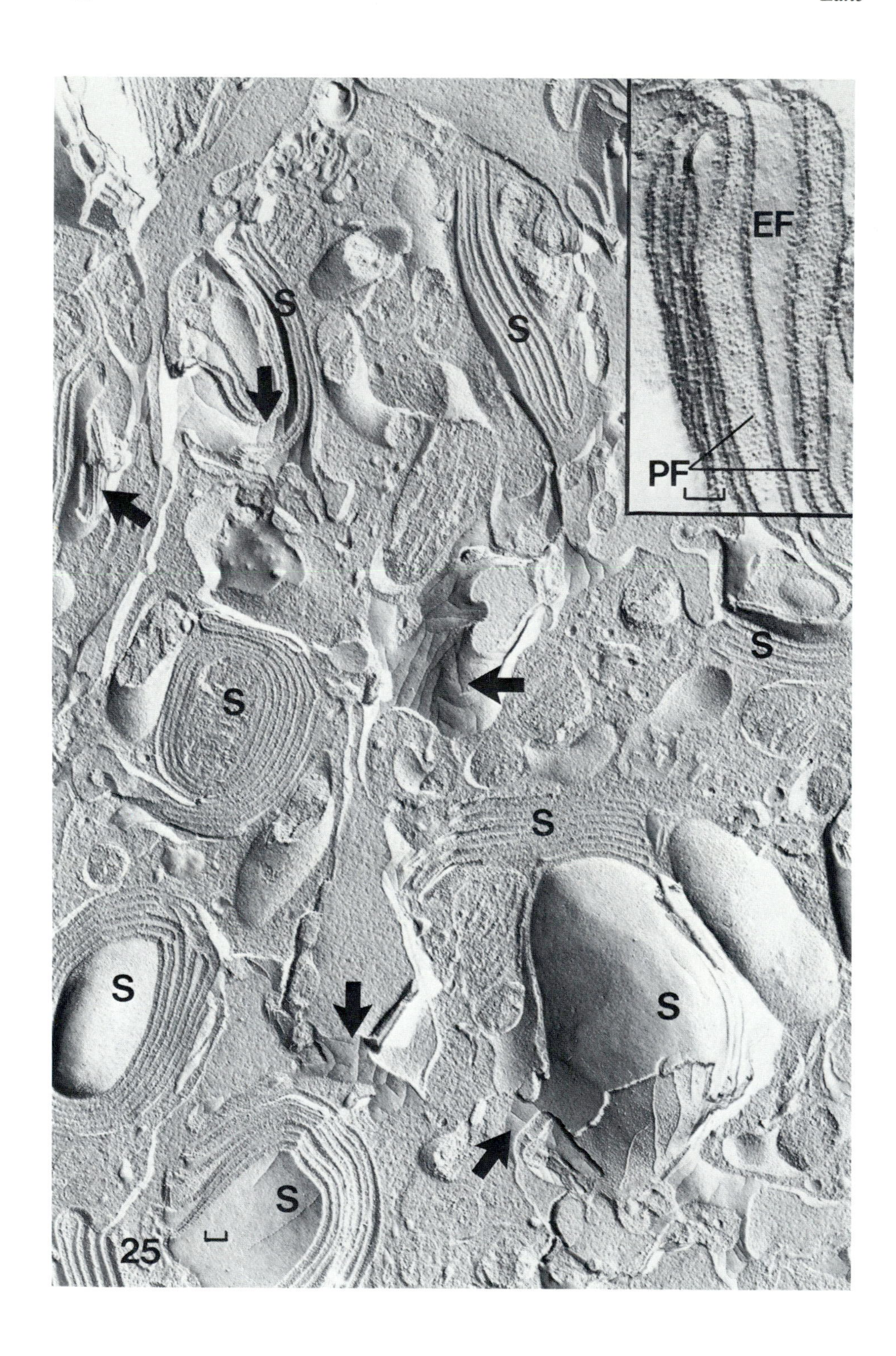
EF
PF
S
S
S
S
S
S
S
S
S
25

ber of parallel junctional strands is held to be important in allowing changes in volume to take place, since a cross-linked network would not permit this (Hull and Staehelin, 1976). While the utility of such expansion is not entirely clear for the CNS of insects, in the case of the rectum (Lane, 1979c) and testis (Toshimori *et al.*, 1979), volume changes certainly occur.

7. *Scalariform Junctions*

7.1. *Location*

Scalariform junctions, first described in the excretory system of the insect *Petrobius maritimus* (Fain-Maurel and Cassier, 1972), are found primarily in tissues such as insect rectal pads and papillae (Lane, 1979c; Noirot-Timothée *et al.*, 1979; Lane and Skaer, 1980; Noirot-Timothée and Noirot, 1980). They occur as autocellular interactions between the self-infoldings of the rectal papillae and as homocellular associations between adjacent rectal pad cells. They have also been found, relatively infrequently, in the CNS (Lane, 1968; Lane and Treherne, 1972; Lane and Treherne, 1980; Lane, 1981c) between glial processes surrounding nerve cell bodies in ganglia. Here the junctions are much less extensive than those in the rectal tissues.

7.2. *Structure*

7.2.1. *Thin Sections*

Scalariform junctions are characterized by a constant 15–20-nm intercellular cleft in which faint cross striations, rather than septa, occur. After incubation with tracers there seems to be no occlusion to penetration, and unstained columns may be observed in *en face* tangential sections (Lane, 1979c).

7.2.2. *Freeze-Fracture Replicas*

In freeze-fracture replicas there appears to be no characteristic intramembranous profile involving ridges of any kind (Figure 25). The junctional regions possess only numerous IMPs of at least two different sizes (insert, Figure 25), of which some may represent the location of pumps and others the insertion sites of the intercellular columns (see Figure 69 in Lane and Skaer, 1980).

←

Figure 25. A freeze-fracture replica through the rectal papillae of *Calliphora erythrocephala*. Note the stacked membrane arrays (S), which represent scalariform junctions. These are characterized by arrays of IMPs of different diameters on the P face (PF in insert). The reticular septate junctions are distinguished by PF ridges and EF grooves (thick arrows) that occur in the membranes that join the stacks together. (The bars equal 0.1 μm.)

7.3. Physiological Role

The scalariform junctions are thought to keep the intercellular cleft open in regions where rapid ion pumping and transport is required, as occurs in rectal pads and papillae (Berridge and Gupta, 1967). Some of the IMPs observed in these regions may be implicated in pumping, since ATPase has been shown to be present in these locations (Berridge and Gupta, 1968). No details are available yet as to their development, since these junctions have no prominent intramembranous structure in which the changes can be followed during junctional assembly.

8. Retinular Junctions

8.1. Location

Retinular junctions (Carlson and Chi, 1979; Lane, 1979d, 1981b,d; Chi and Carlson, 1981) are located in the distal portions of dipteran compound eyes (Figure 26). They occur between the adjacent photoreceptor rhabdomeres, also called retinular axons. Here they are homocellular, but they can also be heterocellular, since they can also occur between retinular axons and glia (Figure 27) (Lane, 1981d). Other homocellular interglial junctions, called marginal glial cell junctions, also exist (Chi and Carlson, 1980b). These exhibit many similarities to the retinular junctions with respect to the size of their intercellular cleft and their freeze-fracture appearance.

8.2. Structure

8.2.1. Thin Sections

These junctions exhibit a fairly constant 15–20-nm intercellular space which may have a faintly striated appearance. In preparations incubated in ionic lanthanum in saline, tangential sectioning reveals cross or longitudinal sections (Figure 27) or oblique sections (insert, Figure 26) of nonopaque columns in an electron-dense background. Therefore, the junctions appear to be patent to tracers, although the incubation procedure may have damaged the system.

8.2.2. Freeze-Fracture Replicas

Freeze-cleaved preparations reveal characteristic discontinuous PF ridge-like structures scattered at angles to each other over the retinular surfaces (Figure 26). The EF grooves that are also seen may in some cases be noncomplementary to their PF ridges (Carlson and Chi, 1979; Lane and Skaer, 1980). No three-dimensional models of these junctions have thus far been drawn up because the relationship between the intramembranous ridges and the intercellular columns is not at all clear.

8.3. Physiological Role

The retinular junctions were at first thought to be related to the orientation of the photoreceptor axons, perhaps enforcing the ommatidial twist observed in the retina (Carlson and Chi, 1979). Since desmosomes are also present in this area, the retinular junctions are probably not primarily adhesive, but they may maintain localized ionic concentration gradients (Chi and Carlson, 1980a). More recent studies show that tracers gain ready access into the system (Chi and Carlson, 1981), and so this latter speculation needs to be modified. At the proximal end of the rhabdomeric axons, the retinular junctions become transformed into septate junctions (Chi *et al.*, 1979; Lane, 1981c,d), so the intercellular columns are replaced by septa (Figure 26), which may have an adhesive role. Both of these become infiltrated after incubation with ionic lanthanum (Lane 1981c, d), indicating that they do not form a completely restricted permeability barrier.

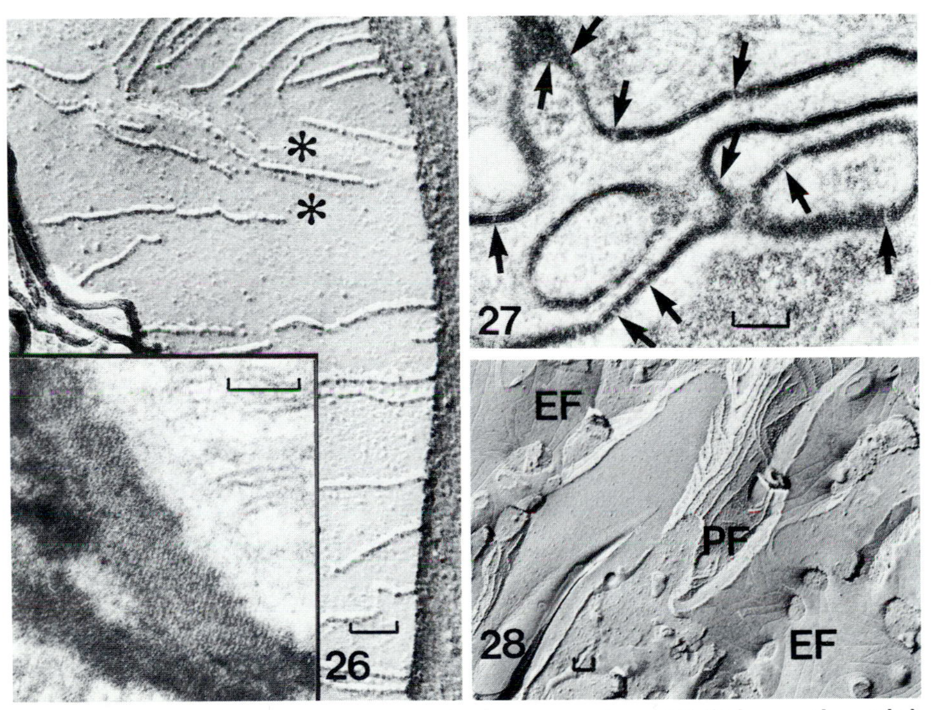

Figures 26 and 27. The photoreceptor region in the compound eye of *Calliphora erythrocephala*. The bars equal 0.1 μm. Figure 26 shows the P face of one of the component retinular axons with the intramembranous particles arrayed in ridges. Between these, on the more proximal surface of the axons, septate junctional particle arrays (*) are interposed. The insert shows an oblique area of the retinular junctions infiltrated with lanthanum revealing the columnar substructure.

Figure 27. Shows intercellular clefts between glial cells and retinular axons, where cross striations or dots (arrows) indicate columns cut longitudinally or transversely. These appear in negative contrast against a lanthanum-stained background. (The bars equal 0.1 μm.)

Figure 28. Rectal papillae of *Calliphora* showing the innerstack regions with reticular "septate" junctions. These exhibit P-face (PF) ridges and E-face (EF) grooves. (The bar equals 0.1 μm.)

Occlusion to tracer entry seems to occur more proximally, where glial tight junctions have been found (Lane, 1981a). Further studies are required to clarify the precise role of these retinular junctions as well as to follow the stages in their development and assembly.

9. *Reticular "Septate" Junctions*

9.1. *Location*

The reticular "septate" junctions have thus far been observed only in the rectal papillae of dipteran flies such as *Musca domestica* (Flower and Walker, 1979) and *Calliphora erythrocephala* (Lane, 1979c). They occur in the intercellular clefts that lie outside the stacked membrane arrays of the scalariform junctions that characterize the papillae (Figure 25). Hence, they occur in the extensible regions that are thought to become distended when water and ion transport occurs across the rectum from the gut lumen to the extracellular hemolymph (Berridge and Gupta, 1967). Here they coexist with gap junctions, while normal septate junctions are found in the apical and basal regions of the same lateral intercellular borders.

9.2. *Structure*

9.2.1. *Thin Sections*

In thin sections, the cleft of the reticular "septate" junctions is usually 15 to 20 nm across and has a striated appearance (Lane, 1979c). These features give the appearance of a variant of conventional septate junctions, and the use of the modifying term "reticular" refers to the characteristic freeze-fracture appearance. However, their intercellular clefts may measure much more than 20 nm, and may be swollen into lacunae. In tracer-infiltrated preparations the occasional region of unstained columns may be seen against the dense background of tracer.

9.2.2. *Freeze-Fracture Replicas*

In freeze-fracture replicas, the regions of the reticular septate junctions are characterized by striking arrays of discontinuous PF ridges and EF grooves (Figures 25, 28) which are not always complementary. The ridges appear to be comprised of 8–10-nm particles that are fused laterally. As the relationship between these ridges and the infrequent intercellular columns is not clear, no useful model can be constructed. Developmental studies have not been made on these structures.

9.3. *Physiological Role*

The function of these reticular septate junctions appears to be related to the even flow of solutes through the clefts in order to maintain adequate hydrostatic pressure. The presumably rigid intramembranous ridges may act to regulate the speed and the extent of cleft distension as water flow occurs across the rectum (Lane, 1979c). It has also been suggested that the "rods" may be associated with the transport of K^+ or water across the plasma membrane (Flower and Walker, 1979).

10. Neuroglial Junctions

The nervous system in insects exhibits a number of different types of neuroglial junctions, which includes interactions identifiable by either thin section or freeze-fracture criteria alone and occasionally by both (Lane and Skaer, 1980; Lane and Treherne, 1980; Lane, 1981c).

10.1. Axonal and Glial Intramembranous Ridges or Particle Arrays

The membrane P face of both axons and glial cells in insects may exhibit discontinuous moniliform PF ridges (Figure 29) and EF grooves oriented parallel to the longitudinal axis of the axons (Lane *et al.*, 1977; Lane, 1978a; 1979a; Lane and Swales, 1978a,b). In other cases, particle plaques may be found, as for example on photoreceptor axons (Gemne, 1969) and in crayfish axoglial interfaces (Peracchia, 1974). Occasional close appositions between the membranes of axons and glia, which could be thin-sectioned correlates of such intramembranous structures, have been observed in the CNS (Lane, 1978a; Lane and Swales, 1978b) and sensory sensilla of insects (Thiele, personal communication), arachnids (Lasansky, 1967; Binnington and Lane, 1980; Harrison and Lane, 1981), and crustaceans (Peracchia, 1974). The physiological significance of these axoglial structures is not known, although they could serve an adhesive or trophic role. The intramembranous ridges could also be involved in axoglial guidance during development (Lane, 1979a; 1981c).

10.2. Capitate Projections

These structures are invaginations of glial cells into the photoreceptor axons in the distal retina of dipteran eyes. In thin sections, they exhibit enhanced electron density in the glial protrusions as well as in patches on the axonal membrane, where synapticlike vesicles also abound (Figure 30) (Carlson and Chi, 1979; Chi and Carlson, 1980a; Lane, 1981c). In freeze-fracture replicas, plaques of intramembranous particles are found where the glial processes abut on the axolemma as well as in axonal patches (Figure 31). The possibility that they are receptor sites cannot be excluded, although they could be a form of communicating junction like the gap junction, permitting the exchange of ions, small molecules, or trophic substances between axons and glial cells.

Other modifications of the photoreceptor axons occur near the extracellular space of the ommatidial cavity (Figure 32) at the distal end. In thin sections, these appear as dense cross striations across both the exposed membrane

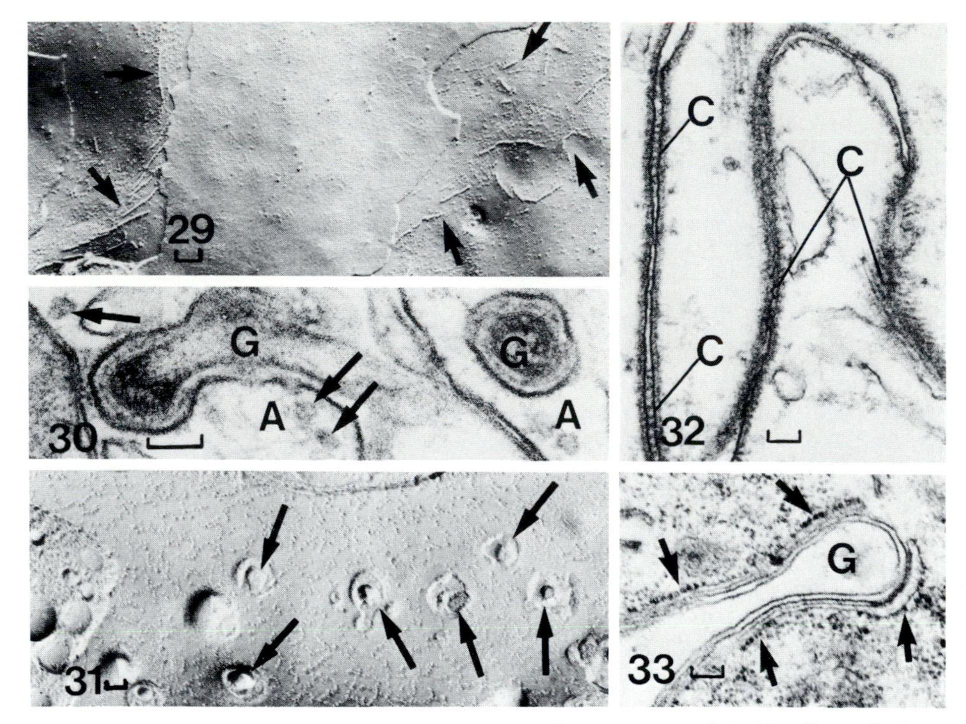

Figures 29–33. These figures are electron micrographs from the CNS of the blowfly, *Calliphora erythrocephala.* (The bars equal 0.1 μm.)
Figure 29. CNS cell membranes showing the intramembranous PF ridges (arrows) that characterize both glial and axonal membranes. These may represent a form of axoglial interaction.
Figure 30. A thin section through capitate glial (G) projections into retinular axons (A). Note the density of the membranes at this axoglial specialization and the vesicles (arrows) in the axoplasm.
Figure 31. Axonal membranes in the compound eye where the glial cells form capitate projections (as in Figure 30) into the axolemma. These sites are identifiable as cross fractures through the glial processes (arrows) that protrude into the axoplasm. IMP arrays decorate the membranes at these axoglial modifications.
Figure 32. Photoreceptor cell membrane near the ommatidial cavity in the compound eye. Note the membrane-associated striations that traverse the intercellular clefts (C).
Figure 33. Subsurface cisernae (arrows) in a nerve cell body associated with glial cell (G) invaginations. Note that the cisternae display ribosomes only on the surface facing away from the area of neuroglial contact.

surfaces and the lateral cell-to-cell associations, some of which may be heterocellular with the adjacent glial cells.

10.3. Subsurface Cisternae

When glial cells send attenuated trophospongial processes into the peripheral cytoplasm of large nerve cells bodies, specialized sites of neuroglial contact are found, called subsurface cisternae. These are close appositions between the glial plasma membrane, the neurolemma, and underlying cisternae of endoplasmic reticulum which have a ribosome-free surface intimately associated with the neurolemma (Figure 33). In some vertebrate systems, comparable structures possess a recognizable freeze-fracture profile of membrane clusters

(Henkart *et al.*, 1976), and the same may be true for insects. These could be sites of exchange of trophic substances.

10.4. Axoglial Modifications in the Insect Eye

In the retina of the dipteran compound eye, interaxonal retinular junctions exist (Chi *et al.*, 1979; Chi and Carlson, 1981) as already mentioned, and there are also axoglial associations in the near vicinity (Lane, 1981c,d). The former are characterized in thin sections by intercellular columns seen as nonopaque dots against a densely stained background in tracer-incubated preparations (Figure 27). These columns may serve to maintain the constant 15-20-nm intercellular cleft. In freeze-fracture replicas these feature PF ridges (Figure 26). These junctions become transformed to axoaxonic and axoglial septate junctions as the retinular axons penetrate more proximally toward the optic neuropile (Figure 26). The intramembranous modifications can be seen both in thin sections (Lane, 1981d) and in freeze-fracture replicas (Figure 26) (Lane, 1981c). The surrounding glial cells also become associated by septate junctions as well as by gap junctions (Lane, 1981d) and ultimately also by tight junctions (Figures 21-23) (Lane, 1981a). These junctions appear to form the basis of the observed insect blood-retinal barrier (Boschek, 1971; Shaw, 1978; Lane and Skaer, 1980).

11. Concluding Remarks

Perhaps after reviewing the nature and location of the different kinds of insect intercellular junctions in this chapter it has become clear that although many of their ultrastructural features have been elucidated, there are actually few cases where a function is unequivocally correlated with structure. For example, desmosomes seem very likely to be responsible for maintaining structural integrity, and tight junctions for creating permeability barriers. However, the coexistence of both of these with septate junctions, believed by some investigators to be capable of carrying out both of these functions, means that no final conclusion can be reached. Similarly, the coexistence of septate junctions with gap junctions has led some investigators to doubt that the gap junctions are the only structures capable of forming the basis of cell-to-cell communication (Gilula *et al.*, 1970; Loewenstein, 1977; Gilula, 1978). The spectrum of variations on the theme of the septate junction seems to be expanding almost yearly. In many cases the subtle distinctions in function that structural differences must signify remain entirely speculative. It is to be hoped that future developments in technology will elucidate the numerous enigmas that currently vex those who are concerned with the structure and physiological significance of insect intercellular junctions.

Acknowledgments

I should like to thank Mr. William Lee, who has kindly prepared the photographic montages for me, Mr. John Rodford, who drew the junctional models, and Mrs. Vanessa Rule, who typed the manuscript.

References

Ashhurst, D. E., 1970, An insect desmosome. *J. Cell Biol.* **46:**421–425.

Berridge, M. J., and Gupta, B. L., 1967, Fine-structural changes in relation to ion and water transport in the rectal papillae of the blowfly, *Calliphora, J. Cell Sci.* **2:**89–112.

Berridge, M. J., and Gupta, B. L., 1968, Fine-structural localization of adenosine trisposphatase in the rectum of *Calliphora, J. Cell Sci.* **3:**17–32.

Binnington, K. C., and Lane, N. J., 1980, Perineurial and glial cells in the tick *Boophilus microplus* (Acarina: Ixodidae): Freeze-fracture and tracer studies, *J. Neurocytol.* **9:**343–362.

Boschek, C. B., 1971, On the fine structure of the peripheral retina and lamina ganglionaria of the fly, *Z. Zellforsch.* **118:**369–409.

Bullivant, S., 1978, The structure of tight junctions. In *Electron Microscopy 1978,* vol. III, *State of the Art, Symposia, Proc. 9th Int. Congress on Electron Microscopy,* edited by J. M. Sturgess, pp. 659-672. Imperial Press Ltd., Toronto, Canada.

Carlson, S. D., and Chi, C., 1979, The functional morphology of the insect photoreceptor, *Annu. Rev. Entomol.* **24:**379–416.

Casper, D. L. D., Goodenough, D. A., Makowski, L., and Phillips, W. C., 1977, Gap junction structures. 1. Correlated electron microscopy and x-ray diffraction. *J. Cell Biol.* **74:**605–628.

Caveney, S., 1978, Intercellular communication in insect development is hormonally controlled, *Science,* **199:**192–195.

Caveney, S., and Podgorski, C., 1975, Intercellular communication in a positional field. Ultrastructural correlates and tracer analysis of communication between insect epidermal cells, *Tissue Cell* **7:**559–574.

Chi, C., and Carlson, S. D., 1980a, Membrane specializations in the first optic neuropile of the housefly, *Musca domestica* L. I. Junctions between neurones, *J. Neurocytol.* **9:**429–449.

Chi, C., and Carlson, S. D., 1980b, Membrane specializations in the first optic neuropile of the housefly, *Musca domestica* L. II. Junctions between glial cells, *J. Neurocytol.* **9:**451–469.

Chi, C., and Carlson, S. D., 1981, Lanthanum and freeze fracture studies on the retinular cell junction in the compound eye of the housefly, *Cell Tissue Res.* **214:**541–552.

Chi, C., Carlson, S. D., and Ste. Marie, R., 1979, Membrane specializations in the peripheral retina of the housefly, *Musca domestica* L. *Cell Tissue Res.* **198:**501–520.

Claude, P., and Goodenough, D. A., 1973, Fracture faces of *zonulae occludentes* from "tight" and "leaky" epithelia, *J. Cell Biol.* **58:**390–400.

Dallai, R., 1976, Septate and continuous junctions associated in the same epithelium, *J. Submicrosc. Cytol.* **8:**163–174.

DeCamilli, P., Peluchetti, D., and Meldolesi, J., 1974, Structural differences between luminal and lateral plasmalemma in pancreatic acinar cells, *Nature* (London) **248:**245–247.

Decker, R. S., 1976, Hormonal regulation of gap junction differentiation, *J. Cell Biol.* **69:**669–685.

Dermietzel, R., Meller, K., Tetzlaff, W., and Waelsch, M., 1977, *In vivo* and *in vitro* formation of the junctional complex in choroid epithelium, *Cell Tissue Res.* **181:**427–441.

DiBona, D. R., 1972, Passive intercellular pathway in amphibian epithelia, *Nature New Biol.* **238:** 179–181.

Eley, S., and Shelton, P. M. J., 1976, Cell junctions in the developing compound eye of the desert locust, *Schistocerca gregaria. J. Embryol. Exp. Morphol.* **36:**409–423.

Elias, P. M., and Friend, D. S., 1976, Vitamin-A-induced mucous metaplasia: An in vitro system for modulating tight and gap junction differentiation, *J. Cell Biol.* **68:**173–188.

Fain-Maurel, M.-A., and Cassier, P., 1972, Un nouveau type de jonctions: Les jonctions scalariformes. Étude ultrastructurale et cytochimique, *J. Ultrastruct. Res.* **39:**222–238.

Fernández-Morán, H., 1958, Fine structure of the light receptors in the compound eyes of insects, *Exp. Cell Res., Suppl.* **5:**586–644.

Flagg-Newton, J. L., and Loewenstein, W. R., 1980, Asymmetrically permeable membrane channels in cell junction, *Science* **207:**771–773.

Flower, N. E., 1972, A new junctional structure in the epithelia of insects of the order Dictyoptera, *J. Cell Sci.* **10:**683–691.

Flower, N. E., and Filshie, B. K., 1975, Junctional structures in the midgut cells of lepidopteran caterpillars, *J. Cell Sci.* **17:**221–239.

Flower, N. E., and Walker, G. D., 1979, Rectal papillae in *Musca domestica*: The cuticle and lateral membranes, *J. Cell Sci.* **39**:167–186.

Gemne, G., 1969, Axon membrane crystallites in insect photoreceptors, in *Symmetry and Function of Biological Systems at the Macromolecular Level*, edited by A. Engstrom and B. Strandberg, pp. 305–309, Almqvist and Wiksell, Stockholm.

Gilula, N. B., 1973, Development of cell junctions, *Amer. Zool.* **13**:1109–1117.

Gilula, N. B. 1978, Structure of intercellular junctions. In *Intercelluar junctions and synapses*, vol. 2, *Receptors and Recognition*, Ser. B, edited by J. Feldman, N. B. Gilula, and J. D. Pitts, pp. 3–22, Chapman and Hall, London.

Gilula, N. B., Branton, D. and Satir, P., 1970, The septate junction: A structural basis for intercellular coupling, *Proc. Nat. Acad. Sci. USA* **67**:213–220.

Goodman, C. S., and Spitzer, N. C., 1979, Embryonic development of identified neurones: Differentiation from neuroblast to neurone, *Nature* **280**:208–214.

Green, C. R., Bergquist, P. R., and Bullivant, S., 1979, An anastomosing septate junction in endothelial cells of the phylum Echinodermata, *J. Ultrastruct. Res.* **68**:72–80.

Green, L. F. B., Bergquist, P. R., and Bullivant, S. 1980, The structure and function of the smooth septate junction in a transporting epithelium: The Malpighian tubules of the New Zealand glow worm *Arachriocampa luminosa*, *Tissue Cell* **12**:365–381.

Harrison, J. B., and Lane, N. J., 1981, Lack of restriction at the blood-brain interface in *Limulus* despite atypical junctional arrangements, *J. Neurocytol.* **10**:233–250.

Henkart, M., Landis, D. M. D., and Reese, T. S., 1976, Similarity of junctions between plasma membranes and endoplasmic reticulum in muscle and neurones, *J. Cell Biol.* **70**:338–347.

Houk, E. J., 1977, Midgut ultrastructure of *Culex tarsalis* (Diptera: culcidae) before and after a bloodmeal, *Tissue Cell* **9**:103–118.

Hull, B. E., and Staehelin, L. A., 1976, Functional significance of the variations in the geometrical organisation of tight junction networks, *J. Cell Biol.* **68**:688–704.

Lane, N. J., 1968, The thoracic ganglia of the grasshopper, *Melanoplus differentialis*: Fine structure of the perineurium and neuroglia with special reference to the intracellular distribution of phosphates, *Z. Zellforsch.* **86**:293–312.

Lane, N. J., 1972, Fine structure of a lepidopteran nervous system and its accessibility to peroxidase and lanthanum, *Z. Zellforsch.* **131**:205–222.

Lane, N. J., 1978a, Intercellular junctions and cell contacts in invertebrates. In *Electron Microscopy 1978*, vol. 3, *State of the Art. Symposia, Proc. 9th Int. Congress on Electron Microscopy*, edited by J. M. Sturgess, pp. 673–691, Imperial Press Ltd., Toronto, Canada.

Lane, N. J., 1978b, Developmental stages in the formation of inverted gap junctions during turnover in the adult horseshoe crab, *Limulus*, *J. Cell Sci.* **32**:293–305.

Lane, N. J., 1979a, Intramembranous particles in the form of bracelets or assemblies in arthropod tissues, *Tissue Cell* **11**:1–18.

Lane, N. J., 1979b, Tight junctions in a fluid-transporting epithelium of an insect, *Science*, **204**: 91–93.

Lane, N. J., 1979c, Freeze-fracture and tracer studies on the intercellular junctions of insect rectal tissues, *Tissue Cell* **11**:481–506.

Lane, N. J., 1979d, A new kind of tight junction-like structure in insect tissues, *J. Cell Biol.* **83**:82a.

Lane, N. J., 1980, Stages in the development of co-existing tight junctional P face ridges and gap junctional E face plaques: Two distinct intramembranous junctional particle populations, *J. Cell Biol.* **87**:198A.

Lane, N. J., 1981a, Vertebrate-like tight junctions in an insect, *Exp. Cell Res.* **132**:482–488.

Lane, N. J., 1981b, Tight junctions in arthropod tissues, *Int. Rev. Cytol.* **73**:243–318.

Lane, N. J., 1981c, Invertebrate neuroglia: Junctional structure and development, *J. Exp. Biol.* **95**:7–33.

Lane, N. J., 1981d, Studies on invertebrate cell interactions: Advantages and limitations of the morphological approach, *Proc. Symp. on Functional Organisation of Animal Tissues, British Soc. Cell Biol.*, edited by J. D. Pitts, Cambridge University Press, Cambridge. In press.

Lane, N. J., 1981e, Evidence for two separate categories of junctional particle during the concurrent formation of tight and gap junctions, *J. Ultrastruct. Res.* **77**:54–65.

Lane, N. J. and Chandler, H. J., 1980, Definitive evidence for the existence of tight junctions in invertebrates, *J. Cell Biol.* **86**:765–774.

Lane, N. J., and Harrison, J. B. 1978, An unusual type of continuous junction in *Limulus, J. Ultrastruct. Res.* **64**:85–97.

Lane, N. J., and Skaer, H. leB., 1980, Intercellular junctions in insect tissues, *Adv. Insect Physiol.* **15**:35–213.

Lane, N. J., and Swales, L. S., 1978a, Changes in the blood-brain barrier of the central nervous system in the blowfly during development, with special reference to the formation and disaggregation of gap and tight junctions. 1. Larval development, *Develop. Biol.* **62**:389–414.

Lane, N. J., and Swales, L. S., 1978b, Changes in the blood-brain barrier of the central nervous system in the blowfly during development, with special reference to the formation and disaggragation of gap and tight junctions. II. Pupal development and adult flies, *Develop. Biol.* **62**:415–431.

Lane, N. J., and Swales, L. S., 1979, Intercellular junctions and the development of the blood-brain barrier in *Manduca sexta, Brain Res.* **169**:227–245.

Lane, N. J., and Swales, L. S., 1980, Dispersal of gap junctional particles, not internalization, during the *in vivo* disappearance of gap junctions, *Cell* **19**:579–586.

Lane, N. J., and Swales, L. S., 1981, Stages in the assembly of pleated and smooth septate junctions in developing insect embryonic tissues, *J. Cell Sci.* in press.

Lane, N. J., and Treherne, J. E., 1971, The distribution of the neural fat body sheath and the accessibility of the extraneural space in the stick insect, *Carausius morosus, Tissue Cell* **3**:589–603.

Lane, N. J., and Treherne, J. E., 1972, Studies on perineurial junctional complexes and the sites of uptake of microperoxidase and lanthanum in the cockroach central nervous system, *Tissue Cell* **4**:427–436.

Lane, N. J., and Treherne, J. E., 1980, Functional organisation of arthropod neuroglia. In *Insect Biology in the Future—VBW 80,* edited by M. Locke and D. S. Smith, pp. 765–795, Academic Press, London.

Lane, N. J., Skaer, H. LeB., and Swales, L. S., 1977, Intercellular junctions in the central nervous system of insects, *J. Cell Sci.* **26**:175–199.

Lane, N. J., Swales, L. S., and Lee, W. M., 1980, Junctional dispersal and reaggregation: IMP reutilization? *Cell Biol. Int. Rep.* **4**:738.

Lane, N. J., Harrison, J. B., and Bowerman, R. F., 1981, A vertebrate-like blood-brain barrier, with intraganglionic blood channels and occluding junctions, in the scorpion, *Tissue Cell,* **13**: 557–576.

Larsen, W. J., 1977, Structural diversity of gap junctions: A review, *Tissue Cell* **9**:373–394.

Lasansky, A., 1967, Cell junctions in ommatidia in *Limulus, J. Cell Biol.* **33**:365–383.

Lawrence, T. S., Beers, W. H., and Gilula, N. B., 1978, Transmission of hormonal stimulation by cell-to-cell communication, *Nature* (London) **272**:501–506.

Loewenstein, W. R., 1977, Permeability of the junctional membrane channel. In *International* Cell Biology, edited by B. R. Brinkley and K. R. Porter, pp. 70–82, Rockefeller University Press, New York.

Loewenstein, W. R., and Kanno, Y., 1964, Studies on an epithelial (gland) cell junction. 1. Modifications of surface membrane permeability, *J. Cell Biol.* **22**:565–586.

Mills, J. W., Lord, B. A. P., and DiBona, D. R., 1976, Osmotic sensitivity of septate junctions in the crayfish midgut, *J. Cell Biol.* **70**:327A.

Noirot, C., and Noirot-Timothée, C., 1967, Un nouveau type de jonction intercellulaire (zonula continua) dans l'intestin moyen des insectes, *C. R. Hebd. Séances Acad. Sci. Ser. D. Sci. Nat.* **264**:2796–2798.

Noirot-Timothée, C., and Noirot, C., 1980, Septate and scalariform junctions in arthropods, *Int. Rev. Cytol.* **63**:97–140.

Noirot-Timothée, C., Smith, D. S., Cayer, M. L., and Noirot, C., 1978, Septate junctions in insects: Comparison between intercellular and intramembranous structures, *Tissue Cell* **10**:125–136.

Noirot-Timothée, C., Noirot, C., Smith, D. S., and Cayer, M. L., 1979, Jonctions et contacts intercellulaires chez les insectes. II. Jonctions scalariformes et complexes formes avec les mitochondries: Étude par coupes fines et cryofracture, *Biol. Cellulaire* **34**:127–136.

Peracchia, C., 1973a, Low resistance junctions in crayfish. I. Two arrays of globules in junctional membranes, *J. Cell Biol.* **57**:54–65.

Peracchia, C., 1973b, Low resistance junctions in crayfish. II. Structural details and further evidence for intercellular channels by freeze fracture and negative staining, *J. Cell Biol.* **57**:66–76.

Peracchia, C., 1974, Excitable membrane ultrastructure. I. Freeze fracture of crayfish axons, *J. Cell Biol.* **61**:107–122.

Peracchia, C., 1980, Structural correlates of gap junction permeation, *Int. Rev. Cytol.* **66**:81–146.

Pichon, Y., Moreton, R. B., and Treherne, J. E., 1971, A quantitative study of the ionic basis of extraneuronal potential changes in the central nervous system of the cockroach (*Periplaneta americana*), *J. Exp. Biol.* **54**:757–777.

Polak-Charcon, S., and Ben-Shaul, Y., 1979, Degradation of tight junctions in HT29, a human colon adenocarcinoma cell line, *J. Cell Sci.* **35**:393–402.

Porvaznik, M., Johnson, R. G., and Sheridan, J. D., 1979, Tight junction development between cultured hepatoma cells: Possible stages in assembly and enhancement with dexamethasone, *J. Supramol. Struct.* **10**:13–30.

Rapaport, S. I., Horn, M., and Klatzo, I., 1971, Reversible osmotic opening of the blood-brain barrier, *Science* **173**:1026–1028.

Raviola, E., Goodenough, D. A., and Raviola, G., 1980, Structure of rapidly frozen gap junctions, *J. Cell Biol.* **87**:273–279.

Schnapp, B., and Mugnaini, E., 1975, The myelin sheath: Electron microscopic studies with thin sections and freeze-fracture. In *Golgi Centennial Symposium Proceedings*, edited by M. Santini, pp. 209–233, Raven Press, New York.

Shaw, S. R., 1978, The extracellular space and blood-eye barrier in an insect retina: An ultrastructural study, *Cell Tissue Res.* **188**:35–61.

Shinowara, N. J., Beutel, W. B., and Revel, J. P., 1980, Comparative analysis of junctions in the myelin sheath of central and peripheral axons of fish, amphibians and mammals: A freeze-fracture study using complementary replicas, *J. Neurocytol.* **9**:15–38.

Simpson, I., Rose, B., and Loewenstein, W. R., 1977, Size limit of molecules permeating the junctional membrane channels, *Science* **195**:294–296.

Skaer, H., leB., Harrison, J. B., and Lee, W. M., 1979, Topographical variations in the structure of the smooth septate junction, *J. Cell Sci.* **37**:373–389.

Skaer, H. leB., Lane, N. J., and Lee, W. M., 1980, Junctional specializations of the digestive system in a range of arthropods, *Eur. J. Cell Biol.* **22**:245.

Swales, L. S., Lane, N. J., and Schofield, P., 1981, Glial and perineurial cells during the development of the blood-brain barrier in embryonic locust CNS, in preparation.

Toshimori, K., Iwashita, T., and Oura, C., 1979, Cell junctions in the cyst envelope in the silkworm testis, *Bombyx mori* Linne, *Cell Tissue Res.* **202**:63–73.

Treherne, J. E., Schofield, P. K., and Lane, N. J., 1973, Experimental disruption of the blood-brain barrier system in an insect (*Periplaneta americana*), *J. Exp. Biol.* **59**:711–723.

Wiley, C. A., and Ellisman, M. H., 1980, Rows of dimeric particles within the axolemma and juxtaposed particles within glia, incorporated into a new model for the paranodal glial-axonal junction at the node of Ranvier, *J. Cell Biol.* **84**:261–280.

Wood, R. L., 1959, Intercellular attachment in the epithelium of *Hydra* as revealed by electron microscopy, *J. Biophys. Biochem. Cytol.* **6**:343–352.

Wood, R. L., and Kuda, A. M., 1980, Formation of junctions in regenerating *Hydra*: Septate junctions, *J. Ultrastruct. Res.* **70**:104–117.

Zampighi, G., Corless, J. M., and Robertson, J. D., 1980, On gap junction structure, *J. Cell Biol.* **86**:190–198.

15

Selectivity in Junctional Coupling between Cells of Insect Tissues

STANLEY CAVENEY AND ROBERT BERDAN

1. *Introduction*

Ionic and metabolic coupling between the cells of compact tissues depends on the presence of a specialized membrane structure—the gap junction. Through this junction inorganic ions and small metabolites may move rapidly from cell to cell, providing a channel for internal tissue homeostasis and synchronization. The gap junction is present in the plasma membranes of nonexcitable cells in most compact animal tissues. The ubiquity of this "ancient hole" (Loewenstein, 1978) is underscored by the early finding that cell lines derived from heterologous tissues, when cocultured *in vitro*, showed little specificity in the formation of the communication channels (Michalke and Loewenstein, 1971). Any "communication-competent cell"—a cell capable of synthesizing and inserting normal gap junctional elements into its plasma membrane—may couple with any other communication-competent cell not physically separated from it. Even differentiated cells from different vertebrate tissues form *heterotypic* gap junctions when cocultured (Lawrence *et al.*, 1978).

Coupling specificity *in vitro* is seen, however, when there is a wide phylogenetic gap between the two cell types being cocultured: insect cells will not couple with vertebrate cells *in vitro* (Epstein and Gilula, 1977). But since insect and vertebrate gap junctions are both structurally and physiologically distinct, this is hardly surprising.

Of greater physiological interest are the few recorded examples of coupling specificity between cells from different insect orders [homopteran cell lines will not couple with a lepidopteran cell line (Epstein and Gilula, 1977)] or between heterologous vertebrate cells [epithelial cells do not couple with fibro-

STANLEY CAVENEY AND ROBERT BERDAN • Department of Zoology, University of Western Ontario, London, Ontario, Canada N6A 5B7.

blasts *in vitro* (Fentiman *et al.*, 1976; Pitts and Burk, 1976)] and the recent detection of directional selectivity in the transfer of molecules between heterotypic cell pairs in culture (Flagg-Newton and Loewenstein, 1980). These findings hint that *in vivo* the cells of complex aggregates may form several discrete communication compartments rather than the single one suggested by the early work. However, it has recently been suggested that specificity in junction formation in vertebrate cells *in vitro* may be largely a property of the established cell lines used rather than the *in vivo* condition (Pitts, 1980). To date, where specificity in junctional coupling has been demonstrated between undifferentiated cells *in vitro*, it is interphyletic, interspecific, or, where intraspecific, between tissue cells derived from different germ layers.

In our opinion, the apparent lack of junctional specificity *in vitro* does not reflect the situation within tissues *in vivo*. Although the cells of several complex tissues in insects do form heterotypic gap junctions and are coupled (Table 1), cells within, say, a complex epithelium may often form select pathways for intercellular ion and metabolite movement. Most mature insect tissues are polymorphic in that they consist of several subpopulations of structurally distinct cells with apparently different functions. If the cells are all in junctional communication, it follows that the metabolic activity of one cell type might directly influence that of the others.

Conceivably, in certain tissues where the activity of the various cell types is synchronous and not very different (such as the different cells that contribute to the cuticle pattern) this may be advantageous, but typical complex tissues are comprised of cells having both temporally and biochemically distinct patterns of activity. Here, ionic and metabolic coupling between different subpopulations may be of little value, although coupling within a subpopulation of cells may be physiologically important. Consequently, *selectivity in junctional communication* may be an important requirement for normal tissue function. Cells within a simple tissue must choose with which of their neighbors, and when, they will form gap junctions to coordinate their metabolism. The communication compartments established by selective junctional coupling could be both temporal and spatial in nature.

Spatial selectivity may result from (1) a partial or complete loss of coupling competence within a subpopulation of tissue cells (a partial loss of coupling—a "leaky" compartment—could result from either a symmetrical or an asymmetrical reduction in the molecular permeability of the coupling channel at the compartment border, whereas the complete loss of coupling could result from one population of cells becoming communication-incompetent); or (2) the ability of different communication-competent cell populations to distinguish between the different cell types with which they are in membrane contact and select the cell type(s) with which they will couple. (If the gap junction phenotype is similar or identical in the different subpopulations, selectivity would likely be based on other cell surface properties.)

Temporal selectivity in junctional coupling is seen when (1) tissue cells form homotypic or heterotypic gap junctions at certain times of development only; or (2) where the gap junctions are temporally stable, their permeability properties may be under physiological (possibly hormonal) control.

Table 1. Heterocellular Coupling in Insect Tissues

Tissue cell types coupled	Species	1. Junctions present[a]	2. Ionic coupling	3. Tracer movement	References
Epidermis					
i. Pit gland cell–general epidermis	*Tenebrio molitor*	und.	+	+	2,3: Own data
ii. Scent gland cell–general epidermis	Several moths	gj (+)	und.	und.	1: Percy, personal communications
iii. Sensilla accessory cell–general epidermis	*Tenebrio molitor*	und.	+	+	2,3: Own data
	Calliphora erythrocephala	gj psj (+)	+	und.	1: Keil, 1978; 2: Thurm and Küppers, 1980
Salivary gland					
Giant cell–flat cell	*Chironomus thummi*	(gj)[b] psj (+ −)	+	+	Rose, 1971
Midgut					
Goblet cell–columnar cell	*Anagasta kühniella*	gj ssj (+)	und.	und.	1: Smith *et al.*, 1969; Flower and Filshie, 1975
	Hyalophora cecropia	gj (+)	+	und.	1: Smith, 1968; 2: Blankemeyer and Harvey, 1978
Rectum					
Principal cell–basal cell	*Periplaneta americana*	gj psj (+ −ff)	+	+	1: Noirot and Noirot-Timothée, 1976
Ovary					
Follicular cell–oocyte	*Locusta migratoria*	gj (+)	+	und.	1: Wollberg *et al.*, 1976
	Hyalophora cecropia	gj (+)	+	+	1: Woodruff, 1979
	Rhodnius prolixus	gj (+ −ff)	und.	+	1: Huebner, 1981 2,3: Own data
Testis					
Cyst envelope–spermatocytes	*Anagasta kühniella*	gj (+)	und.	und.	1: Szöllösi and Marcaillou, 1980

[a] gj, gap junction; psj, pleated septate junction; ssj, smooth septate junction. Determined in positively stained (+) or negatively stained (−) sections or in freeze-fracture replicas (ff). und, undetermined.

[b] It is almost certain that gap junctions occur between these cells, although not detected by this author.

In searching for coupling selectivity in the cells of complex insect tissues, we selected those in which the structure of the membrane junctions is well documented and have attempted to correlate the ultrastructural data with electrophysiological measurements of ionic coupling and the movement of fluorescent tracers injected into the constituent cells. This eclectic approach is not intended to be exhaustive. We have purposely excluded examples from excitable tissues, despite electrophysiological evidence for selective electrotonic synapses in, for example, the photoreceptors in the compound eye (Shaw, 1969; 1979), where the membrane channel for photoreceptor coupling may not be the typical insect (E-face) gap junction (Chi and Carlson, 1980; S. Shaw, personal communication).

Most types of spatial and temporal selectivity in coupling mentioned above are seen in insect tissues *in vivo*. The complex epithelia of integument, midgut, and hindgut are excellent systems in which to study spatial selectivity, while temporal changes in coupling are seen in the ovary, testis, and hemocytes.

2. *Gap Junction Structure and Physiology*

The fundamental importance of a direct cell-to-cell membrane channel in metabolic cooperation and tissue homeostasis, as well as its possible role in growth control and differentiation, is the reason behind the current intense investigation of the gap junction (see reviews by Bennett and Goodenough, 1978; Loewenstein, 1979; Peracchia, 1980). Although most research is focused on the vertebrate gap junction, the structure and physiology (but not the biochemistry) of the insect gap junction is well characterized.

The gap junction, typically disc-shaped or macular in appearance, obtains its name from the presence, in sectioned material, of a small (1–3 nm) extracellular gap that separates the membranes of two tightly apposed cells (Figures 1, 2). This gap distinguishes the gap junction from the tight or occluding junction (see Lane, this volume) with which it has been confused in many early papers on insect junction ultrastructure. In tissues impregnated with the extracellular tracer lanthanum, the gap appears somewhat larger (5 nm in the insect epidermis, for example) and in *en face* view, the gap junction is composed of a clustered array of electron-lucent annular particles called "connexons" (Figure 3). The particles in each membrane protrude into the extracellular gap to link coaxially with those of the other membrane. Each particle pair is thought to contain a central channel that spans the extracellular gap to connect the interiors of the two cells. The vertebrate gap junction particle and that of the electrical synapse in the crayfish septate axon is a hexamer of six identical subunits (Peracchia, 1980; Zampighi, 1980). The symmetry of the insect gap junction is unknown.

When freeze fractured, a major distinction between the arthropod gap junction and that of vertebrates and molluscs is evident. The intramembranous particle clusters of the vertebrate gap junction remain attached to the P (protoplasmic) face of the fracture plane, whereas the particles of the arthropod gap junction adhere to the E (exoplasmic) face (Figure 5). In both instances, pits

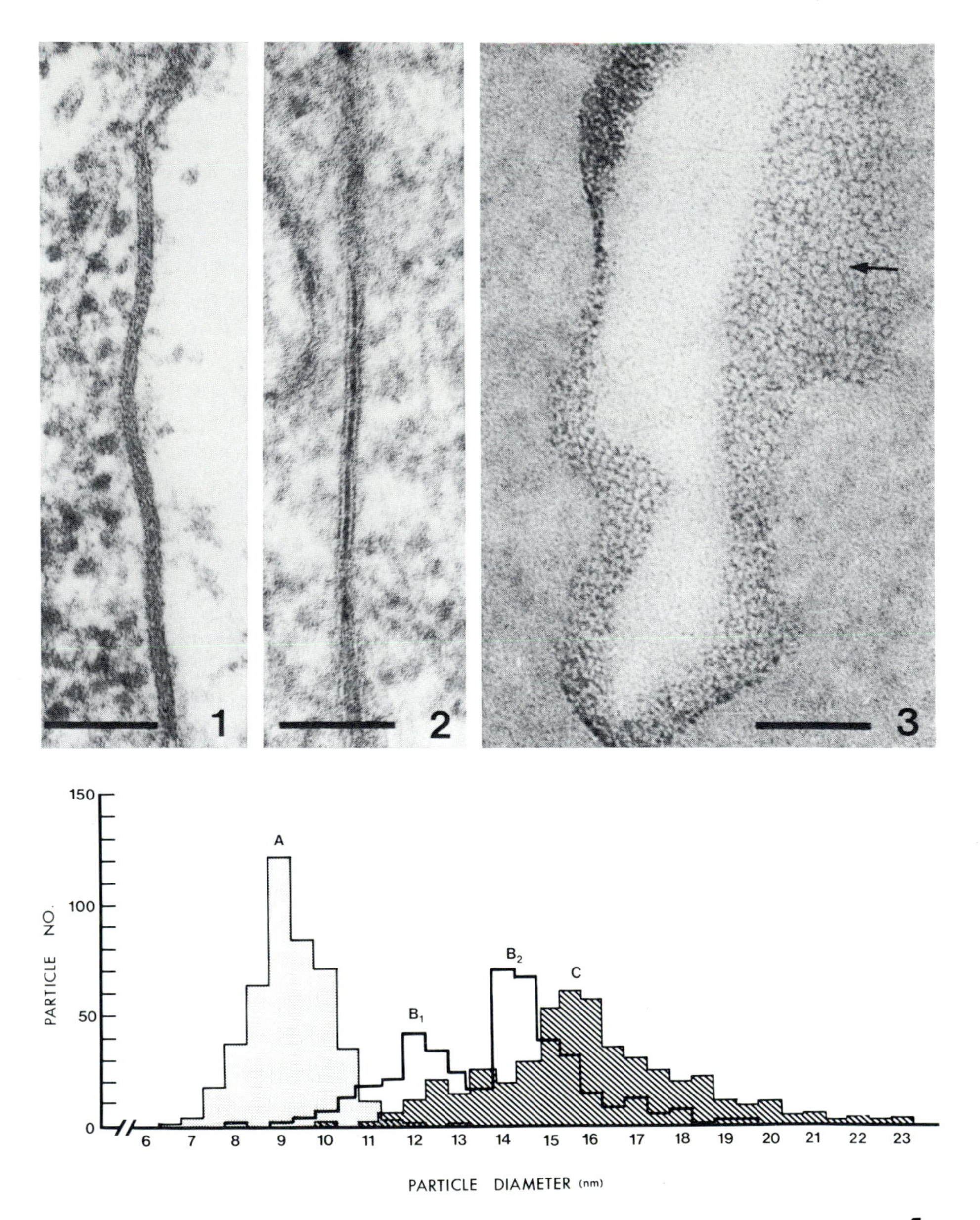

Figures 1–3. The appearance of gap junctions in sectioned material from the epidermis of *Tenebrio molitor*. (Scale bars = 200 nm.)

Figure 1. A transverse section of a gap junction stained with uranyl acetate.

Figure 2. Similar aspect as Figure 1, but the extracellular gap is accentuated by staining with potassium permanganate.

Figure 3. *En face* view of a gap junction after lanthanum has penetrated the extracellular space. The gap junction appears as a noncrystalline array of electron-lucent particles, many with a central electron-dense spot (arrow).

Figure 4. A histogram comparing particle size distribution in freeze-fracture replicas of gap junctions from a vertebrate tissue. A, rabbit liver; B, *Tenebrio* fat body; C, epidermis.

corresponding to the particles are seen in the complementary fracture (Figure 8). Consequently, the gap junctions of arthropods are termed E-type gap junctions, those of vertebrates P-type gap junctions. The gap junctional particles of insect cells are considerably larger (average 13–15 nm) and more variable in size than the 8–10-nm particles characteristic of vertebrate gap junctions (Figure 4). Several particle size classes may coexist in a single gap junction in insect cells (Figures 5, 6). The precise measurement of particle size in these classes is affected by their packing density, since at high density adjacent particles appear to fuse (Figure 7). As packing density rises, the ability to detect smaller classes of particles diminishes as well. Consequently, the more tightly packed particles of the epidermal gap junctions (Figure 7) have a larger (apparent) mean size than those of the more loosely packed fat body gap junctions (Figure 5), which therefore seems to have a higher proportion of small particles. The difference in size of B_2 (fat body) and C (epidermal) particle populations (Figure 4) is not significant. The membranes of both tissues do appear to contain two classes of gap junctional particles, however. (The influence of packing density on measured particle size is also supported by our unpublished data for the apical and basal populations of gap junction in the insect epidermis. The particles of the basal gap junctions are more loosely packed and have smaller mean size.) In both insect and vertebrate cells the packing lattice of the particles is generally noncrystalline, although hexagonal packing has been reported in several cases where the cells have been placed in conditions that ionically uncouple the cells prior to fixation and freeze fracture (Peracchia, 1980).

The structural differences outlined above appear to be supported by physiological studies that compare the permeability properties of the cell-to-cell membrane channel in arthropod and vertebrate cells. The arthropod channel has a greater permeability; if the gap junction is the site of this channel (as generally assumed), the larger arthropod gap junction particle may form a wider channel for the intercellular movement of metabolites than the smaller particle of the vertebrate junction. By injecting fluorochrome-tagged peptides and oligosaccharides of known dimensions and molecular weights into fly salivary gland cells, Rose (1980) has shown the junctional channel in insect cells to have a diameter greater than 2 nm. A molecule with a maximum abaxial dimension less than this value may diffuse directly from cell to cell, and transfer of molecules up to molecular weight 3000 is reported. On the other hand, the mammalian junctional channel has a much smaller cut-off limit. Weakly charged peptide tracers of molecular weight greater than 901 do not leave the injected cell, and the channel is even less permeable to strongly charged peptides (Flagg-Newton, 1980). Because of this charge discrimination, a precise estimate of channel diameter is not available, although it is probably less than 1.4 nm.

Although there is a strong correlation between ionic and dye coupling between cells and the presence of gap junctions in their plasma membranes, the detection of gap junctions between adjoining cells does not necessarily imply that the cells are physiologically coupled. The permeability of both insect and vertebrate channels is modulated intracellularly by the local concentration of one or more inorganic ions. H^+, Ca^{++}, and Mg^{++} have been demonstrated to close the junctional channel. External factors, such as hormones,

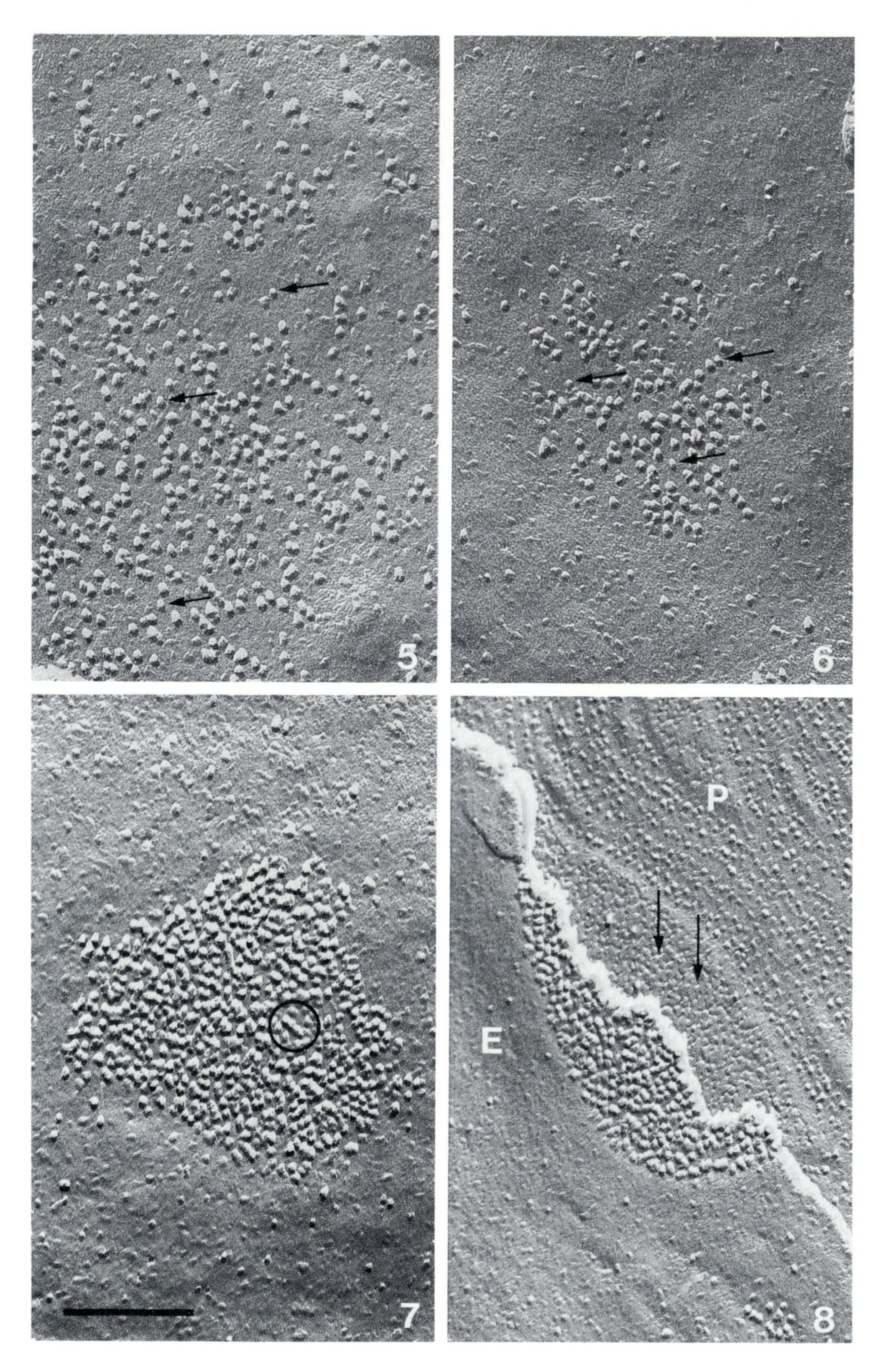
5
6
7
P
E
8

may also regulate junctional conductivity (see section 3). The close interdependence of H^+ and Ca^{++} inside cells complicates any interpretation of which ion has primary control on the junctional channel. The insect channels, however, appear to be primarily sensitive to the concentration of local cytosolic Ca^{++}, since cells may be uncoupled by elevated Ca^{++} levels in the absence of changes in intracellular pH (Rose and Rick, 1978). We mention this because temporal changes in the permeability of the junctional channels may take place without obvious structural changes in the membrane. For this reason, studies on structural coupling between cells should be supplemented with direct evidence for functional coupling. Under normal physiological conditions, the electrophysiological detection of low-resistance ion pathways between cells is the most sensitive indicator of functional coupling. For example, in the tracheal oenocytes strong ionic coupling is recorded between the cells (Figure 9), but our ability to detect tracer movement and gap junctions (Figure 11) in this tissue is far more limited, as tracer movement appears to be absent (Figure 10). In general, this review discusses those insect tissues for which both structural and functional coupling data are available.

3. Spatial Selectivity in Junctional Coupling

3.1. The Epithelium of the Integument

The integument is a complex and versatile insect tissue. A vast array of specialized cell types is incorporated into the basic framework of generalized epidermal cells that secrete the cuticle. These specialized cells, derived from epidermal cells during development, include: various sensilla scattered nonrandomly throughout the epidermis; glandular cells with either direct or indirect access to the outer cuticle surface; clusters of dermal oenocytes; nonennervated scales and hairs; and tendon cells that link skeletal muscle fibers to the rigid cuticle. Many specialized cells are tightly integrated into the epidermal monolayer and synthesize elements of the cuticle pattern as well as contribute to the transepithelial barrier. In other instances, specialization appears to involve the loss of functional coupling with the epidermis, although the cells remain in close proximity.

←

Figures 5–7. The freeze-fracture appearance of three gap junctions from the same piece of larval fat body of *Tenebrio*. Note the considerable variation in gap junction morphology and in particle size. On the exoplasmic (E) face, smaller particles (arrows) are dispersed among a population of larger particles. Scale bar, 200 nm.

Figure 5. A loosely packed cluster of particles.

Figure 6. A small cluster of more tightly packed particles.

Figure 7. A "mature" gap junction. When the particles are tightly packed they often fuse (circle).

Figure 8. An epidermal gap junction, showing both protoplasmic (P) and exoplasmic (E) fracture faces, surrounded by septate junction. The septate junction appears as rows of 11.5-nm particles on the P face; on this face pits corresponding to the E-face particles of the gap junction are also visible (arrows). Scale as in Figures 5 to 7.

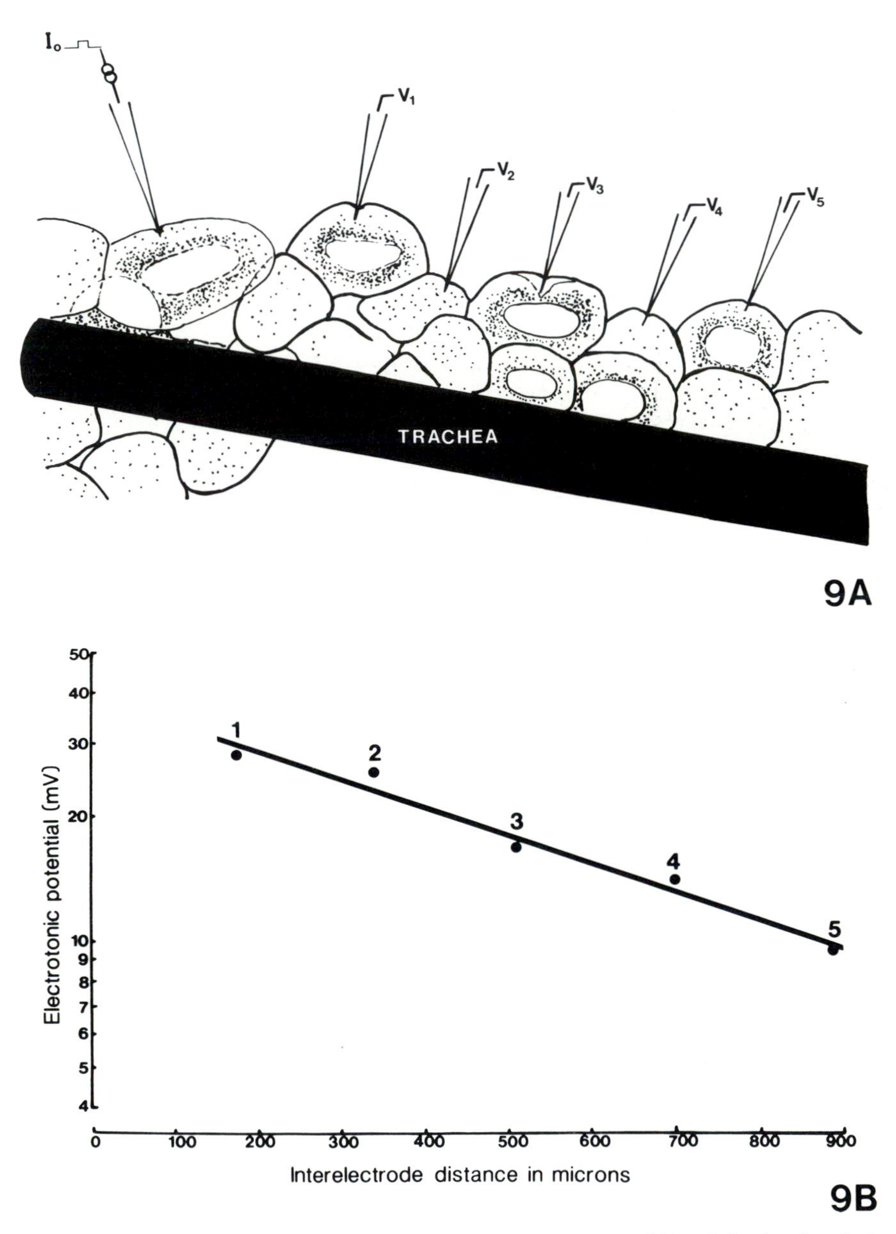

Figure 9. (A.) Analysis of ionic coupling in the tracheal oenocytes of *Tenebrio*. An electrical current injected into the cell on the left (I_0 = 60 nA) causes a change in membrane potential in the other cells along the row. (B.) This electrotonic effect is an exponential function of the distance from the injected cell.

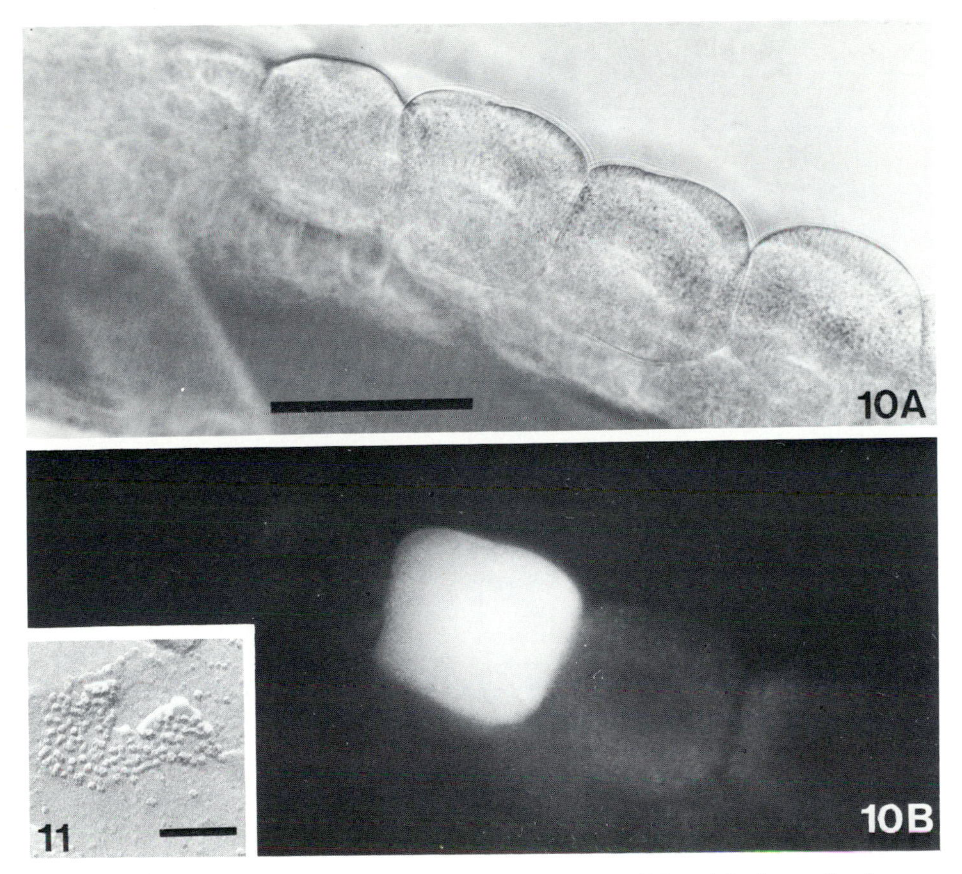

Figure 10. Failure to detect tracer movement between oenocytes. Although ionic coupling between the cells is strong (coupling ratio $V_2/V_1 \geqslant 0.7$), carboxyfluorescein injected into one cell does not appear to pass into adjacent cells. In A the oenocytes are seen in bright field; in B they appear in dark field fluorescence. The bright (green) fluorescence of the tracer fills the cell into which it is injected, the other cells have only a low (red) autofluorescence. 30 min later, tracer fluorescence in the injected cell had dropped considerably, presumably owing to it passing to adjoining cells at levels below our detection threshold. The sensitivity of electrotonic methods for detecting functional coupling between cells is far greater than that of tracer spread. On its own, our inability to detect tracer movement may have led to the erroneous conclusion that the cells were not coupled. Scale for A and B, 100 μm.

Figure 11. An oenocyte gap junction. These junctions are small and found only infrequently. (Scale bar = 100 nm.)

3.1.1. The General Epidermis

The epidermis (Figure 13) is a tightly coupled functional syncytium in which gap junctions are prevalent (Figures 1–3). They are embedded in the apical region of the lateral borders between the cells, surrounded by septate junctions (Caveney and Podgorski, 1975; Lawrence and Green, 1975); in the basal region they are surrounded by extracellular space (Lawrence and Green, 1975; Caveney *et al.*, 1980). Gap junctional content ranges between 6% [*Onco-*

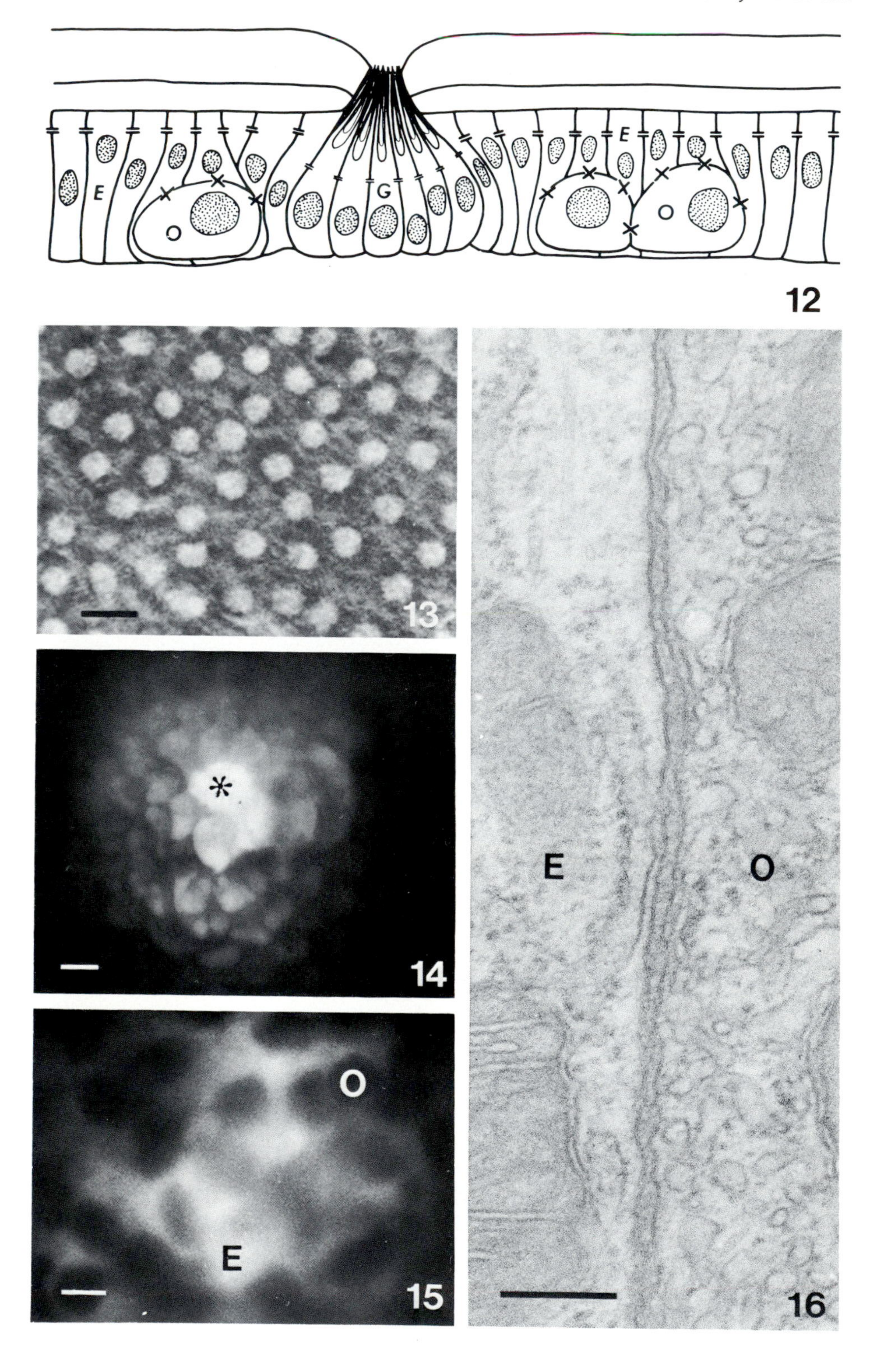
E
G
E
O
O
12
13
*
14
O
E
15
E
O
16

peltus fasciatus, mid-fifth stage nymph (Lawrence and Green, 1975)] and 14% [newly ecdysed *Tenebrio molitor* (Caveney and Podgorski, 1975)] of the total interface between adjoining epidermal cells. Predictably, ionic coupling between epidermal cells is well developed in the larva (Caveney and Blennerhassett, 1980) and this coupling persists through to adult life. That the epidermal cells form a common intracellular ionic compartment is also reflected in the uniformity of their membrane potentials (Caveney, 1974). Fluorescent tracers (MW < 625) injected into the epidermis move away from the electrode tip in a radial fashion within the epidermal plane (Figure 14), showing the cells to be coupled equally as well axially as laterally (Caveney, 1974; Caveney and Podgorski, 1975). Ionic, and presumably metabolic, coupling in the epidermis, fluctuates during larval and prepupal development, but is never lost (Caveney, 1978). Such modulation in ionic coupling may be interpreted in several ways: the amount of gap junction in the membranes may change, or the ionic conductivity of relatively stable membrane junctions may vary. Both mechanisms appear to operate in the developing beetle epidermis. The amount of gap junction in the epidermal membranes in the newly ecdysed *Tenebrio* larva is three times that seen in the midinstar larva. This increase in gap junctional content involves the enlargement of preexisting junctions as well as the formation of new ones (Caveney and Podgorski, 1975). In *Oncopeltus*, too, extra gap junctions develop in the intersegmental region of the epidermis prior to the metamorphic molt (Lawrence and Green, 1975). Yet junctional conductivity in the beetle epidermis may be raised by 66% by exposing the cells to 20-hydroxyecdysone *in vitro* (Caveney and Blennerhassett, 1980), a process that does not appear to involve gap junction growth (Caveney *et al.*, 1980). Thin-section analysis of the junctional membrane revealed that its gap junction content remained constant under a variety of experimental conditions, and recent freeze-fracture data show that the elevated conductivity induced by hormone is not associated with changes in particle size or packing in the gap junction (R. B., unpublished). Our working hypothesis is that 20-hydroxyecdysone either opens preexisting closed gap junction channels, or, if the channels are all open to start with, increases the diameter of the channels. We are currently attempting to distinguish between these two possibilities.

←

Figures 12–16. Spatial selectivity in functional coupling in the epidermis of *Tenebrio molitor*.
Figure 12. A diagram summarizing the experimental findings. A region of adult male integument is shown, in which a multicellular pit gland (G) and dermal oenocytes (O) are embedded in the general epidermis (E). The epidermal cells are stongly coupled (⫩), pit gland cells weakly coupled (⫲) both to each other and to the general epidermis, while the oenocytes lack functional coupling altogether (*).
Figures 13 and 14. Larval epidermis. At this developmental stage, no oenocytes or pit glands are present, and tracer injected into an epidermal cell (*) passes uniformly through the sheet. In Figure 13 the epidermal cells were photographed under phase contrast; Figure 14 shows the dark field fluorescence of transfered dye. (Scale bars = 10 μm.)
Figure 15. In the sternite epidermis of the adult female, tracer injected into the epidermis fails to pass into the oenocytes (O). Consequently they appear as dark areas against a background of fluorescing epidermal cells. (Scale bar = 10 μm.)
Figure 16. No septate or gap junctions are present at the site of contact between oenocyte (O) and epidermal cell (E), although the cells are very close together. (Scale bar = 200 nm.)

As expected of a single-layered epidermis, its secretory product, the cuticle, is of multilaminate construction (see Neville, 1975). It is tempting to suggest that a major metabolic function of the gap junctions in this tissue is to provide a means whereby the activity of the epidermal cells is strongly coordinated and the deposition of the various cuticle layers precisely synchronized. The deposition of the epicuticular layers coincides with a major peak in blood 20-hydroxyecdysone titer in several insects (Dean *et al.*, 1980); our electrophysiological data suggest that junctional conductivity is maximal at this time. Specialized cells involved in the deposition of other elements of the cuticle pattern—accessory cells of the sensillum, tracheal cells, duct cells of certain dermal glands, and tendon cells at apodemes—would, according to this argument, be coupled to the adjoining general epidermis, at least at the time of cuticle secretion.

The converse, that specialized cells in the integument not involved in cuticle secretion be less tightly coupled to he epidermis, need not hold. Yet it is noteworthy that certain classes of glandular cells appear to be poorly coupled to the epidermis, or may even be communication-incompetent.

3.1.2. Dermal Gland and Oenocytes

A vast variety of glandular structures are found in the insect integument, ranging from dispersed to dermal glands to specialized organs such as defensive and scent glands. In spite of the great diversity in structure and function, Noirot and Quennedey (1974) were able to divide glandular units into three classes, based on the relationship of the gland cell to the overlying cuticle and the route of release of the glandular product to the exterior. In so doing, they noted that simple gland cells had the normal junctional specializations seen in the general epidermis, but more complicated glandular units often had much-reduced junctional complexes. Our aim in this section is to attempt to relate the function and structure of the epidermal gland to the quantity of the gap junctions seen between component cells.

The three classes established by Noirot and Quennedey are: Class 1 gland cells, in which the gland cell is directly attached to cuticle that it has previously secreted, and the glandular secretion must cross this barrier; Class 2 gland cells, in which the gland cells are not in direct contact with either the cuticle or the basal lamina, but are surrounded both apically and basally by extensions of the epidermal monolayer; and Class 3 gland cells, in which a cuticular duct, produced by one or more duct cells, provides a canal for the release of the secretion of one or more gland cells to the exterior. Each class of glandular unit may occur in isolation or clustered into single- or multiple-class groupings in the integument.

3.1.2a. Class 1 Gland Cells (*Unicellular Secretory Units*). In this class of epidermal gland, the columnar secretory cells are little more than slightly modified epidermal cells. The membrane junctions are identical to those of the surrounding epidermis. The sex pheromone glands that lie in the abdominal intersegmental membranes of various moths are clusters of Class 1 gland cells. Gap junctions have been reported between the pheromone gland cells in *Chor-*

istoneura fumiferana (Percy, 1974) and are also commonly seen between gland cells and the adjacent intersegmental epidermis in several moths (Percy, personal communication).

The pit glands in the integument of the adult *Tenebrio molitor* (Wigglesworth, 1948) appear to be slightly more complex Class 1 glands (Figure 12). The cuticle above each gland is perforated by a cluster of pores. Each pore is connected to a single pit gland cell. The pit gland cells differentiate in the pupa but do not become maximally active until a week or so after adult emergence, when they release a secretion associated with mating. The pit gland cells grow considerably during the first week of adult life, at a time when the general epidermis is secreting the bulk of the cuticle (Caveney, 1970). This similarity in their periods of synthetic activity may explain why cells of these two types, although producing very different secretions, remain weakly coupled. Carboxyfluorescein injected into epidermis adjacent to a pit gland slowly enters the cells of the gland, but long after the dye has spread widely throughout the epidermis. Ionic and dye coupling suggest that the pit gland cells are poorly coupled, not only to each other but also to the surrounding epidermis (Figure 12). A partial loss of coupling has occurred in association with cell diversification in the epidermis.

3.1.2b. Class 2 Gland Cells and Dermal Oenocytes. This gland type, *sensu strictu*, is only reported from the sternal gland of several termite species (see Noirot and Quennedey, 1974), but because of its similarity to the dermal oenocyte, we shall discuss them together. Both Class 2 gland cells and oenocytes lack specialized ducts; they have no direct contact with the overlying cuticle, being partially or totally enveloped by the surrounding epidermis; they are relatively unpolarized in shape with few morphological indicators of apical-basal polarity; and finally, the plasma membranes in both lack junctions. Because of their location and unpolarized nature, it is possible that these cells may release their secretion for export to the cuticle (and beyond) or into the hemolymph for transfer to internal organs, as required by the insect. Secretion destined for the cuticle is thought to pass through the overlying epidermis; the absence of gap junctions between the Class 2 gland cell and the epidermis removes the possibility of a directed cytoplasmic channel being used. An exocytotic mechanism for secretion release is most likely.

The dermal oenocytes in the larva of *Rhodnius prolixus* release stored granules in a cyclical fashion coincident with lipid deposition in the forming epicuticle and exocuticle; in the reproductive female the granules are released into the blood and pass to the follicular epithelium of the ovary during eggshell formation (Wigglesworth, 1970). Wigglesworth asserts that the dermal oenocytes are the site of lipid and protein carrier synthesis for cuticulin production.

In *Tenebrio*, the dermal oenocytes first appear in the sternite epidermis (from which they are derived) during the early prepupal period (Wigglesworth, 1948; Caveney, 1973). Oenocytes lose their junctional specializations early in their differentiation, and probably because of this they are squeezed away from the apical region of the epidermis to lie largely underneath the epidermis close to, but generally not against, the basal lamina (Figure 12). These primary oenocyte cells undergo a second and occasionally a third round of cell division during the prepupal period to give rise to oenocyte pairs and tetrads that survive well into adult life. Within hours of the epidermal differentiative division that gives rise to the primary oenocyte, no gap junctions or septate junctions are detectable between the oenocytes and the adjacent epidermis (Figure 16). If microelectrodes are placed in the oenocytes and epidermis at this time, no ionic coupling is recorded between these two cell types. Microelectrodes placed into oenocyte pairs indicated that they too, were uncoupled. (Occasionally, apparent coupling was detected between oenocyte pairs, but this we attribute to the presence of persistent cytoplasmic bridges). Carboxyfluorescein does not pass out of an injected oenocyte into either a contiguous oenocyte or the epidermis. Conversely, if the fluroescent dye is injected into an epidermal cell, the oenocytes are silhouetted because of their inability to take up the dye from the surrounding epidermis (Figure 15).

All evidence suggests that Class 2 gland cells and dermal oenocytes are communication-incompetent.

3.1.2c. Class 3 Glandular Units (Multicellular Secretory Units). Glandular units of this class originate from a tetrad of epidermal cells during the molt cycle (Sreng and Quennedey, 1976). Between one and three of these cells degenerate during or after gland differentiation. In the socket gland of *Dysdercus fasciatus,* a duct system is not developed, and only a single secretory cell persists after ecdysis (Lawrence and Staddon, 1975). More commonly, one or two other specialized accessory cells also remain after ecdysis to form elements of the complex duct system (end apparatus, bulb or saccule, duct canal, etc.). A peripheral enveloping cell is sometimes present, separating the glandular unit from the surrounding epidermis. Although the main function of these cells is in the production of the duct system, the duct cell nearest the primary secretory cell may also have a secretory function, as in the dermal glands of *Tenebrio* (Delachambre, 1973). (For general reviews of gland structure and ontogeny, refer to Noirot and Quennedey, 1974; Percy and Weatherston, 1974; Sreng and Quennedey, 1976; and Staddon, 1979).

In spite of many detailed accounts of Class 3 gland ultrastructure, knowledge of the junctional specializations between the component cells is fragmentary. Septate junctions and desmosomes link the secretory cell tightly to its adjacent accessory cell (Noirot and Quennedey, 1974; Barbier, 1975; Sreng and Quennedey, 1976; Lococo and Huebner, 1980), yet with rare exceptions (Filshie and Waterhouse, 1968; Crossley and Waterhouse, 1969) tightly apposed secretory cells of different units are not connected by septate junctions (Quennedey, 1971; Bonnanfant-Jais, 1974; Lococo and Huebner, 1980). This implies that the secretory cells can discriminate among cell types of equal junctional competence (but different ancestry) during septate junction formation—a property that may tell us something about coupling selectivity in gap junction formation. In the milk gland of *Glossina austeni* (Bonnanfant-Jais, 1974) and the honeybee venom gland (Bridges, 1979), spot desmosomes, maculae adhaerentes, connect adjacent secretory cells. Septate junctions are found between the various accessory cells (Delachambre, 1973) and between the peripheral enveloping cell and the general epidermis in most Class 3 dermal glands (Lai-Fook, 1970; Sreng and Quennedey, 1976).

Gap junctions have not been detected in the membranes of the secretory cells of Class 3 units. In common with Class 2 gland cells and dermal oenocytes, these secretory cells may be communication-incompetent. Gap junctions, although rare, do occur in the membranes of the duct cells: In the *Rhodnius* cement gland, autocellular gap junctions connect regions of the same duct cell (Lococo and Heubner, 1980). Adjacent duct cells in parallel in the bee venom gland are also linked by gap junctions (Bridges, 1979). Considering the similarity in origin of the sensillum, where gap junctions have been reported, and the glandular unit, we anticipate that future studies will reveal gap junctional channels between at least all the nonsecretory cells of the gland and between these cells and the adjoining epidermis.

The secretory elements of Class 3 dermal glands and the subdermal oenocyte clearly do not have the typical complement of epithelial membrane junctions. There may be several reasons for this, the most obvious being differences in the metabolic properties and demands of the various secretory cell types in the multicellular gland units, and others relating to the timing of gland activity during the life of the insect.

In the secretory units of the tergal gland of *Dacus tryoni*, a basal secretory cell (trichogen cell) synthesizes a waxy substance, while the adjoining apical secretory cell (tormogen cell) produces the aqueous fraction of the tergal gland secretion (Evans, 1967). In the dermal glands of *Tenebrio molitor* (Delachambre, 1973) and *Bledius mandibularis* (Happ and Happ, 1973) two secretory cells are present, but although the two cells are ultrastructurally distinct, the precise nature of their secretions is unknown. Since it is feasible that their precursor pools and ionic demands are different, the absence of gap junctions in their membranes may be necessary for the optimal functioning of each cell.

The synthetic activity of the cell and the timing of the release of its stored secretion are under hormonal and neurosecretory control. The presence of gap junctions between glandular cells and the unspecialized epidermis may reduce the specificity of the intracellular regulators (such as cAMP, which passes through gap junctions) that are produced in response to specific external factors. Since many cell types have several second-messenger molecules in common, the response to a specific external stimulus needs to be restricted to its target cells. This is particularly important when the activity patterns of contiguous cell types are temporally distinct.

It is true that in many instances the release of the secretion of Class 3 dermal glands coincides with considerable activity in the epidermis, such as at ecdysis (*Rhodnius*—Lai-Fook, 1972; *Tenebrio*—Delachambre, 1973; *Dysdercus*—Lawrence and Staddon, 1975), when the glands are thought to produce a cement layer over the cuticle (Wigglesworth, 1948). However, the glands do not secrete their product while the epidermis is actively depositing preecdysial cuticle, although cuticle stabilization and the production of the cement layer do occur concurrently and could be under the control of the same hormone. The glands are active again at sexual maturity (Delachambre, 1973; Lawrence and Staddon, 1975) and at this time epidermal activity is minimal. A similar pattern of activity is seen in dermal oenocytes, which are not particularly synchronized in their release of stored secretory granules (Wigglesworth, 1970). This could be explained by their physical isolation from one another, and their lack of communication with the epidermis.

3.1.3. The Sensillum

A typical insect sensillum is separated from the surrounding epidermis by up to three nonneuronal accessory cells that in concert secrete the specialized cuticular elements and fluid-filled cavities that enclose the neuronal dendrite (reviewed in Thurm and Küppers, 1980). The accessory cells are arranged concentrically around the cell body of the neuron and its enveloping glial cell. Where three accessory cells are present, the outermost (tormogen) cell, in a trichoid sensillum, forms a cuticular socket for the bristle produced by the middle (trichogen) cell. Both cells secrete a K^+-rich fluid into the lymph space around the receptor dendrite, while the innermost (thecogen) cell secretes fluid into a secondary lymph cavity. Thurm and Küppers (1980) suggest that the accessory cells form an integral part of the epidermal "functional syncytium" (Caveney, 1976) and share a common intracellular ionic compartment. Gap junctions have been reported to occur between accessory cells (Keil, 1978), and they are said to link the outermost accessory cell to the general epidermis (Thurm and Küppers, 1980). Gap junctions have not been detected between the neuron and the glial cell, or between the glial cell and the accessory cells. The presence of gap junctional channels between the epidermis and accessory cells is conceived as important to the recycling of ions from the epidermal compartment across the apical face of the tormogen cell into the lymph space (Küppers and Thurm, 1979).

That the accessory cells of the sensillum are tightly coupled to the general epidermis can be shown for the campaniform sensilla of the larval beetle sternite. By careful placement of intracellular recording electrodes, we have been able to show that the accessory cells (at least two have been detected—see below) are ionically coupled to the epidermis in both the newly ecdysed and the midinstar larva. The membrane potential of the accessory cells is the same as that of the epidermal cells, and analysis of the spread of injected electrical current within the epidermis and accessory cells suggests that they are indistinguishable in their ionic coupling properties. Carboxyfluorescein moves readily from accessory cell to adjacent general epidermis and *vice versa*. Some preliminary data suggest, however, that the physiological control of junctional conductivity between accessory cells in the receptor may differ from that of the general epidermis. In epidermis placed *in vitro* for 24 hr the accessory cells lose their dye coupling, but not ionic coupling, with the surrounding epidermis.

3.2. The Rectal Epithelium

The insect rectum is a major site for the reabsorption of solutes and water (Wall, 1977) and for the absorption of atmospheric moisture (Machin, 1979). The rectal epithelium, the site of fluid transport, can be very complex in structure. In many insects, specialized regions of high fluid transporting activity, the rectal pads or papillae, are surrounded by a simple rectal epithelium. The cockroach *Periplaneta americana* has six rectal pads of thickened epithelium that protrude into the rectal lumen; the fly *Calliphora erythrocephala* has four rectal papillae sunk into the rectal lining. These regions of specialized epithelium may consist of a single layer of columnar epithelial cells, called *principal cells*, as seen in *Thysanura* (Fain-Maurel and Cassier, 1972), or *cortical cells*, as in *Calliphora* (Gupta and Berridge, 1966). The transporting epithelium may be stratified, however, into an inner layer of columnar principal cells and an outer layer of continuous (*Periplaneta*—Noirot and Noirot-Timothée, 1976) or discontinuous (*Blatella orientalis*—Wall and Oschman, 1973) squamous *basal cells*. The cockroach pad epithelium (Figure 17) is separated at its lateral margin from the surrounding rectal epithelium by a multilayered stack of thin *sheath cells* (Wall and Oschman, 1973; Noirot *et al.*, 1979), and in the fly papilla a single row of "junctional" cells is found at this site (Gupta and Berridge, 1966). In larval beetles (Grimstone *et al.*, 1968) and in caterpillars, the sheath cells completely encase the basal surface of the transporting epithelium, as well as the distal ends of the Malpighian tubules, to form a cryptonephridial complex.

The plasma membranes of the three basic types of epithelial cells in the rectal pad—principal (cortical) cells, basal cells, and sheath cells—contain various junctions. Here we are only concerned with the distribution of gap junctions. Adjacent principal cells are coupled by gap junctions (Noirot and Noirot-Timothée, 1976) that are scattered along the extensively folded lateral membranes of the cells (Lane, 1979). Autocellular gap junctions (junctions formed between two regions of the same cell) are seen where the intercellular space between the principal cells is wide (Noirot and Noirot-Timothée, 1976). Although small gap junctions are found between basal cells, they are most conspicuous at the principal–basal cell interface (Wall and Oschman, 1971; Noirot and Noirot-Timothée, 1976). By contrast, gap junctions are rarely seen between the flattened sheath cells that envelop the principal cells, or between the innermost sheath cell and its adjacent principal and basal cells. In those recta containing basal cells, the principal cells have no direct access to the hemolymph (Figure 17).

Can the concept of selective ion pathways suggested from the analysis of the distribution of gap junctions in the rectum be supported by direct electrophysiological evidence? We have recently made such studies on the rectum of the adult male of the cockroach *Periplaneta americana*. The isolated rectum was treated with a collagenase solution (in continually aerated Pringle's Ringer solution) until the circular muscle layer and tracheae could be removed without damage to the underlying epithelium. An open preparation consisting of two pads and intervening simple rectal epithelium was stretched out between two clamps. Microelectrodes were inserted into the cells from the basal surface. The cells of the simple rectal epithelium are clearly visible in phase contrast, but the margins of the thicker principal cells of the pads are somewhat obscured (Figure 19A). It was not possible to discern individual sheath cells, although their location was clear (Figure 18A, 19A). On inserting standard microelectrodes into the cells, three populations of cells could be identified on the basis of three sets of nonoverlapping recorded membrane potentials. The principal–basal cell region of the pad had the highest membrane potentials, ranging from -46 to -50 mV. The principal and basal cell populations could not be separated

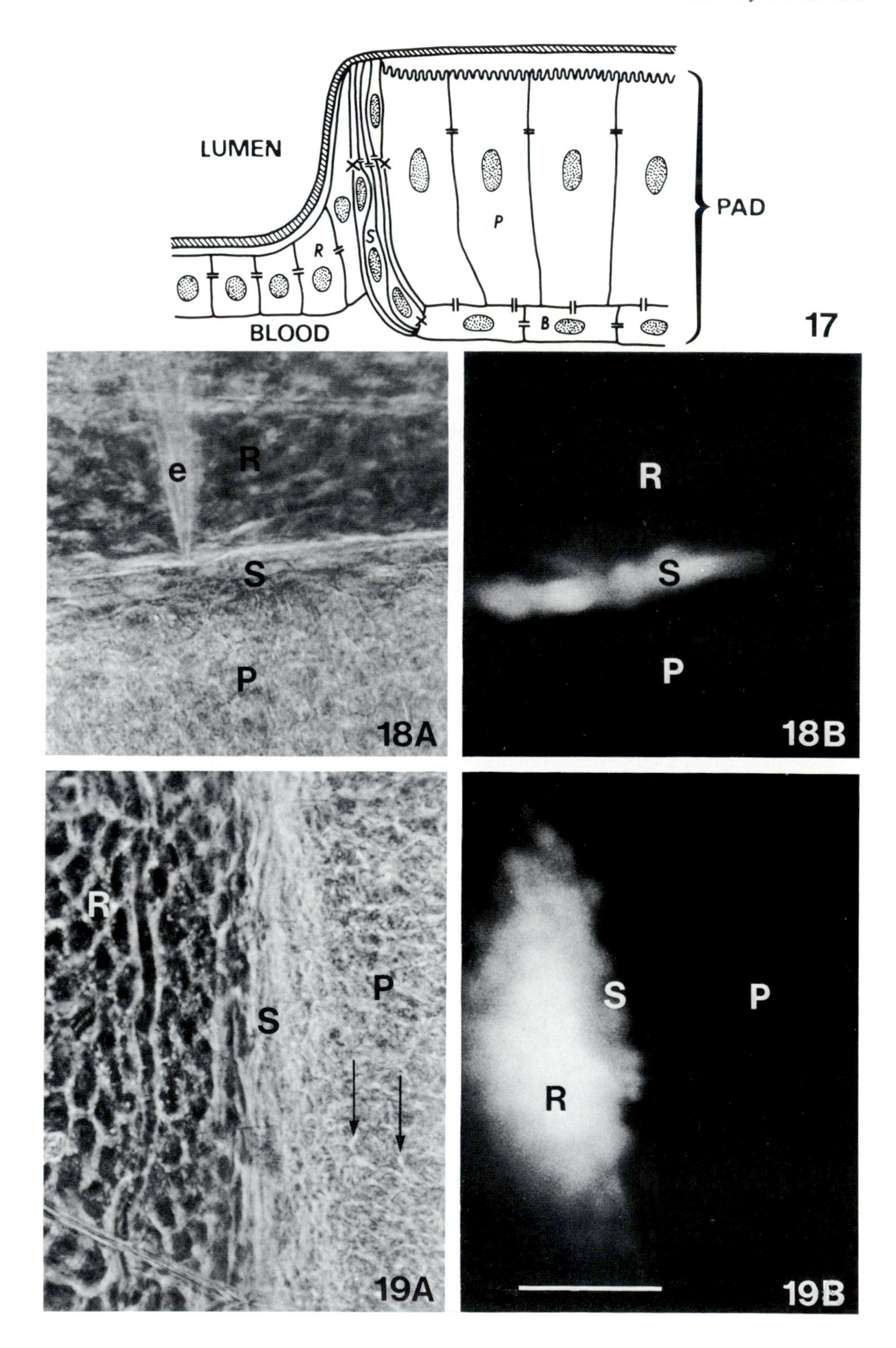
LUMEN
PAD
P
R
S
B
BLOOD
17
e
R
S
P
18A
R
S
P
18B
R
S
P
19A
S
P
R
19B

by membrane potential. Potentials of between -16 and -20 mV were recorded from the simple rectal cells. The lowest membrane potentials were recorded in the sheath region, ranging in value between -9 and -11 mV. On the basis of fine structure, it had been suggested that sheath cells have a low physiological activity (Noirot *et al.*, 1979), and these low membrane potentials support this idea. Furthermore, the presence of three discrete sets of membrane potentials hint that, although the cells may be strongly coupled with members of their own epithelial type, heterocellular ionic coupling is weak or absent (with the exception of principal-basal cell coupling). This conclusion is supported by electrophysiological data on ionic coupling. Simple rectal cells are strongly coupled to one another, as are the principal cells of the pad. The sheath cells, too, are coupled to one another, but this coupling appears to be weak. In no case was ionic coupling detected between the pad epithelium and the surrounding simple rectal cells. Sheath cells are not coupled to the principal cells and only rarely to the simple rectal cells. The physiological data are summarized in Figure 17. The existence of three distinct functional compartments within the rectal epithelium, as diagnosed from coupling data, is further supported by the lack of movement of fluorescent tracer molecules between the compartments (Figures 18, 19).

Coupling via gap junctions in a transporting epithelium has important functional consequences, since the presence of this junction would facilitate the cell-to-cell movement of inorganic ions (Na^+, K^+, Cl^-) required for fluid transport. A key feature of the physiology of rectal pad transport is the recycling of inorganic ions from the hemolymph back to the principal cells for pumping into the intercellular spaces between them (Oschman and Wall, 1969). The presence of extensive gap junctions at the basal-principal cell interface would aid ion recycling. The sheath cells are thought to enhance the efficiency of water and ion transport and its regulation by the pad epithelium (Noirot *et al.*, 1979). The sparsity of gap junctions in the sheath cell membranes would tend to restrict the low-resistance movement of ions between the hemolymph and the principal cells to a route passing through the basal cells (Figure 17). The sheath cells could also serve to isolate the influence of metabolic hormones dependent for their action on intracellular second messengers, such as antidiuretic hormone (Steele and Tolman, 1980), to their target site, the basal cell-principal cell compartment.

←

Figures 17–19. Communication compartments in the rectal pad of the cockroach *Periplaneta*.
Figure 17. A drawing of the edge of a rectal pad, showing the columnar principal cells (P) and squamous basal cells (B) of the pad, separated from the general rectal epithelium (R) by a ring of sheath cells (S). The general rectal cells are strongly coupled to each other (ǂ) as are the cells of the principal cell-basal cell population. The sheath cells, although coupled to one another, are not coupled to the adjoining general rectal cells or to the cells of the rectal pads (*).
Figure 18. (A.) The edge of a pad with an electrode (e) inserted into a sheath cell (phase contrast). (B.) When fluorescent tracer is injected into this cell, its diffusion is restricted to the sheath cell layer (dark field fluorescence).
Figure 19. (A.) The edge of a pad showing the borders of the general rectal cells (R) more clearly, as well as the extracellular spaces between the principal cells (arrows). (Phase contrast). (B.) Tracer injected into a general rectal cell moves radially away from this point, but is blocked from entering the pad regions by the sheath cells (dark field fluorescence). (Scale bar in micrographs = 50 μm.)

3.3. The Midgut Epithelium

In some species, this epithelium consists of a single columnar cell type in which a population of goblet cells is dispersed. Replacement or reserve cells are frequently seen along the basal edge of the epithelium. Both columnar and goblet cells are connected by a smooth septate (formerly "continuous") junction that girdles the apical third of the cells and is thought to form a transepithelial barrier restricting ion and water movement (see Noirot-Timothée and Noirot, 1980, for review).

In spite of its relatively simple construction, the midgut epithelium appears to display the most deviant coupling properties of any insect epithelium studied to date. Gap junctions have been reported in the midgut epithelium of several species. In the collembolan *Orchesella cincta*, gap junctions are numerous along the lateral interface between the columnar cells, though excluded from the region of smooth septate junction (Dallai, 1975); in the midgut of the caterpillar *Anagasta kühniella*, gap junctions are present both between adjacent columnar cells and between goblet and columnar cells (Smith *et al.*, 1969; Flower and Filshie, 1975). The E-type gap junctions of the caterpillar midgut are unusual in that the subunits (in lanthanum-impregnated material) are hexagonally packed, a condition associated with the uncoupled state in certain other tissues (Peracchia, 1980). It is therefore interesting that these gap junctions may not be functional. In *Hyalophora* and *Manduca* caterpillars, ionic coupling is normally absent in the midgut epithelium. Only when the transepithelial potential was short circuited *in vitro* did ionic coupling develop between a goblet cell and its adjoining columnar cells in *Hyalophora*. The columnar cells were not coupled under any of the experimental conditions tested (Blankemeyer and Harvey, 1978). In the caterpillar midgut, the goblet cell pumps K^+ into the gut lumen, and therefore these authors suggested that by opening the gap junctions under short circuit conditions, the goblet cell avails itself of a reserve pool of cytosolic K^+ for transfer into the lumen.

Functional coupling may be absent in the midgut of *Tenebrio* larvae. We were not able to detect ionic coupling between columnar cells (no goblet cells are present), nor does Lucifer yellow (or carboxyfluorescein) spread beyond the cell into which it is injected (Figure 21). Furthermore, analysis of the lateral membranes of the columnar cells in both sectioned and freeze-fractured tissue showed that the regions of cell contact lack gap junctions, although a smooth septate junction forms a broad belt around the apical region of the cells (Figure 20).

Figures 20 and 21. The larval midgut epithelium of *Tenebrio*.

Figure 20. A freeze-fracture micrograph showing the extent of smooth septate junction in the apical region of the cells. This junction appears as rows of particles distributed on both P and E fracture faces (see inset for detail). Gap junctions were never detected in numerous replicas of this type. (mv, microvilli). (Scale bar, 2 μm; inset, 200 nm.)

Figure 21. Carboxyfluorescein was not seen to move beyond the midgut cell into which it was injected (within a 15-minute observation period), supporting the structural and electrophysiological data. (Scale bar = 10 μm.)

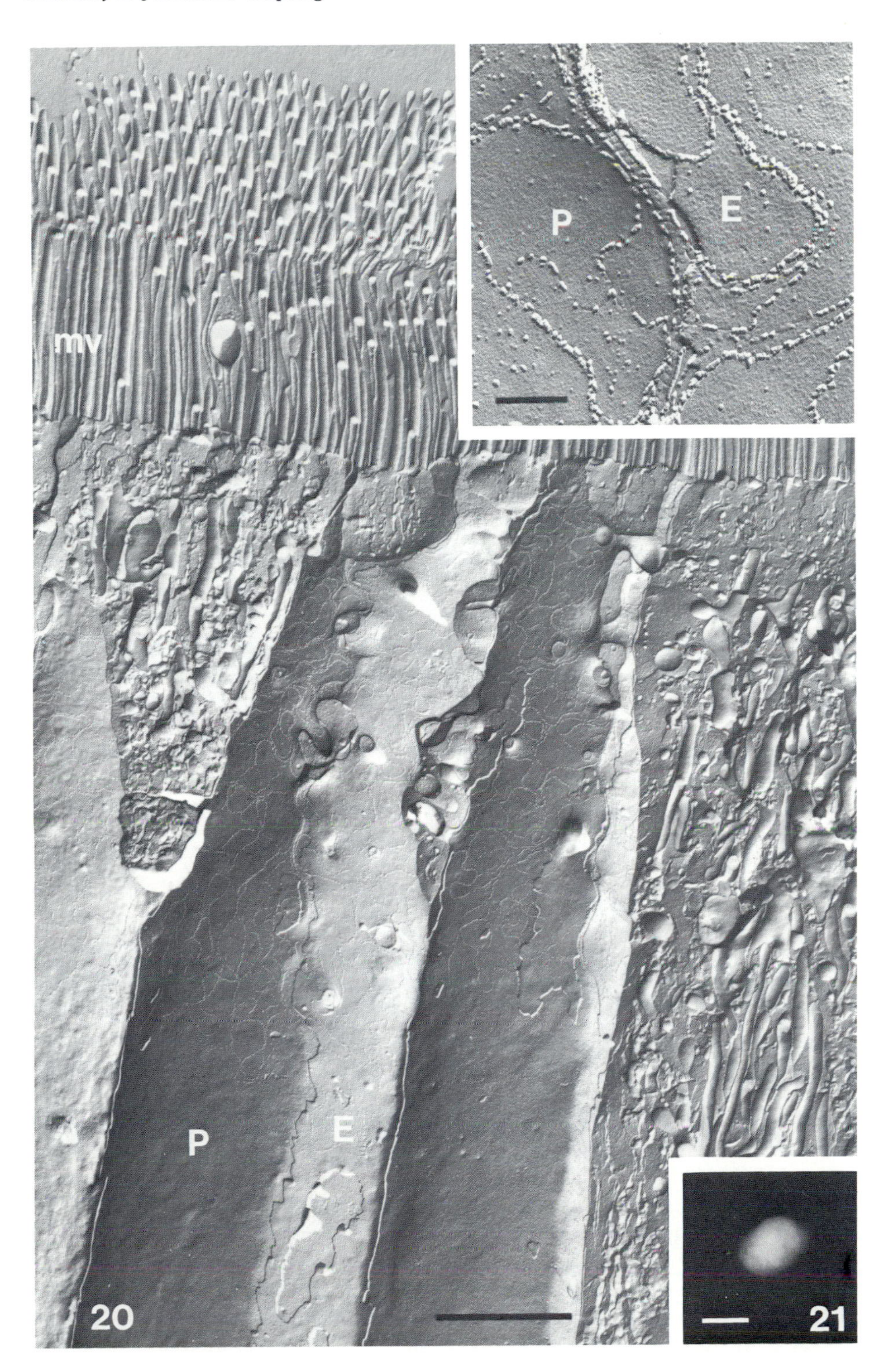
mv
P
E
P
E
20
21

Certainly the coupling properties of the beetle and the lepidopteran midgut are not common to other epithelia studied to date. The physiological control of the rapid opening and closure of the heterocellular junctional channels in the lepidopteran midgut makes it a system worthy of further study.

4. Temporal Selectivity in Junctional Coupling

In attempting to relate the spatial distribution of gap junctions in differentiated tissues to the suspected functions of the component cells, we do not wish to imply that gap junctions are static membrane structures. Rather, gap junctions are dynamic elements of the plasma membrane; their intramembranous subunits may be seen to aggregate, disperse, redistribute, and reassociate within the plane of the membrane, or whole gap junctions may disappear from the membrane by internalization during development and metamorphosis of insect tissues. In the insect central nervous system, gap junction formation is heralded by the appearance of 13-nm particles in the exoplasmic faces of adjacent cell membranes; the particles then align or cluster to form irregular aggregates that rearrange to form macular plaques typical of mature gap junctions (see Lane and Swales, 1980). In the moth testis, gap junctions between germ cells and the cells of the cyst envelope are endocytosed into the cyst cells toward the end of spermiogenesis (Szöllösi and Marcaillou, 1980). Since gap junctions normally form only between cells in close proximity, the stimulus to form gap junctions may be cell contact in cells that aggregate during development. Where tissue cells are already in close contact, gap junction formation or breakdown may be controlled by hormones such as 20-hydroxyecdysone (see section 3.1; also Lane and Swales, 1980).

The rest of this review will describe a case where junctional coupling develops and a case where it is lost during tissue function.

4.1. Oocyte–Follicular Cell Coupling in the Ovary

A simple insect ovary, such as the panoistic ovary of the locust, consists of a series of ovarioles, each with a linear row of oocytes at progressive stages of development as they move towards the oviduct. Each oocyte is surrounded by a single layer of follicular epithelial cells and is joined to adjacent oocytes by a short interfollicular stalk that consists of a group of flattened cells arranged like a stack of pancakes. In the polytrophic ovary of the cecropia silkmoth, a group of seven nurse cells is enclosed along with each oocyte in every follicle, and the interfollicular stalk is longer than that of the locust (see King and Aggarwal, 1965, Figure 9). In *Locusta,* all the oocytes (about fifteen) arranged along a single ovariole are electrotonically coupled (Wollberg *et al.,* 1976); while in a *Hyalophora* ovariole, at least adjacent oocytes are coupled (Woodruff, 1979). Since the adjacent oocytes are not in direct membrane contact, the ionic coupling is indirect and must be through the intervening follicular stalk cells. Support for this conclusion comes from the microinjection of fluorescent tracers into the oocyte. Fluorescence is detected in the follicular epithelium within minutes of injection (Woodruff, 1979; Huebner, 1981) and in 30 min has spread to adjacent follicles through the stalk cells (Woodruff, 1979).

Homocellular gap junctions exist in both the simple follicular epithelium (Huebner and Anderson, 1972; Huebner and Injeyan, 1981) and in the stalk epithelium (Wollberg *et al.*, 1976; Woodruff, 1979; Huebner, 1981). The most detailed studies of the membrane junctions in the ovary are those of Huebner (1981) and Huebner and Injeyan (1981) on the telotrophic ovary of *Rhodnius*. During the previtellogenic phase in this insect, particles seen in the membranes between the follicle cells cluster to form small gap junctions (Figure 22) that grow as extra particles are added (Figures 23, 24) while the oocyte grows (Huebner and Injeyan, 1981). At vitellogenesis, the epithelium changes from a tightly sealed to an open, porous state by the development of extracellular channels that facilitate the transfer of yolk precursors from the blood to the oocyte. During these extensive changes in cell shape, the gap junctions apparently disassemble, to reorganize at localized regions of cell contact across the extracellular space (Huebner and Injeyan, 1981).

Small gap junctions are also present between membrane folds of the developing oocyte and follicle cell processes in *Locusta* (Wollberg *et al.*, 1976), *Rhodnius* (Huebner, 1981), and *Hyalophora* (Woodruff, 1979). Since the ovaries in these three species represent the three major types seen in insects, the presence of gap junctions in this location (Figure 25) is probably a general phenomenon. The junctions are first seen in early previtellogenesis but disappear at the end of vitellogenesis just prior to eggshell formation. During vitellogenesis in *Rhodnius*, as the perivitelline space between the oocyte and the follicle cells enlarges owing to the uptake of yolk precursors, the cells retain gap junctional contact (Figure 26) via bulbous and slender projections (Huebner, 1981). At the end of vitellogenesis the follicular epithelium reverts back to a tightly packed layer of thin cells.

The situation is similar in *Hyalophora*. Here, junctional contact is mediated through bulbous oocyte processes that extend across the vitelline membrane to form gap junctions with depressions in the apical membrane of the follicle cells. At the end of vitellogenesis, the vitelline membrane thickens and these precesses are retracted. Ionic coupling between the oocyte and the follicular epithelium is lost at this time (Woodruff, 1979). A water barrier and subsequently a chorion are laid down by the epithelium around the oocyte during the terminal phase of growth (Telfer and Smith, 1970).

The presence of pathways of intercellular communication along the full length of the ovariole and between the follicular epithelial cells is thought to play an important role in the regulation of oocyte growth and the functional coordination of the cells involved (Woodruff, 1979; Huebner, 1981). In *Rhodnius* and *Locusta* only the terminal oocyte in each ovariole is vitellogenic at any one time. The others are arrested in the previtellogenic state. The gap junctions, by indirectly connecting adjacent follicles, would provide an economic route for the internal diffusion of factors regulating the development of adjacent oocytes in the same ovariole. Secondly, the transition from the previtellogenic to the vitellogenic state is influenced by blood-borne factors. Since the previtellogenic oocyte is effectively isolated from the blood by the follicular

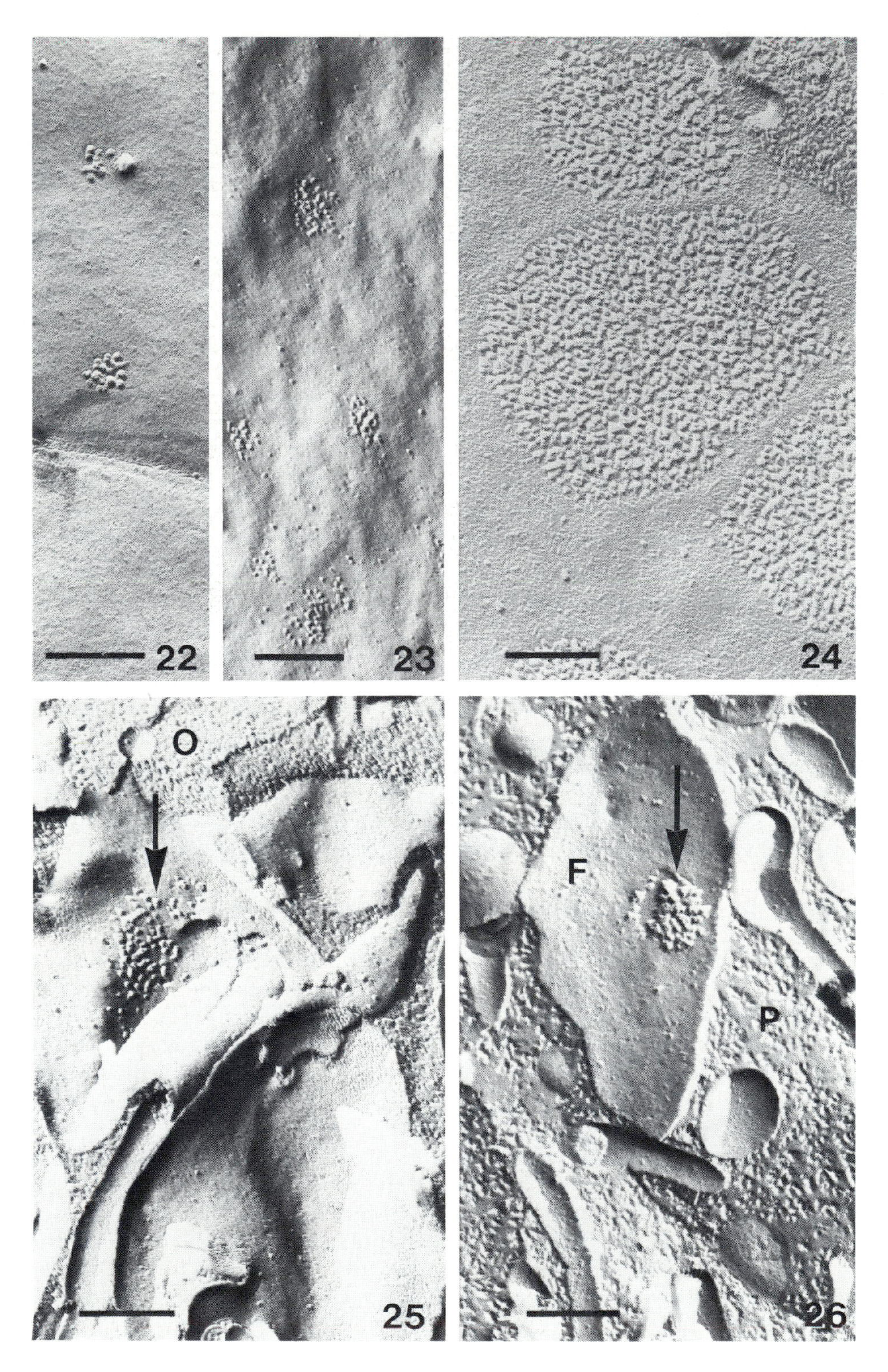
22
23
24
O
25
F
P
26

epithelium, hormonal stimuli at this time (such as by juvenile hormone) might be transmitted to the oocyte either directly, or indirectly via second-messenger systems, through the heterotypic follicle cell–oocyte gap junctions. Both juvenile hormone and cAMP have molecular weights low enough for them to pass through gap junctions. The action of gonadotrophins on mammalian oocyte growth and development is thought to involve cAMP transfer through gap junctional contacts between follicle cells and the oocyte (Gilula *et al.*, 1978). Consequently, the role of the follicle cell in regulating oocyte growth in insects should not be underestimated. The gap junctions between the follicle cells and oocytes also provide a ready pathway for metabolic cooperation during oocyte growth. This route may also be used by signals involved in setting up the anteroposterior and dorsoventral axes of the future embryo, which are known to be under maternal (i.e., ovarian) influence (Nüsslein-Volhard, 1979).

4.2. Hemocyte Capsules

When metazoan parasites, or even plastic implants, are introduced into the insect hemocoel, a cellular defense reaction involving encapsulation of the foreign material by certain classes of hemocytes ensues (Baerwald, 1979). This process involves a drastic transformation in the contact behavior of these hemocytes, since they change from a circulating and dispersed cell population to an immobile and tightly packed cellular capsule. Within 48 hr the capsule may consist of twenty to fifty layers of flattened cells (Baerwald, 1979). As the hemocytes surround the implant, the innermost cells flatten to become tightly apposed, and in many species desmosomes hold them together. Gap junctions have been identified between the cells in the capsules of some species (Grimstone *et al.*, 1967; Baerwald, 1975). In *Periplaneta*, numerous gap junctions are found in the membranes of both inner and peripheral hemocytes; in freeze fracture they appear as typical E-type gap junctions (Baerwald, 1975; 1979).

To test whether these gap junctions are functional structures, we implanted small strips of transparent plastic into the abdomens of midinstar *Periplaneta* nymphs. After 72 hr the plastic strip was removed, placed in saline, and elec-

←

Figures 22–24. Formation of gap junctions in regions of membrane contact between follicular cells during oogenesis in the ovarian follicle of *Rhodnius*. (Scale bar = 200 nm.)
Figure 22. Small clusters of intramembranous E-face particles seen in the follicle cells of the early previtellogenic follicle.
Figure 23. The gap junctions appear to grow during previtellogenesis by the accretion of extra particles.
Figure 24. At late previtellogenesis, large gap junctions are seen between the follicle cells.
Figures 25 and 26. Heterocellular gap junctions between follicle cells and the oocyte in *Rhodnius*. (Scale bar = 200 nm.)
Figure 25. A loose aggregate of gap junctional particles (arrow) located in a membrane fold of the oocyte (O) in late previtellogenesis.
Figure 26. Tightly packed particles of a gap junction (arrow) in the membrane of a bulbous follicle-cell process (F) that extends across the perivitelline space (P) to contact the oocyte during vitellogenesis.
Figures 22–26 kindly supplied by Dr. Erwin Huebner, courtesy of Academic Press.

trotonic coupling between the cells attached to the implant was measured immediately. Under phase contrast, a multilayered stack of flattened cells at least six cells deep was seen to encase the implant. No necrotic cells were seen. On cell penetration with microelectrodes, stable and uniform membrane potentials of between -40 mV and -50 mV were recorded, depending on the incubation time of the capsule *in vitro*. Ionic coupling, present in all cells tested, was detectable between cells as far as 600 μm apart. Microinjected carboxyfluorescein moved readily from cell to cell.

This rapid *de novo* formation of gap junctions probably enhances metabolic interactions within the cells recruited into the capsule and allows global coordination of capsule physiology. Since only the outer capsule cells have direct access to nutrients and hormones in the blood, the gap junctional channels would provide a cell-to-cell trophic pathway to the innermost cells in the capsule, in a manner analogous to the presumed role of gap junctions in the trophic function of perineurial and glial cells in the insect CNS (Lane *et al.*, 1977). Similar analogies exist in the highly coupled vertebrate lens cells (Goodenough *et al.*, 1980) and the mammalian follicular cell-oocyte complex (Gilula *et al.*, 1978), which are not vascularized, and diffusion of nutrients and regulatory molecules through the junctional channels is most expedient. Finally, it should be noted that circulating lymphocytes in vertebrates, when stimulated by phytohemagglutinin, will aggregate and also form gap junctions (see Hooper and Subak-Sharp, 1981).

5. *Implications of Junctional Selectivity*

In general, homologous cells in electrically nonexcitable tissues are functionally coupled by specialized membrane channels (the midgut epithelium is the only known exception). The structural site of these channels is the gap junction. Functional coupling is commonly seen between heterologous cells in complex insect tissues as well. Yet this review suggests that selectivity in junctional coupling in complex tissues such as epithelia may be a widespread and physiologically important phenomenon. The reasons for this selectivity remain largely obscure; we have speculated on what we regard to be rational physiological objectives for the setting up of regionally isolated domains of cells within a tissue, or the isolation of individual cells from an embracing coupled network of cells. Preoccupation with the significance of ionic and metabolic coupling between cells in membrane contact has, over the last fifteen years, dulled our senses to the realization that many cells, or domains of cells, in complex tissues may only function adequately when metabolically isolated from other cell groups in the tissue. In mature tissues, this functional isolation may be relatively stable, or even irreversible, but it is equally likely that junctional coupling between different communication compartments in tissues may occur transiently, as needed, and is under the control of intracellular factors that influence cell metabolism. There are two obvious metabolic reasons for selectivity in functional coupling. First there are *conflicting patterns of activity in tissue cells*. Peak activity in a subpopulation of specialized cells may not coincide with that of supporting cells. It is assumed here that cell activity is under hormonal or neurosecretory control. The presence of gap junctions between, say, secretory gland cells and the general epidermis would reduce the efficiency of intracellular metabolic regulators responsive to external factors. The responses to specific external stimuli need to be isolated to the target cell type, since many cells have several second-messenger molecules in common. Second, there are *conflicting ionic and metabolic demands in cells.* Adjoining heterologous cells with different metabolic functions contain cell-type specific macromolecules and biosynthetic machinery. It may not suffice to argue that the low molecular weight cut-off to the transfer of molecules through gap junctions allows such cells to retain their synthetic autonomy as well as to cooperate metabolically. Different subpopulations of cells in a tissue may have conflicting ionic and metabolic demands. Firstly, in the examples discussed in this review, transporting epithelial cells pump water across the rectal epithelium by setting up intracellular and intercellular standing gradients of certain inorganic ions. For maximum effect, localized epithelial regions of high pumping activity may require ionic isolation from adjacent less active regions. Secondly, contiguous Class 3 glandular cells that produce different secretions (e.g., aqueous *vs.* lipoidal) may have conflicting metabolic demands, as well as the possibility that the first reason given above applies. Multicellular gland units of this type are common in insects, but gap junctions have never been reported between them.

In the main, this review has discussed the metabolic and physiological potential of junctional coupling between the cells of mature tissues. In closing, it is essential to point out that the junctional channel has long been an attractive candidate for the transmission of postulated signal molecules that affect cell growth and cell patterning in tissues (see Loewenstein, 1979). Selectivity in coupling, whether it be permanent, transient, or partial [as seen in directional selectivity between heterologous cell pairs (Flagg-Newton and Loewenstein, 1980)], would enable tissues to set up internal boundaries that restrict junctional communication to subpopulations of cells within the developing tissue. A mechanism of this type could feasibly be involved in the segmentation of the epidermis in the integument and in compartmentation in imaginal discs. Certainly restricted molecular transfer across the boundaries of developmental fields would be an economic means of stabilizing the autonomy of individual domains and may also allow positional signaling systems (e.g., gradients) to be set up.

Acknowledgments

We thank our colleague Dr. Dick Shivers for introducing one of us (R. B.) to the rigors of freeze-fracture electron microscopy; Dr. Erwin Huebner for permission to use some of his micrographs from papers in press; and Drs. Jean Percy and Steve Shaw for providing unpublished data. Dr. Walter Stewart, N.I.H., gave us Lucifer yellow, and Peter Bieman provided cockroach rectal-pad preparations. Shirley Schuurs typed this manuscript. Work in the authors' laboratory was funded by the Natural Sciences and Engineering Research Council of Canada (S.C.) and an Ontario Graduate Scholarship (R.B).

References

Baerwald, R. J., 1975, Inverted gap and other cell junctions in cockroach hemocyte capsules: A thin section and freeze fracture study, *Tissue Cell* **7**:575–85.

Baerwald, R. J., 1979, Fine structure of hemocyte membranes and intercellular junctions formed during hemocyte encapsulation. In *Insect Hemocytes*, edited by A. P. Gupta, pp. 155–188, Cambridge University Press, Cambridge.

Barbier, R., 1975, Differenciation de structures ciliares et mise en plance des canaux au cours de l'organogenèse des glands collétériques de *Galleria mellonella* L. (Lepidoptere, Pyralidae), *J. Microsc.* (Paris) **24**:315–326.

Bennett, M. V. L., and Goodenough, D. A., 1978, Gap junctions, electronic coupling and intercellular communication, *Neuro. Sci. Res. Program Bull.* **16**:377–488.

Blankemeyer, J. T., and Harvey, W. R., 1978, Identification of active cell in potassium transporting epithelium, *J. Exp. Biol.* **77**:1–13.

Bonnanfant-Jais, M. L., 1974, Morphologie de la glande lactée d'une glossine, *Glossina austeni* Newst. au cours du cycle de gestation, *J. Microsc.* (Paris) **19**:265–284.

Bridges, A. R., 1979, Biogenic amines in the honey bee venom gland and venom reservoir. Ph.D. thesis, Department of Zoology, University of Western Ontario, Canada.

Caveney, S., 1970, Juvenile hormone and wound modelling of *Tenebrio* cuticle architecture, *J. Insect Physiol.* **16**:1087–1107.

Caveney, S., 1973, Stability of polarity in the epidermis of a beetle, *Tenebrio molitor* L., *Develop. Biol.* **30**:321–335.

Caveney, S., 1974, Intercellular communication in a positional field: Movement of small ions between insect epidermal cells, *Develop. Biol.* **40**:311–322.

Caveney, S., 1976, The insect epidermis: A functional syncytium. In *The Insect Integument*, edited by H. R. Hepburn, pp. 259–274, Elsevier, Amsterdam.

Caveney, S., 1978, Intercellular communication in insect development is hormonally controlled, *Science* **199**:192–195.

Caveney, S., and Blennerhassett, M. G., 1980, Elevation of ionic conductance between insect epidermal cells by β-ecdysone *in vitro*, *J. Insect. Physiol.* **26**:13–25.

Caveney, S., and Podgorski, C., 1975, Intercellular communication in a positional field: Ultrastructural correlates and tracer analysis of communication between insect epidermal cells, *Tissue Cell* **7**:559–574.

Caveney, S., Berdan, R. C., and McLean, S., 1980, Cell-to-cell ionic communication stimulated by 20-hydroxyecdysone occurs in the absence of protein synthesis and gap junction growth, *J. Insect Physiol.* **26**:557–567.

Chi, C., and Carlson, S. D., 1980, Membrane specializations in the first optic neuropil of the housefly, *Musca domestica* L. I. Junctions between neurons, *J. Neurocytol.* **9**:429–449.

Crossley, A. C., and Waterhouse, D. F., 1969, The ultrastructure of a pheromone-secreting gland in the male scorpion fly *Harpobittacus australis* (Bittacidae: Mecoptera), *Tissue Cell* **1**:273–294.

Dallai, R., 1975, Continuous and gap junction in the midgut of Collembola as revealed by lanthanum tracer and freeze-etching techniques, *J. Submicrosc. Cytol.* **7**:249–257.

Dean, R. L., Bollenbacher, W. E., Locke, M., Smith, S., and Gilbert, L. I., 1980, Haemolymph ecdysteroid levels and cellular events in the intermoult/moult sequence of *Calpodes ethlius*, *J. Insect Physiol.* **26**:267–280.

Delachambre, J., 1973, L'ultrastructure des glandes dermiques de *Tenebrio molitor* L. (Insecta, Coleoptera), *Tissue Cell* **5**:243–257.

Epstein, M. L., and Gilula, N. B., 1977, A study of communication specificity between cells in culture, *J. Cell Biol.* **75**:769–787.

Evans, J. J. T., 1967, The integument of the Queensland fruit fly, *Dacus tryoni* (Frogg). I. The tergal glands, *Z. Zellforsch.* **81**:18–33.

Fain-Maurel, M.-A., and Cassier, P., 1972, Une nouveau type de jonctions; Les jonctions scalariform: Etude ultrastructurale et cytochemique, *J. Ultrastruct. Res.* **39:**222–238.
Fentiman, I., Taylor-Papadimitrou, J., and Stoker, M., 1976, Selective contact-dependent cell communication, *Nature* **264:**760–762.
Filshie, B. K., and Waterhouse, D. F., 1968, The fine structure of the lateral scent glands of the green vegetable bug, *Nezara viridula* (Hemiptera, Pentatomidae), *J. Microsc.* (Oxford) **7:**231–244.
Flagg-Newton, J. L., 1980, Permeability of the cell-to-cell membrane channel and its regulation in mammalian cell junctions, *In Vitro* **16:**1043–1048.
Flagg-Newton, J. L., and Loewenstein, W. R., 1980, Asymmetrically permeable channels in cell junction, *Science* **207:**771–773.
Flower, N. E., and Filshie, B. K., 1975, Junctional structures in the midgut cells of lepidopteran caterpillars, *J. Cell Sci.* **17:**221–239.
Gilula, N. B., Epstein, M. L., and Beers, W. H., 1978, Cell-to-cell communication and ovulation: A study of the cumulus-oocyte complex, *J. Cell Biol.* **78:**58–75.
Goodenough, D. A., Dick, J. S. B., and Lyons, J. E., 1980, Lens metabolic cooperation: A study of mouse lens transport and permeability visualized with freeze-substitution autoradiography and electron microscopy, *J. Cell Biol.* **86:**576–589.
Grimstone, A. V., Rotheram, S., and Salt, G., 1967, An electron-microscopic study of capsule formation by insect blood cells, *J. Cell Sci.* **2:**281–292.
Grimstone, A. V., Mullinger, A. M., and Ramsey, J. A., 1968, Further studies on the rectal complex of the mealworm, *Tenebrio molitor* L. (Coleoptera, Tenebrionidae), *Philos. Trans. R. Soc. London Ser. B.* **253:**343–382.
Gupta, B. L., and Berridge, M. J., 1966, Fine structural organization of the rectum in the blowfly, *Calliphora erythrocephala* (Meig.) with special reference to connective tissue, tracheae and neurosecretory innervation in the rectal papillae, *J. Morphol.* **120:**23–82.
Happ, G. M., and Happ, C. M., 1973, Fine structure of the pygidial glands of *Bledius mandibularis* (Coleoptera: Staphylinidae), *Tissue Cell* **5:**215–231.
Hooper, M. L., and Subak-Sharpe, J. H., 1981, Metabolic cooperation between cells, *Int. Rev. Cytol.* **69:**46–104.
Huebner, E., 1981, Oocyte-follicle cell interaction during normal oogenesis and atresia in an insect, *J. Ultrastruct. Res.* **74:**95–104.
Huebner, E., and Anderson, E., 1972, A cytological study of the ovary of *Rhodnius prolixus*. I. The ontogeny of the follicular epithelium, *J. Morphol.* **136:**459–493.
Huebner, E., and Injeyan, H., 1981, Follicular modulation during oocyte development in an insect: Formation and modification of septate and gap junctions, *Develop. Biol.* **83:**101–113.
Keil, T., 1978, Die Makrochaeten auf dem Thorax von *Calliphora vicina* Robineau-Desvoidy (Calliphoridae, Diptera): Feinstruktur und Morphogenese eines epidermalen Insekten-Mechanoreceptors, *Zoomorphologie* **90:**151–180.
King, R. C., and Aggarwal, S. K., 1965, Oogenesis in *Hyalophora cecropia*, *Growth* **29:**17–83.
Küppers, J., and Thurm, U., 1979, Active ion transport by a sensory epithelium. I. Transepithelial short circuit current, potential difference, and their dependence on metabolism, *J. Comp. Physiol.* **134:**131–136.
Lai-Fook, J., 1970, The fine structure of developing type "B" dermal glands in *Rhodnius prolixus*, *Tissue Cell* **2:**119–138.
Lai-Fook, J., 1972, A comparison between the dermal glands of two insects, *Rhodnius prolixus* (Hemiptera) and *Calpodes ethlius* (Lepidoptera), *J. Morphol.* **136:**495–503.
Lane, N. J., 1979, Freeze-fracture and tracer studies on the intercellular junctions of insect rectal tissues, *Tissue Cell* **11:**481–506.
Lane, N. J., and Swales, L. S., 1980, Dispersal of junctional particles, not internalization, during the *in vivo* disappearance of gap junctions, *Cell* **19:**579–586.
Lane, N. J., Skaer, H.leB., and Swales, L. S., 1977, Intercellular junctions in the central nervous system of insects, *J. Cell Sci.* **26:**175–199.
Lawrence, P. A., and Green, S. M., 1975, The anatomy of a compartment border: The intersegmental boundary in *Oncopeltus*, *J. Cell Biol.* **65:**373–382.
Lawrence, P. A., and Staddon, B. W., 1975, Peculiarities of the epidermal gland system of the cotton stainer *Dysdercus fasciatus* Signoret (Hemiptera: Pyrrhocoridae), *J. Entomol. (A)* **49:**121–130.
Lawrence, T. S., Beers, W. H., and Gilula, N. B., 1978, Hormonal stimulation and cell communication in co-cultures, *Nature* **272:**501–506.

Lococo, D., and Huebner, E., 1980, The ultrastructue of the female accessory gland, the cement gland, in the insect *Rhodnius prolixus*, *Tissue Cell* **12**:557–580.

Loewenstein, W. R., 1978, The cell-to-cell membrane channel in development and growth. In *Differentiation and Development*, edited by F. Ahmad, J. Schultz, T. R. Russell, and R. Werner, pp. 399–409, Academic Press, New York.

Loewenstein, W. R., 1979, Junctional intercellular communication and the control of growth, *Biochem. Biophys. Acta* **560**:1–65.

Machin, J., 1979, Atmospheric water absorption in arthropods, *Adv. Insect Physiol.* **14**:1–48.

Michalke, W., and Loewenstein, W. R., 1971, Communication between cells of different types, *Nature* **232**:121–122.

Neville, A. C., 1975, *The Biology of Arthropod Cuticle*, Springer-Verlag, Berlin.

Noirot, C., and Noirot-Timothée, C., 1976, Fine structure of the rectum in cockroaches (Dictyoptera): General organization and intercellular junctions, *Tissue Cell* **8**:345–368.

Noirot, D., and Quennedey, A., 1974, Fine structure of insect epidermal glands, *Annu. Rev. Entomol.* **19**:61–80.

Noirot, C., Smith, D. S., Cayer, M. L., and Noirot-Timothée, D., 1979, The organization and isolating function of insect rectal tissue cells: A freeze-fracture study, *Tissue Cell* **11**:325–336.

Noirot-Timothée, C., and Noirot, C., 1980, Septate and scalariform junctions in arthropods, *Int. Rev. Cytol.* **63**:97–140.

Nüsslein-Volhard, C., 1979, Maternal effect mutations that alter the spatial coordinates of the embryo of *Drosophila melanogaster*. In *Determinants of Spatial Organization*, edited by S. Subtelny and I. R. Konigsberg, pp. 185–211, Academic Press, New York.

Oschman, J. L., and Wall, B. J., 1969, The structure of the rectal pads of *Periplaneta americana* L. with regard to fluid transport, *J. Morphol.* **127**:475–510.

Peracchia, C., 1980, Structural correlates of gap junction permeation, *Int. Rev. Cytol.* **66**:81–146.

Percy, J. E., 1974, Ultrastructure of sex-pheromone gland cells and cuticle before and during release of pheromone in female eastern spruce budworm, *Choristoneura fumiferana* (Clem.) (Lepidoptera: Tortricidae), *Can. J. Zool.* **52**:695–705.

Percy, J. E., and Weatherston, J., 1974, Gland structure and pheromone production in insects. In *Frontiers of Biology*, vol. 32, *Pheromones*, edited by M. S. Birch, pp. 11–34, American Elsevier, New York.

Pitts, J. D., 1980, Role of junctional communication in animal tissues, *In Vitro* **16**:1049–1056.

Pitts, J. D., and Burk, R. R., 1976, Specificity of junctional communication between animal cells, *Nature* **264**:762–764.

Quennedey, A., 1971, Les glandes exocrines des termites. I. Etude histochimique et ultrastructurale de la gland sternale de *Kalotermes flavicollis* Fab. (Isoptera, Kalotermitidae), *Z. Zellforsch.* **121**:27–47.

Rose, B., 1971, Intercellular communication and some structural aspects of membrane junctions in a simple cell system, *J. Membrane Biol.* **5**:1–19.

Rose, B., 1980, Permeability of the cell-to-cell membrane channel and its regulation in an insect cell junction, *In Vitro* **16**:1029–1042.

Rose, B., and Rick, R., 1978, Intracellular pH, intracellular free Ca, and junctional cell–cell coupling, *J. Membrane Biol.* **44**:337–415.

Shaw, S. R., 1969, Interreceptor coupling in ommatidia of drone honeybee and locust compound eyes, *Vision Res.* **9**:999–1029.

Shaw, S. R., 1979, Signal transmission by graded slow potentials in the arthropod peripheral visual system. In *The Neurosciences: 4th Study Program*, edited by F. O. Schmitt and F. G. Worden, pp. 275–295, MIT Press, Cambridge, Mass.

Smith, D. S., 1968, *Insect Cells: Structure and Function*, Oliver and Boyd, Edinburgh.

Smith, D. S., Compher, K., Janners, M., Lipton, C., and Wittle, L. W., 1969, Cellular organization and ferritin uptake in the mid-gut epithelium of a moth, *Ephestia kühniella*. *J. Morphol.* **127**:41–72.

Sreng, L., and Quennedey, A., 1976, Role of a temporary ciliary structure in the morphogenesis of insect glands: An electron microscope study of the tergal glands of male *Blattella germanica* L. (Dictyoptera, Blattellidae), *J. Ultrastruct. Res.* **56**:78–95.

Staddon, B. W., 1979, The scent glands of Heteroptera, *Adv. Insect Physiol.* **14**:351–418.

Steele, J. E., and Tolman, J. H., 1980, Regulation of water transport in the cockroach rectum by the corpora cardiaca-corpora allata system: The requirement of Na^+, *J. Comp. Physiol.* **138:** 357–365.

Szöllösi, A., and Marcaillou, C., 1980, Gap junctions between germ and somatic cells in the testis of the moth, *Anagasta küehniella* (Insecta: Lepidoptera), *Cell Tissue Res.* **213:**137–147.

Telfer, W. H., and Smith, D. S., 1970, Aspects of egg formation. In *Insect Ultrastructure*, edited by A. C. Neville, pp. 117–134, Blackwell Scientific Publications, Oxford.

Thurm, U., and Küppers, J., 1980, Epithelial physiology of insect sensilla. In *Insect Biology in the Future—VBW 80*, edited by M. Locke and D. S. Smith, pp. 735–763, Academic Press, New York.

Wall, B. J., 1977, Fluid transport in the cockroach rectum. In *Transport of Ions and Water in Animals*, edited by B. L. Gupta, R. B. Moreton, J. L. Oschman, and B. J. Wall, pp. 599–612, Academic Press, New York.

Wall, B. J., and Oschman, J. L., 1973, Structure and function of rectal pads in *Blatella* and *Blaberus* with respect to the mechanisms of water uptake, *J. Morphol.* **140:**105–118.

Wigglesworth, V. B., 1948, The structure and deposition of the cuticle in the adult mealworm, *Tenebrio molitor* L. (Coleoptera), *Quart. J. Microsc. Sci.* **89:**197–216.

Wigglesworth, V. B., 1970, Structural lipids in the insect cuticle and the function of the oenocytes, *Tissue Cell* **2:**155–179.

Wollberg, Z., Cohen, E., and Kalina, M., 1976, Electrical properties of developing oocytes of the migratory locust, *Locusta migratoria*, *J. Cell Physiol.* **88:**145–158.

Woodruff, R. I., 1979, Electrotonic junctions in Cecropia moth ovaries, *Develop. Biol.* **69:**281–295.

Zampighi, G., 1980, On the structure of isolated junctions between communicating cells, *In Vitro* **16:**1018–1028.

Author Index*

*This index contains only textual citations to authors.

Subject Index